ELECTRONIC FILTER DESIGN HANDBOOK

OTHER McGRAW-HILL HANDBOOKS OF INTEREST

American Institute of Physics • American Institute of Physics Handbook
Baumeister • Marks' Standard Handbook for Mechanical Engineers
Beeman • Industrial Power Systems Handbook
Brady and Clauser • Materials Handbook
Burington and May • Handbook of Probability and Statistics with Tables
Condon and Odishaw • Handbook of Physics
Considine • Energy Technology Handbook
Coombs • Basic Electronic Instrument Handbook
Coombs • Printed Circuits Handbook
Croft, Carr, and Watt • American Electricians' Handbook
Dean • Lange's Handbook of Chemistry
Fink • Electronics Engineers' Handbook
Fink and Beaty • Standard Handbook for Electrical Engineers
Giacoletto • Electronics Designers' Handbook
Harper • Handbook of Components for Electronics
Harper • Handbook of Electronic Packaging
Harper • Handbook of Electronic Systems Design
Harper • Handbook of Materials and Processes for Electronics
Harper • Handbook of Thick Film Hybrid Microelectronics
Harper • Handbook of Wiring, Cabling, and Interconnecting for Electronics
Hicks • Standard Handbook of Engineering Calculations
Hunter • Handbook of Semiconductor Electronics
Huskey and Korn • Computer Handbook
Ireson • Reliability Handbook
Jasik • Antenna Engineering Handbook
Juran • Quality Control Handbook
Kaufman and Seidman • Handbook of Electronics Calculations
Kaufman and Seidman • Handbook for Electronics Engineering Technicians
Korn and Korn • Mathematical Handbook for Scientists and Engineers
Kurtz and Shoemaker • The Lineman's and Cableman's Handbook
Machol • System Engineering Handbook
Maissel and Glang • Handbook of Thin Film Technology
Markus • Electronics Dictionary
McPartland • McGraw-Hill's National Electrical Code Handbook
Perry • Engineering Manual
Skolnik • Radar Handbook
Smeaton • Motor Application and Maintenance Handbook
Stout and Kaufman • Handbook of Microcircuit Design and Application
Stout and Kaufman • Handbook of Operational Amplifier Circuit Design
Truxal • Control Engineers' Handbook
Tuma • Engineering Mathematics Handbook
Tuma • Handbook of Physical Calculations
Tuma • Technology Mathematics Handbook
Williams • Electronic Filter Design Handbook

ELECTRONIC FILTER DESIGN HANDBOOK

Arthur B. Williams

Manager of Research and Development
Coherent Communications Systems Corp.

McGraw-Hill Book Company

New York   St. Louis   San Francisco   Auckland
Bogotá   Singapore   Johannesburg   London
Madrid   Mexico   Montreal   New Delhi
Panama   São Paulo   Hamburg
Sydney   Tokyo   Paris
Toronto

To My Family My Wife Ellen
and Children Howard, Bonnie, and Robin
Mrs. Jean Williams and Mr. and Mrs. Marcus Fuhr
for all their Love, Encouragement, and Inspiration

Library of Congress Cataloging in Publication Data

Williams, Arthur Bernard, date.
  Electronic filter design handbook.

  Includes index.
  1. Electric filters.   I. Title.
TK7872.F5W55   621.3815'324   80-11998
ISBN 0-07-070430-9

    6 7 8 9   VBVB   8987

The editors for this book were Harold Crawford and Geraldine
Fahey, the designer was Mark E. Safran, and the production
supervisor was Paul Malchow. It was set in Baskerville by The
Kingsport Press.

Contents

Preface

The design of filters has generally been reserved for specialists, since higher mathematics and specialized knowledge, which can be acquired only from practical experience, are necessary. Further design specialization occurs, since filters fall into one of two general categories, active and passive.

This handbook treats the design of filters, both active and passive, in a practical in-depth manner so that the average engineer can design almost any type of filter with no prior experience.

Emphasis has been placed on the design of filters by using normalized numerical tables. Sophisticated filters can be generated from these tables with basic formulas, rules, and guidelines. This book pays particular attention to elliptic-function filters. Using these tables, this extremely powerful filter type can be easily designed.

A chapter is included on predicting the parameters of the tabulated filters in both the frequency and time domains. This assists design engineers in selecting the optimum design which will satisfy their overall requirements.

Design of almost any filter type is achieved by using the precalculated data. Simple transformations will convert these tabulated normalized low-pass values into high-pass, bandpass, and band-reject designs. Equal emphasis is placed on active and passive filters.

After a suitable paper design has been completed, the design engineer must ensure that the filter will operate satisfactorily after construction. Practical problems have been anticipated in many ways. The effects of component tolerances, magnetic material selection, amplifier limitations, etc., are discussed. Rules and guidelines are established so that good accuracy is obtainable without extensive breadboarding. Parasitics in LC filters, such as low Q and stray capacitance, are discussed and techniques are presented to compensate for these effects. Special methods are described so that a design which contains impractical element values can be modified.

Many topics are covered for the first time in any book. The sections on active bandpass filters, delay lines, and equalizers and the chapter on refinements in LC filter design will be of special interest to the seasoned filter design specialist.

It is felt that this book can be used for the following purposes:

1. As a self-study handbook for practicing engineers and technicians so that they can easily design working filters

2. As a supplementary textbook for a graduate or undergraduate course on the design of filters

3. As a reference book to be used by practicing filter design specialists.

I would like to express my appreciation to A. Bodony for his useful suggestions and assistance. I would also like to thank Joanne Backhaus for her patient typing of the manuscript. Additional thanks are expressed to the individuals and corporations who allowed me to reproduce some of their material. I am especially grateful to Philip Geffe and Dr. Herman Blinchikoff for providing me with some of the pole locations.

Arthur B. Williams

1

Introduction to Modern Network Theory

1.1 DEVELOPMENT OF THE IMAGE-PARAMETER CONCEPT

The image-parameter method of filter design was developed by Zobel and others in the 1920s. Filters obtained by using image-parameter techniques are generally inferior and use more reactive elements than filters designed by using modern network theory. However, because of the historical significance, some basic concepts of image-parameter theory are briefly reviewed in this section.

Image Impedance

Image-parameter filters are constructed by combining filter building blocks, referred to as half sections and full sections, as shown in figure 1-1. Element-value equations are based on the assumption that the network is terminated in impedances that satisfy the following expressions:

$$Z_T = \sqrt{Z_1 Z_2 + \frac{Z_1^2}{4}} \tag{1-1}$$

$$Z_\pi = \frac{Z_1 Z_2}{\sqrt{Z_1 Z_2 + \frac{Z_1^2}{4}}} \tag{1-2}$$

Z_T and Z_π are referred to as image impedances. These expressions also define the input impedance of the network when the network's output is terminated with Z_T or Z_π as in figure 1-1.

Unfortunately, since Z_1 and Z_2 are reactive elements, Z_T and Z_π may change rather abruptly with frequency. It therefore becomes almost impossible to terminate image-parameter filters precisely. This is the basic limitation of the image-parameter concept.

At some frequency within the filter passband the image impedance becomes purely resistive. Terminating the filter in this nominal value R approximates

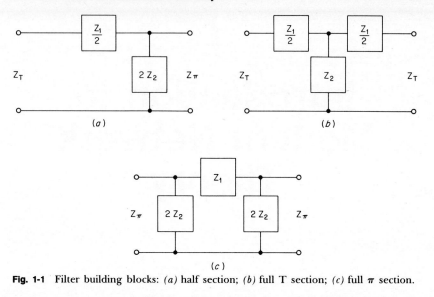

Fig. 1-1   Filter building blocks: *(a)* half section; *(b)* full T section; *(c)* full π section.

the image impedance over much of the passband. Still, the inability to terminate image-parameter filters in their exact image impedance can result in large deviations of filter response from the theoretical predictions.

Constant-k Low-Pass Filters

A constant-*k* low-pass half section and its ideal frequency response are shown in figure 1-2. The element values are computed by

$$L_K = \frac{R}{2\pi f_c} \tag{1-3}$$

$$C_K = \frac{1}{2\pi f_c R} \tag{1-4}$$

where f_c is the cutoff frequency.

The ultimate rate of roll-off of constant-*k* low-pass filters is limited to 6 dB per octave for each reactive element. For instance, a filter consisting of two coils and three capacitors eventually rolls off at 30 dB per octave.

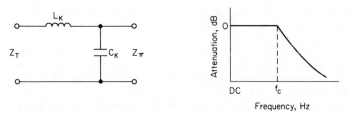

Fig. 1-2   Constant-*K* low-pass filter half section.

The following design example illustrates how constant-k low-pass half sections are combined to obtain increased selectivity.

Example 1-1

REQUIRED: Low-pass filter
Ultimate roll-off of 24 dB per octave
$f_c = 1000$ Hz
$R = 600$ Ω

RESULT: *(a)* Compute L_K and C_K.

$$L_K = \frac{R}{2\pi f_c} = 0.0955 \text{ H} \tag{1-3}$$

$$C_K = \frac{1}{2\pi f_c R} = 0.265 \text{ }\mu\text{F} \tag{1-4}$$

(b) Combine half sections to obtain four reactive elements for a 24 dB per octave roll-off.

The completed low-pass filter design is shown in figure 1-3 *a*. By combining series inductors and parallel capacitors, the final circuit of figure 1-3 *b* is obtained.

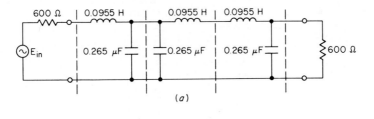

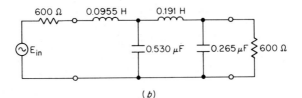

Fig. 1-3 Filter of example 1-1: *(a)* low-pass filter before combining elements; *(b)* final circuit.

The passband behavior of this design may not be suitable for many applications as a result of the inherent deficiencies of the image-parameter concept. This example does illustrate the ease with which element values can be computed and combined, which is one of the few attributes of the image-parameter method.

m-Derived Low-Pass Filters

The m-derived low-pass filters can provide a steeper rate of roll-off than the constant-k type, with fewer elements. An m-derived low-pass filter section has a transmission zero, i.e., a notch or resonance in the stopband having near infinite rejection. This notch can be located wherever desired and is generally placed near the cutoff frequency whenever a very steep rate of roll-off is required.

An m-derived low-pass filter half section and its frequency response are shown in figure 1-4. Although the transmission zero results in a steeper rate of descent from the passband into the stopband, a return lobe referred to as a come-back occurs beyond the notch. The amplitude of this come-back increases as the notch is placed closer to the cutoff frequency.

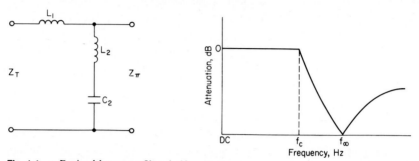

Fig. 1-4 m-Derived low-pass filter half section.

Since constant-k type sections have a monotonic (continuous) roll-off characteristic, m-derived and constant-k filter sections are usually combined to reduce the amplitude of the come-backs.

The design equations of the m-derived half section are

$$m = \sqrt{1 - \frac{f_c^2}{f_\infty^2}} \tag{1-5}$$

$$L_1 = mL_K \tag{1-6}$$

$$L_2 = \frac{1 - m^2}{m} L_K \tag{1-7}$$

$$C_2 = mC_K \tag{1-8}$$

The effect of the design parameter m on the image impedance is plotted in figure 1-5. For $m = 0.6$ the impedance is flat over most of the passband. This

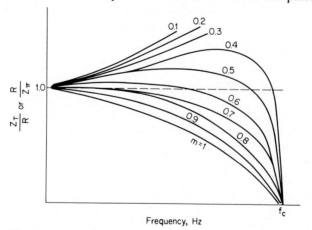

Fig. 1-5   Image impedance vs. m.

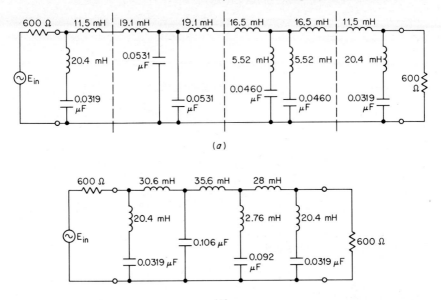

(a)

(b)

Fig. 1-6   Filter of example 1-2: (a) filter with individual sections; (b) combined filter.

value results in a good match to a resistive load; so image-parameter filters are usually terminated with $m = 0.6$ half sections.

It is important to recognize that m-derived sections provide two equally important functions. They provide a suitable match between the image impedance of constant-k sections and a resistive load termination and also increase the rate of descent into the stopband.

The following example demonstrates the design of an image-parameter low-pass filter using both constant-k and m-derived sections.

Example 1-2

REQUIRED: Low-pass filter
$$f_c = 5 \text{ kHz}$$
$$f_\infty = 10 \text{ kHz}$$
$$R = 600 \ \Omega$$

RESULT: Use a constant-k full-T midsection, an m-derived full-T midsection, and m-derived end sections for terminating.

(a)   Compute constant-k midsection.

$$L_K = \frac{R}{2\pi f_c} = 0.0191 \text{ H} \tag{1-3}$$

$$C_K = \frac{1}{2\pi f_c R} = 0.0531 \ \mu\text{F} \tag{1-4}$$

(b)   Compute m-derived midsection.

$$m = \sqrt{1 - \frac{f_c^2}{f_\infty^2}} = 0.866 \tag{1-5}$$

$$L_1 = mL_K = 0.0165 \text{ H} \tag{1-6}$$

$$L_2 = \frac{1-m^2}{m} L_K = 0.00552 \text{ H} \qquad (1\text{-}7)$$

$$C_2 = mC_K = 0.0460 \ \mu\text{F} \qquad (1\text{-}8)$$

(c) Compute m-derived end sections using $m = 0.6$.

$$L_1 = mL_K = 0.0115 \text{ H} \qquad (1\text{-}6)$$

$$L_2 = \frac{1-m^2}{m} L_K = 0.0204 \text{ H} \qquad (1\text{-}7)$$

$$C_2 = mC_K = 0.0319 \ \mu\text{F} \qquad (1\text{-}8)$$

The resulting circuit is shown in figure 1-6a. By combining elements, the filter of figure 1-6b is obtained.

Constant-k and m-derived image-parameter design techniques are generally limited. Filters designed by these methods are inefficient and their response characteristics are difficult to predict. Because of these limitations, the subject will not be carried beyond this point.

1.2 MODERN NETWORK THEORY

A generalized filter is shown in figure 1-7. The filter block may consist of inductors, capacitors, resistors, and possibly active elements such as operational amplifiers and transistors. The terminations shown are a voltage source E_s, a source resistance R_s, and a load resistor R_L.

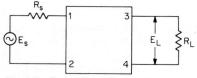

Fig. 1-7 Generalized filter.

The circuit equations for the network of figure 1-7 can be written by using circuit-analysis techniques. Modern network theory solves these equations to determine the network values for optimum performance in some respect.

The Pole-Zero Concept

The frequency response of the generalized filter can be expressed as a ratio of two polynomials in s where $s = j\omega$ ($j = \sqrt{-1}$ and ω, the frequency in radians per second, is $2\pi f$) and is referred to as a transfer function. This can be stated mathematically as

$$T(s) = \frac{E_L}{E_s} = \frac{N(s)}{D(s)} \qquad (1\text{-}9)$$

The roots of the denominator polynomial $D(s)$ are called poles and the roots of the numerator polynomial $N(s)$ are referred to as zeros.

Deriving a network's transfer function could become quite tedious and is beyond the scope of this book. The following discussion explores the evaluation and representation of a relatively simple transfer function.

Analysis of the low-pass filter of figure 1-8a results in the following transfer function:

$$T(s) = \frac{1}{s^3 + 2s^2 + 2s + 1} \tag{1-10}$$

Let us now evaluate this expression at different frequencies after substituting $j\omega$ for s. The result will be expressed as the absolute magnitude of $T(j\omega)$ and the relative attenuation in decibels with respect to the response at DC.

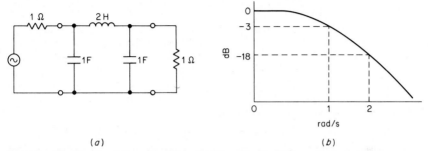

(a)   (b)

Fig. 1-8   All-pole $N = 3$ low-pass filter: (a) filter circuit; (b) frequency response.

$$T(j\omega) = \frac{1}{1 - 2\omega^2 + j(2\omega - \omega^3)} \tag{1-11}$$

ω	$|T(j\omega)|$	$20 \log|T(j\omega)|$
0	1	0 dB
1	0.707	−3 dB
2	0.124	−18 dB
3	0.0370	−29 dB
4	0.0156	−36 dB

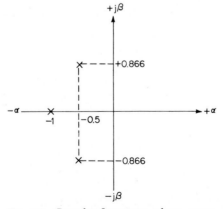

Fig. 1-9   Complex-frequency plane representation of equation (1-10).

The frequency-response curve is plotted in figure 1-8*b*.

Analysis of equation (1-10) indicates that the denominator of the transfer function has three roots or poles and the numerator has none. The filter is therefore called an all-pole type. Since the denominator is a third-order polynomial, the filter is also said to have an $n = 3$ complexity. The denominator poles are $s = -1$, $s = -0.500 + j0.866$, and $s = -0.500 - j0.866$.

These complex numbers can be represented as symbols on a complex-number plane. The abscissa is α, the real component of the root, and the ordinate is β, the imaginary part. Each pole is represented as the symbol X, and a zero is represented as 0. Figure 1-9 illustrates the complex-number plane representation for the roots of equation (1-10).

There are certain mathematical restrictions on the location of poles and zeros in order for the filter to be realizable. They must occur in pairs which are conjugates of each other, except for real-axis poles and zeros, which may occur singly. Poles must also be restricted to the left half plane (i.e., the real coordinate of the pole must be negative). Zeros may occur in either plane.

Synthesis of Filters from Polynomials

Modern network theory has produced families of standard transfer functions that provide optimum filter performance in some desired respect. Synthesis is the process of deriving circuit component values from these transfer functions. Chapter 12 contains extensive tables of transfer functions and their associated component values so that design by synthesis is not required. However, in order to gain some understanding as to how these values have been determined, we will now discuss a few methods of filter synthesis.

Synthesis by Expansion of Driving-Point Impedance
The input impedance to the generalized filter of figure 1-7 is the impedance seen looking into terminals 1 and 2 with terminals 3 and 4 terminated, and is referred to as the driving-point impedance or Z_{11} of the network. If an expression for Z_{11} could be determined from the given transfer function, this expression could then be expanded to define the filter.

A family of transfer functions describing the flattest possible shape and a monotonically increasing attenuation in the stopband is the Butterworth low-pass response. These all-pole transfer functions have denominator polynomial roots which fall on a circle having a radius of unity from the origin of the $j\omega$ axis. The attenuation for this family is 3 dB at 1 rad/s.

The transfer function of equation (1-10) satisfies this criterion. It is evident from figure 1-9 that if a circle were drawn having a radius of 1, with the origin as the center, it would intersect the real root and both complex roots.

If R_s in the generalized filter of figure 1-7 is set to 1 Ω, a driving-point impedance expression can be derived in terms of the Butterworth transfer function as

$$Z_{11} = \frac{D(s) - s^n}{D(s) + s^n} \tag{1-12}$$

where $D(s)$ is the denominator polynomial of the transfer function and n is the order of the polynomial.

After $D(s)$ is substituted into equation (1-12), Z_{11} is expanded using the continuous-fraction expansion. This expansion involves successive division and inversion of a ratio of two polynomials. The final form contains a sequence of terms each alternately representing a capacitor and an inductor and finally the resistive termination. This procedure is demonstrated by the following example.

Example 1-3

REQUIRED: Low-pass LC filter having a Butterworth $n = 3$ response.

RESULT: (a) Use the Butterworth transfer function:

$$T(s) = \frac{1}{s^3 + 2s^2 + 2s + 1} \qquad (1\text{-}10)$$

(b) Substitute $D(s) = s^3 + 2s^2 + 2s + 1$ and $s^n = s^3$ into equation (1-12), which results in

$$Z_{11} = \frac{2s^2 + 2s + 1}{2s^3 + 2s^2 + 2s + 1} \qquad (1\text{-}12)$$

(c) Express Z_{11} so that the denominator is a ratio of the higher-order to the lower-order polynomial:

$$Z_{11} = \frac{1}{\dfrac{2s^3 + 2s^2 + 2s + 1}{2s^2 + 2s + 1}}$$

(d) Dividing the denominator and inverting the remainder results in

$$Z_{11} = \frac{1}{s + 1\dfrac{1}{\dfrac{2s^2 + 2s + 1}{s + 1}}}$$

(e) After further division and inversion, we get as our final expression

$$Z_{11} = \frac{1}{s + 1\dfrac{1}{2s + 1\dfrac{1}{s+1}}} \qquad (1\text{-}13)$$

The circuit configuration of figure 1-10 is called a ladder network, since it consists of alternating series and shunt branches. The input impedance can be expressed as the following continued fraction:

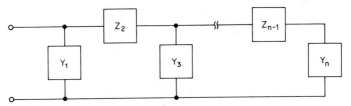

Fig. 1-10 General ladder network.

$$Z_{11} = \frac{1}{Y_1 + \dfrac{1}{Z_2 + \dfrac{1}{Y_3 + \ldots \dfrac{1}{Z_{n-1} + \dfrac{1}{Y_n}}}}} \qquad (1\text{-}14)$$

where $Y = sC$ and $Z = sL$ for the low-pass all-pole ladder except for a resistive termination where $Y_n = sC + 1/R_L$.

Figure 1-11 can then be derived from equations (1-13) and (1-14) by inspection. This can be proved by reversing the process of expanding Z_{11}. By alternately adding admittances and impedances while working toward the input, Z_{11} is verified as being equal to equation (1-13).

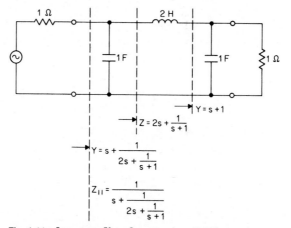

Fig. 1-11 Low-pass filter for equation (1-13).

Synthesis for Unequal Terminations If the source resistor is set equal to 1 Ω and the load resistor is desired to be infinite (unterminated), the impedance looking into terminals 1 and 2 of the generalized filter of figure 1-7 can be expressed as

$$Z_{11} = \frac{D(s \text{ even})}{D(s \text{ odd})} \tag{1-15}$$

$D(s$ even$)$ contains all the even-power s terms of the denominator polynomial and $D(s$ odd$)$ consists of all the odd-power s terms of any realizable all-pole low-pass transfer function. Z_{11} is expanded into a continued fraction as in example 1-3 to define the circuit.

Example 1-4

REQUIRED: Low-pass filter having a Butterworth $n = 3$ response with a source resistance of 1 Ω and an infinite termination.

RESULT: (a) Use the Butterworth transfer function:

$$T(s) = \frac{1}{s^3 + 2s^2 + 2s + 1} \tag{1-10}$$

(b) Substitute $D(s$ even$) = 2s^2 + 1$ and $D(s$ odd$) = s^3 + 2s$ into equation (1-15):

$$Z_{11} = \frac{2s^2 + 1}{s^3 + 2s} \tag{1-15}$$

(c) Express Z_{11} so that the denominator is a ratio of the higher- to the lower-order polynomial:

$$Z_{11} = \frac{1}{\dfrac{s^3 + 2s}{2s^2 + 1}}$$

(d) Dividing the denominator and inverting the remainder results in

$$Z_{11} = \frac{1}{0.5s + \dfrac{1}{\dfrac{2s^2 + 1}{1.5s}}}$$

(e) Dividing and further inverting results in the final continued fraction

$$Z_{11} = \frac{1}{0.5s + \dfrac{1}{1.333s + \dfrac{1}{1.5s}}}$$
(1-16)

The circuit is shown in figure 1-12.

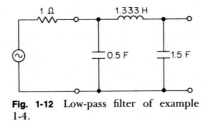

Fig. 1-12 Low-pass filter of example 1-4.

Synthesis by Equating Coefficients An active three-pole low-pass filter is shown in figure 1-13. Its transfer function is given by

$$T(s) = \frac{1}{s^3 A + s^2 B + sC + 1}$$
(1-17)

where

$$A = C_1 C_2 C_3$$
(1-18)

$$B = 2C_3(C_1 + C_2)$$
(1-19)

and

$$C = C_2 + 3C_3$$
(1-20)

If a Butterworth transfer function is desired, we can set equation (1-17) equal to equation (1-10).

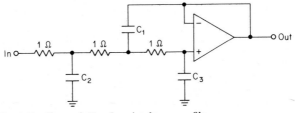

Fig. 1-13 General $N = 3$ active low-pass filter.

$$T(s) = \frac{1}{s^3 A + s^2 B + sC + 1} = \frac{1}{s^3 + 2s^2 + 2s + 1} \qquad (1\text{-}21)$$

By equating coefficients we obtain

$$A = 1$$
$$B = 2$$
$$C = 2$$

Substituting these coefficients in equations (1-18) through (1-20) and solving for C_1, C_2, and C_3 results in the circuit of figure 1-14.

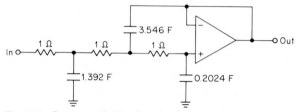

Fig. 1-14   Butterworth $N = 3$ active low-pass filter.

Synthesis of filters directly from polynomials offers an elegant solution to filter design. However, it also may involve laborious computations to determine circuit element values. Design methods have been greatly simplified by the curves, tables, and step-by-step procedures provided in this handbook; so design by synthesis can be left to the advanced specialist.

Active versus Passive Filters

The LC filters of figures 1-11 and 1-12 and the active filter of figure 1-14 all satisfy an $n = 3$ Butterworth low-pass transfer function. The filter designer is frequently faced with the sometimes difficult decision of choosing whether to use an active or LC design. A number of factors must be considered. Some of the limitations and considerations for each filter type will now be discussed.

Frequency Limitations   At subaudio frequencies, LC filter designs require high values of inductance and capacitance along with their associated bulk. Active filters are more practical because they can be designed at higher impedance levels so that capacitor magnitudes are reduced.

Above 50 kHz, most commercial-grade operational amplifiers have insufficient open-loop gain for the average active filter requirement. However, amplifiers are available with extended bandwidth at increased cost so that active filters at frequencies up to 500 kHz are possible. LC filters, on the other hand, are practical at frequencies up to a few hundred megahertz. Beyond this range, filters become impractical to build in lumped form, and so distributed parameter techniques are used.

Size Considerations   Active filters are generally smaller than their LC counterparts, since inductors are not required. Further reduction in size is possible with microelectronic technology. By using deposited RC networks and monolithic operational amplifier chips or with hybrid technology, active filters can be reduced to microscopic proportions.

Economics and Ease of Manufacture LC filters generally cost more than active filters because they use inductors. High-quality coils require efficient magnetic cores. Sometimes, special coil-winding methods are needed. These factors lead to the increased cost of LC filters.

Active filters have the distinct advantage that they can be easily assembled using standard off-the-shelf components. LC filters require coil-winding and coil-assembly skills.

Ease of Adjustment   In critical LC filters, tuned circuits require adjustment to specific resonances. Capacitors cannot be made variable unless they are below a few hundred picofarads. Inductors, however, can easily be adjusted, since most coil structures provide a means for tuning such as an adjustment slug.

Many active filter circuits are not easily adjustable. They may contain RC sections where two or more resistors in each section have to be varied in order to control resonances. These circuits have been avoided. The active filter design techniques presented in this handbook include convenient methods for adjusting resonances where required such as for narrow-band bandpass filters.

REFERENCES

Campbell, G. A., "Physical Theory on the Electric Wave Filter," Bell System Technical Journal, Vol. 1, p. 2, November 1922.

Guillemin, E. A., "Introductory Circuit Theory," John Wiley and Sons, New York, 1953.

ITT, "Reference Data for Radio Engineers," Fourth Edition, International Telephone and Telegraph Corp., pp. 164–185, New York, 1956.

Landee, R. W., et al., "Electronic Designers Handbook," McGraw-Hill Book Company, New York, 1957.

Stewart, J. L., "Circuit Theory and Design," John Wiley and Sons, New York, 1956.

White Electromagnetics, "A Handbook on Electrical Filters," White Electromagnetics Inc., 1963.

Zobel, O. J., "Theory and Design of Uniform and Composite Electric Wave-Filters," Bell System Technical Journal, Vol. 2, p. 1, January 1923.

2

Selecting
the Response
Characteristic

2.1 FREQUENCY-RESPONSE NORMALIZATION

Several parameters are used to characterize a filter's performance. The most commonly specified requirement is frequency response. When given a frequency-response specification, the engineer must select a filter design that meets these requirements. This is accomplished by transforming the required response to a normalized low-pass specification having a cutoff of 1 rad/s. This normalized response is compared with curves of normalized low-pass filters which also have a 1 rad/s cutoff. After a satisfactory low-pass filter is determined from the curves, the tabulated normalized element values of the chosen filter are transformed or denormalized to the final design.

Modern network theory has provided us with many different shapes of amplitude versus frequency which have been analytically derived by placing various restrictions on transfer functions. The major categories of these low-pass responses are:

Butterworth
Chebyshev
Linear phase
Transitional
Synchronously tuned
Elliptic-function

With the exception of the elliptic-function family, these responses are all normalized to a 3-dB cutoff of 1 rad/s.

Frequency and Impedance Scaling

The basis for normalization of filters is the fact that a given filter's response can be scaled (shifted) to a different frequency range by dividing the reactive elements by a frequency-scaling factor (FSF). The FSF is the ratio of a reference frequency of the desired response to the corresponding reference frequency of the given filter. Usually 3-dB points are selected as reference frequencies of low-pass and high-pass filters and the center frequency is chosen as the reference for bandpass filters. The FSF can be expressed as

$$FSF = \frac{\text{desired reference frequency}}{\text{existing reference frequency}} \qquad (2\text{-}1)$$

The FSF must be a dimensionless number; so both the numerator and denominator of equation (2-1) must be expressed in the same units, usually radians per second. The following example demonstrates computation of the FSF and frequency scaling of filters.

Example 2-1

REQUIRED: Low-pass filter either LC or active with $n = 3$ Butterworth transfer function having a 3-dB cutoff at 1000 Hz.

RESULT: Figure 2-1 illustrates the LC and active $n = 3$ Butterworth low-pass filters discussed in chapter 1 and their response.

   (a)   Compute FSF.

$$FSF = \frac{2\pi 1000 \text{ rad/s}}{1 \text{ rad/s}} = 6280 \qquad (2\text{-}1)$$

   (b)   Dividing all the reactive elements by the FSF results in the filters of figure 2-2 a and b and the response of figure 2-2 c.

Note that all points on the frequency axis of the normalized response have been multiplied by the FSF. Also since the normalized filter has its cutoff at 1 rad/s, the FSF can be directly expressed by $2\pi f_c$, where f_c is the desired low-pass cutoff frequency in hertz.

Frequency scaling a filter has the effect of multiplying all points on the frequency axis of the response curve by the FSF. Therefore, a normalized response curve can be directly used to predict the attenuation of the denormalized filter.

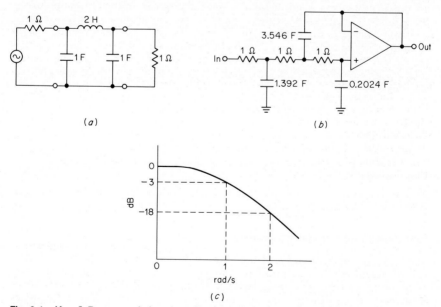

(a)   (b)

(c)

Fig. 2-1 $N = 3$ Butterworth low-pass filter: (a) LC filter; (b) active filter; (c) frequency response.

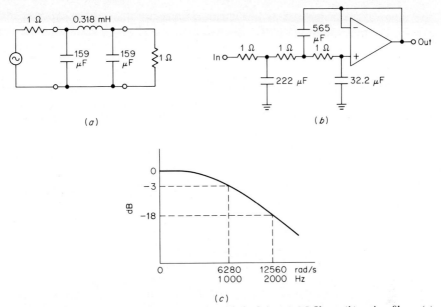

(a)

(b)

(c)

Fig. 2-2 Denormalized low-pass filter of example 2-1: (a) LC filter; (b) active filter; (c) frequency response.

When the filters of figure 2-1 were denormalized to those of figure 2-2, the transfer function changed as well. The denormalized transfer function became

$$T(s) = \frac{1}{4.03 \times 10^{-12}s^3 + 5.08 \times 10^{-9}s^2 + 3.18 \times 10^{-4}s + 1} \tag{2-2}$$

The denominator has the roots: $s = -6280$, $s = -3140 + j5438$, and $s = -3140 - j5438$.

These roots can be obtained directly from the normalized roots by multiplying the normalized root coordinates by the FSF. Frequency scaling a filter also scales the poles and zeros (if any) by the same factor.

The component values of the filters in figure 2-2 are not very practical. The capacitor values are much too large and the $1 - \Omega$ resistor values are not very desirable. This situation can be resolved by impedance scaling. Any linear active or passive network maintains its transfer function if all resistor and inductor values are multiplied by an impedance-scaling factor Z and all capacitors are divided by the same factor Z. This occurs because the Z's cancel in the transfer

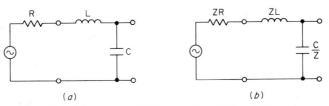

(a)

(b)

Fig. 2-3 Two-pole low-pass LC filter: (a) basic filter; (b) impedance-scaled filter.

function. To prove this, let us investigate the transfer function of the simple two-pole low-pass filter of figure 2-3 a, which is

$$T(s) = \frac{1}{s^2 LC + sCR + 1} \qquad (2\text{-}3)$$

Impedance scaling can be mathematically expressed as

$$R' = ZR \qquad (2\text{-}4)$$

$$L' = ZL \qquad (2\text{-}5)$$

$$C' = \frac{C}{Z} \qquad (2\text{-}6)$$

where the primes denote the values after impedance scaling.

If we impedance-scale the filter, we obtain the circuit of figure 2-3b. The new transfer function becomes

$$T(s) = \frac{1}{s^2 ZL \dfrac{C}{Z} + s \dfrac{C}{Z} ZR + 1} \qquad (2\text{-}7)$$

Clearly, the Z's cancel; so both transfer functions are equivalent.

We can now use impedance scaling to make the values in the filters of figure 2-2 more practical. If we use impedance scaling with a Z of 1000, we obtain the filters of figure 2-4. The values are certainly more suitable.

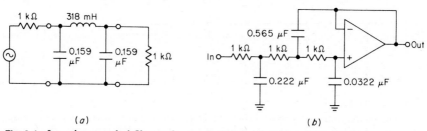

(a)   (b)

Fig. 2-4   Impedance-scaled filters of example 2-1: (a) LC filter; (b) active filter.

Frequency and impedance scaling are normally combined into one step rather than performed sequentially. The denormalized values are then given by

$$R' = R \times Z \qquad (2\text{-}8)$$

$$L' = \frac{L \times Z}{FSF} \qquad (2\text{-}9)$$

$$C' = \frac{C}{FSF \times Z} \qquad (2\text{-}10)$$

where the primed values are both frequency- and impedance-scaled.

Low-Pass Normalization

In order to use normalized low-pass filter curves and tables, a given low-pass filter requirement must first be converted into a normalized requirement. The curves can now be entered to find a satisfactory normalized filter which is then scaled to the desired cutoff.

The first step in selecting a normalized design is to convert the requirement into a steepness factor A_s, which can be defined as

$$A_s = \frac{f_s}{f_c} \tag{2-11}$$

where f_s is the frequency having the minimum required stopband attenuation and f_c is the limiting frequency or cutoff of the passband, usually the 3-dB point. The normalized curves are compared with A_s, and a design is selected that meets or exceeds the requirement. The design is then frequency-scaled so that the selected passband limit of the normalized design occurs at f_c.

If the required passband limit f_c is defined as the 3-dB cutoff, the steepness factor A_s can be directly looked up in radians per second on the frequency axis of the normalized curves.

Suppose that we required a low-pass filter that has a 3-dB point at 100 Hz and more than 30 dB attenuation at 400 Hz. A normalized low-pass filter that has its 3-dB point at 1 rad/s and over 30 dB attenuation at 4 rad/s would meet the requirement if the filter were frequency-scaled so that the 3-dB point occurred at 100 Hz. Then there would be over 30 dB attenuation at 400 Hz, or 4 times the cutoff, because a response shape is retained when a filter is frequency-scaled.

The following example demonstrates normalizing a simple low-pass requirement.

Example 2-2

REQUIRED: Normalize the following specification:
Low-pass filter
3 dB at 200 Hz
30 dB minimum at 800 Hz

RESULT: (a) Compute A_s.

$$A_s = \frac{f_s}{f_c} = \frac{800 \text{ Hz}}{200 \text{ Hz}} = 4 \tag{2-11}$$

(b) Normalized requirement:
3 dB at 1 rad/s
30 dB minimum at 4 rad/s

In the event f_c does not correspond to the 3-dB cutoff, A_s can still be computed and a normalized design found that will meet the specifications. This is illustrated in the following example.

Example 2-3

REQUIRED: Normalize the following specification:
Low-pass filter
1 dB at 200 Hz
30 dB minimum at 800 Hz

RESULT: (a) Compute A_s.

$$A_s = \frac{f_s}{f_c} = \frac{800 \text{ Hz}}{200 \text{ Hz}} = 4 \tag{2-11}$$

(b) Normalized requirement:
1 dB at K rad/s
30 dB minimum at $4K$ rad/s
(where K is arbitrary)

A possible solution to example 2-3 would be a normalized filter which has a 1-dB point at 0.8 rad/s and over 30 dB attenuation at 3.2 rad/s. The fundamental requirement is that the normalized filter makes the transition between the passband and stopband limits within a frequency ratio A_s.

High-Pass Normalization

A normalized $n = 3$ low-pass Butterworth transfer function was given in section 1.2 as

$$T(s) = \frac{1}{s^3 + 2s^2 + 2s + 1} \tag{1-10}$$

and the results of evaluating this transfer function at various frequencies were:

ω	$|T(j\omega)|$	$20 \log|T(j\omega)|$
0	1	0 dB
1	0.707	−3 dB
2	0.124	−18 dB
3	0.0370	−29 dB
4	0.0156	−36 dB

Let us now perform a high-pass transformation by substituting $1/s$ for s in equation (1-10). After some algebraic manipulations the resulting transfer function becomes

$$T(s) = \frac{s^3}{s^3 + 2s^2 + 2s + 1} \tag{2-12}$$

If we evaluate this expression at specific frequencies, we can generate the following table:

ω	$|T(j\omega)|$	$20 \log|T(j\omega)|$
0.25	0.0156	−36 dB
0.333	0.0370	−29 dB
0.500	0.124	−18 dB
1	0.707	−3 dB
∞	1	0 dB

The response is clearly that of a high-pass filter. It is also apparent that the low-pass attenuation values now occur at high-pass frequencies that are exactly the reciprocals of the corresponding low-pass frequencies. A high-pass transformation of a normalized low-pass filter transposes the low-pass attenuation values to reciprocal frequencies and retains the 3-dB cutoff at 1 rad/s. This relationship is evident in figure 2-5, where both filter responses are compared.

The normalized low-pass curves could be interpreted as normalized high-pass curves by reading the attenuation as indicated and taking the reciprocals of the frequencies. However, it is much easier to convert a high-pass specification into a normalized low-pass requirement and use the curves directly.

To normalize a high-pass filter specification calculate A_s, which in the case of high-pass filters is given by

$$A_s = \frac{f_c}{f_s} \qquad (2\text{-}13)$$

Since the A_s for high-pass filters is defined as the reciprocal of the A_s for low-pass filters, equation (2-13) can be directly interpreted as a low-pass requirement.

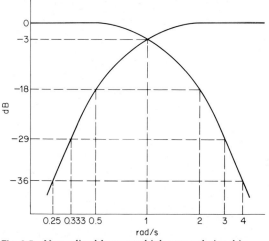

Fig. 2-5 Normalized low-pass high-pass relationship.

A normalized low-pass filter can then be selected from the curves. A high-pass transformation is performed on the corresponding low-pass filter, and the resulting high-pass filter is scaled to the desired cutoff frequency.

The following example shows the normalization of a high-pass filter requirement.

Example 2-4

REQUIRED: Normalize the following requirement:
            High-pass filter
            3 dB at 200 Hz
            30 dB minimum at 50 Hz

RESULT: (a) Compute A_s.

$$A_s = \frac{f_c}{f_s} = \frac{200 \text{ Hz}}{50 \text{ Hz}} = 4 \qquad (2\text{-}13)$$

(b) Normalized equivalent low-pass requirement:
      3 dB at 1 rad/s
      30 dB minimum at 4 rad/s

Bandpass Normalization

Bandpass filters fall into two categories, narrow-band and wide-band. If the ratio of the upper cutoff frequency to the lower cutoff frequency is over 2 (an octave), the filter is considered a wide-band type.

Wide-Band Bandpass Filters   Wide-band filter specifications can be separated into individual low-pass and high-pass requirements which are treated independently. The resulting low-pass and high-pass filters are then cascaded to meet the composite response.

Example 2-5

REQUIRED: Normalize the following specification:

> Bandpass filter
> 3 dB at 500 and 1000 Hz
> 40 dB minimum at 200 and 2000 Hz

RESULT: (a)   Determine the ratio of upper cutoff to lower cutoff.

$$\frac{1000 \text{ Hz}}{500 \text{ Hz}} = 2$$

wide-band type

(b)   Separate requirement into individual specifications.

High-pass filter:	Low-pass filter:
3 dB at 500 Hz	3 dB at 1000 Hz
40 dB minimum at 200 Hz	40 dB minimum at 2000 Hz
$A_s = 2.5$ (2-13)	$A_s = 2.0$ (2-11)

(c)   Normalized high-pass and low-pass filters are now selected, scaled to the required cutoff frequencies, and cascaded to meet the composite requirements. Figure 2-6 shows the resulting circuit and response.

Narrow-Band Bandpass Filters   Narrow-band bandpass filters have a ratio of upper cutoff frequency to lower cutoff frequency of approximately 2 or less

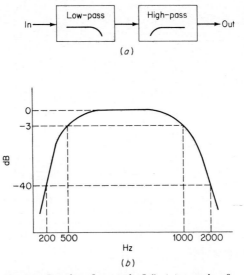

Fig. 2-6   Results of example 2-5: (a) cascade of low-pass and high-pass filters; (b) frequency response.

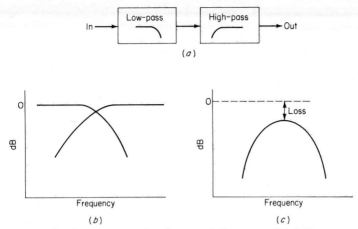

Fig. 2-7 Limitation of wide-band approach for narrow-band filters: (a) cascade of low-pass and high-pass filters; (b) composite response; (c) algebraic sum of attenuation.

and cannot be designed as separate low-pass and high-pass filters. The major reason for this is evident from figure 2-7. As the ratio of upper cutoff to lower cutoff decreases, the loss at center frequency will increase, and it may become prohibitive for ratios near unity.

If we substitute $s + 1/s$ for s in a low-pass transfer function, a bandpass filter results. The center frequency occurs at 1 rad/s, and the frequency response of the low-pass filter is directly transformed into the bandwidth of the bandpass filter at points of equivalent attenuation. In other words, the attenuation bandwidth ratios remain unchanged. This is shown in figure 2-8, which shows the relationship between a low-pass filter and its transformed bandpass equivalent. Each pole and zero of the low-pass filter is transformed into a *pair* of poles and zeros in the bandpass filter.

In order to design a bandpass filter, the following sequence of steps is involved.

1. Convert the given bandpass filter requirement into a normalized low-pass specification.
2. Select a satisfactory low-pass filter from the normalized frequency-response curves.
3. Transform the normalized low-pass parameters into the required bandpass filter.

The response shape of a bandpass filter is shown in figure 2-9 along with some basic terminology. The center frequency is defined as

$$f_0 = \sqrt{f_L f_u} \tag{2-14}$$

where f_L is the lower passband limit and f_u is the upper passband limit, usually the 3-dB attenuation frequencies. For the more general case

$$f_0 = \sqrt{f_1 f_2} \tag{2-15}$$

where f_1 and f_2 are any two frequencies having equal attenuation. These relationships imply geometric symmetry; that is, the entire curve below f_0 is the mirror image of the curve above f_0 when plotted on a *logarithmic* frequency axis.

An important parameter of bandpass filters is the filter selectivity factor or Q, which is defined as

$$Q = \frac{f_0}{\text{BW}} \qquad (2\text{-}16)$$

where BW is the passband bandwidth or $f_u - f_L$.

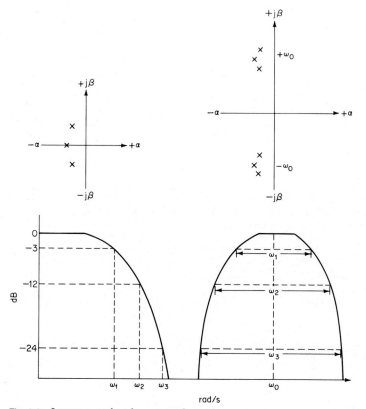

Fig. 2-8 Low-pass to bandpass transformation.

As the filter Q increases, the response shape near the passband approaches the arithmetically symmetrical condition, i.e., mirror-image symmetry near the center frequency, when plotted using a *linear* frequency axis. For Q's of 10 or more the center frequency can be redefined as the arithmetic mean of the passband limits; so we can replace equation (2-14) with

$$f_0 = \frac{f_L + f_u}{2} \qquad (2\text{-}17)$$

In order to utilize the normalized low-pass filter frequency-response curves, a given narrow-band bandpass filter specification must be transformed into a normalized low-pass requirement. This is accomplished by first manipulating the specification to make it geometrically symmetrical. At equivalent attenuation points, corresponding frequencies above and below f_0 must satisfy

$$f_1 f_2 = f_0^2 \qquad (2\text{-}18)$$

which is an alternate form of equation (2-15) for geometric symmetry. The given specification is modified by calculating the corresponding opposite geome-

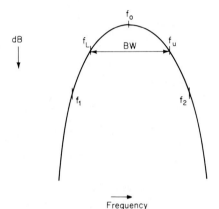

Fig. 2-9 General bandpass filter response shape.

tric frequency for each stopband frequency specified. Each pair of stopband frequencies will result in two new frequency pairs. The pair having the lesser separation is retained, since it represents the more severe requirement.

A bandpass filter steepness factor can now be defined as

$$A_s = \frac{\text{stopband bandwidth}}{\text{passband bandwidth}} \qquad (2\text{-}19)$$

This steepness factor is used to select a normalized low-pass filter from the frequency-response curves that makes the passband to stopband transition within a frequency ratio of A_s.

The following example illustrates the normalization of a bandpass filter requirement.

Example 2-6

REQUIRED: Normalize the following bandpass filter requirement:
Bandpass filter
Center frequency of 100 Hz
3 dB at ±15 Hz (85 Hz, 115 Hz)
40 dB at ±30 Hz (70 Hz, 130 Hz)

RESULT: (a) First compute center frequency f_0.

$$f_0 = \sqrt{f_L f_u} = \sqrt{85 \times 115} = 98.9 \text{ Hz} \qquad (2\text{-}14)$$

(b) Compute two geometrically related stopband frequency pairs for each pair of stopband frequencies given.

Let $f_1 = 70$ Hz.

$$f_2 = \frac{f_0^2}{f_1} = \frac{(98.9)^2}{70} = 139.7 \text{ Hz} \qquad (2\text{-}18)$$

Let $f_2 = 130$ Hz.

$$f_1 = \frac{f_0^2}{f_2} = \frac{(98.9)^2}{130} = 75.2 \text{ Hz} \qquad (2\text{-}18)$$

The two pairs are

$$f_1 = 70 \text{ Hz}, f_2 = 139.7 \text{ Hz } (f_2 - f_1 = 69.7 \text{ Hz})$$

and $\qquad f_1 = 75.2 \text{ Hz}, f_2 = 130 \text{ Hz } (f_2 - f_1 = 54.8 \text{ Hz})$

Retain the second frequency pair, since it has the lesser separation. Figure 2-10 compares the specified filter requirement and the geometrically symmetrical equivalent.

   (c) Calculate A_s.

$$A_s = \frac{\text{stopband bandwidth}}{\text{passband bandwidth}} = \frac{54.8 \text{ Hz}}{30 \text{ Hz}} = 1.83 \qquad (2\text{-}19)$$

   (d) A normalized low-pass filter can now be selected from the normalized curves. Since the passband limit is the 3-dB point, the normalized filter is required to have over 40 dB of rejection at 1.83 rad/s or 1.83 times the 1-rad/s cutoff.

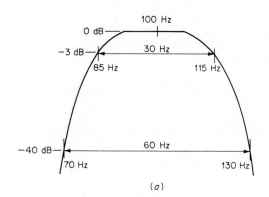

(a)

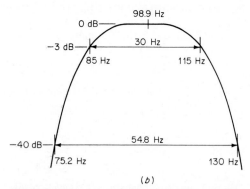

(b)

Fig. 2-10 Frequency-response requirements of example 2-6: (a) given filter requirement; (b) geometrically symmetrical requirement.

Bandpass filter requirements are not always specified in an arithmetically symmetrical manner as in the previous example. Multiple stopband attenuation requirements may also exist. The design engineer is still faced with the basic problem of converting the given parameters into geometrically symmetrical characteristics so that a steepness factor or factors can be determined. The following example demonstrates conversion of a specification somewhat more complicated than the previous example.

Example 2-7

REQUIRED: Normalize the following bandpass filter specification:
Bandpass filter
1 dB passband limits of 12 and 14 kHz
20 dB minimum at 6 kHz
30 dB minimum at 4 kHz
40 dB minimum at 56 kHz

RESULT: (a) First compute the center frequency.

$$f_L = 12 \text{ kHz} \qquad f_u = 14 \text{ kHz}$$
$$f_0 = 12.96 \text{ kHz} \qquad\qquad\qquad (2\text{-}14)$$

(b) Compute the corresponding geometric frequency for each stopband frequency given, using equation (2-18).

$$f_1 f_2 = f_0^2 \qquad\qquad\qquad (2\text{-}18)$$

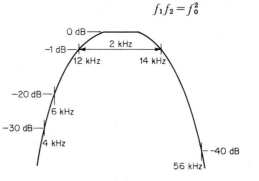

(a)

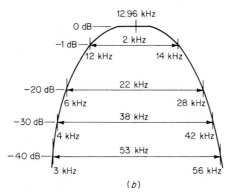

(b)

Fig. 2-11 Given and transformed response of example 2-7: (a) given requirement; (b) geometrically symmetrical response.

f_1	f_2
6 kHz	28 kHz
4 kHz	42 kHz
3 kHz	56 kHz

Figure 2-11 illustrates the comparison between the given requirement and the corresponding geometrically symmetrical equivalent response.

(c) Calculate the steepness factor for each stopband bandwidth in figure 2-11b.

20 dB:
$$A_s = \frac{22 \text{ kHz}}{2 \text{ kHz}} = 11 \qquad (2\text{-}19)$$

30 dB:
$$A_s = \frac{38 \text{ kHz}}{2 \text{ kHz}} = 19$$

40 dB:
$$A_s = \frac{53 \text{ kHz}}{2 \text{ kHz}} = 26.5$$

(d) Select a low-pass filter from the normalized tables. A filter is required that has over 20, 30, and 40 dB of rejection at, respectively, 11, 19, and 26.5 times its 1-dB cutoff.

Band-Reject Normalization

Wide-Band Band-Reject Filters   Normalizing a band-reject filter requirement proceeds along the same lines as a bandpass filter. If the ratio of the upper cutoff frequency to the lower cutoff frequency is an octave or more, a band-reject filter requirement can be classified as wide-band and separated into individual low-pass and high-pass specifications. The resulting filters are paralleled at the input and combined at the output. The following example demonstrates normalization of a wide-band band-reject filter requirement.

Example 2-8

REQUIRED: Band-reject filter
3 dB at 200 and 800 Hz
40 dB minimum at 300 and 500 Hz

RESULT: (a)   Determine ratio of upper cutoff to lower cutoff.

$$\frac{800 \text{ Hz}}{200 \text{ Hz}} = 4$$

wide-band type

(b) Separate requirements into individual low-pass and high-pass specifications.

Low-pass filter:	High-pass filter:
3 dB at 200 Hz	3 dB at 800 Hz
40 dB minimum at 300 Hz	40 dB minimum at 500 Hz
$A_s = 1.5$ (2-11)	$A_s = 1.6$ (2-13)

(c) Select appropriate filters from the normalized curves and scale the normalized low-pass and high-pass filters to cutoffs of 200 and 800 Hz, respectively. Figure 2-12 shows the resulting circuit and response.

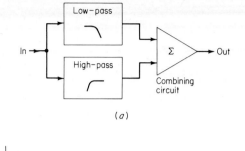

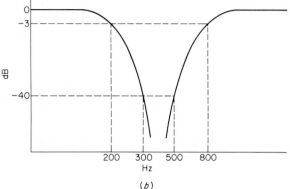

Fig. 2-12 Results of example 2-8: *(a)* combined low-pass and high-pass filters; *(b)* frequency response.

The basic assumption of the previous example is that when the filter outputs are combined, the resulting response is the superimposed individual response of both filters. This is a valid assumption if each filter has sufficient rejection in the band of the other filter so that there is no interaction when the outputs are combined. Figure 2-13 shows the case where inadequate separation exists.

The requirement for a minimum separation between cutoffs of an octave or more is by no means rigid. Sharper filters can have their cutoffs placed closer together with minimal interaction.

Narrow-Band Band-Reject Filters The normalized transformation described for bandpass filters where $s + 1/s$ is substituted into a low-pass transfer function can instead be applied to a high-pass transfer function to obtain a band-reject filter. Figure 2-14 shows the direct equivalence between a high-pass filter's frequency response and the transformed band-reject filter's bandwidth.

The design method for narrow-band band-reject filters can be defined as follows:

1. Convert the band-reject requirement directly into a normalized low-pass specification.
2. Select a low-pass filter from the normalized curves that meets the normalized requirements.

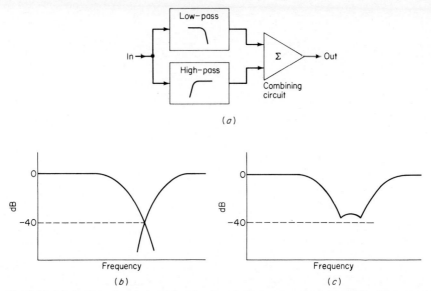

(a)

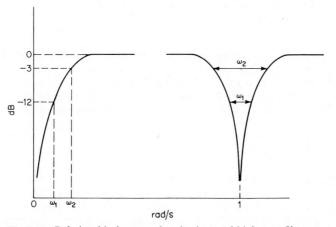

(b) (c)

Fig. 2-13 Limitation of wide-band band-reject design approach: (a) combined low-pass and high-pass filters; (b) composite response; (c) combined response by summation of outputs.

Fig. 2-14 Relationship between band-reject and high-pass filters.

3. Transform the normalized low-pass parameters into the required band-reject filter. This may involve designing the intermediate high-pass filter, or the transformation may be direct.

The band-reject response has geometric symmetry just as bandpass filters have. Figure 2-15 defines this response shape. The parameters shown have the same relationship to each other as they do for bandpass filters. The attenuation

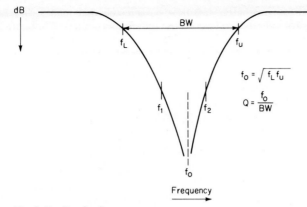

Fig. 2-15   Band-reject response.

at center frequency is theoretically infinite, since the response of a high-pass filter at DC has been transformed to the center frequency.

The geometric center frequency can be defined as

$$f_0 = \sqrt{f_L f_u} \qquad (2\text{-}14)$$

where f_L and f_u are usually the 3-dB frequencies, or for the more general case:

$$f_0 = \sqrt{f_1 f_2} \qquad (2\text{-}15)$$

The selectivity factor Q is defined as

$$Q = \frac{f_0}{\text{BW}} \qquad (2\text{-}16)$$

where BW is $f_u - f_L$. For Q's of 10 or more, the response near center frequency approaches the arithmetically symmetrical condition; so we can then state

$$f_0 = \frac{f_L + f_u}{2} \qquad (2\text{-}17)$$

To use the normalized curves for the design of a band-reject filter, the response requirement must be converted to a normalized low-pass filter specification. In order to accomplish this, the band-reject specification should first be made geometrically symmetrical; that is, each pair of frequencies having equal attenuation should satisfy

$$f_1 f_2 = f_0^2 \qquad (2\text{-}18)$$

which is an alternate form of equation (2-15). When two frequencies are specified at a particular attenuation level, two frequency pairs will result from calculating the corresponding opposite geometric frequency for each frequency specified. Retain the pair having the wider separation, since it represents the more severe requirement. In the bandpass case the pair having the lesser separation represented the more difficult requirement.

The band-reject filter steepness factor is defined by

$$A_s = \frac{\text{passband bandwidth}}{\text{stopband bandwidth}} \qquad (2\text{-}20)$$

A normalized low-pass filter can now be selected that makes the transition from the passband attenuation limit to the minimum required stopband attenuation within a frequency ratio A_s.

The following example demonstrates the normalization procedure for a band-reject filter.

Example 2-9

REQUIRED: Band-reject filter
Center frequency of 1000 Hz
3 dB at ±300 Hz (700 Hz, 1300 Hz)
40 dB at ±200 Hz (800 Hz, 1200 Hz)

RESULT: (a)   First compute center frequency f_0.

$$f_0 = \sqrt{f_L f_u} = \sqrt{700 \times 1300} = 954 \text{ Hz} \qquad (2\text{-}14)$$

(b)   Compute two geometrically related stopband frequency pairs for each pair of stopband frequencies given:
Let $f_1 = 800$ Hz

$$f_2 = \frac{f_0^2}{f_1} = \frac{(954)^2}{800} = 1138 \text{ Hz} \qquad (2\text{-}18)$$

Let $f_2 = 1200$ Hz

$$f_1 = \frac{f_0^2}{f_2} = \frac{(954)^2}{1200} = 758 \text{ Hz} \qquad (2\text{-}18)$$

the two pairs are

$$f_1 = 800 \text{ Hz}, f_2 = 1138 \text{ Hz} \ (f_2 - f_1 = 338 \text{ Hz})$$
and
$$f_1 = 758 \text{ Hz}, f_2 = 1200 \text{ Hz} \ (f_2 - f_1 = 442 \text{ Hz})$$

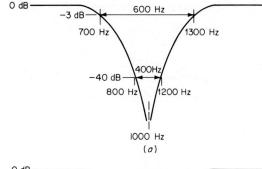

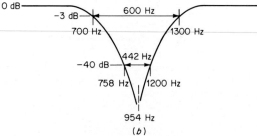

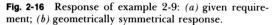

Fig. 2-16   Response of example 2-9: (a) given requirement; (b) geometrically symmetrical response.

Retain the second pair, since it has the *wider* separation and represents the more severe requirement. The given response requirement and the geometrically symmetrical equivalent are compared in figure 2-16.

(c) Calculate A_s.

$$A_s = \frac{\text{passband bandwidth}}{\text{stopband bandwidth}} = \frac{600 \text{ Hz}}{442 \text{ Hz}} = 1.36 \quad (2\text{-}20)$$

(d) Select a normalized low-pass filter from the normalized curves that makes the transition from the 3-dB point to the 40-dB point within a frequency ratio of 1.36. Since these curves are all normalized to 3 dB, a filter is required with over 40 dB of rejection at 1.36 rad/s.

2.2 TRANSIENT RESPONSE

In our previous discussions of filters we have restricted our interest to frequency-domain parameters such as frequency response. The input forcing function was a sine wave. In real-world applications of filters, input signals consist of a variety of complex waveforms. The response of filters to these nonsinusoidal inputs is called "transient response."

A filter's transient response is best evaluated in the time domain, since we are usually dealing with input signals which are functions of time such as pulses or amplitude steps. The frequency- and time-domain parameters of a filter are directly related through the Fourier or Laplace transforms.

Effect of Nonuniform Time Delay

Evaluating a transfer function as a function of frequency results in both a magnitude and phase characteristic. Figure 2-17 shows the amplitude and phase response of a normalized $n = 3$ Butterworth low-pass filter. Butterworth low-pass filters have a phase shift of exactly n times $-45°$ at the 3-dB frequency. The phase shift continuously increases as the transition is made into the stopband and eventually approaches n times $-90°$ at frequencies far removed from the passband. Since the filter described by figure 2-17 has a complexity of $n = 3$, the phase shift is $-135°$ at the 3-dB cutoff and approaches $-270°$ in the stopband. Frequency scaling will transpose the phase characteristics to a new frequency range as determined by the FSF.

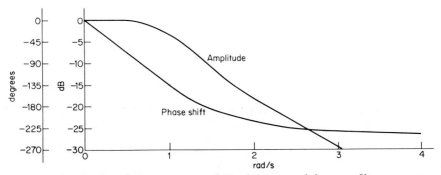

Fig. 2-17 Amplitude and phase response of $N = 3$ Butterworth low-pass filter.

It is well known that a square wave can be represented by a Fourier series of odd harmonic components as indicated in figure 2-18. Since the amplitude of each harmonic is reduced as the harmonic order increases, only the first

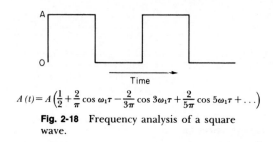

$$A(t) = A\left(\frac{1}{2} + \frac{2}{\pi}\cos\omega_1\tau - \frac{2}{3\pi}\cos 3\omega_1\tau + \frac{2}{5\pi}\cos 5\omega_1\tau + \ldots\right)$$

Fig. 2-18 Frequency analysis of a square wave.

few harmonics are of significance. If a square wave is applied to a filter, the fundamental and its significant harmonics must have the proper relative amplitude relationship at the filter's output in order to retain the square waveshape. In addition, these components must not be displaced in time with respect to each other. Let us now consider the effect of a low-pass filter's phase shift on a square wave.

If we assume that a low-pass filter has a linear phase shift between 0° at DC and n times $-45°$ at the cutoff, we can express the phase shift in the passband as

$$\phi = -\frac{45nf_x}{f_c} \tag{2-21}$$

where f_x is any frequency in the passband and f_c is the 3-dB cutoff frequency.

A phase-shifted sine wave appears displaced in time from the input waveform. This displacement is called "phase delay" and can be computed by determining the time interval represented by the phase shift, using the fact that a full period contains 360°. Phase delay can then be computed by

$$T_{pd} = \frac{\phi}{360}\frac{1}{f_x} \tag{2-22}$$

or, as an alternate form,

$$T_{pd} = -\frac{\beta}{\omega} \tag{2-23}$$

where β is the phase shift in radians (1 rad = $360/2\pi$ or 57.3°) and ω is the input frequency expressed in radians per second ($\omega = 2\pi f_x$).

Example 2-10

REQUIRED: Compute the phase delay of the fundamental and the third, fifth, seventh, and ninth harmonics of a 1-kHz square wave applied to an $n = 3$ Butterworth low-pass filter having a 3-dB cutoff of 10 kHz. Assume a linear phase shift with frequency in the passband.

RESULT: Using formulas (2-21) and (2-22), the following table can be computed:

Frequency	ϕ	T_{pd}
1 kHz	−13.5°	37.5 μs
3 kHz	−40.5°	37.5 μs
5 kHz	−67.5°	37.5 μs
7 kHz	−94.5°	37.5 μs
9 kHz	−121.5°	37.5 μs

The phase delays of the fundamental and each of the significant harmonics in example 2-10 are identical. The output waveform would then appear nearly equivalent to the input except for a delay of 37.5 μs. If the phase shift is not linear with frequency, the ratio ϕ/f_x in equation (2-22) is not constant; so each significant component of the input square wave would undergo a different delay. This displacement in time of the spectral components, with respect to each other, introduces a distortion of the output waveform. Figure 2-19 shows some typical effects of nonlinear phase shift upon a square wave. Most filters have nonlinear phase versus frequency characteristics; so some waveform distortion will usually occur for complex input signals.

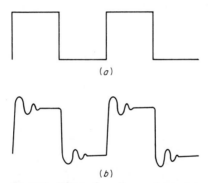

Fig. 2-19 Effect of nonlinear phase: *(a)* ideal square wave; *(b)* distorted square wave.

Not all complex waveforms have harmonically related spectral components. An amplitude-modulated signal, for example, consists of a carrier and two sidebands, each sideband separated from the carrier by the modulating frequency. If a filter's phase characteristic is linear with frequency and intersects zero phase shift at zero frequency (DC), both the carrier and the two sidebands will have the same delay in passing through the filter; so the output will be a delayed replica of the input. If these conditions are not satisfied, the carrier and both sidebands will be delayed by different amounts. The carrier delay will be in accordance with the equation for phase delay

$$T_{pd} = -\frac{\beta}{\omega} \qquad (2\text{-}23)$$

(The terms carrier delay and phase delay are used interchangeably.)

A new definition is required for the delay of the sidebands. This delay is commonly called "group delay" and is defined as the derivative of phase versus frequency, which can be expressed as

$$T_{gd} = -\frac{d\beta}{d\omega} \tag{2-24}$$

Linear phase shift results in constant group delay, since the derivative of a linear function is a constant. Figure 2-20 illustrates a low-pass filter phase shift which is nonlinear in the vicinity of a carrier ω_c and the two sidebands $\omega_c - \omega_m$ and $\omega_c + \omega_m$. The phase delay at ω_c is the negative slope of a line drawn

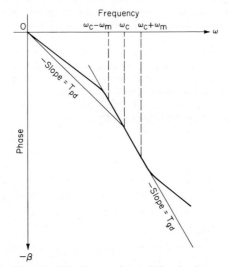

Fig. 2-20 Nonlinear phase shift of a low-pass filter.

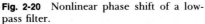
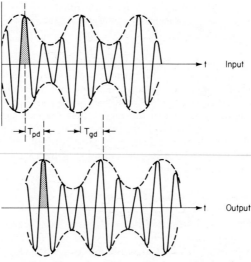

Fig. 2-21 Effect of nonlinear phase on AM signal.

from the origin to the phase shift corresponding to ω_c which is in agreement with equation (2-23). The group delay at ω_c is shown as the negative slope of a line which is tangent to the phase response at ω_c. This can be mathematically expressed as

$$T_{gd} = -\frac{d\beta}{d\omega}\bigg|_{\omega=\omega_c}$$

If the two sidebands are restricted to a region surrounding ω_c having a constant group delay, the envelope of the modulated signal will be delayed by T_{gd}. Figure 2-21 compares the input and output waveforms of an amplitude-modulated signal applied to the filter depicted by figure 2-20. Note that the carrier is delayed by the phase delay while the envelope is delayed by the group delay. For this reason group delay is sometimes called "envelope delay."

If the group delay is not constant over the bandwidth of the modulated signal, waveform distortion will occur. Narrow-bandwidth signals are more likely to encounter constant group delay than signals having a wider spectrum. It is common practice to use group-delay variation as a criterion to evaluate phase nonlinearity and subsequent waveform distortion. The absolute magnitude of the nominal delay is usually of little consequence.

Step Response of Networks

If we were to define a hypothetical ideal low-pass filter, it would have the response shown in figure 2-22. The amplitude response is unity from DC to the cutoff frequency ω_c and zero beyond the cutoff. The phase shift is a linearly increasing function in the passband, where n is the order of the ideal filter. The group delay is constant in the passband and zero in the stopband. If a unity amplitude step were applied to this ideal filter at $t = 0$, the output would be in accordance

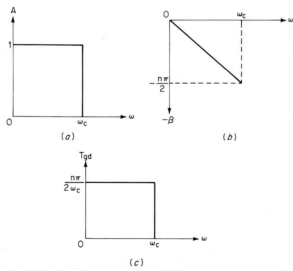

Fig. 2-22 Ideal low-pass filter: (a) frequency response; (b) phase shift; (c) group delay.

with figure 2-23. The delay of the half-amplitude point would be $n\pi/2\omega_c$ and the rise time, which is defined as the interval required to go from zero amplitude to unity amplitude with a slope equal to that at the half-amplitude point, would be equal to π/ω_c. Since rise time is inversely proportional to ω_c, a wider filter results in reduced rise time. This proportionality is in agreement with a fundamental rule of thumb relating rise time to bandwidth, which is

$$T_r \approx \frac{0.35}{f_c} \tag{2-25}$$

where T_r is the rise time in seconds and f_c is the 3-dB cutoff in hertz.

A 9% overshoot exists on the leading edge. Also a sustained oscillation occurs having a period of $2\pi/\omega_c$ which eventually decays and unity amplitude is estab-

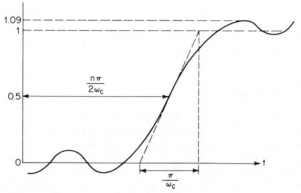

Fig. 2-23   Step response of ideal low-pass filter.

lished. This oscillation is called "ringing." Overshoot and ringing occur in an ideal low-pass filter even though we have linear phase. This is because of the abrupt amplitude roll-off at cutoff. Therefore, both linear phase and a prescribed roll-off are required for minimum transient distortion.

Overshoot and prolonged ringing are both very undesirable if the filter is required to pass pulses with minimum waveform distortion. The step-response curves provided for the different families of normalized low-pass filters can be very useful for evaluating the transient properties of these filters.

Impulse Response

A unit impulse is defined as a pulse which is infinitely high and infinitesimally narrow, and has an area of unity. The response of the ideal filter of figure 2-22 to a unit impulse is shown in figure 2-24. The peak output amplitude is ω_c/π, which is proportional to the filter's bandwidth. The pulse width, $2\pi/\omega_c$, is inversely proportional to bandwidth.

An input signal having the form of a unit impulse is physically impossible. However, a narrow pulse of finite amplitude will represent a reasonable approximation; so the impulse response of normalized low-pass filters can be useful in estimating the filter's response to a relatively narrow pulse.

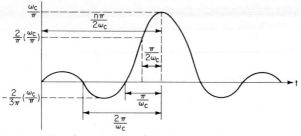

Fig. 2-24 Impulse response of ideal low-pass filter.

Estimating Transient Characteristics

Group-delay, step-response, and impulse-response curves are given for the nor-malized low-pass filters discussed in the latter section of this chapter. These curves are useful for estimating filter responses to nonsinusoidal signals. If the input waveforms are steps or pulses, the curves may be used directly. For more complex inputs we can use the method of superposition, which permits the representation of a complex signal as the sum of individual components. If we find the filter's output for each individual input signal, we can combine these responses to obtain the composite output.

Group Delay of Low-Pass Filters   When a normalized low-pass filter is fre-quency-scaled, the delay characteristics are frequency-scaled as well. The follow-ing rules can be applied to derive the resulting delay curve from the normalized response:

1. Divide the delay axis by $2\pi f_c$, where f_c is the filter's 3-dB cutoff.
2. Multiply all points on the frequency axis by f_c.

The following example demonstrates the denormalization of a low-pass curve.

Example 2-11

REQUIRED: Using the normalized delay curve of an $n = 3$ Butterworth low-pass filter given in figure 2-25a, compute the delay at DC and the delay variation in the passband if the filter is frequency-scaled to a 3-dB cutoff of 100 Hz.
RESULT: To denormalize the curve, divide the delay axis by $2\pi f_c$ and multiply the frequency axis by f_c where f_c is 100 Hz. The resulting curve is shown in figure 2-25b. The delay at DC is 3.2 ms, and the delay variation in the passband is 1.3 ms.

The nominal delay of a low-pass filter at frequencies well below the cutoff can be estimated by the following formula:

$$T \approx \frac{125n}{f_c} \tag{2-26}$$

where T is the delay in milliseconds, n is the order of the filter, and f_c is the 3-dB cutoff in hertz. Equation (2-26) is an approximation which usually is accurate to within 25%.

Group Delay of Bandpass Filters   When a low-pass filter is transformed to a narrow-band bandpass filter, the delay is transformed to a nearly symmetrical curve mirrored about the center frequency. As the bandwidth increases from the narrow-bandwidth case, the symmetry of the delay curve is distorted approximately in proportion to the filter's bandwidth.

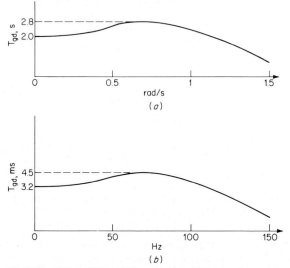

Fig. 2-25   Delay of $N = 3$ Butterworth low-pass filter: (a) normalized delay; (b) delay with $f_c = 100$ Hz.

For the narrow-band condition the bandpass delay curve can be approximated by implementing the following rules:

1. Divide the delay axis of the normalized delay curve by πBW where BW is the 3-dB bandwidth in hertz.
2. Multiply the frequency axis by BW/2.
3. A delay characteristic symmetrical around the center frequency can now be formed by generating the mirror image of the curve obtained by implementing steps 1 and 2. The total 3-dB bandwidth becomes BW.

The following example demonstrates the approximation of a narrow-band bandpass filter's delay curve.

Example 2-12

REQUIRED: Estimate the group delay at the center frequency and the delay variation over the passband of a bandpass filter having a center frequency of 1000 Hz and a 3-dB bandwidth of 100 Hz. The bandpass filter is derived from a normalized $n = 3$ Butterworth low-pass filter.

RESULT: The delay of the normalized filter is shown in figure 2-25a. If we divide the delay axis by πBW and multiply the frequency axis by BW/2, where BW = 100 Hz, we obtain the delay curve of figure 2-26a. We can now reflect this delay curve on both sides of the center frequency of 1000 Hz to obtain figure 2-26b. The delay at center frequency is 6.4 ms and the delay variation over the passband is 2.6 ms.

The technique used in example 2-12 to approximate a bandpass delay curve is valid for bandpass filter Q's of 10 or more (f_0/BW $\geqq$ 10). As the fractional bandwidth increases, the delay becomes less symmetrical and peaks toward the low side of center frequency as shown in figure 2-27.

The delay at center frequency of a bandpass filter can be estimated by

$$T \approx \frac{250\,n}{\text{BW}} \qquad (2\text{-}27)$$

where T is the delay in milliseconds. This approximation is usually accurate within 25%.

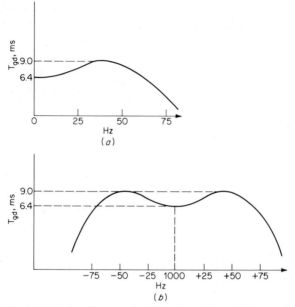

Fig. 2-26 Delay of narrow-band bandpass filter: (a) low-pass delay; (b) bandpass delay.

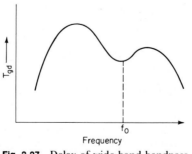

Fig. 2-27 Delay of wide-band bandpass filter.

Comparison of figures 2-25*b* and 2-26*b* indicates that a bandpass filter has twice the delay of the equivalent low-pass filter of the same bandwidth. This results from the low-pass to bandpass transformation where a low-pass filter transfer function of order n always results in a bandpass filter transfer function having an order $2n$. However, a bandpass filter is conventionally referred to as having the same order n as the low-pass filter it was derived from.

Step Response of Low-Pass Filters Delay distortion usually cannot be directly used to determine the extent of the distortion of a modulated signal. A more direct parameter would be the step response, especially where the modulation consists of an amplitude step or pulse.

The two essential parameters of a filter's step response are overshoot and ringing. Overshoot should be minimized for accurate pulse reproduction. Ringing should decay as rapidly as possible to prevent interference with subsequent pulses. Rise time and delay are usually less important considerations.

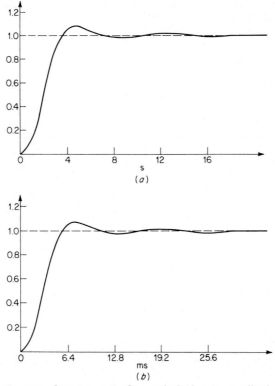

Fig. 2-28 Step response of example 2-13: *(a)* normalized step response; *(b)* denormalized step response.

Step-response curves for standard normalized low-pass filters are provided in the latter part of this chapter. These responses can be denormalized by dividing the time axis by $2\pi f_c$ where f_c is the 3-dB cutoff of the filter. Denormalization of the step response is shown in the following example.

Example 2-13

REQUIRED: Determine the amount of overshoot of an $n = 3$ Butterworth low-pass filter having a 3-dB cutoff of 100 Hz. Also determine the approximate time required for the ringing to decay substantially, i.e., the settling time.
RESULT: The step response of the normalized low-pass filter is shown in figure 2-28a. If the time axis is divided by $2\pi f_c$, where $f_c = 100$ Hz, the step response of figure 2-28b is obtained. The overshoot is slightly under 10%. After 25 ms the amplitude has almost completely settled.

If the input signal to a filter is a pulse rather than a step, the step-response curves can still be used to estimate the transient response provided that the pulse width is greater than the settling time.

Example 2-14

REQUIRED: Estimate the output waveform of the filter of example 2-13 if the input is the pulse of figure 2-29a.
RESULT: Since the pulse width is in excess of the settling time, the step response can be used to estimate the transient response. The leading edge is determined by the shape of the denormalized step response of figure 2-28b. The trailing edge can be derived by inverting the denormalized step response. The resulting waveform is shown in figure 2-29b.

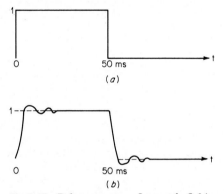

Fig. 2-29 Pulse response of example 2-14:
(a) input pulse; (b) output pulse.

Step Response of Bandpass Filters   The envelope of the response of a narrow bandpass filter to a step of the center frequency is almost identical to the step response of the equivalent low-pass filter having half the bandwidth. To determine this envelope shape, denormalize the low-pass step response by dividing the time axis by πBW, where BW is the 3-dB bandwidth of the bandpass filter. The previous discussions of overshoot, ringing, etc., can be applied to carrier envelope.

Example 2-15

REQUIRED: Determine the envelope of the response to a 1000-Hz step for an $n = 3$ Butterworth bandpass filter having a center frequency of 1000 Hz and a 3-dB bandwidth of 100 Hz.
RESULT: Using the normalized step response of figure 2-28a, divide the time axis by πBW, where BW = 100 Hz. The results are shown in figure 2-30.

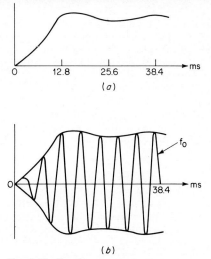

Fig. 2-30 Bandpass response to center frequency step: *(a)* denormalized low-pass step response; *(b)* bandpass envelope response.

Impulse Response of Low-Pass Filters If the duration of a pulse applied to a low-pass filter is much less than the rise time of the filter's step response, the filter's impulse response will provide a reasonable approximation to the shape of the output waveform.

Impulse-response curves are provided for the different families of low-pass filters. These curves are all normalized to correspond to a filter having a 3-dB cutoff of 1 rad/s and have an area of unity. To denormalize the curve, multiply the amplitude by the FSF and divide the time axis by the same factor.

It is desirable to select a normalized low-pass filter having an impulse response whose peak is as high as possible. The ringing which occurs after the trailing edge should also decay rapidly to avoid interference with subsequent pulses.

Example 2-16

REQUIRED: Determine the approximate output waveform if a 100-μs pulse is applied to an $n = 3$ Butterworth low-pass filter having a 3-dB cutoff of 100 Hz.

RESULT: The denormalized step response of the filter is given in figure 2-28*b*. The rise time is well in excess of the given pulse width of 100 μs; so the impulse response curve should be used to approximate the output waveform.

The impulse response of a normalized $n = 3$ Butterworth low-pass filter is shown in figure 2-31*a*. If the time axis is divided by the FSF and the amplitude is multiplied by this same factor, the curve of figure 2-31*b* results.

Since the input pulse amplitude of example 2-16 is certainly not infinite, the amplitude axis is in error. However, the pulse shape is retained at a lower amplitude. As the input pulse width is reduced in relation to the filter rise time, the output amplitude will decrease and eventually the output pulse will vanish.

Impulse Response of Bandpass Filters The envelope of the response of a narrow-band bandpass filter to a short tone burst of center frequency can be found by denormalizing the low-pass impulse response. This approximation is valid if the burst width is much less than the rise time of the denormalized step response of the bandpass filter. Also the center frequency should be high enough so that many cycles occur during the burst interval.

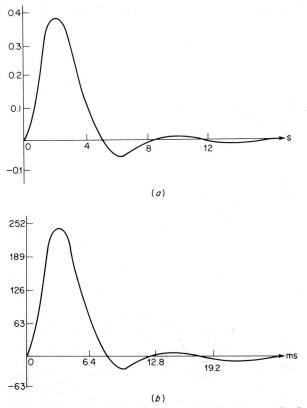

(a)

(b)

Fig. 2-31 Impulse response for example 2-16: (a) normalized response; (b) denormalized response.

To transform the impulse-response curve, multiply the amplitude axis by πBW and divide the time axis by this same factor, where BW is the 3-dB bandwidth of the bandpass filter. The resulting curve defines the shape of the envelope of the filter's response to the tone burst.

Example 2-17

REQUIRED: Determine the approximate shape of the response of an $n = 3$ Butterworth bandpass filter having a center frequency of 1000 Hz and a 3-dB bandwidth of 10 Hz to a tone burst of center frequency having a duration of 10 ms.

RESULT: The step response of a normalized $n = 3$ Butterworth low-pass filter is shown in figure 2-28a. To determine the rise time of the bandpass step response, divide the normalized low-pass rise time by πBW, where BW is 10 Hz. The resulting rise time is approximately 120 ms, which well exceeds the burst duration. Also, 10 cycles of center frequency occur during the burst interval; so the impulse response can be used to approximate the output envelope. To denormalize the impulse response, multiply the amplitude axis by πBW and divide the time axis by the same factor. The results are shown in figure 2-32.

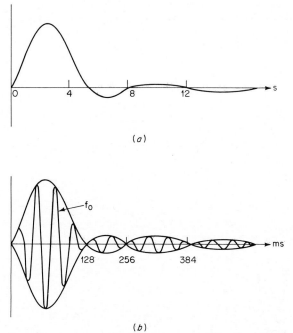

(a)

(b)

Fig. 2-32   Results of example 2-17: (a) normalized low-pass impulse response; (b) impulse response of bandpass filter.

Effective Use of the Group-Delay, Step-Response, and Impulse-Response Curves

Many signals consist of complex forms of modulation rather than pulses or steps; so the transient response curves cannot be directly used to estimate the amount of distortion introduced by the filters. However, the curves are useful as a figure of merit, since networks having desirable step- or impulse-response behavior introduce minimal distortion to most forms of modulation.

Examination of the step- and impulse-response curves in conjunction with group delay indicates that a necessary condition for good pulse transmission is a flat group delay. A gradual transition from the passband to the stopband is also required for low transient distortion but is highly undesirable from a frequency-attenuation point of view.

In order to obtain a rapid pulse rise time the higher-frequency spectral components should not be delayed with respect to the lower frequencies. The curves indicate that low-pass filters which do have sharply increasing delay at higher frequencies have an impulse response which comes to a peak at a later time.

When a low-pass filter is transformed to a high-pass, a band-reject, or a wide-band bandpass filter the transient properties are not preserved. Lindquist and

Zverev (see references) provide computational methods for the calculation of these responses.

2.3 BUTTERWORTH MAXIMALLY FLAT AMPLITUDE

The Butterworth approximation to an ideal low-pass filter is based on the assumption that a flat response at zero frequency is more important than the response at other frequencies. The normalized transfer function is an all-pole type having roots which all fall on a unit circle. The attenuation is 3 dB at 1 rad/s.

The attenuation of a Butterworth low-pass filter can be expressed by

$$A_{dB} = 10 \log \left[1 + \left(\frac{\omega_x}{\omega_c} \right)^{2n} \right] \tag{2-28}$$

where ω_x/ω_c is the ratio of the given frequency ω_x to the 3-dB cutoff frequency ω_c and n is the order of the filter.

For the more general case,

$$A_{dB} = 10 \log (1 + \Omega^{2n}) \tag{2-29}$$

where Ω is defined by the following table:

Filter Type	Ω
Low-pass	ω_x/ω_c
High-pass	ω_c/ω_x
Bandpass	$BW_x/BW_{3\,dB}$
Band-reject	$BW_{3\,dB}/BW_x$

The value Ω is a dimensionless ratio of frequencies or normalized frequency. $BW_{3\,dB}$ is the 3-dB bandwidth and BW_x is the bandwidth of interest. At high values of Ω the attenuation increases at a rate of $6n$ dB per octave, where an octave is defined as a frequency ratio of 2 for the low-pass and high-pass cases and a *bandwidth* ratio of 2 for bandpass and band-reject filters.

The pole positions of the normalized filter all lie on a unit circle and can be computed by

$$-\sin \frac{(2K-1)\pi}{2n} + j \cos \frac{(2K-1)\pi}{2n}, \qquad K = 1, 2, \ldots, n \tag{2-30}$$

and the element values for an *LC* normalized low-pass filter operating between equal 1-Ω terminations can be calculated by

$$L_K \text{ or } C_K = 2 \sin \frac{(2K-1)\pi}{2n}, \qquad K = 1, 2, \ldots, n \tag{2-31}$$

where $(2K-1)\pi/2n$ is in radians.

Equation (2-31) is exactly equal to twice the real part of the pole positions of equation (2-30) except that the sign is positive.

Example 2-18

REQUIRED: Calculate the frequency response at 1, 2, and 4 rad/s, the pole positions, and the *LC* element values of a normalized $n = 5$ Butterworth low-pass filter.

RESULT: *(a)* Using equation (2-29) with $n = 5$, the following frequency-response table can be derived:

Ω	Attenuation
1	3 dB
2	30 dB
4	60 dB

(b) The pole positions are computed using equation (2-30) as follows:

K	$-\sin\dfrac{(2K-1)\pi}{2n}$	$j\cos\dfrac{(2K-1)\pi}{2n}$
1	-0.309	$+j\,0.951$
2	-0.809	$+j\,0.588$
3	-1	
4	-0.809	$-j\,0.588$
5	-0.309	$-j\,0.951$

(c) The element values can be computed by equation (2-31) and have the following values:

$$
\begin{array}{lcl}
L_1 = 0.618\text{ H} & & C_1 = 0.618\text{ F} \\
C_2 = 1.618\text{ F} & & L_2 = 1.618\text{ H} \\
L_3 = 2\text{ H} & \text{or} & C_3 = 2\text{ F} \\
C_4 = 1.618\text{ F} & & L_4 = 1.618\text{ H} \\
L_5 = 0.618\text{ H} & & C_5 = 0.618\text{ F}
\end{array}
$$

The results of example 2-18 are shown in figure 2-33.

Chapter 12 provides pole locations and element values for both LC and active Butterworth low-pass filters having complexities up to $n = 10$.

The Butterworth approximation results in a class of filters which have moderate attenuation steepness and acceptable transient characteristics. Their element values are more practical and less critical than those of most other filter types. The rounding of the frequency response in the vicinity of cutoff may make these filters undesirable where a sharp cutoff is required, but nevertheless they should be used wherever possible because of their favorable characteristics.

Figures 2-34 through 2-37 indicate the frequency response, group delay, impulse response, and step response for the Butterworth family of low-pass filters normalized to a 3-dB cutoff of 1 rad/s.

2.4 CHEBYSHEV RESPONSE

If the poles of the normalized Butterworth low-pass transfer function were moved to the right by multiplying the real parts of the pole positions by a constant k_c where $k_c < 1$, the poles would now lie on an ellipse instead of a unit circle. The frequency response would ripple evenly and have a 3-dB cutoff of 1 rad/s. As the real part of the poles is decreased by lowering k_c, the ripples will grow in magnitude. The resulting response is called the Chebyshev or equiripple function.

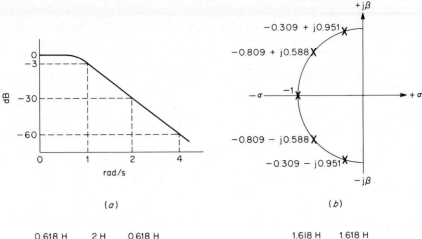

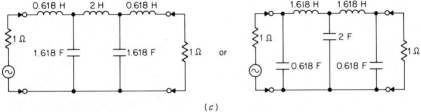

Fig. 2-33 Butterworth low-pass filter of example 2-18: (a) frequency response; (b) pole locations; (c) circuit configuration.

The Chebyshev approximation to an ideal filter has a much more rectangular frequency response in the region near cutoff than the Butterworth family of filters. This is accomplished at the expense of allowing ripples in the passband.

The factor k_c can be computed by

$$k_c = \tanh A \qquad (2\text{-}32)$$

The parameter A is given by

$$A = \frac{1}{n} \sinh^{-1} \frac{1}{\epsilon} \qquad (2\text{-}33)$$

where

$$\epsilon = \sqrt{10^{R_{dB}/10} - 1} \qquad (2\text{-}34)$$

and R_{db} is the ripple in decibels.

Figure 2-38 compares the voltage response of an $n = 3$ Butterworth normalized low-pass filter and the Chebyshev filter generated by multiplying the real parts of the roots by k_c. Both filters have half-power (3-dB) bandwidths of 1 rad/s. The ripple bandwidth of the Chebyshev filter is $1/\cosh A$.

The attenuation of Chebyshev filters can be expressed as

$$A_{dB} = 10 \log [1 + \epsilon^2 C_n^2(\Omega)] \qquad (2\text{-}35)$$

where $C_n(\Omega)$ is a Chebyshev polynomial whose magnitude oscillates between ± 1 for $\Omega \leq 1$. Table 2-1 lists the Chebyshev polynomials up to order $n = 10$.

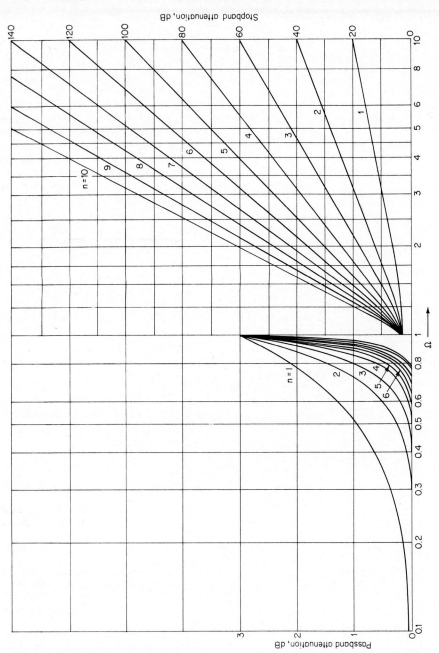

Fig. 2-34 Attenuation characteristics for Butterworth filters. *(From Anatol I. Zverev, Handbook of Filter Synthesis, John Wiley and Sons, Inc., New York, 1967. By permission of the publishers.)*

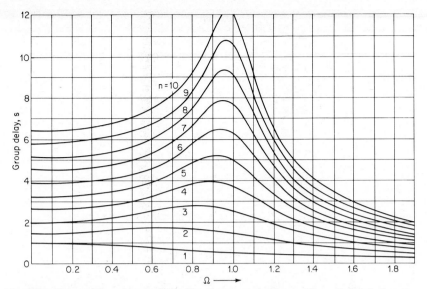

Fig. 2-35 Group-delay characteristics for Butterworth filters. (*From Anatol I. Zverev, Handbook of Filter Synthesis, John Wiley and Sons, Inc., New York, 1967. By permission of the publishers.*)

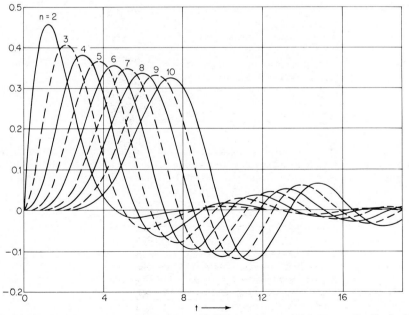

Fig. 2-36 Impulse response for Butterworth filters. (*From Anatol I. Zverev, Handbook of Filter Synthesis, John Wiley and Sons, Inc., New York, 1967. By permission of the publishers.*)

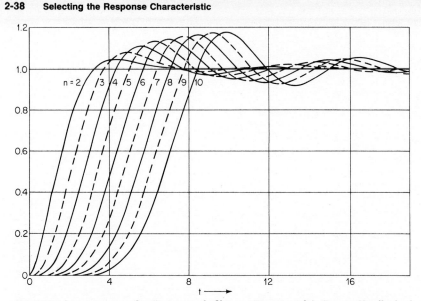

Fig. 2-37 Step response for Butterworth filters. *(From Anatol I. Zverev, Handbook of Filter Synthesis, John Wiley and Sons, Inc., New York, 1967. By permission of the publishers.)*

TABLE 2-1   Chebyshev Polynomials

1. Ω
2. $2\Omega^2 - 1$
3. $4\Omega^3 - 3\Omega$
4. $8\Omega^4 - 8\Omega^2 + 1$
5. $16\Omega^5 - 20\Omega^3 + 5\Omega$
6. $32\Omega^6 - 48\Omega^4 + 18\Omega^2 - 1$
7. $64\Omega^7 - 112\Omega^5 + 56\Omega^3 - 7\Omega$
8. $128\Omega^8 - 256\Omega^6 + 160\Omega^4 - 32\Omega^2 + 1$
9. $256\Omega^9 - 576\Omega^7 + 432\Omega^5 - 120\Omega^3 + 9\Omega$
10. $512\Omega^{10} - 1280\Omega^8 + 1120\Omega^6 - 400\Omega^4 + 50\Omega^2 - 1$

At $\Omega = 1$, Chebyshev polynomials have a value of unity; so the attenuation defined by equation (2-35) would be equal to the ripple. The 3-dB cutoff is slightly above $\Omega = 1$ and is equal to cosh A. In order to normalize the response equation so that 3 dB of attenuation occurs at $\Omega = 1$, the Ω of equation (2-35) is computed by using the following table:

Filter Type	Ω
Low-pass	(cosh A) ω_x/ω_c
High-pass	(cosh A) ω_c/ω_x
Bandpass	(cosh A) $BW_x/BW_{3\,dB}$
Band-reject	(cosh A) $BW_{3\,dB}/BW_x$

Figure 2-39 compares the ratios of 3-dB bandwidth to ripple bandwidth (cosh A) for Chebyshev low-pass filters ranging from $n = 2$ through $n = 10$.

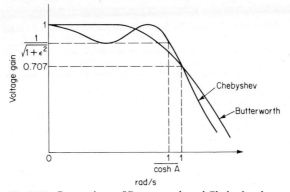

Fig. 2-38 Comparison of Butterworth and Chebyshev low-pass filters.

n	0.001 dB	0.005 dB	0.01 dB	0.05 dB
2	5.7834930	3.9027831	3.3036192	2.2685899
3	2.6427081	2.0740079	1.8771819	1.5120983
4	1.8416695	1.5656920	1.4669048	1.2783955
5	1.5155888	1.3510908	1.2912179	1.1753684
6	1.3495755	1.2397596	1.1994127	1.1207360
7	1.2531352	1.1743735	1.1452685	1.0882424
8	1.1919877	1.1326279	1.1106090	1.0673321
9	1.1507149	1.1043196	1.0870644	1.0530771
10	1.1215143	1.0842257	1.0703312	1.0429210

n	0.10 dB	0.25 dB	0.50 dB	1.00 dB
2	1.9432194	1.5981413	1.3897437	1.2176261
3	1.3889948	1.2528880	1.1674852	1.0948680
4	1.2130992	1.1397678	1.0931019	1.0530019
5	1.1347180	1.0887238	1.0592591	1.0338146
6	1.0929306	1.0613406	1.0410296	1.0234422
7	1.0680005	1.0449460	1.0300900	1.0172051
8	1.0519266	1.0343519	1.0230107	1.0131638
9	1.0409547	1.0271099	1.0181668	1.0103963
10	1.0331307	1.0219402	1.0147066	1.0084182

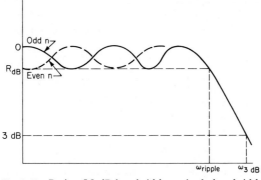

Fig. 2-39 Ratio of 3-dB bandwidth to ripple bandwidth.

Odd-order Chebyshev *LC* filters have zero relative attenuation at DC. Even-order filters, however, have a loss at DC equal to the passband ripple. As a result the even-order networks must operate between unequal source and load resistances, whereas for odd n's, the source and load may be equal. However, a mathematical transformation can alter even-order networks for operation between equal terminations (see $\theta = T$ in table 12-56). The result is a Chebyshev-like behavior in the passband and a slightly diminished rate of roll-off when compared with a comparable unaltered network.

The element values for an *LC* normalized low-pass filter operating between equal 1-Ω terminations and having an odd n can be calculated from the following series of relations:

$$G_1 = \frac{2A_1 \cosh A}{Y} \tag{2-36}$$

$$G_k = \frac{4A_{k-1}A_k \cosh^2 A}{B_{k-1} G_{k-1}} \qquad k = 2, 3, 4, \ldots, n \tag{2-37}$$

where

$$Y = \sinh \frac{\beta}{2n} \tag{2-38}$$

$$\beta = \ln \left(\coth \frac{R_{dB}}{17.37} \right) \tag{2-39}$$

$$A_k = \sin \frac{(2k-1)\,\pi}{2n} \qquad k = 1, 2, 3, \ldots, n \tag{2-40}$$

$$B_k = Y^2 + \sin^2 \left(\frac{k\,\pi}{n} \right) \qquad k = 1, 2, 3, \ldots, n \tag{2-41}$$

Coefficients G_1 through G_n are the element values.

An alternate form of determining *LC* element values is by synthesis of the driving-point impedance directly from the transfer function. This method includes both odd- and even-order n's.

Example 2–19

REQUIRED: Compute the pole positions, the frequency response at 1, 2, and 4 rad/s, and the element values of a normalized $n = 5$ Chebyshev low-pass filter having a ripple of 0.5 dB.

RESULT: (*a*)   To compute the pole positions, first solve for k_c as follows:

$$\epsilon = \sqrt{10^{R_{dB}/10} - 1} = 0.349 \tag{2-34}$$

$$A = \frac{1}{n} \sinh^{-1} \frac{1}{\epsilon} = 0.355 \tag{2-33}$$

$$k_c = \tanh A = 0.340 \tag{2-32}$$

Multiplication of the real parts of the normalized Butterworth poles of example 2-18 by k_c results in the following new pole positions:

$-0.105 \pm j0.951$
$-0.275 \pm j0.588$
-0.34

(*b*)   To calculate the frequency response, substitute a fifth-order Chebyshev polynomial and $\epsilon = 0.349$ into equation (2-35). The following results are obtained:

Ω	A_{dB}
1.0	3 dB
2.0	45 dB
4.0	77 db

(c) The element values are computed as follows:

$$A_1 = 0.309 \qquad (2\text{-}40)$$
$$\beta = 3.55 \qquad (2\text{-}39)$$
$$Y = 0.363 \qquad (2\text{-}38)$$
$$G_1 = 1.81 \qquad (2\text{-}36)$$
$$G_2 = 1.30 \qquad (2\text{-}37)$$
$$G_3 = 2.69 \qquad (2\text{-}37)$$
$$G_4 = 1.30 \qquad (2\text{-}37)$$
$$G_5 = 1.81 \qquad (2\text{-}37)$$

Coefficients G_1 through G_5 represent the element values of a normalized Chebyshev low-pass filter having a 0.5-dB ripple and a 3-dB cutoff of 1 rad/s.

Figure 2-40 shows the results of this example.

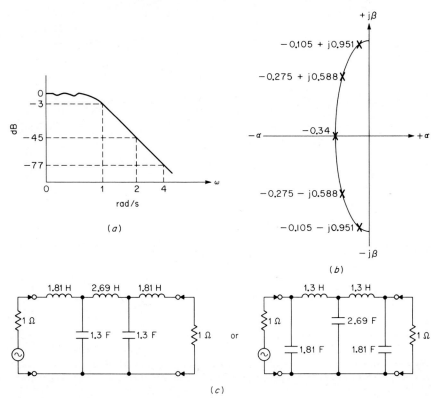

Fig. 2-40 Chebyshev low-pass filter of example 2-19: (a) frequency response; (b) pole locations; (c) circuit configuration.

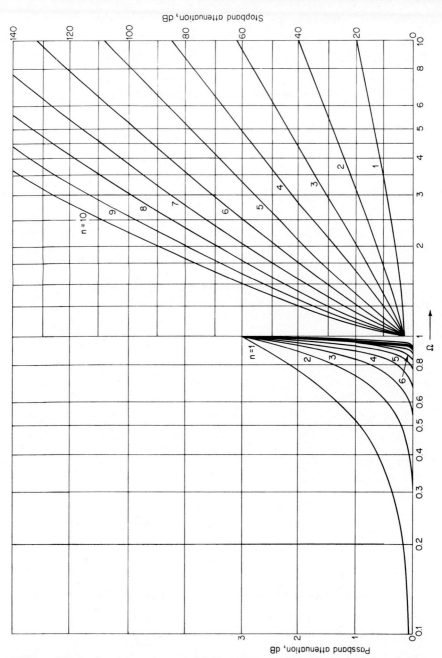

Fig. 2-41 Attenuation characteristics for Chebyshev filters with 0.01-dB ripple. (*From Anatol I. Zverev, Handbook of Filter Synthesis, John Wiley and Sons, Inc., New York, 1967. By permission of the publishers.*)

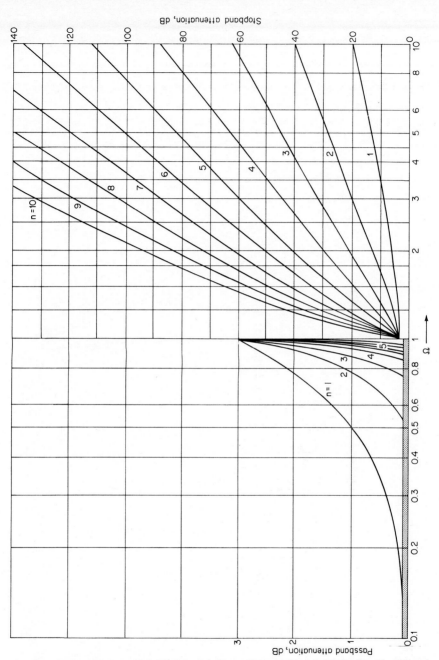

Fig. 2-42 Attenuation characteristics for Chebyshev filters with 0.1-dB ripple. (*From Anatol I. Zverev, Handbook of Filter Synthesis, John Wiley and Sons, Inc., New York, 1967. By permission of the publishers.*)

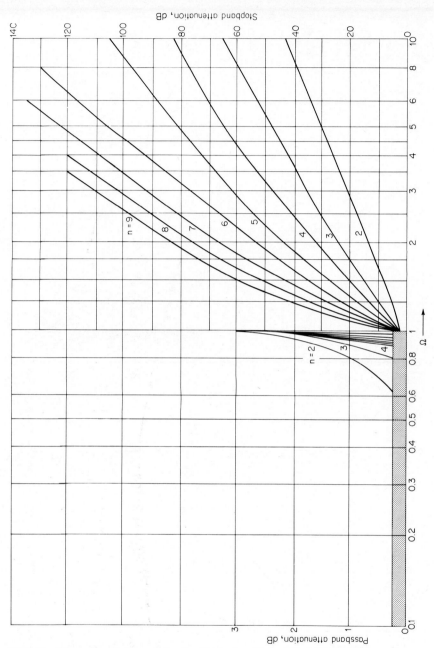

Fig. 2-43 Attenuation characteristics for Chebyshev filters with 0.25-dB ripple.

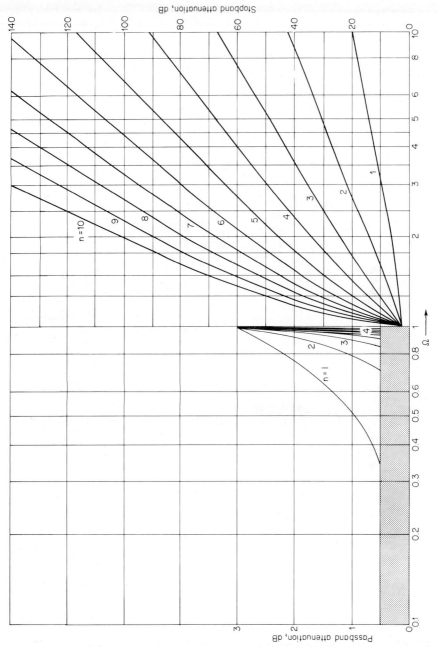

Fig. 2-44 Attenuation characteristics for Chebyshev filters with 0.5-dB ripple. *(From Anatol I. Zverev, Handbook of Filter Synthesis, John Wiley and Sons, Inc., New York, 1967. By permission of the publishers.)*

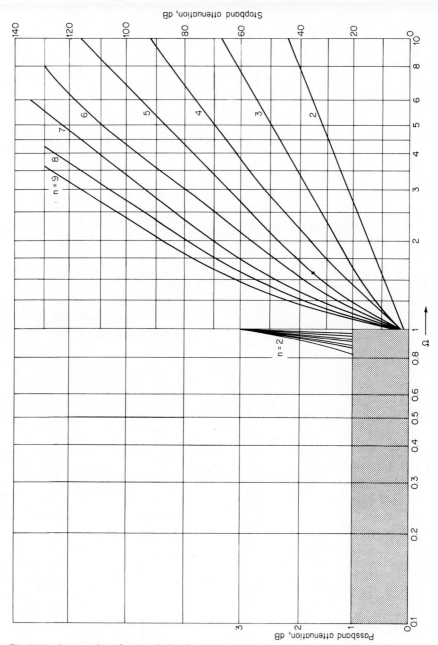

Fig. 2-45 Attenuation characteristics for Chebyshev filters with 1-dB ripple.

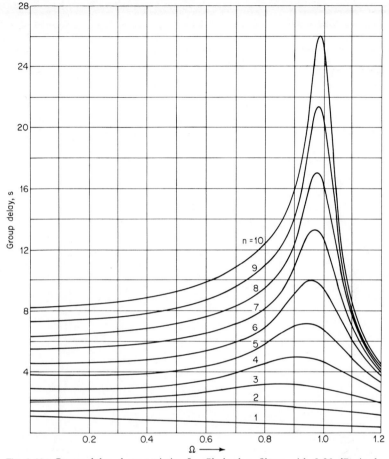

Fig. 2-46 Group-delay characteristics for Chebyshev filters with 0.01-dB ripple. *(From Anatol I. Zverev, Handbook of Filter Synthesis, John Wiley and Sons, Inc., New York, 1967. By permission of the publishers.)*

Chebyshev filters have a narrower transition region between the passband and stopband than Butterworth filters but have more delay variation in their passband. As the passband ripple is made larger, the rate of roll-off increases, but the transient properties rapidly deteriorate. If no ripples are permitted, the Chebyshev filter degenerates to a Butterworth.

The Chebyshev function is useful where frequency response is the major consideration. It provides the maximum theoretical rate of roll-off of any all-pole transfer function for a given order. It does not have the mathematical simplicity of the Butterworth family which should be evident from comparing examples 2-19 and 2-18. Fortunately the computation of poles and element values is not required, since this information is provided in chapter 12.

Figures 2-41 through 2-54 show the frequency and time-domain parameters of Chebyshev low-pass filters for ripples of 0.01, 0.1, 0.25, 0.5, and 1 dB all normalized for a 3-dB cutoff of 1 rad/s.

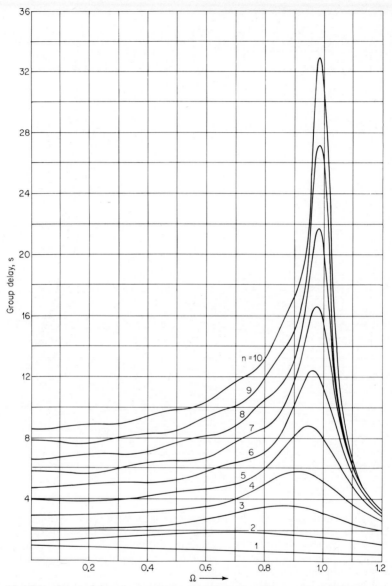

Fig. 2-47 Group-delay characteristics for Chebyshev filters with 0.1-dB ripple. *(From Anatol I. Zverev, Handbook of Filter Synthesis, John Wiley and Sons, Inc., New York, 1967. By permission of the publishers.)*

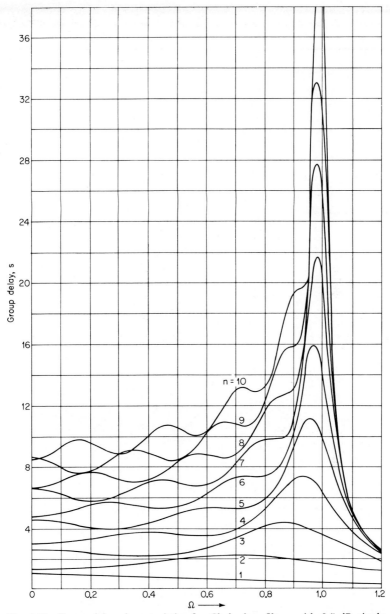

Fig. 2-48 Group-delay characteristics for Chebyshev filters with 0.5-dB ripple. *(From Anatol I. Zverev, Handbook of Filter Synthesis, John Wiley and Sons, Inc., New York, 1967. By permission of the publishers.)*

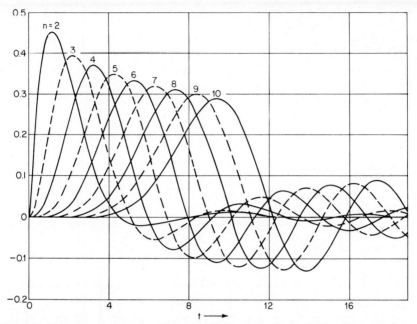

Fig. 2-49 Impulse response for Chebyshev filters with 0.01-dB ripple. (*From Anatol I. Zverev, Handbook of Filter Synthesis, John Wiley and Sons, Inc., New York, 1967. By permission of the publishers.*)

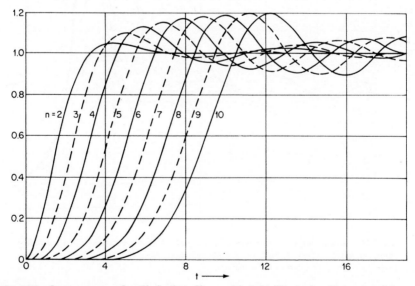

Fig. 2-50 Step response for Chebyshev filters with 0.01-dB ripple. (*From Anatol I. Zverev, Handbook of Filter Synthesis, John Wiley and Sons, Inc., New York, 1967. By permission of the publishers.*)

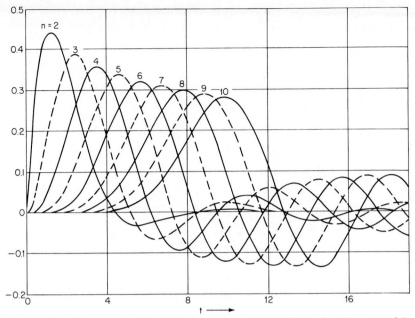

Fig. 2-51 Impulse response for Chebyshev filters with 0.1-dB ripple. (*From Anatol I. Zverev, Handbook of Filter Synthesis, John Wiley and Sons, Inc., New York, 1967. By permission of the publishers.*)

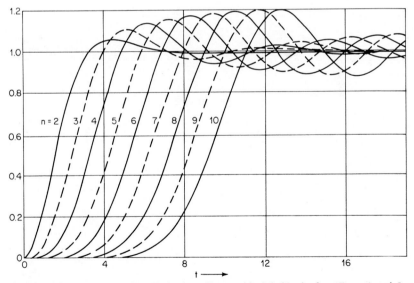

Fig. 2-52 Step response for Chebyshev filters with 0.1-dB ripple. (*From Anatol I. Zverev, Handbook of Filter Synthesis, John Wiley and Sons, Inc., New York, 1967. By permission of the publishers.*)

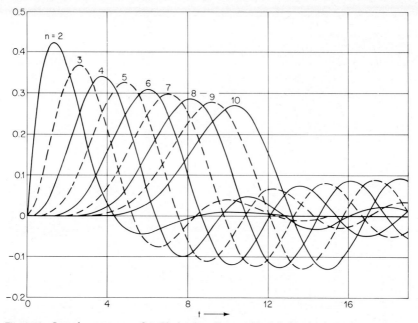

Fig. 2-53 Impulse response for Chebyshev filters with 0.5-dB ripple. *(From Anatol I. Zverev, Handbook of Filter Synthesis, John Wiley and Sons, Inc., New York, 1967. By permission of the publishers.)*

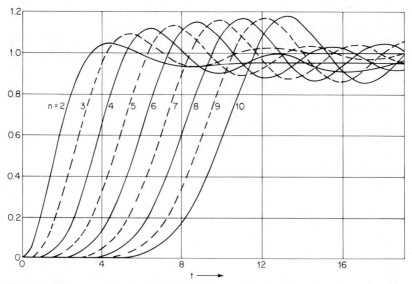

Fig. 2-54 Step response for Chebyshev filters with 0.5-dB ripple. *(From Anatol I. Zverev, Handbook of Filter Synthesis, John Wiley and Sons, Inc., New York, 1967. By permission of the publishers.)*

2.5 BESSEL MAXIMALLY FLAT DELAY

Butterworth filters have fairly good amplitude and transient characteristics. The Chebyshev family of filters offers increased selectivity but poor transient behavior. Neither approximation to an ideal filter is directed toward obtaining a constant delay in the passband.

The Bessel transfer function has been optimized to obtain a linear phase, i.e., a maximally flat delay. The step response has essentially no overshoot or ringing and the impulse response lacks oscillatory behavior. However, the frequency response is much less selective than in the other filter types.

The low-pass approximation to a constant delay can be expressed as the following general transfer function:

$$T(s) = \frac{1}{\sinh s + \cosh s} \tag{2-42}$$

If a continued-fraction expansion is used to approximate the hyperbolic functions and the expansion is truncated at different lengths, the Bessel family of transfer functions will result.

A crude approximation to the pole locations can be found by locating all the poles on a circle and separating their imaginary parts by $2/n$, as shown in figure 2-55. The vertical spacing between poles is equal, whereas in the Butterworth case the angles were equal.

The relative attenuation of a Bessel low-pass filter can be approximated by

$$A_{dB} = 3\left(\frac{\omega_x}{\omega_c}\right)^2 \tag{2-43}$$

This expression is reasonably accurate for ω_x/ω_c ranging between 0 and 2.

Figures 2-56 through 2-59 indicate that as the order n is increased, the region of flat delay is extended farther into the stopband. However, the steepness of roll-off in the transition region does not improve significantly. This restricts the use of Bessel filters to applications where the transient properties are the major consideration.

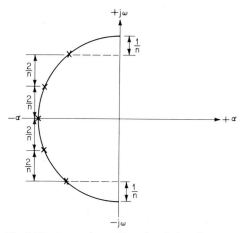

Fig. 2-55 Approximate Bessel pole locations.

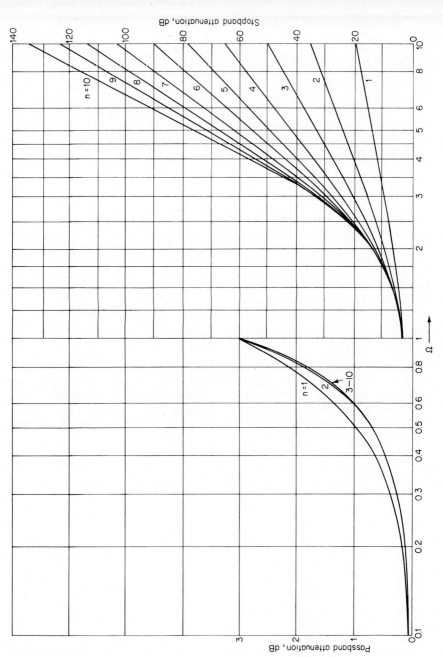

Fig. 2-56 Attenuation characteristics for maximally flat delay (Bessel) filters. *(From Anatol I. Zverev, Handbook of Filter Synthesis, John Wiley and Sons, Inc., New York, 1967. By permission of the publishers.)*

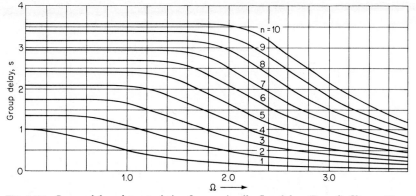

Fig. 2-57 Group-delay characteristics for maximally flat delay (Bessel) filters. *(From Anatol I. Zverev, Handbook of Filter Synthesis, John Wiley and Sons, Inc., New York, 1967. By permission of the publishers.)*

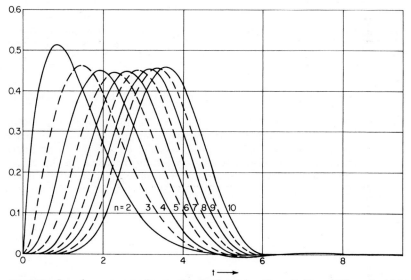

Fig. 2-58 Impulse response for maximally flat delay (Bessel) filters. *(From Anatol I. Zverev, Handbook of Filter Synthesis, John Wiley and Sons, Inc., New York, 1967. By permission of the publishers.)*

A similar family of filters is the Gaussian type. However, the Gaussian phase response is not as linear as the Bessel for the same number of poles, and the selectivity is not as sharp.

2.6 LINEAR PHASE WITH EQUIRIPPLE ERROR

The Chebyshev (equiripple amplitude) function is a better approximation of an ideal amplitude curve than the Butterworth. Therefore, it stands to reason

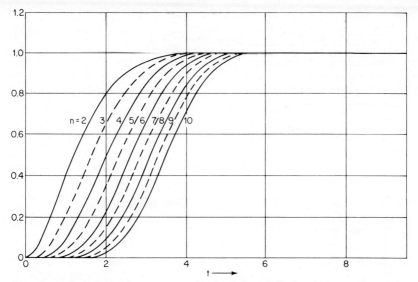

Fig. 2-59 Step response for maximally flat delay (Bessel) filters. *(From Anatol I. Zverev, Handbook of Filter Synthesis, John Wiley and Sons, Inc., New York, 1967. By permission of the publishers.)*

that an equiripple approximation of a linear phase will be more efficient than the Bessel family of filters.

Figure 2-60 illustrates how a linear phase can be approximated to within a given ripple of ϵ degrees. For the same n the equiripple-phase approximation results in a linear phase and consequently a constant delay over a larger interval than the Bessel approximation. Also the amplitude response is superior far from cutoff. In the transition region and below cutoff both approximations have nearly identical responses.

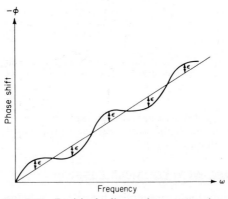

Fig. 2-60 Equiripple linear-phase approximation.

As the phase ripple ϵ is increased, the region of constant delay is extended farther into the stopband. However, the delay develops ripples. The step response has slightly more overshoot than Bessel filters.

A closed-form method for computation of the pole positions is not available. The pole locations tabulated in chapter 12 were developed by iterative techniques. Values are provided for phase ripples of 0.05° and 0.5°, and the associated frequency and time-domain parameters are given in figures 2-61 through 2-68.

2.7 TRANSITIONAL FILTERS

The Bessel filters discussed in section 2.5 have excellent transient properties but poor selectivity. Chebyshev filters, on the other hand, have steep roll-off characteristics but poor time-domain behavior. A transitional filter offers a compromise between a gaussian filter, which is similar to the Bessel family, and Chebyshev filters.

Transitional filters have a near linear phase shift and smooth amplitude roll-off in the passband. Outside the passband a sharp break in the amplitude characteristics occurs. Beyond this breakpoint the attenuation increases quite abruptly in comparison with Bessel filters, especially for the higher n's.

In the tables in chapter 12 transitional filters are provided which have gaussian characteristics to both 6 and 12 dB. The transient properties of the gaussian to 6-dB filters are somewhat superior to those of the Butterworth family. Beyond the 6-dB point, which occurs at approximately 1.5 rad/s, the attenuation characteristics are nearly comparable with Butterworth filters. The gaussian to 12-dB filters have time-domain parameters far superior to those of Butterworth filters. However, the 12-dB breakpoint occurs at 2 rad/s, and the attenuation characteristics beyond this point are inferior to those of Butterworth filters.

The transitional filters tabulated in chapter 12 were generated by mathematical techniques which involve interpolation of pole locations. Figures 2-69 through 2-76 indicate the frequency and time-domain properties of both the gaussian to 6-dB and gaussian to 12-dB transitional filters.

2.8 SYNCHRONOUSLY TUNED FILTERS

Synchronously tuned filters are the most basic filter type and are the easiest to construct and align. They consist of identical multiple poles. A typical application is in the case of a bandpass amplifier, where a number of stages are cascaded with each stage having the same center frequency and Q.

The attenuation of a synchronously tuned filter can be expressed as

$$A_{dB} = 10n \log[1+(2^{1/n} - 1)\Omega^2] \qquad (2\text{-}44)$$

Equation (2-44) is normalized so that 3 dB of attenuation occurs at $\Omega = 1$.

The individual section Q can be defined in terms of the composite circuit Q requirement by the following relationship:

$$Q_{section} = Q_{overall}\sqrt{2^{1/n} - 1} \qquad (2\text{-}45)$$

Alternately we can state that the 3-dB bandwidth of the individual sections is reduced by the shrinkage factor $(2^{1/n} - 1)^{1/2}$. The individual section Q is less than the overall Q, whereas in the case of nonsynchronously tuned filters the section Q's may be required to be much higher than the composite Q.

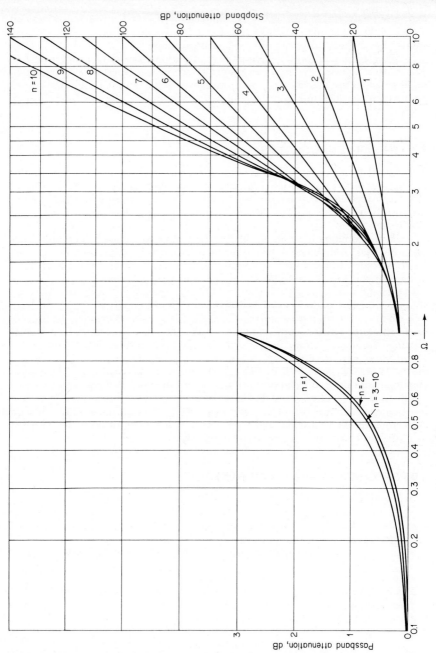

Fig. 2-61 Attenuation characteristics for linear phase with equiripple error filters (phase error = 0.05°). *(From Anatol I. Zverev, Handbook of Filter Synthesis, John Wiley and Sons, Inc., New York, 1967. By permission of the publishers.)*

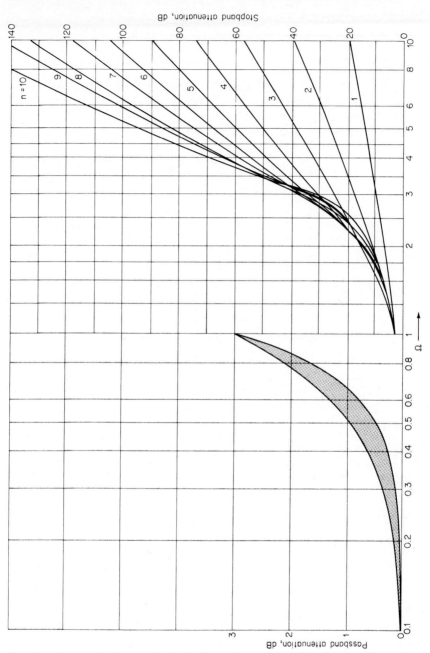

Fig. 2-62 Attenuation characteristics for linear phase with equiripple error filters (phase error = 0.5°). *(From Anatol I. Zverev, Handbook of Filter Synthesis, John Wiley and Sons, Inc., New York, 1967. By permission of the publishers.)*

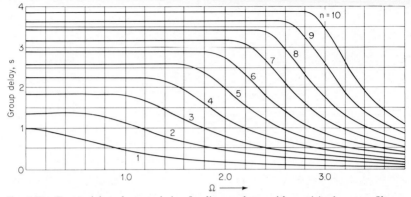

Fig. 2-63 Group-delay characteristics for linear phase with equiripple error filters (phase error = 0.05°). *(From Anatol I. Zverev, Handbook of Filter Synthesis, John Wiley and Sons, Inc., New York, 1967. By permission of the publishers.)*

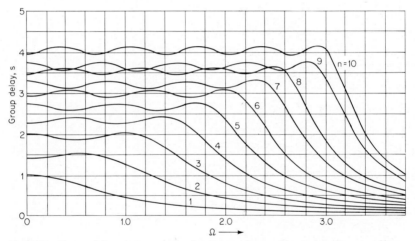

Fig. 2-64 Group-delay characteristics for linear phase with equiripple error filters (phase error = 0.5°). *(From Anatol I. Zverev, Handbook of Filter Synthesis, John Wiley and Sons, Inc., New York, 1967. By permission of the publishers.)*

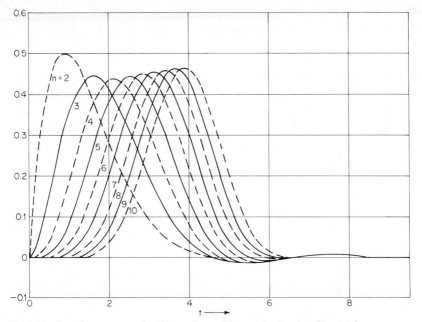

Fig. 2-65   Impulse response for linear phase with equiripple error filters (phase error =
0.05°). *(From Anatol I. Zverev, Handbook of Filter Synthesis, John Wiley and Sons, Inc., New
York, 1967. By permission of the publishers.)*

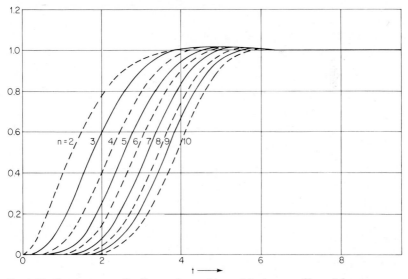

Fig. 2-66   Step response for linear phase with equiripple error filters (phase error =
0.05°). *(From Anatol I. Zverev, Handbook of Filter Synthesis, John Wiley and Sons, Inc.,
New York, 1967. By permission of the publishers.)*

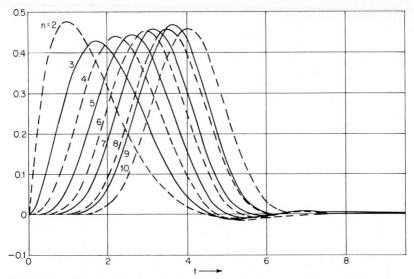

Fig. 2-67 Impulse response for linear phase with equiripple error filters (phase error = 0.5°). *(From Anatol I. Zverev, Handbook of Filter Synthesis, John Wiley and Sons, Inc., New York, 1967. By permission of the publishers.)*

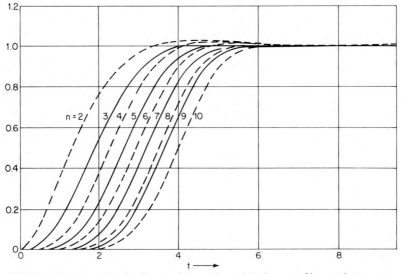

Fig. 2-68 Step response for linear phase with equiripple error filters (phase error = 0.5°). *(From Anatol I. Zverev, Handbook of Filter Synthesis, John Wiley and Sons, Inc., New York, 1967. By permission of the publishers.)*

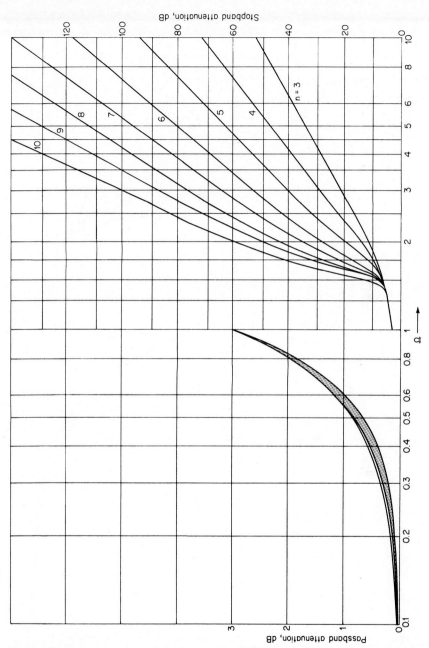

Fig. 2-69 Attenuation characteristics for transitional filters (gaussian to 6 dB). *(From Anatol I. Zverev, Handbook of Filter Synthesis, John Wiley and Sons, Inc., New York, 1967. By permission of the publishers.)*

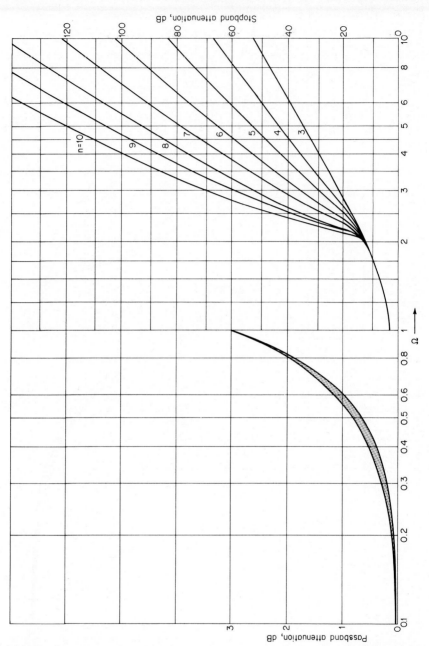

Fig. 2-70 Attenuation characteristics for transitional filters (gaussian to 12 dB). *(From Anatol I. Zverev, Handbook of Filter Synthesis, John Wiley and Sons, Inc., New York, 1967. By permission of the publishers.)*

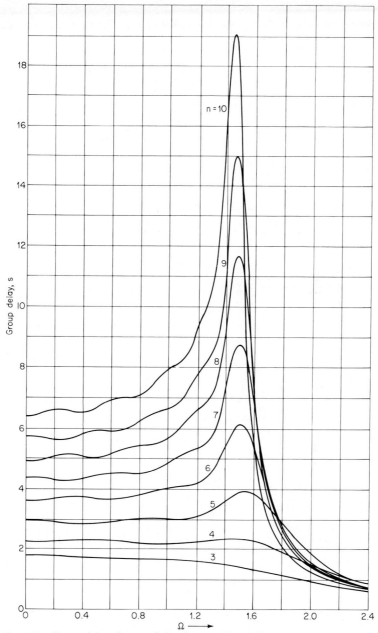

Fig. 2-71 Group-delay characteristics for transitional filters (gaussian to 6 dB). *(From Anatol I. Zverev, Handbook of Filter Synthesis, John Wiley and Sons, Inc., New York, 1967. By permission of the publishers.)*

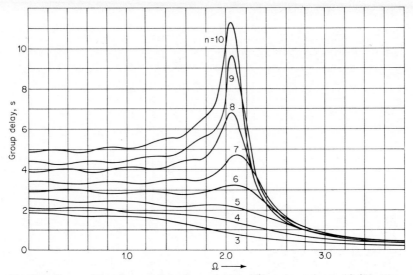

Fig. 2-72 Group-delay characteristics for transitional filters (gaussian to 12 dB). *(From Anatol I. Zverev, Handbook of Filter Synthesis, John Wiley and Sons, Inc., New York, 1967. By permission of the publishers.)*

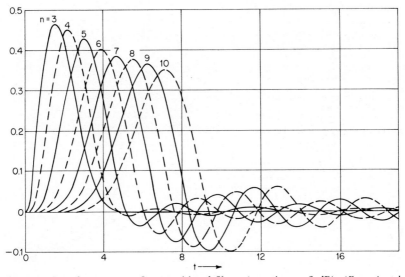

Fig. 2-73 Impulse response for transitional filters (gaussian to 6 dB). *(From Anatol I. Zverev, Handbook of Filter Synthesis, John Wiley and Sons, Inc., New York, 1967. By permission of the publishers.)*

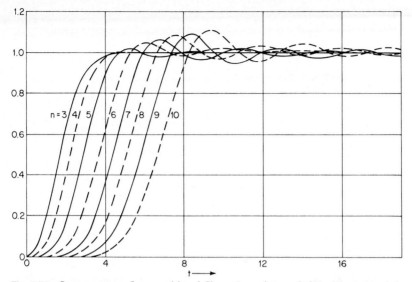

Fig. 2-74 Step response for transitional filters (gaussian to 6 dB). *(From Anatol I. Zverev, Handbook of Filter Synthesis, John Wiley and Sons, Inc., New York, 1967. By permission of the publishers.)*

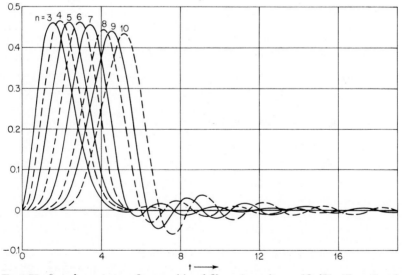

Fig. 2-75 Impulse response for transitional filters (gaussian to 12 dB). *(From Anatol I. Zverev, Handbook of Filter Synthesis, John Wiley and Sons, Inc., New York, 1967. By permission of the publishers.)*

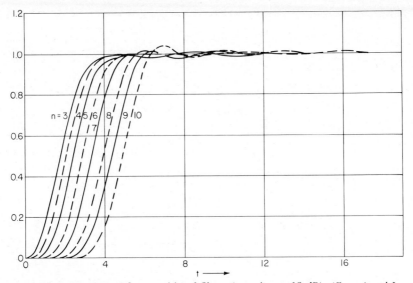

Fig. 2-76 Step response for transitional filters (gaussian to 12 dB). *(From Anatol I. Zverev, Handbook of Filter Synthesis, John Wiley and Sons, Inc., New York, 1967. By permission of the publishers.)*

Example 2-20

REQUIRED: A three-section synchronously tuned bandpass filter is required to have a center frequency of 10 kHz and a 3-dB bandwidth of 100 Hz. Determine the attenuation corresponding to a bandwidth of 300 Hz, and calculate the Q of each section.

RESULT: (a)   The attenuation at the 300-Hz bandwidth can be computed as

$$A_{dB} = 10n \log[1 + (2^{1/n} - 1)\Omega^2] = 15.7 \text{ dB} \qquad (2\text{-}44)$$

where $n = 3$ and Ω, the bandwidth ratio, is 300 Hz/100 Hz, or 3. (Since the filter is a narrow-band type, conversion to a geometrically symmetrical response requirement was not necessary.)

(b)   The Q of each section is

$$Q_{\text{section}} = Q_{\text{overall}}\sqrt{2^{1/n} - 1} = 51 \qquad (2\text{-}45)$$

where Q_{overall} is 10 kHz/100 Hz, or 100.

The synchronously tuned filter of example 2-20 has only 15.7 dB of attenuation at a normalized frequency ratio of 3 and for $n = 3$. Even the gradual roll-off characteristics of the Bessel family provide better selectivity than synchronously tuned filters for equivalent complexities.

The transient properties, however, are near optimum. The step response exhibits no overshoot at all and the impulse response lacks oscillatory behavior.

The poor selectivity of synchronously tuned filters limits their application to circuits requiring modest attenuation steepness and simplicity of alignment. The frequency and time-domain characteristics are illustrated in figures 2-77 through 2-80.

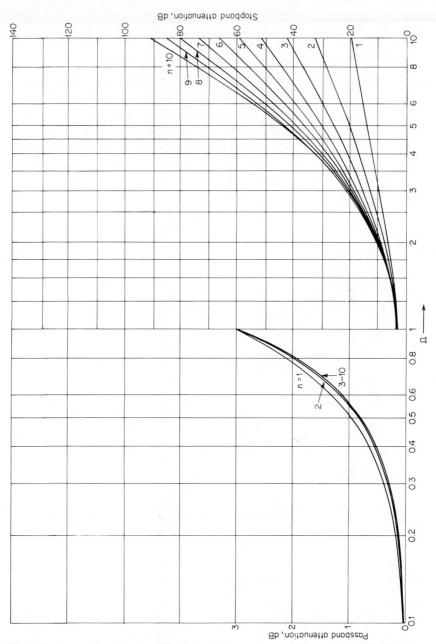

Fig. 2-77 Attenuation characteristics for synchronously tuned filters. *(From Anatol I. Zverev, Handbook of Filter Synthesis, John Wiley and Sons, Inc., New York, 1967. By permission of the publishers.)*

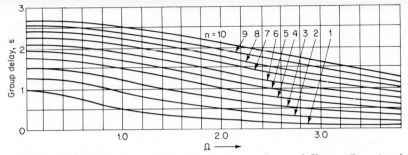

Fig. 2-78   Group-delay characteristics for synchronously tuned filters. (*From Anatol I. Zverev, Handbook of Filter Synthesis, John Wiley and Sons, Inc., New York, 1967. By permission of the publishers.*)

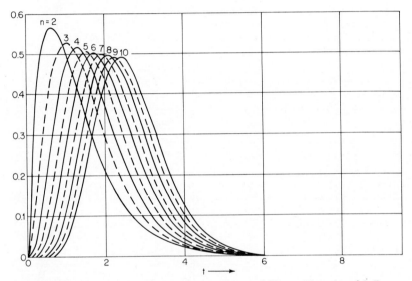

Fig. 2-79   Impulse response for synchronously tuned filters. (*From Anatol I. Zverev, Handbook of Filter Synthesis, John Wiley and Sons, Inc., New York, 1967. By permission of the publishers.*)

2.9   ELLIPTIC-FUNCTION FILTERS

All the previous filter types discussed are all-pole networks. They exhibit infinite rejection only at the extremes of the stopband. Elliptic-function filters have zeros as well as poles at finite frequencies. The location of the poles and zeros creates equiripple behavior in the passband similar to Chebyshev filters. Finite transmission zeros in the stopband reduce the transition region so that extremely sharp roll-off characteristics can be obtained. The introduction of these transmission zeros allows the steepest rate of descent theoretically possible for a given number of poles.

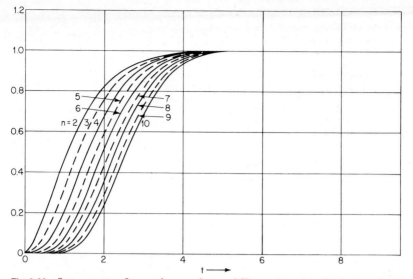

Fig. 2-80 Step response for synchronously tuned filters. *(From Anatol I. Zverev, Handbook of Filter Synthesis, John Wiley and Sons, Inc., New York, 1967. By permission of the publishers.)*

Figure 2-81 compares a five-pole Butterworth, a 0.1-dB Chebyshev, and a 0.1-dB elliptic-function filter having two transmission zeros. Clearly the elliptic-function filter has a much more rapid rate of descent in the transition region than the other filter types.

Improved performance is obtained at the expense of return lobes in the stopband. Elliptic-function filters are also more complex than all-pole networks. Return lobes usually are acceptable to the user, since a minimum stopband attenuation is required and the chosen filter will have return lobes that meet

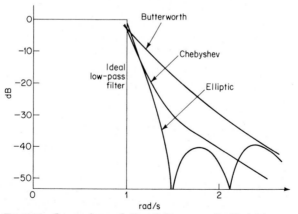

Fig. 2-81 Comparison of $N = 3$ Butterworth, Chebyshev, and elliptic-function filters.

this requirement. Also, even though each filter section is more complex than all-pole filters, fewer sections are required.

The following definitions apply to normalized elliptic-function low-pass filters and are illustrated in figure 2-82:

R_{dB} = passband ripple
A_{min} = minimum stopband attenuation in decibels
Ω_s = lowest stopband frequency at which A_{min} occurs

The response in the passband is similar to Chebyshev filters except that the attenuation at 1 rad/s is equal to the passband ripple instead of 3 dB. The

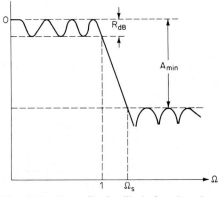

Fig. 2-82 Normalized elliptic-function low-pass filter response.

stopband has transmission zeros, with the first zero occurring slightly beyond Ω_s. All returns (come-backs) in the stopband are equal to A_{min}.

The attenuation of elliptic filters can be expressed as

$$A_{dB} = 10 \log[1 + \epsilon^2 Z_n^2 (\Omega)] \qquad (2\text{-}46)$$

where ϵ is determined by the ripple [equation (2-34)] and $Z_n (\Omega)$ is an elliptic function of the nth order. Elliptic functions have both poles and zeros and can be expressed as

$$Z_n (\Omega) = \frac{\Omega(a_2^2 - \Omega^2)(a_4^2 - \Omega^2) \ \ldots \ (a_m^2 - \Omega^2)}{(1 - a_2^2\Omega^2)(1 - a_4^2\Omega^2) \ \ldots \ (1 - a_m^2\Omega^2)} \qquad (2\text{-}47)$$

where n is odd and $m = (n - 1)/2$, or

$$Z_n (\Omega) = \frac{(a_2^2 - \Omega)(a_4^2 - \Omega^2) \ \ldots \ (a_m^2 - \Omega^2)}{(1 - a_2^2\Omega^2)(1 - a_4^2\Omega^2) \ \ldots \ (1 - a_m^2\Omega^2)} \qquad (2\text{-}48)$$

where n is even and $m = n/2$.

The zeros of Z_n are $a_2, a_4, \ldots, a_m$, whereas the poles are $1/a_2, 1/a_4, \ldots, 1/a_m$. The reciprocal relationship between the poles and zeros of Z_n results in equiripple behavior in both the stopband and the passband.

The values for a_2 through a_m are derived from the elliptic integral, which is defined as

$$K_e = \int_0^{\pi/2} \frac{d\theta}{\sqrt{1 - k^2 \sin^2\theta}} \qquad (2\text{-}49)$$

Numerical evaluation may be somewhat difficult. Glowatski (see references) contains tables specifically intended for determining the poles and zeros of Z_n (Ω).

Elliptic-function filters have been extensively tabulated by Saal and Zverev (see references). The basis for these tabulations was the order n and the parameters θ (degrees) and reflection coefficient ρ (percent).

Elliptic-function filters are sometimes called "Cauer filters" in honor of network theorist Professor Wilhelm Cauer. They are tabulated in chapter 12 and are commonly classified using the following convention:

$$C\, n\, \rho\, \theta$$

where C represents Cauer, n is the filter order, ρ is the reflection coefficient, and θ is the modular angle. A fifth-order filter having a ρ of 15% and a θ of 29° would be described as CO5 15 $\theta = 29°$. This convention will be used throughout the handbook.

The angle θ determines the steepness of the filter and is defined as

$$\theta = \sin^{-1}\frac{1}{\Omega_s} \qquad (2\text{-}50)$$

or alternately we can state

$$\Omega_s = \frac{1}{\sin\theta} \qquad (2\text{-}51)$$

Table 2-2 gives some representative values of θ and Ω_s.

TABLE 2-2 Ω_s vs. θ

θ, degrees	Ω_s
0	∞
10	5.759
20	2.924
30	2.000
40	1.556
50	1.305
60	1.155
70	1.064
80	1.015
90	1.000

The parameter ρ, the reflection coefficient, can be derived from

$$\rho = \frac{\text{VSWR} - 1}{\text{VSWR} + 1} = \sqrt{\frac{\epsilon^2}{1 + \epsilon^2}} \qquad (2\text{-}52)$$

where VSWR is the standing-wave ratio and ϵ is the ripple factor (see section 2.4 on the Chebyshev response). The passband ripple and reflection coefficient are related by

$$R_{\text{dB}} = -10 \log(1 - \rho^2) \qquad (2\text{-}53)$$

Table 2-3 interrelates these parameters for some typical values of reflection coefficient, where ρ is expressed as a percentage.

TABLE 2-3 ρ vs. R_{dB}, VSWR, and ϵ

ρ, %	R_{dB}	VSWR	ϵ (ripple factor)
1	0.0004343	1.0202	0.0100
2	0.001738	1.0408	0.0200
3	0.003910	1.0619	0.0300
4	0.006954	1.0833	0.0400
5	0.01087	1.1053	0.0501
8	0.02788	1.1739	0.08026
10	0.04365	1.2222	0.1005
15	0.09883	1.3529	0.1517
20	0.1773	1.5000	0.2041
25	0.2803	1.6667	0.2582
50	1.249	3.0000	0.5774

As the parameter θ approaches 90°, the edge of the stopband Ω_s approaches unity. For θ's near 90° extremely sharp roll-offs are obtained. However, for a fixed n, the stopband attenuation A_{min} is reduced as the steepness increases. Figure 2-83 shows the frequency response of an $n = 3$ elliptic filter for a fixed ripple of 1 dB ($\rho \approx 50\%$) and different values of θ.

For a given θ and order n, the stopband attenuation parameter A_{min} increases as the ripple is made larger. Since the poles of elliptic-function filters are approximately located on an ellipse, the delay curves behave in a similar manner to those of the Chebyshev family. Figure 2-84 compares the delay characteristics of $n = 3$, 4, and 5 elliptic filters all having A_{min} of 60 dB. The delay variation tends to increase sharply with increasing ripple and filter order n.

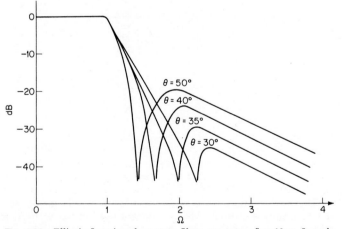

Fig. 2-83 Elliptic-function low-pass filter response for $N = 3$ and $R_{dB} = 1$ dB.

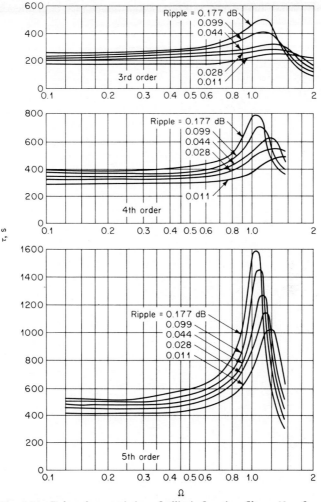

Fig. 2-84 Delay characteristics of elliptic-function filters $N = 3$, 4, 5 and with A_{min} of 60 dB. *(From Claude S. Lindquist, Active Network Design, Steward and Sons, California, 1977.)*

The factor ρ determines the input impedance variation with frequency of *LC* elliptic filters as well as the passband ripple. As ρ is reduced, a better match is achieved between the resistive terminations and the filter impedance. Figure 2-85 illustrates the input impedance variation with frequency of a normalized $n = 5$ elliptic-function low-pass filter. At DC, the input impedance is 1 Ω resistive. As the frequency increases, both positive and negative reactive components appear. All maximum values are within the diameter of a circle whose radius is proportional to the reflection coefficient ρ. As the complexity of the filter is increased, more gyrations occur within the circle.

The relationship between ρ and filter input impedance is defined by

$$|\rho|^2 = \left| \frac{R - Z_{11}}{R + Z_{11}} \right|^2 \tag{2-54}$$

where R is the resistive termination and Z_{11} is the filter input impedance.

The closeness of matching between R and Z_{11} is frequently expressed in decibels as return loss, which is defined as

$$A_\rho = 20 \log \left| \frac{1}{\rho} \right| \tag{2-55}$$

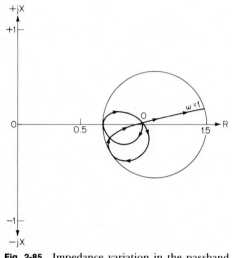

Fig. 2-85 Impedance variation in the passband of a normalized $N = 5$ elliptic-function low-pass filter.

Estimating the Required Order

To estimate the required order of an elliptic-function filter, the curves of figure 2-86 are used. This diagram shows the relationship between Ω_s, filter order n, reflection coefficient ρ, and A_{min}, the minimum stopband rejection.

The abscissa is the stopband limit Ω_s. The ordinate consists of the terms $A_{min} + A_\rho$. To use these curves, first choose ρ. Convert ρ to A_ρ using the table provided. Determine the filter order n to meet or exceed the desired $A_{min} + A_\rho$ at the required Ω_s. A_{min} and A_ρ are given both in nepers (Np) and decibels (dB), where 1 Np = 8.69 dB.

The passband ripple behavior of elliptic-function filters is similar to that of the Chebyshev family, as shown in figure 2-39. Even-order filters are required to have a loss at DC equal to the passband ripple. This is accomplished by providing unequal terminations, and these filters are designated by b in figure 2-86. If the terminations are to be equal, a transformation is applied to the filter which results in the filters designated by c in the curves.

The following example demonstrates the selection of an elliptic-function filter.

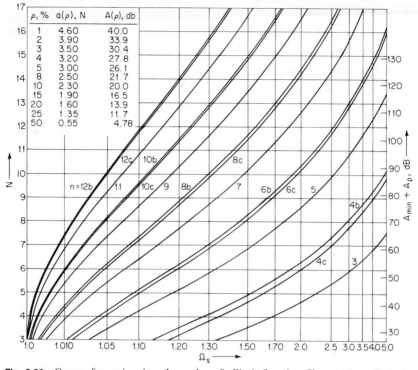

ρ, %	a(ρ), N	A(ρ), db
1	4.60	40.0
2	3.90	33.9
3	3.50	30.4
4	3.20	27.8
5	3.00	26.1
8	2.50	21.7
10	2.30	20.0
15	1.90	16.5
20	1.60	13.9
25	1.35	11.7
50	0.55	4.78

Fig. 2-86 Curves for estimating the order of elliptic-function filters. (*From R. Saal, Der Entwurf von Filtern mit Hilfe des Kataloges Normierter Tiefpasse, Telefunken GMBH, Backnang, West Germany, 1963. Reprinted by permission.*)

Example 2-21

REQUIRED: Determine the order of an elliptic-function filter having a passband ripple less than 0.2 dB up to 1000 Hz and a minimum rejection of 60 dB at 1300 Hz and above.

RESULT: (*a*) Compute the low-pass steepness factor A_s:

$$A_s = \frac{f_s}{f_c} = \frac{1300 \text{ Hz}}{1000 \text{ Hz}} = 1.3 \tag{2-11}$$

(*b*) Select a normalized low-pass filter having an A_{min} of at least 60 dB with a stopband limit Ω_s not exceeding 1.3. Let us choose a ρ of 20% which corresponds to a ripple of 0.18 dB from table 2-3. Using the table of figure 2-86, a ρ of 20% corresponds to an $A_ρ$ of 13.9 dB. Therefore,

$$A_{min} + A_ρ = 60 \text{ dB} + 13.9 \text{ dB} = 73.9 \text{ dB}$$

The curve of figure 2-86 indicates that at an Ω_s of 1.3, a filter of $n = 7$ provides the required attenuation.

In comparison, a twenty-seventh-order Butterworth low-pass filter would be required to meet the requirements of example 2-21; so the elliptic-function family is a *must* for steep filter requirements.

REFERENCES

Glowatski, E., "Sechsstellige Tafel der Cauer-Parameter," Verlag der Bayr, Akademie der Wissenschaften, Munich, Germany, 1955.

Lindquist, C. S., "Active Network Design," Steward and Sons, California, 1977.

Saal, R., "Der Entwurf von Filtern mit Hilfe des Kataloges Normierter Tiefpasse," Telefunken GMBH, Backnang, West Germany, 1963.

White Electromagnetics, "A Handbook on Electrical Filters," White Electromagnetics Inc., 1963.

Zverev, A. I., "Handbook of Filter Synthesis," John Wiley and Sons, New York, 1967.

3

Low-Pass
Filter Design

3.1 LC LOW-PASS FILTERS

All-Pole Filters

LC low-pass filters can be directly designed from the tables provided in chapter 12. A suitable filter must first be selected using the guidelines established in chapter 2. The chosen design is then frequency- and impedance-scaled to the desired cutoff and impedance level.

Example 3-1

REQUIRED: *LC* low-pass filter
3 dB at 1000 Hz
20 dB minimum at 2000 Hz
$R_s = R_L = 600 \ \Omega$

RESULT: (a) To normalize the low-pass requirement compute A_s.

$$A_s = \frac{f_s}{f_c} = \frac{2000 \text{ Hz}}{1000 \text{ Hz}} = 2 \qquad (2\text{-}11)$$

(b) Choose a normalized low-pass filter from the curves of chapter 2 having at least 20 dB of attenuation at 2 rad/s.
Examination of the curves indicates that an $n = 4$ Butterworth or third-order 0.1-dB Chebyshev satisfies this requirement. Let us select the latter, since fewer elements are required.

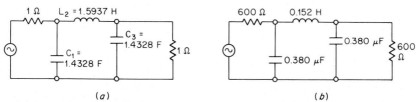

(a)   (b)

Fig. 3-1 Results of example 3-1: (a) normalized filter from table 12-28; (b) frequency- and impedance-scaled filter.

(c) Table 12-28 contains element values for normalized 0.1-dB Chebyshev LC filters ranging from $n = 2$ through $n = 10$. The circuit corresponding to $n = 3$ and equal source and load resistors $(R_s = 1\ \Omega)$ is shown in figure 3-1a.

(d) Denormalize the filter using a Z of 600 and a frequency-scaling factor (FSF) of $2\pi f_c$ or 6280.

$$R_s' = R_L' = 600\ \Omega \qquad (2\text{-}8)$$

$$L_2' = \frac{L \times Z}{\text{FSF}} = \frac{1.5937 \times 600}{6280} = 0.152\ \text{H} \qquad (2\text{-}9)$$

$$C_1' = C_3' = \frac{C}{\text{FSF} \times Z} = \frac{1.4328}{6280 \times 600} = 0.380\ \mu\text{F} \qquad (2\text{-}10)$$

The resulting filter is shown in figure 3-1b.

The normalized filter used in example 3-1 was shown in the table as having a current source input with a parallel resistor of 1 Ω. The reader will recall that Thévenin's theorems permit the replacement of this circuit with a voltage source having an equivalent series source resistance.

Elliptic-Function Filters

Normalized elliptic-function LC low-pass filters are presented in tabular form in table 12-56. They are classified using the convention outlined in section 2.9 and arranged in order of increasing n.

The following sample section of table 12-56 corresponds to C07 20:

Capacitors C_1 through C_7 and inductors L_2, L_4, and L_6 are defined by the table and are located directly above the branch numbers shown in the schematic. Branches 2, 4, and 6 consist of parallel resonant circuits comprised of L_2C_2, L_4C_4, and L_6C_6. The branch resonant frequencies correspond to the transmission zeros and are respectively given in the tables as Ω_2, Ω_4, and Ω_6. All inductors are in henrys, capacitors are in farads, and resonant frequencies are in radians per second. The source and load terminations are both shown to be 1 Ω.

When an elliptic-function low-pass filter is frequency-scaled, the transmission zeros are scaled by the same factor; so the tabulated zeros can be used to determine the resulting branch resonances. Where very steep filters are required, precise adjustment of these resonances to compensate for component tolerances is a necessity; so these frequencies are very useful.

The following example illustrates the design of an elliptic-function low-pass filter using the normalized tables.

Example 3-2

REQUIRED: LC low-pass filter
0.25 dB maximum ripple DC to 100 Hz
60 dB minimum at 132 Hz
$R_s = R_L = 900\ \Omega$

RESULT: (a) To normalize the given requirement, compute the low-pass steepness factor A_s.

$$A_s = \frac{f_s}{f_c} = \frac{132}{100} = 1.32 \qquad (2\text{-}11)$$

(b) Figure 2-86 indicates that for a ρ of 20% (0.18-dB ripple) a filter of $n = 7$ is required. Using the table corresponding to C07 20, select a filter having an Ω_s not exceeding 1.32 and an $A_{\min}$ of 60 dB or more.

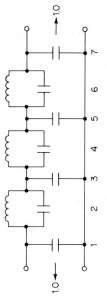

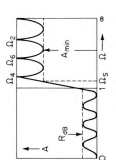

θ	Ω_s	A_min	C_1	C_2	L_2	Ω_2	C_3	C_4	L_4	Ω_4	C_5	C_6	L_6	Ω_6	C_7	θ
46	1.390 164	68.2	1.251	0.1000	1.285	2.789 476	1.808	0.4828	1.035	1.414 728	1.657	0.3428	1.048	1.668 286	1.053	46
47	1.367 327	66.7	1.247	0.1051	1.280	2.725 881	1.789	0.5093	1.015	1.391 016	1.633	0.3617	1.033	1.636 211	1.040	47
48	1.345 633	65.2	1.243	0.1105	1.275	2.664 770	1.770	0.5370	0.9944	1.368 471	1.608	0.3814	1.017	1.605 563	1.027	48
49	1.325 013	63.7	1.238	0.1160	1.269	2.605 984	1.751	0.5661	0.9736	1.347 026	1.583	0.4020	1.001	1.576 255	1.013	49
50	1.305 407	62.3	1.234	0.1217	1.264	2.549 377	1.731	0.5965	0.9525	1.326 618	1.557	0.4235	0.9850	1.548 208	0.9992	50
51	1.286 760	60.9	1.229	0.1277	1.258	2.494 813	1.711	0.6286	0.9310	1.307 190	1.531	0.4462	0.9684	1.521 349	0.9848	51
52	1.269 018	59.5	1.224	0.1339	1.252	2.442 167	1.690	0.6622	0.9093	1.288 687	1.504	0.4699	0.9514	1.495 612	0.9699	52
53	1.252 136	58.1	1.219	0.1404	1.246	2.391 323	1.669	0.6977	0.8872	1.271 063	1.477	0.4948	0.9340	1.470 934	0.9547	53
54	1.236 068	56.8	1.213	0.1471	1.239	2.342 170	1.648	0.7351	0.8648	1.254 270	1.450	0.5211	0.9163	1.447 259	0.9391	54
55	1.220 775	55.4	1.208	0.1541	1.232	2.294 610	1.626	0.7745	0.8420	1.238 269	1.422	0.5487	0.8981	1.424 533	0.9230	55

Let us choose the design which corresponds to $\theta = 50°$, since $\Omega_s = 1.305$ and $A_{min} = 62.3$ dB. The circuit of the C07 20 $\theta = 50°$ filter is shown in figure 3-2a and corresponds to the underlined values of the sample table.

(c)   Denormalize the filter using a Z of 900 and a frequency-scaling factor (FSF) of $2\pi f_c$ or 628.

$$C_1' = \frac{C}{\text{FSF} \times Z} = \frac{1.234}{628 \times 900} = 2.1822 \ \mu\text{F} \qquad (2\text{-}10)$$

$C_2' = 0.2152 \ \mu\text{F}$
$C_3' = 3.061 \ \mu\text{F}$
$C_4' = 1.055 \ \mu\text{F}$
$C_5' = 2.753 \ \mu\text{F}$
$C_6' = 0.7489 \ \mu\text{F}$
$C_7' = 1.767 \ \mu\text{F}$

$$L_2 = \frac{L \times Z}{\text{FSF}} = \frac{1.264 \times 900}{628} = 1.811 \ \text{H} \qquad (2\text{-}9)$$

$L_4' = 1.365$ H
$L_6' = 1.412$ H

To compute the denormalized resonant frequencies of each parallel tuned circuit, multiply the design cutoff ($f_c = 100$ Hz) by Ω_2, Ω_4, and Ω_6 to obtain

$f_2 = 254.9$ Hz
$f_4 = 132.7$ Hz
$f_6 = 154.8$ Hz

The resulting filter is given in figure 3-2b having the frequency response shown in figure 3-2c.

The capacitor magnitudes in the tuned circuits can be reduced by using tapped inductors. This technique is explained in detail in the discussion in chapter 8.

Duality and Reciprocity

A network and its dual have identical response characteristics. Each LC filter tabulated in section 12 has an equivalent dual network. The circuit configuration shown at the bottom of each table and the bottom set of nomenclature corresponds to the dual of the upper filter.

Any ladder-type network can be transformed into its dual by implementing the following rules:

1.   Convert every series branch into a shunt branch and every shunt branch into a series branch.
2.   Convert circuit branch elements in series to elements in parallel and vice versa.
3.   Transform each inductor into a capacitor and vice versa. The values remain unchanged; i.e., 4 H becomes 4 F.
4.   Replace each resistance with a conductance; i.e., 3 Ω becomes 3 mhos or $\frac{1}{3} \Omega$.
5.   Change a voltage source into a current source and vice versa.

Figure 3-3 shows a network and its dual.

The theorem of reciprocity states that if a voltage located at one point of a linear network produces a current at any other point, the same voltage acting at the second point results in the same current at the first point. Alternately,

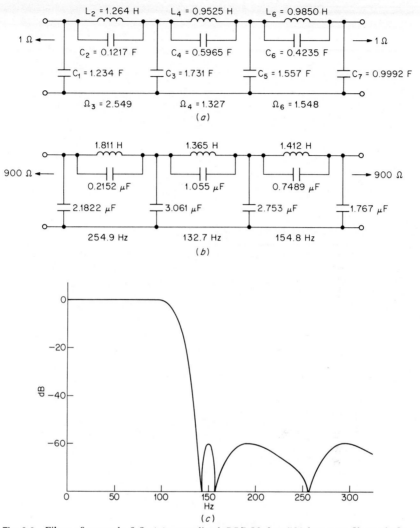

Fig. 3-2   Filter of example 3-2: (*a*) normalized CO7 20 $\theta = 50°$ low-pass filter; (*b*) frequency- and impedance-scaled network; (*c*) frequency response.

if a current source at one point of a linear network results in a voltage measured at a different point, the same current source at the second point produces the same voltage at the first point. As a result, the response of an *LC* filter is the same regardless of which direction the signal flows in, except for a constant multiplier. It is perfectly permissible to turn a filter schematic completely around with regard to its driving source provided that the source and load resistive terminations are also interchanged.

The laws of duality and reciprocity are used to manipulate a filter to satisfy termination requirements or to force a desired configuration.

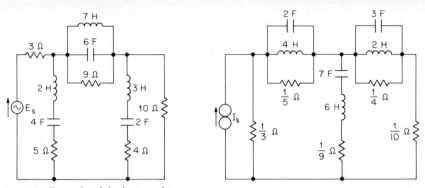

Fig. 3-3 Example of dual networks.

Designing for Unequal Terminations

Tables of LC element values are provided in chapter 12 for both equally terminated and unequally terminated networks. A number of different ratios of source
to load resistance are tabulated, including the impedance extremes of infinity
and zero.

   To design an unequally terminated filter, first determine the desired ratio
of R_s/R_L. Select a normalized filter from the tables that satisfies this ratio.
The reciprocity theorem can be applied to turn a network around end for end
and the source and load resistors can be interchanged. The tabulated impedance
ratio is inverted if the dual network given by the lower schematic is used. The
chosen filter is then frequency- and impedance-scaled.

Example 3-3

   REQUIRED: LC low-pass filter
         1 dB at 900 Hz
         20 dB minimum at 2700 Hz
         $R_s = 1$ kΩ
         $R_L = 5$ kΩ
   RESULT: (a) Compute A_s.

$$A_s = \frac{f_s}{f_c} = \frac{2700 \text{ Hz}}{900 \text{ Hz}} = 3 \qquad (2\text{-}11)$$

   (b) Normalized requirement:
       1 dB at X rad/s
       20 dB minimum at $3X$ rad/s
       (where X is arbitrary)

   (c) Select a normalized low-pass filter that makes the transition from
       1 dB to at least 20 dB over a frequency ratios of 3:1. A Butterworth
       $n = 3$ design will satisfy these requirements, since figure 2-34 indi-
       cates that the 1-dB point occurs at 0.8 rad/s and that more than
       20 dB of attenuation is obtained at 2.4 rad/s. Table 12-2 provides
       element values for normalized Butterworth low-pass filters for a
       variety of impedance ratios. Since the ratio of R_s/R_L is 1:5, we
       will select a design for $n = 3$ corresponding to $R_s = 0.2$ Ω and
       use the upper schematic. (Alternately, we could have selected the
       lower schematic corresponding to $R_s = 5$ Ω and turned the network
       end for end, but an additional inductor would have been required.)

(d) The normalized filter from table 12-2 is shown in figure 3-4a. Since the 1-dB point is required to be 900 Hz, the FSF is calculated by

$$\text{FSF} = \frac{\text{desired reference frequency}}{\text{existing reference frequency}}$$

$$= \frac{2\,\pi 900\ \text{rad/s}}{0.8\ \text{rad/s}} = 7069 \qquad (2\text{-}1)$$

Using a Z of 5000 and an FSF of 7069, the denormalized component values are

$$R_s' = R \times Z = 1\ k\Omega \qquad (2\text{-}8)$$
$$R_L' = 5\ k\Omega$$

$$C_1' = \frac{C}{\text{FSF} \times Z} = \frac{2.6687}{7069 \times 5000} = 0.0755\ \mu\text{F} \qquad (2\text{-}10)$$

$$C_3' = 0.22\mu\text{F}$$

$$L_2 = \frac{L \times Z}{\text{FSF}} = \frac{0.2842 \times 5000}{7069} = 0.201\ \text{H} \qquad (2\text{-}9)$$

The scaled filter is shown in figure 3-4b.

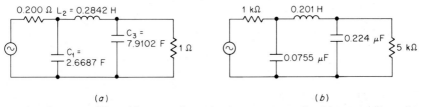

(a)   (b)

Fig. 3-4 Low-pass filter with unequal terminations: (a) normalized low-pass filter; (b) frequency- and impedance-scaled filter.

If an infinite termination is required, a design having an R_s of infinity is selected. When the input is a current source, the configuration is used as given. For an infinite load impedance, the entire network is turned end for end.

If the design requires a source impedance of 0 Ω, the dual network is used corresponding to $1/R_s$ of infinity or $R_s = 0\ \Omega$.

In practice, impedance extremes of near zero or infinity are not always possible. However, for an impedance ratio of 20 or more, the load can be considered infinite in comparison with the source, and the design for an infinite termination is used. Alternately the source may be considered zero with respect to the load and the dual filter corresponding to $R_s = 0\ \Omega$ may be used. When n is odd, the configuration having the infinite termination has one less inductor than its dual.

An alternate method of designing filters to operate between unequal terminations involves partitioning the source or load resistor between the filter and the termination. For example, a filter designed for a 1-kΩ source impedance could operate from a 250-Ω source if a 750-Ω resistor were placed within the filter network in series with the source. However, this approach would result in higher insertion loss.

Bartlett's Bisection Theorem   A filter network designed to operate between equal terminations can be modified for unequal source and load resistors if the circuit is symmetrical. Bartlett's bisection theorem states that if a symmetrical network is bisected and one half is impedance-scaled including the termination, the response shape will not change. All tabulated odd-order Butterworth and Chebyshev filters having equal terminations satisfy the symmetry requirement.

Example 3-4

REQUIRED: LC low-pass filter
3 dB at 200 Hz
15 dB minimum at 400 Hz
$R_s = 1$ kΩ $R_L = 1.5$ kΩ

RESULT: (a)   Compute A_s.

$$A_s = \frac{f_s}{f_c} = \frac{400}{200} = 2 \qquad (2\text{-}11)$$

(b)   Figure 2-34 indicates that an $n = 3$ Butterworth low-pass filter provides 18 dB rejection at 2 rad/s. Normalized LC values for Butterworth low-pass filters are given in table 12-2. The circuit corresponding to $n = 3$ and equal terminations is shown in figure 3-5 a.

(c)   Since the circuit of figure 3-5 a is symmetrical, it can be bisected into two equal halves as shown in figure 3-5 b. The requirement specifies a ratio of load to source resistance of 1.5 (1.5 kΩ/1

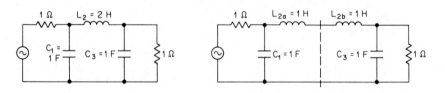

(a)   (b)

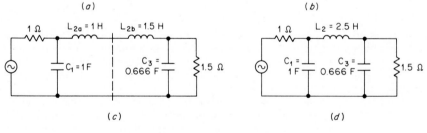

(c)   (d)

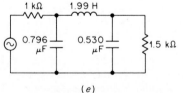

(e)

Fig. 3-5   Example of Bartlett's bisection theorem: (a) normalized filter having equal terminations; (b) bisected filter; (c) impedance-scaled right half section; (d) recombined filter; (e) final scaled network.

$k\Omega$); so we must impedance-scale the right half of the circuit by a factor of 1.5. The circuit of figure 3-5c is obtained.

(d) The recombined filter of figure 3-5d can now be frequency- and impedance-scaled using an FSF of $2\pi 200$ or 1256 and a Z of 1000.

$$R'_s = 1 \ k\Omega$$
$$R'_L = 1.5 \ k\Omega$$

$$C'_1 = \frac{C}{\text{FSF} \times Z} = \frac{1}{1256 \times 1000} = 0.796 \ \mu\text{F} \qquad (2\text{-}10)$$

$$C'_3 = 0.530 \ \mu\text{F}$$

$$L'_2 = \frac{L \times Z}{\text{FSF}} = \frac{2.5 \times 1000}{1256} = 1.99 \ \text{H} \qquad (2\text{-}9)$$

The final filter is shown in figure 3-5e.

Effects of Dissipation

Filters designed using the tables of LC element values in chapter 12 require lossless coils and capacitors to obtain the theoretical responses predicted in chapter 2. In the practical world, capacitors are usually obtainable that have low losses, but inductors are generally lossy, especially at low frequencies. Losses can be defined in terms of Q, the figure of merit or quality factor of a reactive component.

If a lossy coil or capacitor is resonated in parallel with a lossless reactance, the ratio of resonant frequency to 3-dB bandwidth of the resonant circuit's impedance, i.e., the band over which the magnitude of the impedance remains within 0.707 of the resonant value, is given by

$$Q = \frac{f_0}{\text{BW}_{3\,\text{dB}}} \qquad (3\text{-}1)$$

Figure 3-6 gives the low-frequency equivalent circuits for practical inductors and capacitors. Their Q's can be calculated by

Inductors: $$Q = \frac{\omega L}{R_L} \qquad (3\text{-}2)$$

Capacitors: $$Q = \omega C R_c \qquad (3\text{-}3)$$

where ω is the frequency of interest, in radians per second.

Using elements having a finite Q in a design intended for lossless reactances has the following mostly undesirable effects:

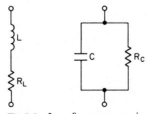

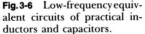

Fig. 3-6 Low-frequency equivalent circuits of practical inductors and capacitors.

1. At the passband edge the response shape becomes more rounded. Within the passband, the ripples are diminished and may completely vanish.
2. The insertion loss of the filter is increased. The loss in the stopband is maintained (except in the vicinity of transmission zeros); so the relative attenuation between the passband and the stopband is reduced.

Figure 3-7 shows some typical examples of these effects.

The most critical problem caused by finite element Q is the effect on the response shape near cutoff. Estimating the extent of this effect is somewhat

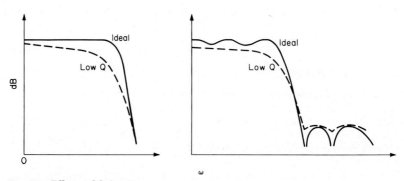

Fig. 3-7   Effects of finite Q.

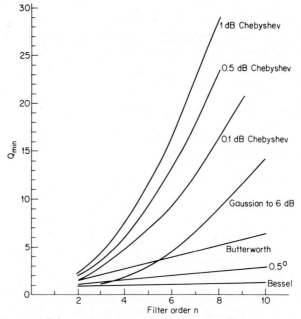

Fig. 3-8   Minimum Q requirements for low-pass filters.

difficult without extensive empirical data. The following variations in the filter design parameters will cause *increased* rounding of the frequency response near cutoff for a fixed Q:

1. Going to a larger passband ripple
2. Increasing the filter order n
3. Decreasing the transition region of elliptic-function filters

Changing these parameters in the opposite direction of course reduces the effects of dissipation.

Filters can be designed to have the responses predicted by modern network theory using finite element Q's. Figure 3-8 shows the minimum Q's required at cutoff for different low-pass responses. If elements are used having Q's slightly above the minimum values given in figure 3-8, the desired response can be obtained provided that certain predistorted element values are used. However, the insertion loss will be prohibitive. It is therefore highly desirable that element Q's be several times higher than the values indicated.

The effect of low Q on the response near cutoff can usually be compensated for by going to a higher-order network or a steeper filter and using a larger design bandwidth to allow for rounding. However, this design approach does not always result in satisfactory results, since the Q requirement may also increase. A method of compensating for low Q by using amplitude equalization is discussed in section 8.4.

The insertion loss of low-pass filters can be computed by replacing the reactive elements with resistances corresponding to their Q's, since at DC the inductors become short circuits and capacitors become open, which leaves the resistive elements only.

Figure 3-9 a shows a normalized third-order 0.1-dB Chebyshev low-pass filter where each reactive element has a Q of 10 at the 1-rad/s cutoff. The series and shunt resistors for the coils and capacitors are calculated using equations (3-2) and (3-3), respectively. At 1 rad/s these equations can be simplified and reexpressed as

$$R_L = \frac{L}{Q} \tag{3-4}$$

$$R_c = \frac{Q}{C} \tag{3-5}$$

The equivalent circuit at DC is shown in figure 3-9b. The insertion loss is 1.9 dB. The actual loss calculated was 7.9 dB, but the 6-dB loss due to the source and load terminations is normally not considered as part of the filter's insertion loss, since it would also occur in the event the filter was completely lossless.

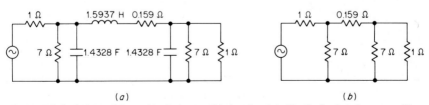

(*a*)   (*b*)

Fig. 3-9 Calculation of insertion loss: (*a*) third-order 0.1-dB Chebyshev low-pass filter with $Q = 10$; (*b*) equivalent circuit at DC.

Using Predistorted Designs

The effect of finite element Q on an LC filter transfer function is to increase the real components of the pole positions by an amount equal to the dissipation factor d, where

$$d = \frac{1}{Q} \tag{3-6}$$

Figure 3-10 shows this effect. All poles are displaced to the left by an equal amount.

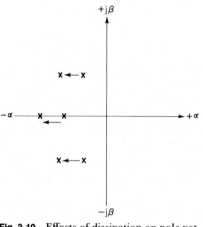

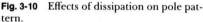

Fig. 3-10 Effects of dissipation on pole pattern.

If the desired poles were first shifted to the right by an amount equal to d, the introduction of the appropriate losses into the corresponding LC filter would move the poles back to the desired locations. This technique is called "predistortion." Predistorted filters are obtained by predistorting the required transfer function for a desired Q and then synthesizing an LC filter from the resulting transfer function. When the reactive elements of the filter have the required losses added, the response shape will correspond to the original transfer function.

The maximum amount that a group of poles can be displaced to the right in the process of predistortion is equal to the smallest real part among the poles, since further movement corresponds to locating a pole in the right half plane, which is an unstable condition. The minimum Q therefore is determined by the highest Q pole (i.e., the pole having the smallest real component). The Q's shown in figure 3-8 correspond to $1/d$, where d is the real component of the highest Q pole.

Tables are provided in chapter 12 for all-pole predistorted low-pass filters. These designs are all singly terminated with a source resistor of 1 Ω and an infinite termination. Their duals turned end for end can be used with a voltage source input and a 1-Ω termination.

Two types of predistorted filters are tabulated for various d's. The uniform dissipation networks require uniform losses in both the coils and the capacitors.

The second type are the Butterworth lossy-L filters, where only the inductors have losses, which closely agrees with practical components. It is important for both types that the element Q's are closely equal to $1/d$ at the cutoff frequency. In the case of the uniform dissipation networks, losses must usually be added to the capacitors.

Example 3-5

REQUIRED: LC low-pass filter
            3 dB at 500 Hz
            24 dB minimum at 1200 Hz
            $R_s = 600\ \Omega$ $R_L = 100\ k\Omega$ minimum
            Inductor Q's of 5 at 500 Hz
            Lossless capacitors

RESULT: (a)   Compute A_s.

$$A_s = \frac{1200}{500} = 2.4 \qquad (2\text{-}11)$$

(b)   The curves of figure 2-34 indicate that an $n = 4$ Butterworth low-pass filter has over 24 dB of rejection at 2.4 rad/s. Table 12-14 contains the element lossy values for Butterworth lossy-L networks where $n = 4$. The circuit corresponding to $d = 0.2$ ($d = 1/Q$) is shown in figure 3-11a.

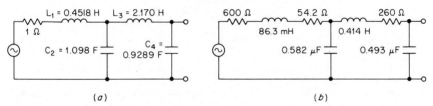

(a)   (b)

Fig. 3-11   Lossy-L low-pass filter of example 3-5: (a) normalized filter; (b) scaled filter.

(c)   The normalized filter can now be frequency- and impedance-scaled using an FSF of $2\pi500 = 3142$ and a Z of 600.

$$R_s' = 600\ \Omega$$

$$L_1' = \frac{L \times Z}{\text{FSF}} = \frac{0.4518 \times 600}{3142} = 86.3\ \text{mH} \qquad (2\text{-}9)$$

$$L_3' = 0.414\ \text{H}$$

$$C_2' = \frac{C}{\text{FSF} \times Z} = \frac{1.098}{3142 \times 600} = 0.582\ \mu\text{F} \qquad (2\text{-}10)$$

$$C_4' = 0.493\ \mu\text{F}$$

(d)   The resistive coil losses are

$$R_1 = \frac{\omega L}{Q} = 54.2\ \Omega \qquad (3\text{-}2)$$

and $R_3 = 260\ \Omega$

where $\omega = 2\pi f_c = 3142$

The final filter is given in figure 3-11b.

Example 3-6

REQUIRED: LC low-pass filter
3 dB at 100 Hz
58 dB minimum at 300 Hz
$R_s = 1$ kΩ $R_L = 100$ kΩ minimum
Inductor Q's of 11 at 100 Hz
Lossless capacitors

RESULT: (a)   Compute A_s.

$$A_s = \frac{300}{100} = 3 \tag{2-11}$$

(b)   Figure 2-42 indicates that a fifth-order 0.1-dB Chebyshev has about 60 dB of rejection at 3 rad/s. Table 12-32 provides LC element values for 0.1-dB Chebyshev uniform dissipation networks. The available inductor Q of 11 corresponds to a d of 0.091 ($d = 1/Q$). Values are tabulated for an $n = 5$ network having a d of 0.0881, which is sufficiently close to the requirement. The corresponding circuit is shown in figure 3-12a.

(c)   The normalized filter is frequency- and impedance-scaled using an FSF of $2\pi 100 = 628$ and a Z of 1000.

$$C_1' = \frac{C}{\text{FSF} \times Z} = \frac{1.1449}{628 \times 1000} = 1.823 \ \mu\text{F} \tag{2-10}$$

$$C_3' = 3.216 \ \mu\text{F}$$
$$C_5' = 1.453 \ \mu\text{F}$$

$$L_2' = \frac{L \times Z}{\text{FSF}} = \frac{1.8416 \times 1000}{628} = 2.932 \ \text{H} \tag{2-9}$$

$$L_4' = 2.681 \ \text{H}$$

(d)   The shunt resistive losses for capacitors C_1', C_3', and C_5' are

$$R_1' = \frac{Q}{\omega C} = 9.91 \ \text{k}\Omega \tag{3-3}$$

$$R_3' = 5.62 \ \text{k}\Omega$$
$$R_5' = 12.44 \ \text{k}\Omega$$

The series resistive inductor losses are

$$R_2' = \frac{\omega L}{Q} = 162 \ \Omega \tag{3-2}$$

$$R_4' = 148 \ \Omega$$

where $$Q = \frac{1}{d} = \frac{1}{0.0881} = 11.35$$

and $$\omega = 2\pi f_c = 2\pi 100 = 628$$

The resulting circuit including all losses is shown in figure 3-12b. This circuit can be turned end for end so that the requirement for a 1-kΩ source resistance is met. The final filter is given in figure 3-12c.

It is important to remember that uniform dissipation networks require the presence of losses in both the coils and the capacitors; so resistors must usually be added. Component Q's within 20% of $1/d$ are usually sufficient for satisfactory results.

Resistors can sometimes be combined to eliminate components. In the circuit of figure 3-12c, the 1-kΩ source and the 12.44-kΩ resistor can be combined,

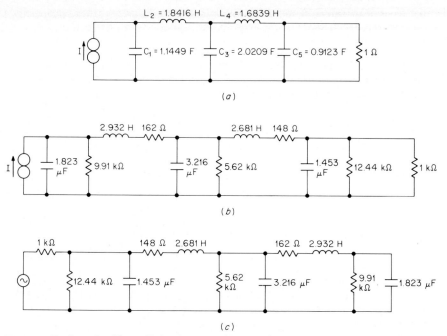

Fig. 3-12 Design of uniform dissipation network of example 3-6: (a) normalized fifth-order 0.1-dB Chebyshev with $d = 0.0881$; (b) frequency- and impedance-scaled filter including losses; (c) final network.

which results in a 926-Ω equivalent source resistance. The network can then be impedance-scaled to restore a 1-kΩ source.

3.2 ACTIVE LOW-PASS FILTERS

Active low-pass filters are designed using a sequence of operations similar to the design of LC filters. The specified low-pass requirement is first normalized and a particular filter type of the required complexity is selected using the response characteristics given in chapter 2. Normalized tables of active filter component values are provided in chapter 12 for each associated transfer function. The corresponding filter is denormalized by frequency and impedance scaling.

Active filters can also be designed directly from the poles and zeros. This approach sometimes offers some additional degrees of freedom and will also be covered.

All-Pole Filters

The transfer function of a passive RC network has poles that lie only on the negative real axis of the complex frequency plane. In order to obtain the complex poles required by the all-pole transfer functions of chapter 2, active elements must be introduced. Integrated circuit operational amplifiers are readily available that have nearly ideal properties such as high gain. However, these properties are limited to frequencies below a few hundred kilohertz; so active filters beyond this range are difficult.

Unity-Gain Single-Feedback Realization   Figure 3-13 shows two active low-pass filter configurations. The two-pole section provides a pair of complex conjugate poles, whereas the three-pole section produces a pair of complex conjugate

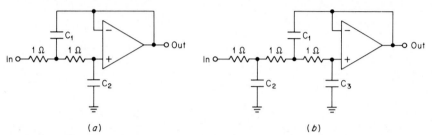

(a)   *(b)*

Fig. 3-13   Unity-gain active low-pass configurations: *(a)* two-pole section; *(b)* three-pole section.

poles and a single real-axis pole. The operational amplifier is configured in the voltage-follower configuration, which has a closed-loop gain of unity, very high input impedance, and nearly zero output impedance.

The two-pole section has the transfer function

$$T(s) = \frac{1}{C_1 C_2 s^2 + 2 C_2 s + 1} \tag{3-7}$$

A second-order low-pass transfer function can be expressed in terms of the pole locations as

$$T(s) = \frac{1}{\dfrac{1}{\alpha^2 + \beta^2} s^2 + \dfrac{2\,\alpha}{\alpha^2 + \beta^2} s + 1} \tag{3-8}$$

Equating coefficients and solving for the capacitors results in

$$C_1 = \frac{1}{\alpha} \tag{3-9}$$

$$C_2 = \frac{\alpha}{\alpha^2 + \beta^2} \tag{3-10}$$

where α and β are the real and imaginary coordinates of the pole pair.

The transfer function of the normalized three-pole section was discussed in section 1.2 and was given by

$$T(S) = \frac{1}{s^3 A + s^2 B + s C + 1} \tag{1-17}$$

where $A = C_1 C_2 C_3 \tag{1-18}$

$B = 2 C_3 (C_1 + C_2) \tag{1-19}$

and $C = C_2 + 3 C_3 \tag{1-20}$

Solution of these equations to find the values of C_1, C_2, and C_3 in terms of the poles is somewhat laborious and is best accomplished with a digital computer.

If the filter order n is an even order, $n/2$ two-pole filter sections are required. Where n is odd, $(n - 3)/2$ two-pole sections and a single three-pole section are necessary. This occurs because even-order filters have complex poles only, whereas an odd-order transfer function has a single real pole in addition to the complex poles.

At DC, the capacitors become open circuits; so the circuit gain becomes equal to that of the amplifier, which is unity. This can also be determined analytically from the transfer functions given by equations (3-7) and (1-17). At DC, $s = 0$ and $T(s)$ reduces to 1. Within the passband of a low-pass filter, the response of individual sections may have sharp peaks and some corresponding gain.

All resistors are 1 Ω in the two normalized filter circuits of figure 3-13. Capacitors C_1, C_2, and C_3 are tabulated in chapter 12. These values result in the normalized all-pole transfer functions of chapter 2 where the 3-dB cutoff occurs at 1 rad/s.

To design a low-pass filter, a filter type is first selected from chapter 2. The corresponding active low-pass filter values are then obtained from chapter 12. The normalized filter is denormalized by dividing all the capacitor values by FSF × Z, which is identical to the denormalization formula for LC filters, i.e.,

$$C' = \frac{C}{\text{FSF} \times Z} \tag{2-10}$$

where FSF is the frequency-scaling factor $2\pi f_c$ and Z is the impedance-scaling factor. The resistors are multiplied by Z, which results in equal resistors throughout of Z Ω.

The factor Z does not have to be the same for each filter section, since the individual circuits are isolated by the operational amplifiers. The value of Z can be independently chosen for each section so that practical capacitor values occur, but the FSF must be the same for all sections. The sequence of the sections can be rearranged if desired.

The frequency response obtained from active filters is usually very close to theoretical predictions provided that the component tolerances are small and that the amplifier has satisfactory properties. The effects of low Q which occur in LC filters do not apply; so the filters have no insertion loss and the passband ripples are well defined.

Example 3-7

REQUIRED: Active low-pass filter
3 dB at 100 Hz
70 dB minimum at 350 Hz

RESULT: (a) Compute low-pass steepness factor A_s.

$$A_s = \frac{f_s}{f_c} = \frac{350}{100} = 3.5 \tag{2-11}$$

(b) The response curve of figure 2-44 indicates that a fifth-order 0.5-dB Chebyshev low-pass filter meets the 70-dB requirement at 3.5 rad/s.

(c) The normalized values can be found in table 12-39. The circuit consists of a three-pole section followed by a two-pole section and is shown in figure 3-14a.

(d) Let us arbitrarily select an impedance-scaling factor of 5×10^4. Using an FSF of $2\pi f_c$ or 628, the resulting new values are

Three-pole section:

$$C_1' = \frac{C}{\text{FSF} \times Z} = \frac{6.842}{628 \times 5 \times 10^4} = 0.218 \ \mu\text{F} \qquad (2\text{-}10)$$

$$C_2' = 0.106 \ \mu\text{F}$$
$$C_3' = 0.00966 \ \mu\text{F}$$

Two-pole section:

$$C_1' = \frac{C}{\text{FSF} \times Z} = \frac{9.462}{628 \times 5 \times 10^4} = 0.301 \ \mu\text{F}$$

$$C_2' = 0.00364 \ \mu\text{F}$$

The resistors in both sections are multiplied by Z, resulting in equal resistors throughout of 50 kΩ. The denormalized circuit is given in figure 3-14b having the frequency response of figure 3-14c.

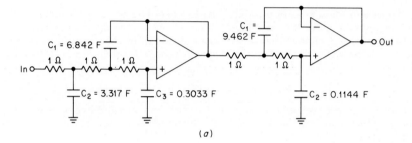

(a)

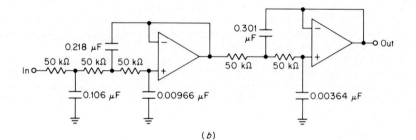

(b)

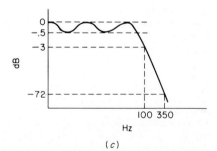

(c)

Fig. 3-14 Low-pass filter of example 3-7: (a) normalized fifth-order 0.5-dB Chebyshev low-pass filter; (b) denormalized filter; (c) frequency response.

The first section of the filter should be driven by a voltage source having a source impedance much less than the first resistor of the section. The input must have a DC return to ground if a blocking capacitor is present. Since the filter's output impedance is low, the frequency response is independent of the terminating load, provided that the operational amplifier has sufficient driving capability.

Real-Pole Configurations   All odd-order low-pass transfer functions have a single real-axis pole. This pole is realized as part of the $n = 3$ section of figure 3-13b when the tables of active low-pass values in chapter 12 are used. If an odd-order filter is designed directly from the tabulated poles, the normalized real-axis pole can be generated using one of the configurations given in figure 3-15.

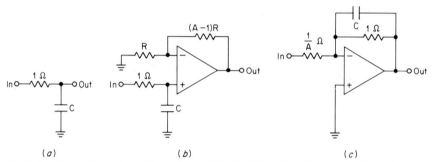

(a)   (b)   (c)

Fig. 3-15   First-order pole configurations: (a) basic RC section; (b) noninverting gain configuration; (c) inverting gain circuit.

The most basic form of a real pole is the circuit of figure 3-15a. The capacitor C is defined by

$$C = \frac{1}{\alpha_0} \qquad (3\text{-}11)$$

where α_0 is the normalized real-axis pole. The circuit gain is unity with a high-impedance termination.

If gain is desirable, the circuit of figure 3-15a can be followed by a noninverting amplifier as in figure 3-15b, where A is the required gain. When the gain must be inverting, the circuit of figure 3-15c is used.

The chosen circuit is frequency- and impedance-scaled in a manner similar to the rest of the filter. The value R in figure 3-15b is arbitrary, since only the ratio of the two feedback resistors determines the gain of the amplifier.

Example 3-8

REQUIRED: Active low-pass filter
3 dB at 75 Hz
15 dB minimum at 150 Hz
Gain of 40 dB ($A = 100$)

RESULT: (a) Compute the steepness factor.

$$A_s = \frac{f_s}{f_c} = \frac{150}{75} = 2 \qquad (2\text{-}11)$$

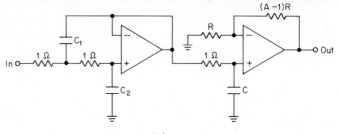

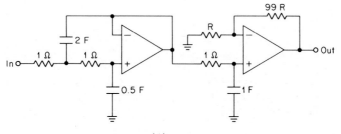

(a)

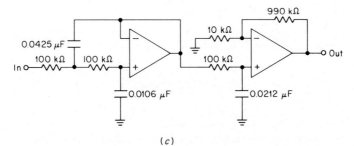

(b)

(c)

Fig. 3-16 Low-pass filter of example 3-8: (a) circuit configuration; (b) normalized circuit; (c) scaled filter.

(b) Figure 2-34 indicates that an $n = 3$ Butterworth low-pass response satisfies the attenuation requirement. Since a gain of 100 is required, we will use the $n = 2$ section of figure 3-13a followed by the $n = 1$ section of figure 3-15b, which provides the gain. The circuit configuration is shown in figure 3-16a.

(c) The following pole locations of a normalized $n = 3$ Butterworth low-pass filter are obtained from table 12-1:
Complex pole $\alpha = 0.5000$ $\beta = 0.8660$
Real pole $\alpha_0 = 1.0000$
The component values for the $n = 2$ section are

$$C_1 = \frac{1}{\alpha} = \frac{1}{0.5} = 2 \text{ F} \tag{3-9}$$

$$C_2 = \frac{\alpha}{\alpha^2 + \beta^2} = \frac{0.5}{0.5^2 + 0.866^2} = 0.5 \text{ F} \tag{3-10}$$

The capacitor in the $n = 1$ circuit is computed by

$$C = \frac{1}{\alpha_0} = \frac{1}{1.0} = 1 \text{ F} \qquad (3\text{-}11)$$

Since $A = 100$, the feedback resistor is $99R$ in the normalized circuit shown in figure 3-16b.

(d) Using an FSF of $2\pi f_c$ or 471 and selecting an impedance-scaling factor of 10^5 the denormalized capacitor values are

n = 2 section:

$$C_1' = \frac{C}{\text{FSF} \times Z} = \frac{2}{471 \times 10^5} = 0.0425 \ \mu\text{F} \qquad (2\text{-}10)$$
$$C_2' = 0.0106 \ \mu\text{F}$$

n = 1 section:

$$C' = 0.0212 \ \mu\text{F}$$

The value R for the $n = 1$ section is arbitrarily selected at 10 kΩ. The final circuit is given in figure 3-16c.

Although these real pole sections are intended to be part of odd-order low-pass filters, they can be independently used as an $n = 1$ low-pass filter. They have the transfer function

$$T(s) = K \frac{1}{sC + 1} \qquad (3\text{-}12)$$

where $K = 1$ for figure 3-15a, $K = A$ for figure 3-15b, and $K = -A$ for figure 3-15c. If $C = 1$ F, the 3-dB cutoff occurs at 1 rad/s.

The attenuation of a first-order filter can be expressed as

$$A_{\text{dB}} = 10 \log \left[1 + \left(\frac{\omega_x}{\omega_c} \right)^2 \right] \qquad (3\text{-}13)$$

where ω_x/ω_c is the ratio of a given frequency to the cutoff frequency. The normalized frequency response corresponds to the $n = 1$ curve of the Butterworth low-pass filter response curves of figure 2-34. The step response has no overshoot and the impulse response does not have any oscillatory behavior.

Example 3-9

REQUIRED: Active low-pass filter
3 dB at 60 Hz
12 dB minimum attenuation at 250 Hz
Gain of 20 dB with inversion

RESULT: (a) Compute A_s.

$$A_s = \frac{f_s}{f_c} = \frac{250}{60} = 4.17 \qquad (2\text{-}11)$$

(b) Figure 2-34 indicates that an $n = 1$ filter provides over 12 dB attenuation at 4.17 rad/s. Since an inverting gain of 20 dB is required, the configuration of figure 3-15c will be used. The normalized circuit is shown in figure 3-17a, where $C = 1$ F and $A = 10$ corresponding to a gain of 20 dB.

(c) Using an FSF of $2\pi 60$ or 377 and an impedance-scaling factor of 10^6, the denormalized capacitor is

$$C' = \frac{C}{\text{FSF} \times Z} = \frac{1}{377 \times 10^6} = 0.00265 \ \mu\text{F} \qquad (2\text{-}10)$$

The input and output feedback resistors are 100 kΩ and 1 MΩ, respectively. The final circuit is shown in figure 3-17b.

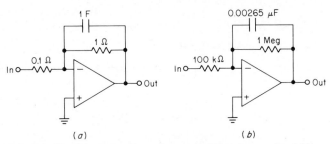

(a)   (b)

Fig. 3-17 $N = 1$ low-pass filter of example 3-9: (a) normalized filter; (b) frequency- and impedance-scaled filter.

Second-Order Section with Gain   If an active low-pass filter is required to have a gain higher than unity and the order is even, the $n = 1$ sections of figure 3-15 cannot be used, since a real pole is not contained in the transfer function.

The circuit of figure 3-18 realizes a pair of complex poles and provides a gain of $-A$. The element values are computed using the following formulas:

$$C_1 = (A+1)\left(1 + \frac{\beta^2}{\alpha^2}\right) \qquad (3\text{-}14)$$

$$R_1 = \frac{\alpha}{A(\alpha^2 + \beta^2)} \qquad (3\text{-}15)$$

$$R_2 = \frac{AR_1}{A+1} \qquad (3\text{-}16)$$

$$R_3 = AR_1 \qquad (3\text{-}17)$$

This section is used in conjunction with the $n = 2$ section of figure 3-13a to realize even-order low-pass filters with gain. This is shown in the following example:

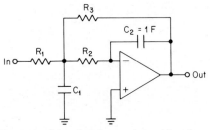

Fig. 3-18   Second-order section with gain.

Example 3-10

REQUIRED: Active low-pass filter
3 dB at 200 Hz
30 dB minimum at 800 Hz
No step-response overshoot
Gain of 6 dB with inversion ($A = 2$)

RESULT: (a) Compute A_s.

$$A_s = \frac{f_s}{f_c} = \frac{800}{200} = 4 \qquad (2\text{-}11)$$

(b) Since no overshoot is permitted, a Bessel filter type will be used. Figure 2-56 indicates that a fourth-order network provides over 30 dB of rejection at 4 rad/s. Since an inverting gain of 2 is required and $n = 4$, the circuit of figure 3-18 will be used followed by the two-pole section of figure 3-13a. The basic circuit configuration is given in figure 3-19a.

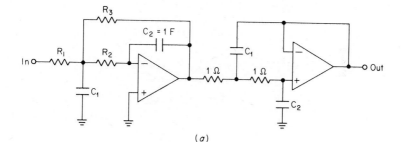

(a)

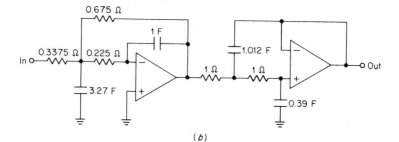

(b)

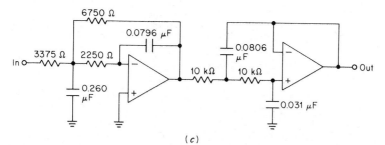

(c)

Fig. 3-19 $N = 4$ Bessel low-pass filter of example 3-10: (a) circuit configuration; (b) normalized filter; (c) frequency- and impedance-scaled filter.

(c) The following pole locations of a normalized $n = 4$ Bessel low-pass filter are obtained from table 12-41:

$$\alpha = 1.3596 \qquad \beta = 0.4071$$
and
$$\alpha = 0.9877 \qquad \beta = 1.2476$$

The normalized component values for the first section are determined by the following formulas, where $\alpha = 1.3596$, $\beta = 0.4071$, and $A = 2$:

$$C_1 = (A + 1)\left(1 + \frac{\beta^2}{\alpha^2}\right) = 3\left(1 + \frac{0.4071^2}{1.3596^2}\right) = 3.27 \text{ F} \tag{3-14}$$

$$R_1 = \frac{\alpha}{A(\alpha^2 + \beta^2)} = \frac{1.3596}{2(1.3596^2 + 0.4071^2)} = 0.3375 \text{ }\Omega \tag{3-15}$$

$$R_2 = \frac{AR_1}{A + 1} = \frac{2 \times 0.3375}{3} = 0.225 \text{ }\Omega \tag{3-16}$$

$$R_3 = AR_1 = 2 \times 0.3375 = 0.675 \text{ }\Omega \tag{3-17}$$

The remaining pole pair of $\alpha = 0.9877$ and $\beta = 1.2476$ is used to compute the component values of the second section.

$$C_1 = \frac{1}{\alpha} = \frac{1}{0.9877} = 1.012 \text{ F} \tag{3-9}$$

$$C_2 = \frac{\alpha}{\alpha^2 + \beta^2} = \frac{0.9877}{0.9877^2 + 1.2476^2} = 0.39 \text{ F} \tag{3-10}$$

The normalized low-pass filter is shown in figure 3-19b.

(d) Using an FSF of $2\pi f_c$ or 1256 and an impedance-scaling factor of 10^4 for both sections, the denormalized values are

n = 2 section with A = 2:

$$R_1' = R \times Z = 0.3375 \times 10^4 = 3375 \text{ }\Omega \tag{2-8}$$

$$R_2' = 2250 \text{ }\Omega$$
$$R_3' = 6750 \text{ }\Omega$$

$$C_1' = \frac{C}{\text{FSF} \times Z} = \frac{3.27}{1256 \times 10^4} = 0.260 \text{ }\mu\text{F} \tag{2-10}$$

$$C_2' = 0.0796 \text{ }\mu\text{F}$$

n = 2 section having unity gain:

$$R' = 10 \text{ k}\Omega$$

$$C_1' = \frac{C}{\text{FSF} \times Z} = \frac{1.012}{1256 \times 10^4} = 0.0806 \text{ }\mu\text{F} \tag{2-10}$$

$$C_2' = 0.0310 \text{ }\mu\text{F}$$

The final circuit is shown in figure 3-19c.

Elliptic-Function VCVS Filters

Elliptic-function filters were first discussed in section 2.9. They contain zeros as well as poles. The zeros begin just outside the passband and force the response to decrease rapidly as Ω_s is approached. (Refer to figure 2-82 for frequency-response definitions.)

Because of these finite zeros, the active filter circuit configurations of the previous section cannot be used, since they are restricted to the realization of poles only.

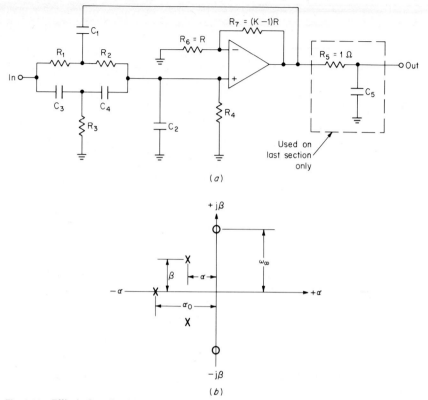

Fig. 3-20 Elliptic-function low-pass filter section: (a) VCVS circuit configuration for $K > 1$; (b) pole-zero pattern.

The schematic of an elliptic-function low-pass filter section is shown in figure 3-20a. This section provides a pair of complex conjugate poles and a pair of imaginary zeros as shown in figure 3-20b. The complex pole pair has a real component of α and an imaginary coordinate of β. The zeros are located at $\pm j\omega_\infty$. The RC section consisting of R_5 and C_5 introduces a real pole at α_0.

The configuration contains a voltage-controlled voltage source (VCVS) as the active element and is frequently referred to as a VCVS realization. Although this structure requires additional elements when compared with other VCVS configurations, it has been found to yield more reliable results and has lower sensitivity factors.[1]

The normalized element values are determined by the following relations: First calculate

$$a = \frac{2\alpha}{\sqrt{\alpha^2 + \beta^2}} \qquad (3\text{-}18)$$

$$b = \frac{\omega_\infty^2}{\alpha^2 + \beta^2} \qquad (3\text{-}19)$$

[1] W. J. Kerwin and L. P. Huelsman, "The Design of High Performance Active RC Band-pass Filters," IEEE International Convention Record, Vol. 14, part 10, March 1966.

where α, β, and ω_∞ are the pole-zero coordinates as defined in figure 3-20 b. The element values are computed by

$$R_1 = \frac{b+1}{3\,b} \tag{3-20}$$

$$R_2 = 2\,R_1 \tag{3-21}$$

$$R_3 = \frac{b+1}{4.5} \tag{3-22}$$

$$R_4 = 4.5\,R_3 \tag{3-23}$$

$$C_1 = \frac{4.5\,b}{(b+1)\sqrt{\alpha^2 + \beta^2}} \tag{3-24}$$

$$C_2 = \frac{C_1}{4.5} \tag{3-25}$$

$$C_3 = \frac{C_1}{1.5\,b} \tag{3-26}$$

$$C_4 = \frac{C_3}{2} \tag{3-27}$$

$$K = \frac{(2.5 - a)(b+1)}{1.5\,b} \tag{3-28}$$

$$\text{Section gain} = \frac{bK}{b+1} \tag{3-29}$$

Capacitor C_5 is determined by the real pole as follows:

$$C_5 = \frac{1}{\alpha_0} \tag{3-30}$$

Zverev (see references) has extensively tabulated design data for elliptic-function filters including pole-zero locations. Component values for active elliptic-function low-pass filters are conveniently given in table 12-57 based on Zverev's data; so the computation of the normalized element values is not required. Table 12-57 is restricted to odd-order filters in order to provide the most efficient utilization of a fixed amount of amplifiers.

Since the circuit of figure 3-20 provides a single pole pair (along with a pair of zeros), the total number of sections required for a filter is determined by $(n - 1)/2$, where n is the order of the filter. Because an odd-order transfer function has a single real pole, R_5 and C_5 appear on the output section only.

In the absence of a detailed analysis, it is a good rule of thumb to pair poles with their nearest zeros when allocating poles and zeros to each active section. This applies to high-pass, bandpass, and band-reject filters as well.

Alongside the normalized component values in table 12-57, the poles and zeros are listed for each section using the nomenclature of figure 3-20b. For the real pole section consisting of R_5 and C_5, no imaginary part exists; so only the real component α_0 is given.

Each section has a gain at DC which is also provided alongside the element values. For a filter consisting of multiple sections, the composite gain is the product of all the individual gains tabulated. The real pole section could be realized using one of the circuits of figure 3-15 if some additional flexibility in determining gain is desired.

A controlled amplification of K is required between the noninverting amplifier input and the section output. Since the gain of a noninverting operational amplifier is the ratio of the feedback resistors plus 1, R_6 and R_7 are R and $(K - 1) R$, respectively, where R can be any convenient value. The value for K is also tabulated for each section in table 12-57.

In the event that K is less than 1, the amplifier is reconfigured as a voltage follower and R_4 is split into two resistors, R_{4a} and R_{4b}, where

$$R_{4a} = (1 - K) R_4 \qquad\qquad (3\text{-}31)$$

$$R_{4b} = KR_4 \qquad\qquad (3\text{-}32)$$

The modified circuit is shown in figure 3-21.

The elliptic-function active filters given in table 12-57 are categorized in the same manner as the LC filters. The filters are described by $C n \rho \theta$, where C represents Cauer, n is the filter order, ρ is the reflection coefficient, and θ is the modular angle.

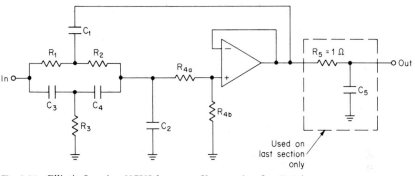

Fig. 3-21 Elliptic-function VCVS low-pass filter section for $K < 1$.

The design of active elliptic-function filters proceeds in a manner similar to the design of the all-pole types. A low-pass requirement is first converted to a steepness factor A_s. A normalized low-pass filter is selected from table 12-57 that meets the required passband to stopband transition steepness by having an Ω_s not in excess of A_s. The passband ripple as determined by the reflection coefficient ρ and the stopband attenuation A_{min} should also satisfy the specification. The filter complexity can first be estimated using the curves of figure 2-86 and a satisfactory design selected from the table. Alternately the tables can be approached directly and a filter can be chosen based on the tabulated design parameters.

The selected filter is scaled to the required cutoff frequency f_c by denormalization. The capacitors are divided by $Z \times$ FSF, and the resistors are multiplied by Z except for R_6 and R_7, where the impedance-scaling factor can be any desired value. Each section may be denormalized by a different value of Z, but the FSF must be constant throughout. The frequency of infinite attenuation of each section is calculated as follows:

$$f_\infty = \omega_\infty \times f_c \qquad (3\text{-}33)$$

The first section should be driven by a voltage source whose source impedance is much less than the input resistor R_1. The load impedance should be high in comparison with R_5 unless the circuit of figure 3-15b or c is used for the real pole, in which case the termination can be any practical value within the driving capability of the operational amplifier.

Example 3-11

REQUIRED: Active low-pass filter
1 dB maximum ripple below 1000 Hz
38 dB minimum rejection above 2950 Hz

RESULT: (*a*) Compute the steepness factor.

$$A_s = \frac{f_s}{f_c} = \frac{2950}{1000} = 2.95 \qquad (2\text{-}11)$$

(*b*) A low-pass filter is required that makes the transition from less than 1 dB to over 38 dB of attenuation within a frequency ratio of 2.95. Table 12-57 indicates that an elliptic filter corresponding to C03 25 $\theta = 20°$ has the following parameters:

$$n = 3$$
$$R_{dB} = 0.28 \text{ dB (from table 2-3 for } p = 25\%)$$
$$\Omega_s = 2.924$$
$$A_{min} = 39.48 \text{ dB}$$

The normalized circuit from table 12-57 is shown in figure 3-22a, where

$$R_7 = (K-1)R = 0.410R \qquad \text{since } K = 1.410$$

(*c*) Frequency- and impedance-scale the normalized filter, where FSF $= 2\,\pi f_c = 6280$ and Z is arbitrarily chosen at 5×10^3. The resulting element values are

$$R_1' = R \times Z = 0.3719 \times 5 \times 10^3 = 1860\ \Omega \qquad (2\text{-}8)$$
$$R_2' = 3719\ \Omega$$
$$R_3' = 10.71\ \text{k}\Omega$$
$$R_4' = 48.19\ \text{k}\Omega$$
$$R_5' = 5\ \text{k}\Omega$$

$$C_1' = \frac{C}{\text{FSF} \times Z} = \frac{3.538}{6280 \times 5 \times 10^3} = 0.113\ \mu\text{F} \qquad (2\text{-}10)$$

$$C_2' = 0.0250\ \mu\text{F}$$
$$C_3' = 8690\ \text{pF}$$
$$C_4' = 4345\ \text{pF}$$
$$C_5' = 0.0408\ \mu\text{F}$$

(*d*) Using an R of 10 kΩ, the feedback resistors are

$$R_6 = 10\ \text{k}\Omega$$
$$R_7 = 4100\ \Omega$$

The section zero is computed by

$$f_\infty = \omega_\infty \times f_c = 3.35 \times 1000 \text{ Hz} = 3350 \text{ Hz} \qquad (3\text{-}33)$$

The resulting filter is shown in figure 3-22b. The corresponding frequency response is given in figure 3-22c.

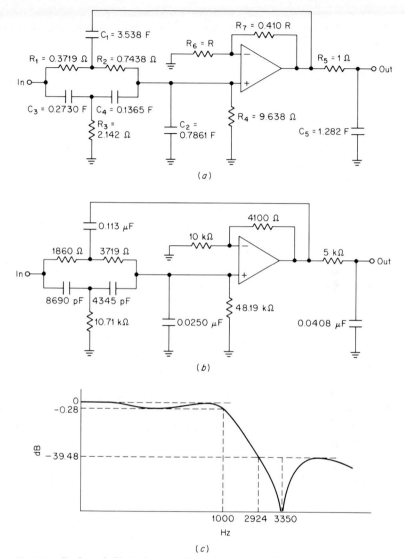

Fig. 3-22 Design of elliptic-function filter of example 3-11: (a) normalized circuit from table 12-57; (b) denormalized filter; (c) frequency response.

State-Variable Low-Pass Filters

The poles and zeros of the previously discussed active filter configurations cannot be easily adjusted because of the interaction of circuit elements. For most industrial requirements, sufficient accuracy is obtained by specifying 1% resistors and 1 or 2% capacitors. In the event greater precision is required, the state-variable approach features independent adjustment of the pole and zero coordinates. Also, the state-variable configuration has a lower sensitivity to many of the inadequacies of operational amplifiers such as finite bandwidth and gain.

All-Pole Configuration The circuit of figure 3-23 realizes a single pair of complex poles. The low-pass transfer function is given by

$$T(s) = \frac{1}{R_2 R_4 C^2} \frac{1}{s^2 + \dfrac{1}{R_1 C} s + \dfrac{1}{R_2 R_3 C^2}} \tag{3-34}$$

If we equate equation (3-34) to the second-order low-pass transfer function expressed by equation (3-8) and solve for the element values, after some algebraic manipulation we obtain the following design equations:

$$R_1 = \frac{1}{2\alpha C} \tag{3-35}$$

$$R_2 = R_3 = R_4 = \frac{1}{C\sqrt{\alpha^2 + \beta^2}} \tag{3-36}$$

where α and β are the real and imaginary components, respectively, of the pole locations and C is arbitrary. The value of R in figure 3-23 is also optional.

The element values computed by equations (3-35) and (3-36) result in a DC gain of unity. If a gain of A is desired, R_4 can instead be defined by

$$R_4 = \frac{1}{AC\sqrt{\alpha^2 + \beta^2}} \tag{3-37}$$

Sometimes it is desirable to design a filter directly at its cutoff frequency instead of calculating the normalized values and then frequency- and impedance-scaling the normalized network. Equations (3-35) and (3-36) result in the denormalized values if α and β are first denormalized by the frequency-scaling factor FSF as follows:

$$\alpha' = \alpha \times \text{FSF} \tag{3-38}$$

$$\beta' = \beta \times \text{FSF} \tag{3-39}$$

Direct design of the denormalized filter is especially advantageous when the design formulas permit the arbitrary selection of capacitors and all network capacitors are equal. A standard capacitance value can then be chosen.

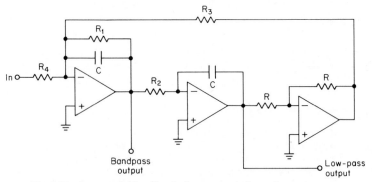

Fig. 3-23 State-variable all-pole low-pass configuration.

Figure 3-23 indicates that a bandpass output is also provided. Although a discussion of bandpass filters will be deferred until chapter 5, this output is useful for tuning of the low-pass filter. To adjust the low-pass real and imaginary pole coordinates, first compute the bandpass resonant frequency:

$$f_0 = \frac{\sqrt{(\alpha')^2 + (\beta')^2}}{2\pi} \tag{3-40}$$

Trim the value of R_3 until resonant conditions occur at the bandpass output with f_0 applied. Resonance can be determined by exactly 180° of phase shift between input and output or by peak output amplitude. The 180° phase shift method normally results in more accuracy and resolution. By connecting the vertical channel of an oscilloscope to the section input and the horizontal channel to the bandpass output, a "Lissajous pattern" is obtained. This pattern is an ellipse that will collapse to a straight line (at a 135° angle) when the phase shift is 180°.

For the final adjustment trim R_1 for a bandpass Q (i.e., f_0/3-dB bandwidth) equal to

$$Q = \frac{\pi f_0}{\alpha'} \tag{3-41}$$

Resistor R_1 can be adjusted for a measured bandpass output gain at f_0 equal to the computed ratio of R_1/R_4. Amplifier phase shift creates a "Q enhancement" effect where the Q of the section is increased. This effect also increases the gain at the bandpass output; so adjustment of R_1 for the calculated gain will usually restore the desired Q. Alternately the 3-dB bandwidth can be measured and the Q computed. Although the Q measurement approach is the more accurate method, it certainly is slower than a simple gain adjustment.

Example 3-12

REQUIRED: Active low-pass filter
3 dB $\pm$ 0.25 dB at 500 Hz
40 dB minimum at 1375 Hz

RESULT: (a) Compute A_s.

$$A_s = \frac{1375}{500} = 2.75 \tag{2-11}$$

(b) Figure 2-42 indicates that a fourth-order 0.1-dB Chebyshev low-pass filter has over 40 dB of rejection at 2.75 rad/s. Since a precise cutoff is required, we will use the state-variable approach so that the filter parameters can be adjusted if necessary.

The pole locations for a normalized fourth-order 0.1-dB Chebyshev low-pass filter are obtained from table 12-23 and are as follows:

$$\alpha = 0.2183 \qquad \beta = 0.9262$$
and $$\alpha = 0.5271 \qquad \beta = 0.3836$$

(c) Two sections of the circuit of figure 3-23 will be cascaded. The denormalized filter will be designed directly. The capacitor value C is chosen to be 0.01 μF, and R is arbitrarily selected to be 10 kΩ.

Section 1:

$$\alpha = 0.2183 \qquad \alpha' = \alpha \times \text{FSF} = 685.5 \tag{3-38}$$
$$\beta = 0.9262 \qquad \beta' = \beta \times \text{FSF} = 2908 \tag{3-39}$$

where $\text{FSF} = 2\pi f_c = 2\pi 500 = 3140$

$$R_1 = \frac{1}{2\,\alpha'\,C} = 72.94 \text{ k}\Omega \qquad (3\text{-}35)$$

$$R_2 = R_3 = R_4 = \frac{1}{C\sqrt{(\alpha')^2 + (\beta')^2}} = 33.47 \text{ k}\Omega \quad (3\text{-}36)$$

Section 2:

$$\alpha = 0.5271 \qquad \alpha' = \alpha \times \text{FSF} = 1655 \qquad (3\text{-}38)$$
$$\beta = 0.3836 \qquad \beta' = \beta \times \text{FSF} = 1205 \qquad (3\text{-}39)$$

where $\text{FSF} = 3140$

$$R_1 = \frac{1}{2\alpha'\,C} = 30.21 \text{ k}\Omega \qquad (3\text{-}35)$$

$$R_2 = R_3 = R_4 = \frac{1}{C\sqrt{(\alpha')^2 + (\beta')^2}} = 48.85 \text{ k}\Omega \quad (3\text{-}36)$$

(d) The resulting filter is shown in figure 3-24. The resistor values have been modified so that standard 1% resistors are used and adjustment capability is provided.

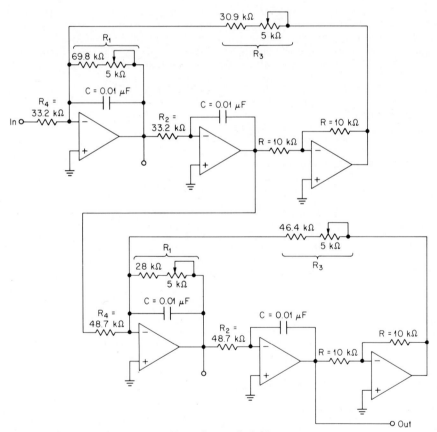

Fig. 3-24   State-variable low-pass filter of example 3-12.

The bandpass resonant frequency and Q are

Section 1:

$$f_0 = \frac{\sqrt{(\alpha')^2 + (\beta')^2}}{2\pi} = 476 \text{ Hz} \quad (3\text{-}40)$$

$$Q = \frac{\pi f_0}{\alpha'} = 2.18 \quad (3\text{-}41)$$

Section 2:

$$f_0 = 326 \text{ Hz} \quad (3\text{-}40)$$
$$Q = 0.619 \quad (3\text{-}41)$$

Elliptic-Function Configuration   When precise control of the parameters of elliptic-function filters is required, a state-variable elliptic-function approach is necessary. This is especially true in the case of very sharp filters where the location of the poles and zeros is highly critical.

The circuit of figure 3-25 has the transfer function

$$T(s) = -\frac{R_6}{R} \frac{s^2 + \dfrac{1}{R_2 R_3 C^2}\left(1 + \dfrac{R_3 R}{R_4 R_5}\right)}{s^2 + \dfrac{1}{R_1 C}s + \dfrac{1}{R_2 R_3 C^2}} \quad (3\text{-}42)$$

where $R_1 = R_4$ and $R_2 = R_3$. The numerator roots result in a pair of imaginary zeros, and the denominator roots determine a pair of complex poles. Since both the numerator and denominator are second-order, this transfer function form is frequently referred to as "biquadratic" and the circuit is called a "biquad." The zeros are restricted to frequencies beyond the pole locations, i.e., the stopband of elliptic-function low-pass filters. If R_5 in figure 3-25 were con-

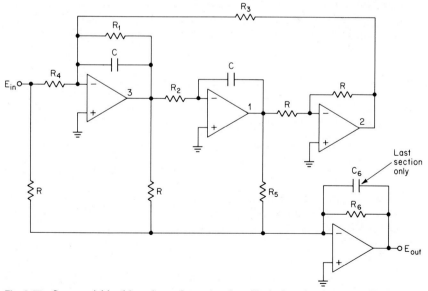

Fig. 3-25   State-variable (biquad) configuration for elliptic-function low-pass filters.

nected to node 2 instead of node 1, the zeros will occur below the poles as in high-pass elliptic-function filters.

Table 12-57 provides the poles and zeros for normalized elliptic-function low-pass filters alongside the element values. The component values of the denormalized biquad section can be directly computed from these poles and zeros.

First denormalize the tabulated pole coordinates α and β and the zero ω_∞ as follows:

$$\alpha' = \alpha \times \text{FSF} \qquad (3\text{-}43)$$

$$\beta' = \beta \times \text{FSF} \qquad (3\text{-}44)$$

$$\omega'_\infty = \omega_\infty \times \text{FSF} \qquad (3\text{-}45)$$

where FSF is the desired frequency-scaling factor.

Compute the bandpass resonant frequency in radians per second:

$$\omega'_0 = \sqrt{(\alpha')^2 + (\beta')^2} \qquad (3\text{-}46)$$

The component values are

$$R_1 = R_4 = \frac{1}{2\alpha' C} \qquad (3\text{-}47)$$

$$R_2 = R_3 = \frac{1}{\omega'_0 C} \qquad (3\text{-}48)$$

$$R_5 = \frac{2\alpha' \omega'_0 R}{(\omega'_\infty)^2 - (\omega'_0)^2} \qquad (3\text{-}49)$$

$$R_6 = \left(\frac{\omega'_0}{\omega'_\infty}\right)^2 AR \qquad (3\text{-}50)$$

where C and R are arbitrary and A is the desired low-pass gain at DC.

Since odd-order elliptic-function filters contain a real pole, the last section of a cascade of biquads should contain capacitor C_6 in parallel with R_6. To compute C_6, first denormalize the real pole α_0 as follows:

$$\alpha'_0 = \alpha_0 \times \text{FSF} \qquad (3\text{-}51)$$

then
$$C_6 = \frac{1}{\alpha'_0 R_6} \qquad (3\text{-}52)$$

The poles and zeros of the biquad configuration of figure 3-25 can be adjusted by implementing the following sequence of steps:

1. *Resonant frequency:* The bandpass resonant frequency is defined by

$$f_0 = \frac{\omega'_0}{2\pi} \qquad (3\text{-}53)$$

If R_3 is made adjustable, the section resonant frequency can be tuned to f_0 by monitoring the bandpass output at node 3. The 180° phase shift method is preferred for the determination of resonance.

2. *Q adjustment:* The bandpass Q is given by

$$Q = \frac{\pi f_0}{\alpha'} \qquad (3\text{-}54)$$

Adjustment of R_1 for unity gain at f_0 measured between the section input and the bandpass output at node 3 will usually compensate for any "Q enhancement" resulting from amplifier phase shift.

3.   *Notch frequency:* The notch frequency was given by

$$f_\infty = \omega_\infty \times f_c \qquad (3\text{-}33)$$

Adjustment of f_∞ usually is not required if the circuit is first tuned to f_0, since f_∞ will then fall in. However, if independent tuning of the notch frequency is desired, R_5 should be made variable. The notch frequency is measured by determining the input frequency where E_{out} is nulled.

Example 3-13

REQUIRED: Active low-pass filter
0.5 dB maximum ripple below 1000 Hz
18 dB minimum rejection above 1600 Hz

RESULT: (*a*) Compute the steepness factor.

$$A_s = \frac{f_s}{f_c} = \frac{1600}{1000} = 1.6 \qquad (2\text{-}11)$$

(*b*) Select a normalized elliptic-function low-pass filter from table 12-57 that makes the transition from less than 0.5 dB to over 18 dB of rejection within a frequency ratio of 1.6. A filter corresponding to C03 25 $\theta = 40°$ has the following parameters:

$$n = 3$$
$$R_{dB} = 0.28 \text{ dB}$$
$$\Omega_s = 1.556$$
$$A_{min} = 20.58 \text{ dB}$$

The normalized complex poles and zeros are

$$\alpha = 0.2643 \qquad \beta = 1.100$$
$$\omega_\infty = 1.742$$

and the real pole is located at

$$\alpha_0 = 0.9142$$

(*c*) A state-variable approach will be used employing the biquad configuration of figure 3-25. The frequency-scaling factor is

$$\text{FSF} = 2\pi f_c = 2\pi 1000 = 6280$$

The poles and zeros are denormalized as follows:

Complex poles and zeros:

$$\alpha = 0.2643 \qquad \alpha' = \alpha \times \text{FSF} = 1660 \qquad (3\text{-}43)$$
$$\beta = 1.100 \qquad \beta' = \beta \times \text{FSF} = 6908 \qquad (3\text{-}44)$$
$$\omega_\infty = 1.742 \qquad \omega'_\infty = \omega_\infty \times \text{FSF} = 10{,}940 \qquad (3\text{-}45)$$

Real pole:

$$\alpha_0 = 0.9142 \qquad \alpha'_0 = \alpha_0 \times \text{FSF} = 5741 \qquad (3\text{-}51)$$

(*d*) A single section is required. Let $R = 100$ kΩ and $C = 0.1$ μF, and let the gain equal unity ($A = 1$). The values are computed as follows:

$$\omega'_0 = \sqrt{(\alpha')^2 + (\beta')^2} = 7105 \qquad (3\text{-}46)$$

$$R_1 = R_4 = \frac{1}{2\alpha' C} = 3012 \ \Omega \qquad (3\text{-}47)$$

$$R_2 = R_3 = \frac{1}{\omega_0' C} = 1408 \ \Omega \qquad (3\text{-}48)$$

$$R_5 = \frac{2\alpha' \omega_0' R}{(\omega_\infty')^2 - (\omega_0')^2} = 34.09 \ \text{k}\Omega \qquad (3\text{-}49)$$

$$R_6 = \left(\frac{\omega_0'}{\omega_\infty'}\right)^2 AR = 42.17 \ \text{k}\Omega \qquad (3\text{-}50)$$

Since a real pole is also required, C_6 is introduced in parallel with R_6 and is calculated by

$$C_6 = \frac{1}{\alpha_0' R_6} = 4130 \ \text{pF} \qquad (3\text{-}52)$$

(e)   The resulting filter is shown in figure 3-26. The resistor values are modified so that standard 1% values are used and the circuit is adjustable. The section f_0 and Q are computed by

$$f_0 = \frac{\omega_0'}{2\pi} = 1131 \ \text{Hz} \qquad (3\text{-}53)$$

$$Q = \frac{\pi f_0}{\alpha'} = 2.14 \qquad (3\text{-}54)$$

The frequency of infinite attenuation is given by

$$f_\infty = \omega_\infty \times f_c = 1742 \ \text{Hz} \qquad (3\text{-}33)$$

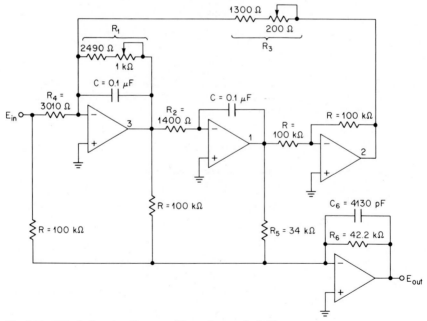

Fig. 3-26   Elliptic-function low-pass filter of example 3-13.

Generalized Impedance Converters

The circuit of figure 3-27 is known as a generalized impedance converter (GIC). The driving-point impedance can be expressed as

$$Z_{11} = \frac{Z_1 Z_3 Z_5}{Z_2 Z_4} \tag{3-55}$$

By substituting RC combinations of up to two capacitors for Z_1 through Z_5, a variety of impedances can be simulated. If, for instance, Z_4 consists of a capacitor having an impedance $1/sC$, where $s = j\omega$ and all other elements are resistors, the driving-point impedance is given by

$$Z_{11} = \frac{sCR_1 R_3 R_5}{R_2} \tag{3-56}$$

The impedance is proportional to frequency and is therefore identical to an inductor having a value of

$$L = \frac{CR_1 R_3 R_5}{R_2} \tag{3-57}$$

as shown in figure 3-28.

If two capacitors are introduced for Z_1 and Z_3 and Z_2, Z_4, and Z_5 are resistors, the resulting driving-point impedance expression can be expressed in the form of

$$Z_{11} = \frac{R_5}{s^2 C^2 R_2 R_4} \tag{3-58}$$

An impedance proportional to $1/s^2$ is called a D element whose driving-point impedance is given by

$$Z_{11} = \frac{1}{s^2 D} \tag{3-59}$$

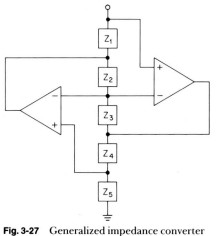

Fig. 3-27   Generalized impedance converter (GIC).

Equation (3-58) therefore defines a D element having the value

$$D = \frac{C^2 R_2 R_4}{R_5} \tag{3-60}$$

If we let $C = 1$ F, $R_2 = R_5 = 1 \ \Omega$, and $R_4 = R$, equation (3-60) simplifies to $D = R$.

In order to gain some insight into the nature of this element, let us substitute $s = j\omega$ into equation (3-58). The resulting expression is

$$Z_{11} = -\frac{R_5}{\omega^2 C^2 R_2 R_4} \tag{3-61}$$

Equation (3-61) corresponds to a frequency-dependent negative resistor (FDNR).

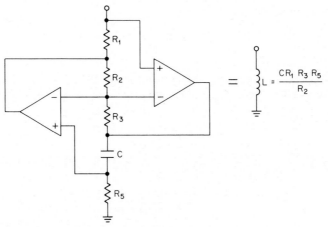

Fig. 3-28   GIC inductor simulation.

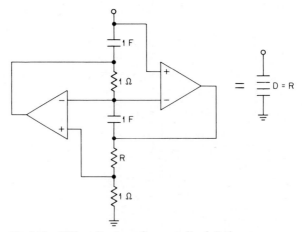

Fig. 3-29   GIC realization of a normalized D element.

A GIC in the form of a normalized D element and its schematic designation are shown in figure 3-29. Bruton (see references) has shown how the FDNR or D element can be used to generate active filters directly from the LC normalized low-pass prototype values. This technique will now be described.

The 1/s Transformation If all the impedances of an LC filter network are multiplied by $1/s$, the transfer function remains unchanged. This operation is equivalent to impedance scaling a filter by the factor $1/s$ and should not be confused with the high-pass transformation which involves the substitution of $1/s$ for s. Section 2.1 under Frequency and Impedance Scaling demonstrated that the impedance scaling of a network by any factor Z does not change the frequency response, since the Z's cancel in the transfer function; so the validity of this transformation should be apparent.

When the elements of a network are impedance-scaled by $1/s$, they undergo a change in form. Inductors are transformed into resistors, resistors into capacitors, and capacitors into D elements as summarized in table 3-1. Clearly, this design technique is extremely powerful. It enables us to design active filters directly from the passive LC circuits, which are extensively tabulated. Knowledge of the pole and zero locations is unnecessary.

The design method proceeds by first selecting a normalized low-pass LC filter. All capacitors must be restricted to the shunt arms only, since they will be transformed into D elements which are connected to ground. The dual LC filter defined by the lower schematic in the tables of chapter 12 is usually chosen to be transformed. The circuit elements are modified by the $1/s$ transformation, and the D elements are realized using the normalized GIC circuit of figure 3-29. The transformed filter is then frequency- and impedance-scaled in the conventional manner. The following example demonstrates the design of an all-pole active low-pass filter using the $1/s$ impedance transformation and the GIC.

TABLE 3-1   The 1/s Impedance Transformation

Element	Impedance	Transformed Element	Transformed Impedance
L	sL	L	L
C	$\dfrac{1}{sC}$	C	$\dfrac{1}{s^2 C}$
R	R	$\dfrac{1}{R}$	$\dfrac{R}{s}$

Example 3-14

REQUIRED: Active low-pass filter
3 dB to 400 Hz
20 dB minimum at 1200 Hz
Minimal ringing and overshoot

RESULT: (a) Compute the steepness factor.

$$A_s = \frac{f_s}{f_c} = \frac{1200}{400} = 3 \tag{2-11}$$

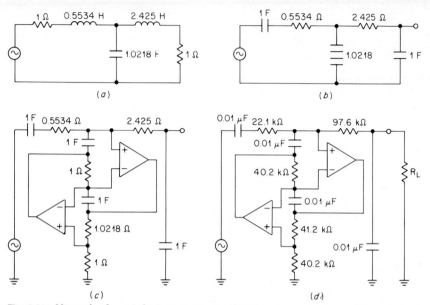

Fig. 3-30 Network of example 3-14: *(a)* normalized low-pass prototype; *(b)* normalized circuit after $1/s$ transformation; *(c)* realization of D element; *(d)* final circuit.

(b) Since low transient distortion is desired, a linear phase filter with a phase error of 0.5° will be selected. The curves of figure 2-62 indicate that a filter complexity of $n = 3$ provides over 20 dB of attenuation at 3 rad/s.

    The $1/s$ transformation and a GIC realization will be used.

(c) The normalized LC low-pass filter from table 12-47 corresponding to $n = 3$ is shown in figure 3-30a. The dual circuit has been selected so that only a single D element will be required.

(d) The normalized filter is transformed in accordance with table 3-1, resulting in the circuit of figure 3-30b. The D element is realized using the normalized GIC configuration of figure 3-29 as shown in figure 3-30c.

(e) Since all normalized capacitors are 1 F, it would be desirable if they were all denormalized to a standard value such as 0.01 μF. Using an FSF of $2\pi f_c$ or 2513 and a C' of 0.01 μF, the required impedance-scaling factor can be found by solving equation (2-10) for Z as follows:

$$Z = \frac{C}{FSF \times C'} = \frac{1}{2513 \times 0.01 \times 10^{-6}} = 39{,}800 \quad (2\text{-}10)$$

Using an FSF of $2\pi f_c$ or 2513 and an impedance-scaling factor Z of 39.8×10^3, the normalized filter is scaled by dividing all capacitors by $Z \times FSF$ and multiplying all resistors by Z. The final circuit is given in figure 3-30d. The resistor values were modified for standard 1% values. The filter loss is 6 dB, corresponding to the loss due to the resistive source and load terminations of the LC filter.

    The D elements are usually realized with dual operational amplifiers which are available as a matched pair in a single package. In order to provide a bias current for the noninverting input of the upper amplifier, a resistive termination to ground must be provided. This resistor will cause a low-frequency roll-off; so true DC-coupled operation will not be possible. However, if the termination is made much larger than the nominal resistor values of the circuit, the low-frequency roll-off can be made to occur well below the frequency range of interest. If a low output impedance is required, the filter can be followed by a voltage follower or an amplifier if gain is also desired. The filter input should be driven by a source impedance much less than the input resistor of the filter. A voltage follower or amplifier could be used for input isolation.

Elliptic-Function Low-Pass Filters Using the GIC The $1/s$ transformation and GIC realization are particularly suited for the realization of active high-order elliptic-function low-pass filters. These circuits exhibit low sensitivity to component tolerances and amplifier characteristics. They can be made tunable, are less complex than the state-variable configurations, and can be designed directly from the LC filter tables. The following example illustrates the design of an elliptic-function low-pass filter using the GIC as a D element.

Example 3-15

REQUIRED: Active low-pass filter
0.5 dB maximum at 260 Hz
60 dB minimum at 270 Hz

RESULT: (a) Compute the steepness factor.

$$A_s = \frac{f_s}{f_c} = \frac{270}{260} = 1.0385 \qquad (2\text{-}11)$$

(b) An extremely sharp filter is required; so an elliptic-function type will be selected. A filter corresponding to C11 20 $\theta = 75°$ is chosen from table 12-56 and has the following parameters:

$$n = 11$$
$$R_{dB} = 0.18 \text{ dB}$$
$$\Omega_s = 1.0353$$
$$A_{min} = 60.8 \text{ dB}$$

The normalized circuit corresponding to the dual filter is shown in figure 3-31a.

(c) The $1/s$ impedance transformation modifies the elements in accordance with table 3-1, resulting in the circuit of figure 3-31b. The D elements are realized using the GIC of figure 3-29, as shown in figure 3-31c.

(d) The normalized circuit can now be frequency- and impedance-scaled. Since all normalized capacitors are equal, it would be desirable if they could all be scaled to a standard value such as 0.1 μF. The required impedance-scaling factor can be determined from equation (2-10) by using an FSF of $2\pi f_c$ or 1634 corresponding to a cutoff of 260 Hz. Therefore,

$$Z = \frac{C}{\text{FSF} \times C'} = \frac{1}{1634 \times 0.1 \times 10^{-6}} = 6120 \qquad (2\text{-}10)$$

Frequency and impedance scaling by dividing all capacitors by $Z \times \text{FSF}$ and multiplying all resistors by Z results in the final filter circuit of figure 3-31d having the frequency response of figure 3-31e. The resistor values have been modified so that 1% values are used and the transmission zeros can be adjusted. The frequency of each zero was computed by multiplying Ω_2, Ω_4, Ω_6, Ω_8, and Ω_{10} by f_c. A termination is provided so that bias current can be supplied to the amplifiers.

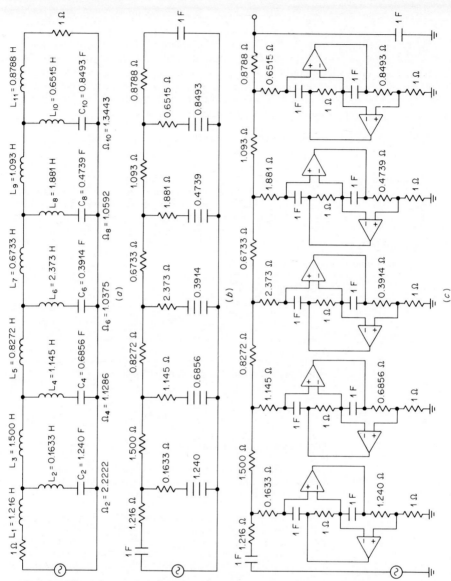

Fig. 3-31 Filter of example 3-15: (a) normalized low-pass filter; (b) circuit after $1/s$ transformation; (c) normalized configuration using GICs for D elements; (d) denormalized filter; (e) frequency response.

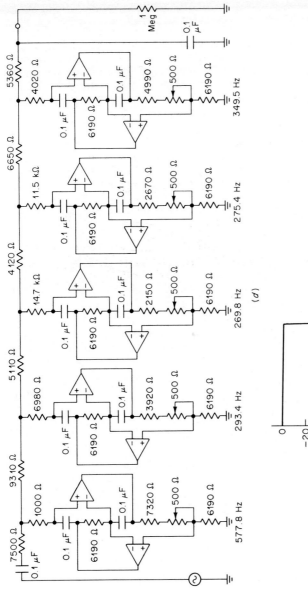

(d)

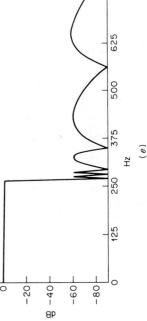

(e)

The transmission zeros generated by the circuit of figure 3-31d occur because at specific frequencies the value of each FDNR is equal to the positive resistor in series, therefore creating cancellations or nulls in the shunt branches. By adjusting each D element, these nulls can be tuned to the required frequencies. These adjustments are usually sufficient to obtain satisfactory results. State-variable filters permit the adjustment of the poles and zeros directly for greater accuracy. However, the realization is more complex; for instance, the filter of example 3-15 would require twice as many amplifiers, resistors, and potentiometers if the state-variable approach were used.

VCVS Uniform Capacitor Structures

The unity-gain n-2 all-pole configuration of figure 3-13a and the VCVS elliptic-function configuration of figure 3-20a both require unequal capacitor values and noninteger capacitor ratios. This inconvenience usually results in either the use of nonstandard capacitor values or the paralleling of two or more standard values.

Alternate configurations are given in this section. These structures feature equal capacitors for the all-pole case and some additional degrees of freedom for capacitor selection in the elliptic-function circuit. However, the circuit sensitivities are somewhat higher than the previously discussed configurations. Nevertheless the more convenient capacitor values may justify their use in many instances where higher sensitivities are tolerable.

All-Pole Configuration   The $n = 2$ low-pass circuit of figure 3-32 features equal capacitors and a gain of 2. The element values are computed as follows:
   Select C.

Then
$$R_1 = \frac{1}{2\alpha' C} \tag{3-62}$$

and
$$R_2 = \frac{2\alpha'}{C(\alpha'^2 + \beta'^2)} \tag{3-63}$$

where α' and β' are the denormalized real and imaginary pole coordinates as given by equations (3-38) and (3-39), respectively. R may be conveniently chosen.

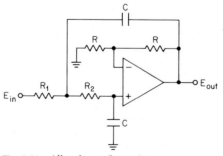

Fig. 3-32   All-pole configuration.

Example 3-16

REQUIRED: Design a fourth-order 0.1-dB Chebyshev active low-pass filter for a 3-dB cutoff of 100 Hz using 0.01-μF capacitors throughout.

RESULT: (a) The pole locations for a normalized 0.1-dB Chebyshev low-pass filter are obtained from table 12-23 and are as follows:

$$\alpha = 0.2177 \quad \beta = 0.9254$$
$$\text{and} \quad \alpha = 0.5257 \quad \beta = 0.3833$$

(b) Two sections of the filter of figure 3-32 will be cascaded. The value of C is 0.01 μF, and R is chosen at 10 kΩ.

Section 1:

$$\alpha = 0.2177 \quad \alpha' = \alpha \times \text{FSF} = 136.8 \quad \text{(3-38)}$$
$$\beta = 0.9254 \quad \beta' = \beta \times \text{FSF} = 581.4 \quad \text{(3-39)}$$

where FSF $= 2\pi f_c = 628.3$

$$R_1 = \frac{1}{2\alpha' C} = 365.5 \text{ k}\Omega \quad \text{(3-62)}$$

$$R_2 = \frac{2\alpha'}{C(\alpha'^2 + \beta'^2)} = 76.7 \text{ k}\Omega \quad \text{(3-63)}$$

Section 2:

$$\alpha = 0.5257 \quad \alpha' = 330.3 \quad \text{(3-38)}$$
$$\beta = 0.3833 \quad \beta' = 240.8 \quad \text{(3-39)}$$
$$R_1 = 151.4 \text{ k}\Omega \quad \text{(3-62)}$$
$$R_2 = 395.4 \text{ k}\Omega \quad \text{(3-63)}$$

The final filter is shown in figure 3-33. The gain is 2^2, or 4.

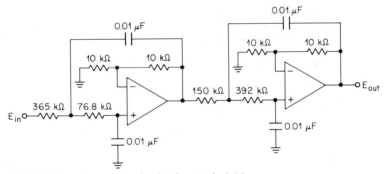

Fig. 3-33 Equal capacitor circuit of example 3-16.

Elliptic-Function Configuration The elliptic-function filter circuit of section 3.2 is repeated in figure 3-34. However, by use of an alternate design procedure, an additional degree of freedom in capacitor selection is possible. The method proceeds as follows:

First compute

$$a = \frac{2a'}{\sqrt{\alpha'^2 + \beta'^2}} \tag{3-64}$$

$$b = \frac{\omega'^2_\infty}{\alpha'^2 + \beta'^2} \tag{3-65}$$

$$c = \sqrt{\alpha'^2 + \beta'^2} \tag{3-66}$$

where α', β', and ω'_∞ are the denormalized pole-zero coordinates obtained from the normalized values of table 12-57 which are then multiplied by the FSF.

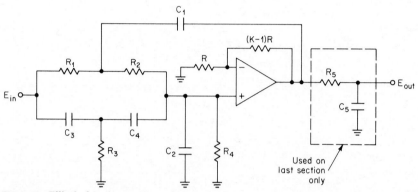

Fig. 3-34 Elliptic-function configuration.

The element values are computed as follows:
Select C.

Then

$$C_1 = C \tag{3-67}$$

$$C_3 = C_4 = \frac{C_1}{2} \tag{3-68}$$

let

$$C_2 \geqq \frac{C_1(b-1)}{4} \tag{3-69}$$

$$R_3 = \frac{1}{cC_1\sqrt{b}} \tag{3-70}$$

$$R_1 = R_2 = 2R_3 \tag{3-71}$$

$$R_4 = \frac{4\sqrt{b}}{cC_1(1-b) + 4cC_2} \tag{3-72}$$

$$K = 2 + \frac{2C_2}{C_1} - \frac{a}{2\sqrt{b}} + \frac{2}{C_1\sqrt{b}}\left(\frac{1}{cR_4} - aC_2\right) \tag{3-73}$$

$$\text{Section gain} = \frac{bKC_1}{4C_2 + C_1} \tag{3-74}$$

Capacitor C_5 is determined from the denormalized real pole by

$$C_5 = \frac{1}{R_5 a_0'} \qquad (3\text{-}75)$$

where both R and R_5 can be arbitrarily chosen and a_0' is $a_0 \times$ FSF.

Example 3-17

REQUIRED: Design an active elliptic-function low-pass filter corresponding to CO3 25 $\theta = 20°$ from table 12-57 for a cutoff of 100 Hz using the VCVS structure of figure 3-34.

RESULT: (a) The following parameters are obtained from table 12-57:

$$n = 3$$
$$R_{dB} = 0.28 \text{ dB}$$
$$\Omega_s = 2.924$$
$$A_{min} = 39.48 \text{ dB}$$

The normalized poles and zeros are given as

$$\alpha = 0.3449 \qquad \beta = 1.0860 \qquad \omega_\infty = 3.350$$
$$\alpha_0 = 0.7801$$

(b) The poles and zeros are denormalized as follows:

$$\alpha' = \alpha \times \text{FSF} = 216.7 \qquad (3\text{-}43)$$
$$\beta' = \beta \times \text{FSF} = 682.4 \qquad (3\text{-}44)$$
$$\omega_\infty' = \omega_\infty \times \text{FSF} = 2105 \qquad (3\text{-}45)$$
$$\alpha_0' = \alpha_0 \times \text{FSF} = 490.2 \qquad (3\text{-}51)$$

where FSF $= 2\pi f_c = 628.3$

(c) The element values are computed as follows:

$$a = 0.6053 \qquad (3\text{-}64)$$
$$b = 8.6437 \qquad (3\text{-}65)$$
$$c = 716 \qquad (3\text{-}66)$$

Select $\qquad C_1 = C = 0.1 \ \mu\text{F} \qquad (3\text{-}67)$

$$C_3 = C_4 = 0.05 \ \mu\text{F} \qquad (3\text{-}68)$$
$$C_2 \geq 0.191 \ \mu\text{F} \qquad (3\text{-}69)$$

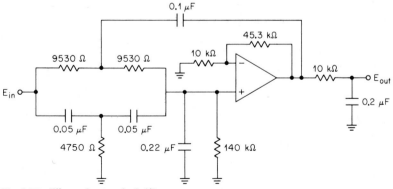

Fig. 3-35 Filter of example 3-17.

Let
$$C_2 = 0.22 \ \mu F$$
$$R_3 = 4751 \ \Omega \tag{3-70}$$
$$R_1 = R_2 = 9502 \ \Omega \tag{3-71}$$
$$R_4 = 142 \ k\Omega \tag{3-72}$$
$$K = 5.458 \tag{3-73}$$

Let
$$R = R_5 = 10 \ k\Omega$$

then
$$C_5 = 0.204 \ \mu F \tag{3-75}$$

The resulting circuit is shown in figure 3-35.

REFERENCES

Bruton, L. T., "Active Filter Design Using Generalized Impedance Converters," EDN, February 1973.

Bruton, L. T., "Network Transfer Functions Using the Concept of Frequency-Dependent Negative Resistance," IEEE Transactions on Circuit Theory, Vol. CT-16, pp. 406–408, August 1969.

Christian, E., and Eisenmann, E., "Filter Design Tables and Graphs," John Wiley and Sons, New York, 1966.

Geffe, P., "Simplified Modern Filter Design," John F. Rider, New York, 1963.

Huelsman, L. P., "Theory and Design of Active RC Circuits," McGraw-Hill Book Company, 1968.

Saal, R., and Ulbrich, E., "On the Design of Filters by Synthesis," IRE Transactions on Circuit Theory, December 1958.

Saal, R. "Der Entwurf von Filtern mit Hilfe des Kataloges Normierter Tiefpasse," Telefunken GMBH, Backnang, West Germany, 1963.

Shepard, B. R., "Active Filters Part 12," Electronics, pp. 82–91, August 18, 1969.

Thomas, L. C., "The Biquad: Part I—Some Practical Design Considerations," IEEE Transactions on Circuit Theory, Vol. CT-18, pp. 350–357, May 1971.

Tow, J., "A Step-by-Step Active Filter Design," IEEE Spectrum, Vol. 6, pp. 64–68, December 1969.

Williams, A. B., "Design Active Elliptic Filters Easily from Tables," Electronic Design, Vol. 19, No. 21, pp. 76–79, October 14, 1971.

Williams, A. B., "Active Filter Design," Artech House, Dedham, Mass., 1975.

Zverev, A. I., "Handbook of Filter Synthesis," John Wiley and Sons, New York, 1967.

4

High-Pass
Filter Design

4.1 LC HIGH-PASS FILTERS

The Low-Pass to High-Pass Transformation

If $1/s$ is substituted for s in a normalized low-pass transfer function, a high-pass response is obtained. The low-pass attenuation values will now occur at high-pass frequencies which are the reciprocal of the corresponding low-pass frequencies. This was demonstrated in section 2.1.

A normalized LC low-pass filter can be transformed into the corresponding high-pass filter by simply replacing each coil with a capacitor and vice versa, using reciprocal element values. This can be expressed as

$$C_{hp} = \frac{1}{L_{Lp}} \qquad (4\text{-}1)$$

and
$$L_{hp} = \frac{1}{C_{Lp}} \qquad (4\text{-}2)$$

The source and load resistive terminations are unaffected.

The transmission zeros of a normalized elliptic-function low-pass filter are also reciprocated when the high-pass transformation occurs. Therefore,

$$\omega_\infty(\text{hp}) = \frac{1}{\omega_\infty(\text{Lp})} \qquad (4\text{-}3)$$

To minimize the number of inductors in the high-pass filter, the dual low-pass circuit defined by the lower schematic in the tables of chapter 12 is usually chosen to be transformed except for even-order all-pole filters, where either circuit may be used.

After the low-pass to high-pass transformation, the normalized high-pass filter is frequency- and impedance-scaled to the required cutoff frequency. The following two examples demonstrate the design of high-pass filters.

Example 4-1

REQUIRED: *LC* high-pass filter
3 dB at 1 MHz
28 dB minimum at 500 kHz
$R_s = R_L = 300\ \Omega$

RESULT: (a) To normalize the requirement, compute the high-pass steepness factor A_s.

$$A_s = \frac{f_c}{f_s} = \frac{1\ \text{MHz}}{500\ \text{kHz}} = 2 \qquad (2\text{-}13)$$

(b) Select a normalized low-pass filter that offers over 28 dB of attenuation at 2 rad/s.

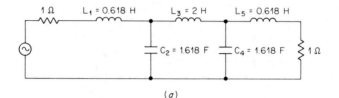

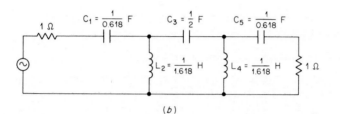

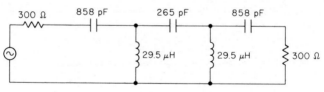

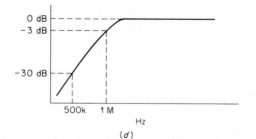

Fig. 4-1 High-pass filter of example 4-1: (a) normalized low-pass filter; (b) high-pass transformation; (c) frequency- and impedance-scaled filter; (d) frequency response.

Inspection of the curves of chapter 2 indicates that a normalized $n = 5$ Butterworth low-pass filter provides the required attenuation. Table 12-2 contains element values for the corresponding network. The normalized low-pass filter for $n = 5$ and equal terminations is shown in figure 4-1a. The dual circuit as defined by the lower schematic of table 12-2 was chosen.

(c) To transform the normalized low-pass circuit to a high-pass configuration, replace each coil with a capacitor and vice versa using reciprocal element values as shown in figure 4-1b.

(d) Denormalize the high-pass filter using a Z of 300 and a frequency-scaling factor (FSF) of $2\pi f_c$ or 6.28×10^6.

$$C_1' = \frac{C}{FSF \times Z} = \frac{\dfrac{1}{0.618}}{6.28 \times 10^6 \times 300} = 858 \text{ pF} \quad (2\text{-}10)$$

$$C_3' = 265 \text{ pF}$$
$$C_5' = 858 \text{ pF}$$

$$L_2' = \frac{L \times Z}{FSF} = \frac{\dfrac{1}{1.618} \times 300}{6.28 \times 10^6} = 29.5 \ \mu\text{H} \quad (2\text{-}9)$$

$$L_4' = 29.5 \ \mu\text{H}$$

The final filter is given in figure 4-1c having the frequency response shown in figure 4-1d.

Example 4-2

REQUIRED: *LC* high-pass filter
2 dB maximum at 3220 Hz
52 dB minimum at 3020 Hz
$R_s = R_L = 300 \ \Omega$

RESULT: (a) Compute the high-pass steepness factor A_s.

$$A_s = \frac{f_c}{f_s} = \frac{3220 \text{ Hz}}{3020 \text{ Hz}} = 1.0662 \quad (2\text{-}13)$$

(b) Since the filter requirement is very steep, an elliptic-function filter will be selected. The curves of figure 2-86 indicate that for a ρ of 20% (0.18-dB passband ripple), a filter of $n = 9$ is required. Using table 12-56, a ninth-order low-pass filter is selected that makes the transition from the passband to over 52 dB of attenuation in the stopband within a frequency ratio of 1.0662. The design corresponding to CO9 20 $\theta = 70°$ has the following parameters:

$$n = 9$$
$$R_{dB} = 0.18 \text{ dB}$$
$$\Omega_s = 1.0642$$
$$A_{min} = 53.6 \text{ dB}$$

The normalized low-pass filter corresponding to the lower schematic of the dual filter is shown in figure 4-2a.

(c) To transform the normalized low-pass circuit into a high-pass configuration, convert inductors into capacitors and vice versa using reciprocal values. The transformed high-pass filter is illustrated in figure 4-2b. The transmission zeros occurring at Ω_2, Ω_4, Ω_6, and Ω_8 are also transformed by conversion to reciprocal values.

(d) Denormalize the high-pass filter using a Z of 300 and a frequency-

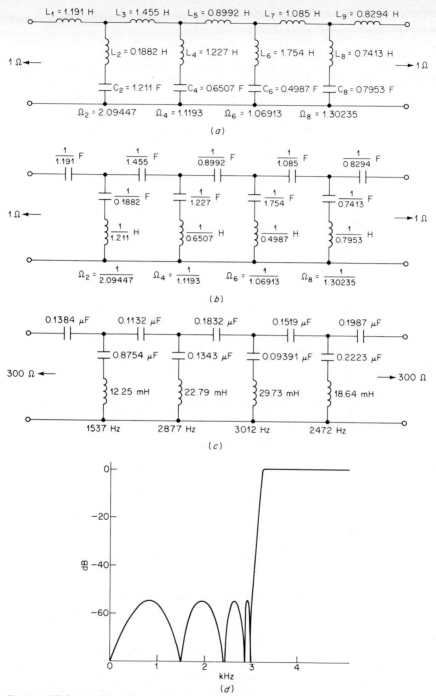

Fig. 4-2 High-pass filter of example 4-2: (a) normalized CO9 20 $\theta = 70°$ low-pass filter; (b) transformed high-pass filter; (c) frequency- and impedance-scaled high-pass filter; (d) frequency response.

scaling factor of $2\pi f_c$ or 20,232. The denormalized elements are computed by

$$L' = \frac{L \times Z}{\text{FSF}} \tag{2-9}$$

and

$$C' = \frac{C}{\text{FSF} \times Z} \tag{2-10}$$

The resulting denormalized high-pass filter is illustrated in figure 4-2c. The stopband peaks were obtained by multiplying Ω_2, Ω_4, Ω_6, and Ω_8 by the design cutoff frequency of $f_c = 3220$ Hz. The frequency response is given in figure 4-2d.

The T to Pi Capacitance Conversion

When the elliptic-function high-pass filters are designed for audio frequencies and at low impedance levels, the capacitor values tend to be large. The T to pi capacitance conversion will usually restore practical capacitor values.

The two circuits of figure 4-3 have identical terminal behavior and are therefore equivalent if

$$C_a = \frac{C_1 C_2}{\Sigma C} \tag{4-4}$$

$$C_b = \frac{C_1 C_3}{\Sigma C} \tag{4-5}$$

$$C_c = \frac{C_2 C_3}{\Sigma C} \tag{4-6}$$

where $\Sigma C = C_1 + C_2 + C_3$. The following example demonstrates the effectiveness of this transformation in reducing large capacitances.

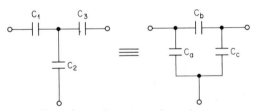

Fig. 4-3 T to pi capacitance transformation.

Example 4-3

REQUIRED: The high-pass filter of figure 4-2c contains a 0.8754-μF capacitor in the first shunt branch. Use the T to pi transformation to provide some relief.
RESULT: The circuit can be redrawn where the second series capacitor is split into two series capacitors each having twice the original value. The modified circuit is shown in figure 4-4a. A T has been formed which includes the undesirable 0.8754-μF capacitor. The T to pi transformation results in

$$C_a = \frac{C_1 C_2}{\Sigma C} = 0.09769 \ \mu\text{F} \tag{4-4}$$

$$C_b = \frac{C_1 C_3}{\Sigma C} = 0.02527 \ \mu\text{F} \tag{4-5}$$

and
$$C_c = \frac{C_2 C_3}{\Sigma C} = 0.1598 \ \mu F \qquad (4\text{-}6)$$

where $C_1 = 0.1384 \ \mu F$, $C_2 = 0.8754 \ \mu F$, and $C_3 = 0.2264 \ \mu F$. The transformed circuit is given in figure 4-4b, where the maximum capacitor value has undergone a 5:1 reduction.

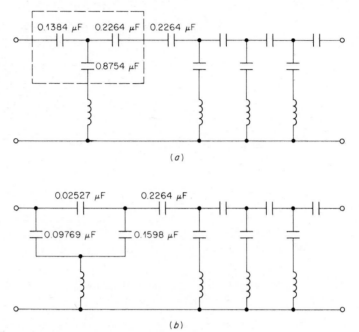

Fig. 4-4   T to pi transformation of example 4-3: (a) high-pass filter of example 4-2; (b) modified configuration.

4.2 ACTIVE HIGH-PASS FILTERS

The Low-Pass to High-Pass Transformation

Active high-pass filters can be derived directly from the normalized low-pass configurations by a suitable transformation in a similar manner to LC high-pass filters. To make the conversion, replace each resistor by a capacitor having the reciprocal value and vice versa as follows:

$$C_{hp} = \frac{1}{R_{Lp}} \qquad (4\text{-}7)$$

$$R_{hp} = \frac{1}{C_{Lp}} \qquad (4\text{-}8)$$

It is important to recognize that only the resistors that are part of the low-pass RC networks are transformed into capacitors by equation (4-7). Feedback resistors that strictly determine operational amplifier gain, such as R_6 and R_7 in figure 3-20a, are omitted from the transformation.

After the normalized low-pass configuration is transformed into a high-pass filter, the circuit is frequency- and impedance-scaled in the same manner as in the design of low-pass filters. The capacitors are divided by $Z \times$ FSF and the resistors are multiplied by Z. A different Z can be used for each section, but the FSF must be uniform throughout the filter.

All-Pole High-Pass Filters   Active two-pole and three-pole low-pass filter sections were shown in figure 3-13 and correspond to the normalized active low-pass values tabulated in chapter 12. These circuits can be directly transformed into high-pass filters by replacing the resistors with capacitors and vice versa using reciprocal element values and then frequency- and impedance-scaling the filter network. The filter gain is unity at frequencies well into the passband corresponding to unity gain at DC for low-pass filters. The source impedance of the driving source should be much less than the reactance of the capacitors of the first filter section at the highest passband frequency of interest. The following example demonstrates the design of an all-pole high-pass filter.

Example 4-4

REQUIRED: Active high-pass filter
3 dB at 100 Hz
75 dB minimum at 25 Hz

RESULT: (a)   Compute the high-pass steepness factor.

$$A_s = \frac{f_c}{f_s} = \frac{100}{25} = 4 \qquad (2\text{-}13)$$

(b)   A normalized low-pass filter must first be selected that makes the transition from 3 to 75 dB within a frequency ratio of $4:1$. The curves of figure 2-44 indicate that a fifth-order 0.5-dB Chebyshev filter is satisfactory. The corresponding active filter consists of a three-pole section and a two-pole section whose values are obtained from table 12-39 and is shown in figure 4-5a.

(c)   To transform the normalized low-pass filter into a high-pass filter, replace each resistor with a capacitor and vice versa using reciprocal element values. The normalized high-pass filter is given in figure 4-5b.

(d)   Since all normalized capacitors are equal, the impedance-scaling factor Z will be computed so that all capacitors become 0.015 μF after denormalization. Since the cutoff frequency is 100 Hz, the FSF is $2\pi f_c$ or 628, so that

$$Z = \frac{C}{\text{FSF} \times C'} = \frac{1}{628 \times 0.015 \times 10^{-6}} = 106.1 \times 10^3 \qquad (2\text{-}10)$$

If we frequency- and impedance-scale the normalized high-pass filter by dividing all capacitors by $Z \times$ FSF and multiplying all resistors by Z, the circuit of figure 4-5c is obtained. The resistors were rounded off to standard 1% values.

Elliptic-Function High-Pass Filters   High-pass elliptic-function filters can be designed directly from the normalized elliptic-function low-pass element values contained in table 12-57. By replacing each resistor with a capacitor and each capacitor with a resistor using reciprocal values, the normalized low-pass filter is transformed into an elliptic-function high-pass filter.

Figure 4-6a shows the normalized high-pass configuration for K greater than 1. Where K is less than 1, the circuit of figure 4-6b is used. These circuits are transformed from the low-pass filter sections of figures 3-20 and 3-21 using

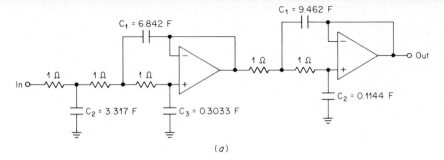

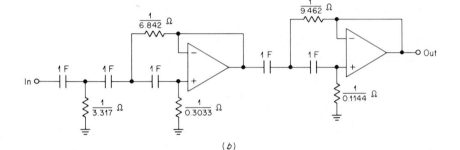

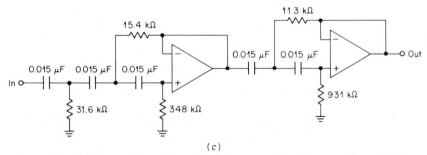

Fig. 4-5   All-pole high-pass filter of example 4-4: *(a)* normalized low-pass filter; *(b)* high-pass transformation; *(c)* frequency- and impedance-scaled high-pass filter.

the corresponding element values given in table 12-57. The low-pass transmission zeros of the normalized low-pass filter are also transformed to reciprocal values by the low-pass to high-pass transformation.

The normalized high-pass filter is frequency- and impedance-scaled to the required cutoff in the conventional manner. The filter gain is calculated by computing the product of the tabulated gains of all the sections. The following example illustrates the design of an elliptic-function high-pass filter by transformation from the low-pass circuit:

Example 4-5

REQUIRED: Active high-pass filter
0.5 dB maximum at 3000 Hz
35 dB minimum rejection at 1000 Hz

RESULT: *(a)* Compute the high-pass steepness factor.

$$A_s = \frac{f_c}{f_s} = \frac{3000}{1000} = 3 \tag{2-13}$$

(b) Select a normalized low-pass filter which provides over 35 dB of rejection within a frequency ratio of 3 and has a passband ripple of less than 0.5 dB. An elliptic-function filter type will be used. The curves of figure 2-86 indicate that for a ρ of 25% ($R_{dB} = 0.28$ dB) a third-order filter provides the required attenuation. The following parameters are found in table 12-57 for a filter corresponding to CO3 25 $\theta = 20°$:

$$n = 3$$
$$R_{dB} = 0.28 \text{ dB}$$
$$\Omega_s = 2.924$$
$$A_{min} = 39.48 \text{ dB}$$

The normalized low-pass filter from table 12-57 is given in figure 4-7*a* using the circuit defined by figure 3-20.

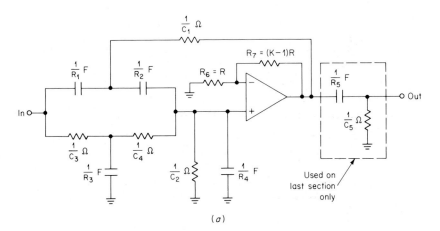

(a)

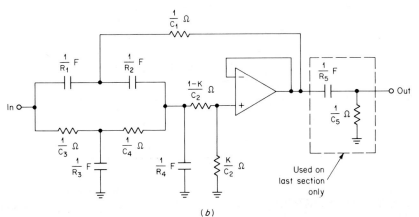

(b)

Fig. 4-6 Elliptic-function normalized high-pass filter section: *(a)* VCVS circuit for $K > 1$; *(b)* VCVS circuit for $K < 1$.

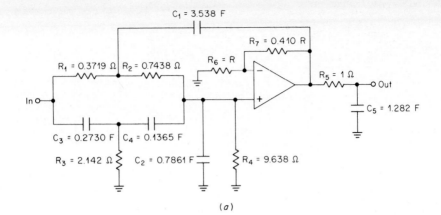

(a)

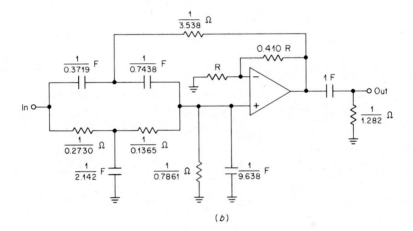

(b)

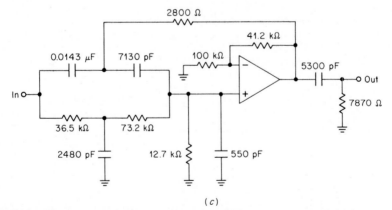

(c)

Fig. 4-7 Elliptic-function high-pass filter of example 4-5: (a) normalized low-pass filter from table 12-57; (b) transformed high-pass filter; (c) denormalized filter.

(c) The normalized low-pass filter can be transformed directly into the high-pass configuration by reciprocating the element values as illustrated in figure 4-6a. The resulting normalized high-pass filter is shown in figure 4-7b.

(d) To denormalize the circuit, frequency- and impedance-scale the filter. The capacitors are all divided by $Z \times$ FSF, where Z is arbitrarily chosen at 10^4 and FSF is $2\pi f_c$ or 18.85×10^3. The resistors are multiplied by Z. R is arbitrarily set equal to 100 kΩ. The final circuit is given by figure 4-7c using standard 1% resistor values.

State-Variable High-Pass Filters

The all-pole and elliptic-function active high-pass filters of section 4.2 cannot be easily adjusted. If the required degree of accuracy results in unreasonable component tolerances, the state-variable or biquad approach will permit independent adjustment of the filter's pole and zero coordinates. Another feature of this circuit is the reduced sensitivity of the response to many of the amplifier limitations such as finite bandwidth and gain.

All-Pole Configuration   In order to design a state-variable all-pole high-pass filter the normalized low-pass poles must first undergo a low-pass to high-pass transformation. Each low-pass pole pair consisting of a real part α and imaginary part β is transformed into a normalized high-pass pole pair as follows:

$$\alpha_{hp} = \frac{\alpha}{\alpha^2 + \beta^2} \tag{4-9}$$

$$\beta_{hp} = \frac{\beta}{\alpha^2 + \beta^2} \tag{4-10}$$

The transformed high-pass pole pair can now be denormalized by

$$\alpha'_{hp} = \alpha_{hp} \times \text{FSF} \tag{4-11}$$

$$\beta'_{hp} = \beta_{hp} \times \text{FSF} \tag{4-12}$$

where FSF is the frequency-scaling factor $2\pi f_c$.

The circuit of figure 4-8 realizes a high-pass second-order biquadratic transfer function. The element values for the all-pole case can be computed in terms of the high-pass pole coordinates as follows:

First compute

$$\omega'_0 = \sqrt{(\alpha'_{hp})^2 + (\beta'_{hp})^2} \tag{4-13}$$

The component values are

$$R_1 = R_4 = \frac{1}{2\alpha'_{hp}C} \tag{4-14}$$

$$R_2 = R_3 = \frac{1}{\omega'_0 C} \tag{4-15}$$

$$R_5 = \frac{2\alpha'_{hp}}{\omega'_0} R \tag{4-16}$$

$$R_6 = AR \tag{4-17}$$

where C and R are arbitrary and A is the section gain.

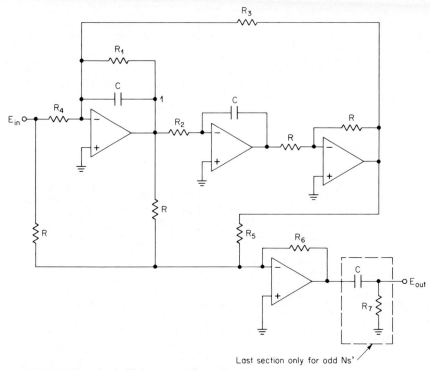

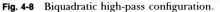

Fig. 4-8 Biquadratic high-pass configuration.

If the transfer function is of an odd order, a real pole must be realized. To transform the normalized low-pass real pole α_0, compute

$$\alpha_{0,\mathrm{hp}} = \frac{1}{\alpha_0} \qquad (4\text{-}18)$$

Then denormalize the high-pass real pole by

$$\alpha'_{0,\mathrm{hp}} = \alpha_{0,\mathrm{hp}} \times \mathrm{FSF} \qquad (4\text{-}19)$$

The last section of the filter is followed by an RC network as shown in figure 4-8. The value of R_7 is computed by

$$R_7 = \frac{1}{\alpha'_{0,\mathrm{hp}} C} \qquad (4\text{-}20)$$

where C is arbitrary.

A bandpass output is provided at node 1 for tuning purposes. The bandpass resonant frequency is given by

$$f_0 = \frac{\omega'_0}{2\pi} \qquad (4\text{-}21)$$

R_3 can be made adjustable and the circuit tuned to resonance by monitoring the phase shift between E_{in} and node 1 and adjusting R_3 for 180° of phase shift at f_0 using a Lissajous pattern.

The bandpass Q can then be monitored at node 1 and is given by

$$Q = \frac{\pi f_0}{\alpha'_{hp}} \qquad (4\text{-}22)$$

R_1 controls the Q and can be adjusted until either the computed Q is obtained or more conveniently the gain is unity between E_{in} and node 1 with f_0 applied.

The following example demonstrates the design of an all-pole high-pass filter using the biquad configuration.

Example 4-6

REQUIRED: Active high-pass filter
3 ± 0.1 dB at 300 Hz
30 dB minimum at 120 Hz
Gain of 2

RESULT: (a) Compute the high-pass steepness factor.

$$A_s = \frac{f_c}{f_s} = \frac{300}{120} = 2.5 \qquad (2\text{-}13)$$

(b) The curves of figure 2-45 indicate that a normalized third-order 1-dB Chebyshev low-pass filter has over 30 dB of attenuation at 2.5 rad/s. Since an accuracy of 0.1 dB is required at the cutoff frequency, a state-variable approach will be used so that adjustment capability is provided.

(c) The low-pass pole locations are found in table 12-26 and are as follows:

Complex pole $\alpha = 0.2257$ $\beta = 0.8822$
Real pole $\alpha_0 = 0.4513$

Complex pole-pair realization:

The complex low-pass pole pair is transformed to a high-pass pole pair as follows:

$$\alpha_{hp} = \frac{\alpha}{\alpha^2 + \beta^2} = \frac{0.2257}{0.2257^2 + 0.8822^2} = 0.2722 \qquad (4\text{-}9)$$

and
$$\beta_{hp} = \frac{\beta}{\alpha^2 + \beta^2} = \frac{0.8822}{0.2257^2 + 0.8822^2} = 1.0639 \qquad (4\text{-}10)$$

The transformed pole pair is then denormalized by

$$\alpha'_{hp} = \alpha_{hp} \times \text{FSF} = 513 \qquad (4\text{-}11)$$

$$\beta'_{hp} = \beta_{hp} \times \text{FSF} = 2005 \qquad (4\text{-}12)$$

where FSF is $2\pi f_c$ or 1885 since $f_c = 300$ Hz. If we choose $R = 10$ kΩ and $C = 0.01$ μF, the component values are calculated by

$$\omega'_0 = \sqrt{(\alpha'_{hp})^2 + (\beta'_{hp})^2} = 2070 \qquad (4\text{-}13)$$

then
$$R_1 = R_4 = \frac{1}{2\alpha'_{hp}C} = 97.47 \text{ k}\Omega \qquad (4\text{-}14)$$

$$R_2 = R_3 = \frac{1}{\omega'_0 C} = 48.31 \text{ k}\Omega \qquad (4\text{-}15)$$

$$R_5 = \frac{2 \, \alpha'_{hp}}{\omega'_0} R = 4957 \ \Omega \qquad (4\text{-}16)$$

$$R_6 = AR = 20 \text{ k}\Omega \qquad (4\text{-}17)$$

where $A = 2$.

The bandpass resonant frequency and Q are

$$f_0 = \frac{\omega_0'}{2\pi} = \frac{2070}{2\pi} = 329 \text{ Hz} \tag{4-21}$$

$$Q = \frac{\pi f_0}{\alpha_{hp}'} = \frac{\pi 329}{513} = 2.015 \tag{4-22}$$

Real-pole realization:

Transform the real pole:

$$\alpha_{0,hp} = \frac{1}{\alpha_0} = \frac{1}{0.4513} = 2.216 \tag{4-18}$$

To denormalize the transformed pole compute

$$\alpha_{0,hp}' = \alpha_{0,hp} \times \text{FSF} = 4177 \tag{4-19}$$

Using $C = 0.01 \ \mu\text{F}$ the real pole section resistor is given by

$$R_7 = \frac{1}{\alpha_{0,hp}' C} = 23.94 \text{ k}\Omega \tag{4-20}$$

The final filter configuration is shown in figure 4-9. The resistors were rounded off to standard 1% values, and R_1 and R_3 were made adjustable.

Elliptic-Function Configuration The biquadratic configuration of figure 4-8 can also be applied to the design of elliptic-function high-pass filters. The

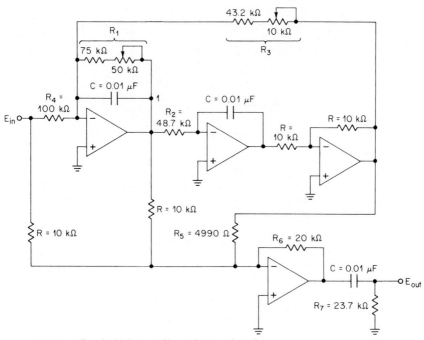

Fig. 4-9 All-pole high-pass filter of example 4-6.

component values are computed by using the same formulas as for the all-pole case except for R_5, which is given by

$$R_5 = \frac{2 \, \alpha'_{hp} \, \omega'_0}{(\omega'_0)^2 - [\omega'_\infty(hp)]^2} R \qquad (4\text{-}23)$$

where $\omega'_\infty(hp)$ is the denormalized high-pass transmission zero which is obtained from

$$\omega'_\infty(hp) = \omega_\infty(hp) \times FSF \qquad (4\text{-}24)$$

As in the all-pole circuit the bandpass resonant frequency f_0 is controlled by R_3 and the bandpass Q is determined by R_1. In addition the section notch can be adjusted if R_5 is made variable. However, this adjustment is usually not required if the circuit is first tuned to f_0, since the notch will then usually fall in.

The following example illustrates the design of an elliptic-function high-pass filter using the biquad configuration of figure 4-8.

Example 4-7

REQUIRED: Active high-pass filter
0.3 dB maximum ripple above 1000 Hz
20 dB minimum at 635 Hz

RESULT: (a) Compute the high-pass steepness factor.

$$A_s = \frac{f_c}{f_s} = \frac{1000}{635} = 1.575 \qquad (2\text{-}13)$$

(b) An elliptic-function filter type will be used. A normalized low-pass filter corresponding to CO3 25 $\theta = 40°$ can be found in table 12-57 and has the following parameters:

$$n = 3$$
$$R_{dB} = 0.28 \text{ dB}$$
$$\Omega_s = 1.556$$
$$A_{min} = 20.58 \text{ dB}$$

The normalized complex poles and zeros are given as

$$\alpha = 0.2643 \qquad \beta = 1.100$$
$$\omega_\infty = 1.742$$

and a real pole is located at

$$\alpha_0 = 0.9142$$

The biquad configuration of figure 4-8 will be used.

(c) First transform the normalized complex low-pass poles and zeros to the high-pass values.

$$\alpha_{hp} = \frac{\alpha}{\alpha^2 + \beta^2} = \frac{0.2643}{0.2643^2 + 1.100^2} = 0.2065 \qquad (4\text{-}9)$$

$$\beta_{hp} = \frac{\beta}{\alpha^2 + \beta^2} = \frac{1.100}{0.2643^2 + 1.100^2} = 0.8594 \qquad (4\text{-}10)$$

$$\omega_\infty(hp) = \frac{1}{\omega_\infty(Lp)} = \frac{1}{1.742} = 0.5741 \qquad (4\text{-}3)$$

The poles and zeros are denormalized as follows:

$$\alpha'_{hp} = \alpha_{hp} \times FSF = 1297 \qquad (4\text{-}11)$$

$$\beta'_{hp} = \beta_{hp} \times FSF = 5400 \tag{4-12}$$

and $$\omega'_\infty(hp) = \omega_\infty(hp) \times FSF = 3607 \tag{4-24}$$

where $FSF = 2\pi f_c$ or 6283.

If we arbitrarily choose $C = 0.01\ \mu F$ and $R = 100\ k\Omega$, the component values can be obtained by

$$\omega'_0 = \sqrt{(\alpha'_{hp})^2 + (\beta'_{hp})^2} = 5554 \tag{4-13}$$

then $$R_1 = R_4 = \frac{1}{2\alpha'_{hp}C} = 38.55\ k\Omega \tag{4-14}$$

$$R_2 = R_3 = \frac{1}{\omega'_0 C} = 18.01\ k\Omega \tag{4-15}$$

$$R_5 = \frac{2\alpha'_{hp}\omega'_0}{(\omega'_0)^2 - [\omega'_\infty(hp)]^2}\,R = 80.77\ k\Omega \tag{4-23}$$

and $$R_6 = AR = 100\ k\Omega \tag{4-17}$$

where the gain A is unity.

The bandpass resonant frequency and Q are determined from

$$f_0 = \frac{\omega'_0}{2\pi} = 884\ Hz \tag{4-21}$$

and $$Q = \frac{\pi f_0}{\alpha'_{hp}} = 2.14 \tag{4-22}$$

The notch frequency occurs at $\omega_\infty(hp) \times f_c$ or 574 Hz.

(d) The normalized real low-pass pole is transformed to a high-pass pole:

$$\alpha_{0,hp} = \frac{1}{\alpha_0} = \frac{1}{0.9142} = 1.0939 \tag{4-18}$$

and is then denormalized by

$$\alpha'_{0,hp} = \alpha_{0,hp} \times FSF = 6873 \tag{4-19}$$

Resistor R_7 is found by

$$R_7 = \frac{1}{\alpha'_{0,hp}C} = 14.55\ k\Omega \tag{4-20}$$

where $C = 0.01\ \mu F$.

The final circuit is given in figure 4-10a using standard 1% values with R_1 and R_3 made adjustable. The frequency response is illustrated in figure 4-10b.

High-Pass Filters Using the GIC

The generalized impedance converter (GIC) was first introduced in section 3.2. This versatile device is capable of simulating a variety of different impedance functions. The circuit of figure 3-28 simulated an inductor whose magnitude was given by

$$L = \frac{CR_1R_3R_5}{R_2} \tag{3-57}$$

If we set R_1 through R_3 equal to 1 Ω and $C = 1$ F, a normalized inductor is obtained where $L = R_5$. This circuit is shown in figure 4-11.

An active realization of a grounded inductor is particularly suited for the

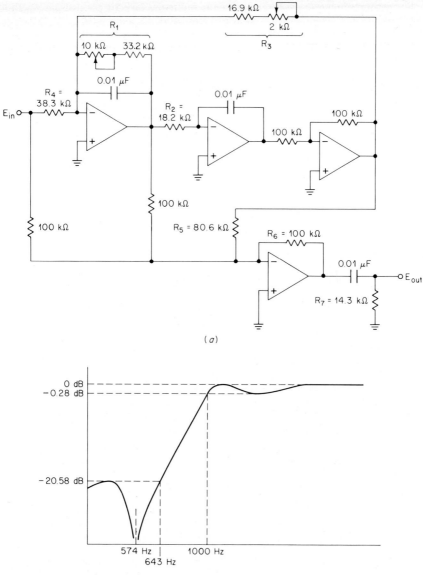

(a)

(b)

Fig. 4-10 Elliptic-function high-pass filter of example 4-7: (a) filter using biquad configuration; (b) frequency response.

design of active high-pass filters. If a passive *LC* low-pass configuration is transformed into a high-pass filter, shunt inductors to ground are obtained which can be implemented using the GIC. The resulting normalized filter can then be frequency- and impedance-scaled. If R_5 is made variable, the equivalent inductance can be adjusted. This feature is especially desirable in the case of steep elliptic-function high-pass filters, since the inductors directly control the location of the critical transmission zeros in the stopband.

The following example illustrates the design of an active all-pole high-pass filter directly from the *LC* element values using the GIC as a simulated inductance.

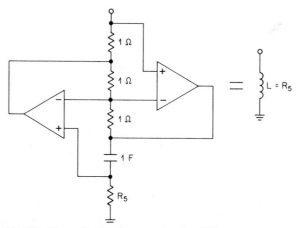

Fig. 4-11   Normalized inductor using the GIC.

Example 4-8

REQUIRED: Active high-pass filter
3 dB at 1200 Hz
35 dB minimum at 375 Hz

RESULT: (*a*)   Compute the high-pass steepness factor.

$$A_s = \frac{f_c}{f_s} = \frac{1200}{375} = 3.2 \qquad (2\text{-}13)$$

(*b*)   The curves of figure 2-45 indicate that a third-order 1-dB Chebyshev low-pass filter provides over 35 dB of attenuation at 3.2 rad/s. For this example we will use a GIC to simulate the inductor of an $n = 3$ *LC* high-pass configuration.

(*c*)   The normalized low-pass filter is obtained from table 12-31 and is shown in figure 4-12*a*. The dual filter configuration is used to minimize the number of inductors in the high-pass filter.

(*d*)   To transform the normalized low-pass filter into a high-pass configuration, replace the inductors with capacitors and vice versa using reciprocal element values. The normalized high-pass filter is shown in figure 4-12*b*. The inductor can now be replaced by the GIC of figure 4-11, resulting in the high-pass filter of figure 4-12*c*.

(*e*)   The filter is frequency- and impedance-scaled. Using an FSF of $2\pi f_c$ or 7540 and a Z of 10^4, divide all capacitors by Z × FSF and multiply all resistors by Z. The final configuration is shown in figure 4-12*d* using standard 1% resistor values.

Active elliptic-function high-pass filters can also be designed directly from the tables of *LC* element values using the GIC. This approach is much less complex than a high-pass configuration involving biquads and still permits ad-

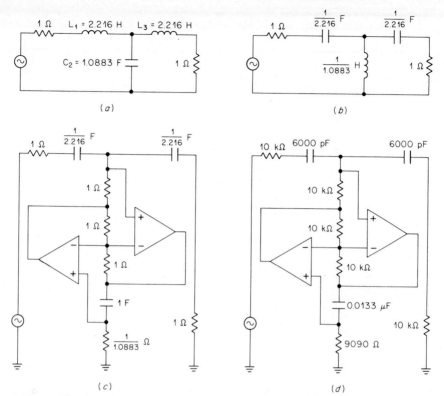

Fig. 4-12   All-pole high-pass filter using the GIC: *(a)* normalized low-pass filter; *(b)* transformed high-pass filter; *(c)* active inductor realization; *(d)* final network after scaling.

justment of the transmission zeros. Design of an elliptic-function high-pass filter is demonstrated in the following example.

Example 4-9

REQUIRED: Active high-pass filter
0.5 dB maximum at 2500 Hz
60 dB minimum at 1500 Hz

RESULT: *(a)*   Compute the high-pass steepness factor.

$$A_s = \frac{f_c}{f_s} = \frac{2500}{1500} = 1.667 \qquad (2\text{-}13)$$

(b)   Since the requirement is rather steep, we will choose an elliptic-function filter type. Using a ρ of 20% (0.18-dB ripple), the curves of figure 2-86 indicate that an $n = 6$ filter is required. A C06 20c $\theta = 40°$ filter is selected from table 12-56 and has the following parameters:

$$n = 6$$
$$R_{dB} = 0.18 \text{ dB}$$
$$\Omega_s = 1.6406$$
$$A_{min} = 62.8 \text{ dB}$$

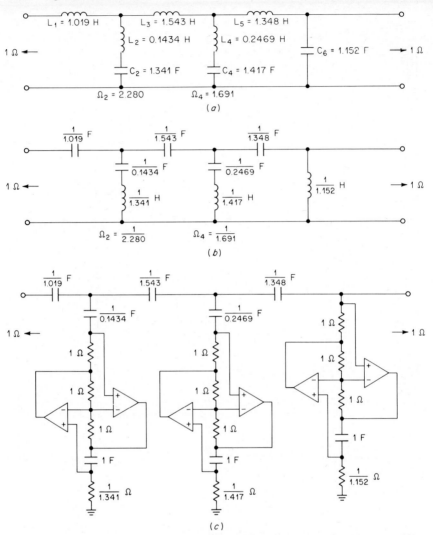

Fig. 4-13 Elliptic-function high-pass filter of example 4-9: (a) normalized low-pass filter; (b) transformed high-pass filter; (c) high-pass filter using GIC; (d) frequency- and imped-ance-scaled network.

The normalized filter using the dual circuit configuration is shown in figure 4-13a.

(c) To transform the network into a normalized high-pass filter, re-place each inductor with a capacitor having a reciprocal value and vice versa. The zeros are also reciprocated. The normalized high-pass filter is given in figure 4-13b.

(d) The inductors can be replaced using the GIC inductor simulation of figure 4-11, resulting in the circuit of figure 4-13c. To scale the network, divide all capacitors by Z × FSF and multiply all

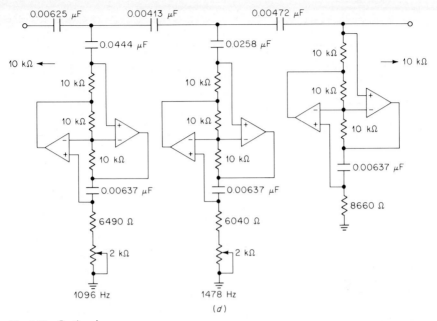

Fig. 4-13 *Continued*

resistors by Z, where Z is arbitrarily chosen at 10^4 and FSF is $2\pi f_c$ or 15,708. The final filter is shown in figure 4-13d. The stop-band peaks were computed by multiplying each normalized high-pass transmission zero by $f_c = 2500$ Hz, resulting in the frequencies indicated.

REFERENCES

Bruton, L. T., "Active Filter Design Using Generalized Impedance Converters," EDN, February 1973.

Geffe, P., "Simplified Modern Filter Design," John F. Rider, New York, 1963.

Williams, A. B., "Active Filter Design," Artech House, Dedham, Mass., 1975.

Williams, A. B., "Design Active Elliptic Filters Easily from Tables," Electronic Design, Vol. 19, No. 21, pp. 76–79, October 14, 1971.

5

Bandpass Filters

5.1 LC BANDPASS FILTERS

Bandpass filters were classified in section 2.1 as either narrow-band or wide-band. If the ratio of upper cutoff frequency to lower cutoff frequency is over an octave, the filter is considered a wide-band type. The specification is separated into individual low-pass and high-pass requirements and is simply treated as a cascade of low-pass and high-pass filters.

The design of narrow-band filters become somewhat more difficult. The circuit configuration must be appropriately chosen and suitable transformations may have to be applied to avoid impractical element values. In addition, as the filter becomes narrower the element Q requirements increase and component tolerances and stability become more critical.

Wide-Band Filters

Wide-band bandpass filters are obtained by cascading a low-pass filter and a high-pass filter. The validity of this approach is based on the assumption that the filters maintain their individual responses even though they are cascaded.

The impedance observed at the input or output terminals of an LC low-pass or high-pass filter approaches the resistive termination at the other end at frequencies well in the passband. This is apparent from the equivalent circuit of the low-pass filter at DC and the high-pass filter at infinite frequency. At DC the inductors become short circuits and capacitors become open circuits, and at infinite frequency the opposite conditions occur. If a low-pass and high-pass filter are cascaded and both filters are designed to have equal source and load terminations and identical impedances, the filters will each be properly terminated in their passband if the cutoff frequencies are separated by at least one or two octaves.

If the separation between passband is insufficient, the filters will interact because of impedance variations. This effect can be minimized by isolating the two filters through an attenuator. Usually 3 dB of loss is sufficient. Further attenuation provides increased isolation. Table 5-1 contains values for T and π attenuators ranging from 1 to 10 dB at an impedance level of 500 Ω. These networks can be impedance-scaled to the filter impedance level R if each resistor value is multiplied by $R/500$.

TABLE 5-1   T and π Attenuators

dB	R_1	R_2	R_a	R_b
1	28.8	4330	8700	57.7
2	57.3	2152	4362	116
3	85.5	1419	2924	176
4	113	1048	2210	239
5	140	822	1785	304
6	166	669	1505	374
7	191	558	1307	448
8	215	473	1161	528
9	238	406	1050	616
10	260	351	963	712

Example 5-1

REQUIRED: LC bandpass filter
3 dB at 500 Hz and 2000 Hz
40 dB minimum at 100 Hz and 4000 Hz
$R_s = R_L = 600\ \Omega$

RESULT: (a) Since the ratio of upper cutoff frequency to lower cutoff frequency is 4:1, a wide-band approach will be used. The requirement is first separated into individual low-pass and high-pass specifications:

High-pass filter:	**Low-pass filter:**
3 dB at 500 Hz	3 dB at 2000 Hz
40 dB minimum at 100 Hz	40 dB minimum at 4000 Hz

(b) The low-pass and high-pass filters are designed independently using the design methods outlined in sections 3.1 and 4.1 as follows:

Low-pass filter:

Compute the low-pass steepness factor.

$$A_s = \frac{f_s}{f_c} = \frac{4000\ \text{Hz}}{2000\ \text{Hz}} = 2 \tag{2-11}$$

Figure 2-43 indicates that a fifth-order 0.25-dB Chebyshev normalized low-pass filter provides over 40 dB of attenuation at 2 rad/s. The normalized low-pass filter is obtained from table 12-29 and is shown in figure 5-1a. The filter is frequency- and impedance-scaled by multiplying all inductors by Z/FSF and dividing all capacitors by $Z \times$ FSF, where Z is 600 and the frequency-scaling

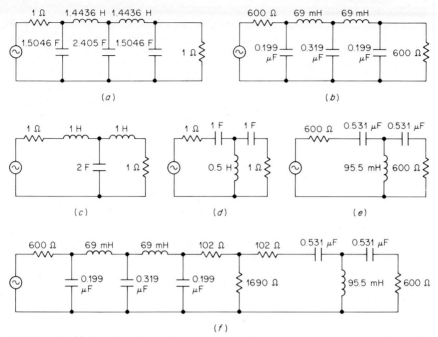

Fig. 5-1   *LC* wide-band bandpass filter of example 5-1: *(a)* normalized low-pass filter; *(b)* scaled low-pass filter; *(c)* normalized low-pass filter for high-pass requirement; *(d)* transformed high-pass filter; *(e)* scaled high-pass filter; *(f)* combined network.

factor FSF is $2\pi f_c$ or 12,560. The denormalized low-pass filter is shown in figure 5-1*b*.

High-pass filter:

Compute the high-pass steepness factor:

$$A_s = \frac{f_c}{f_s} = \frac{500 \text{ Hz}}{100 \text{ Hz}} = 5 \qquad (2\text{-}13)$$

Using figure 2-34, an $n = 3$ Butterworth normalized low-pass filter is selected to meet the attenuation requirement. The normalized filter values are found in table 12-2 and shown in figure 5-1*c*. Since the low-pass filter is to be transformed into a high-pass filter, the dual configuration was selected. By reciprocating element values and replacing inductors with capacitors and vice versa, the normalized high-pass filter of figure 5-1*d* is obtained. The network is then denormalized by multiplying all inductors by Z/FSF and dividing all capacitors by $Z \times \text{FSF}$, where Z is 600 Ω and FSF is 3140. The denormalized high-pass filter is illustrated in figure 5-1*e*.

(*c*) The low-pass and high-pass filters can now be combined. A 3-dB T pad will be used to provide some isolation between filters, since the separation of cutoffs is only 2 octaves. The pad values are obtained by multiplying the resistances of table 5-1 corresponding to 3 dB by 600 Ω/500 Ω or 1.2 and rounding off to standard 1% values. The final circuit is shown in figure 5-1*f*.

Narrow-Band Filters

Narrow-band bandpass filter terminology was introduced in section 2.1 using the concept of bandpass Q, which was defined by

$$Q_{bp} = \frac{f_0}{BW_{3\,dB}} \tag{2-16}$$

where f_0 is the geometric center frequency and BW is the 3-dB bandwidth. The geometric center frequency was given by

$$f_0 = \sqrt{f_L f_u} \tag{2-14}$$

where f_L and f_u are the lower and upper 3-dB limits.

Bandpass filters obtained by transformation from a low-pass filter exhibit geometric symmetry, that is,

$$f_0 = \sqrt{f_1 f_2} \tag{2-15}$$

where f_1 and f_2 are any two frequencies having equal attenuation. Geometric symmetry must be considered when normalizing a bandpass specification. For each stopband frequency specified, the corresponding geometric frequency is calculated and a steepness factor is computed based on the more severe requirement.

For bandpass Q's of 10 or more the passband response approaches arithmetic symmetry. The center frequency then becomes the average of the 3-dB points, i.e.,

$$f_0 = \frac{f_L + f_u}{2} \tag{2-17}$$

The stopband will also become arithmetically symmetrical as the Q increases even further.

The Low-Pass to Bandpass Transformation   A bandpass transfer function can be obtained from a low-pass transfer function by replacing the frequency variable by a new variable which is given by

$$f_{bp} = f_0 \left(\frac{f}{f_0} - \frac{f_0}{f} \right) \tag{5-1}$$

When f is equal to f_0, the bandpass center frequency, the response corresponds to that at DC for the low-pass filter.

If the low-pass filter has a 3-dB cutoff of f_c, the corresponding bandpass frequency f can be found by solving

$$\pm f_c = f_0 \left(\frac{f}{f_0} - \frac{f_0}{f} \right) \tag{5-2}$$

The $\pm$ signs occur because a low-pass filter has a mirrored response at negative frequencies in addition to the normal response. Solving equation (5-2) for f, we obtain

$$f = \pm \frac{f_c}{2} \pm \sqrt{\left(\frac{f_c}{2} \right)^2 + f_0^2} \tag{5-3}$$

Equation (5-3) implies that the bandpass response has two positive frequencies corresponding to the low-pass response at $\pm f_c$ as well as two negative frequencies with identical responses. These frequencies can be obtained from equation (5-3) and are given by

$$f_L = f_0 \left[\sqrt{1 + \left(\frac{f_c}{2 f_0}\right)^2} - \frac{f_c}{2 f_0} \right] \tag{5-4}$$

and

$$f_u = f_0 \left[\sqrt{1 + \left(\frac{f_c}{2 f_0}\right)^2} + \frac{f_c}{2 f_0} \right] \tag{5-5}$$

The bandpass 3-dB bandwidth is

$$\text{BW}_{3\ \text{dB}} = f_u - f_L = f_c \tag{5-6}$$

The correspondence between a low-pass filter and the transformed bandpass filter is shown in figure 5-2. The response of a low-pass filter to positive frequen-

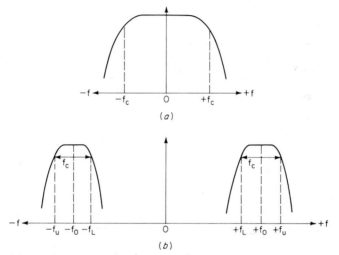

Fig. 5-2 Low-pass to bandpass transformation: (a) low-pass filter; (b) transformed bandpass filter.

cies is transformed into the response of the bandpass filter at an equivalent bandwidth. Therefore, a bandpass filter can be obtained by first designing a low-pass filter that has the required response corresponding to the desired bandwidth characteristics of the bandpass filter. The low-pass filter is then transformed into the bandpass filter.

The reactance of a capacitor in the low-pass filter is given by

$$X_c = \frac{1}{j \omega C} \tag{5-7}$$

where $\omega = 2\pi f$. If we replace the frequency variable f by the expression of equation (5-1), the impedance expression becomes

$$Z = \frac{1}{j \omega C + \dfrac{1}{j \dfrac{\omega}{\omega_0^2 C}}} \tag{5-8}$$

where $\omega_0 = 2\pi f_0$. This is the impedance of a parallel resonant LC circuit where the capacitance is still C and the inductance is $1/\omega_0^2 C$. The resonant frequency is ω_0.

The reactance of an inductor in the low-pass filter is

$$X_L = j\omega L \tag{5-9}$$

If we again replace the frequency variable using equation (5-1), the resulting impedance expression becomes

$$Z = j\omega L + \cfrac{1}{j\cfrac{\omega}{\omega_0^2 L}} \tag{5-10}$$

This corresponds to a series resonant LC circuit where the inductance L is unchanged and C is $1/\omega_0^2 L$. The resonant frequency is ω_0.

We can summarize these results by stating that an LC low-pass filter can be transformed into a bandpass filter having the equivalent bandwidth by resonating each capacitor with a parallel inductor and each inductor with a series capacitor. The resonant frequency is f_0, the bandpass filter center frequency. Table 5-2 shows the circuits which result from the low-pass to bandpass transformation.

TABLE 5-2   The Low-Pass to Bandpass Transformation

Low-Pass Branch	Bandpass Configuration	Circuit Values
Type I		$L = \dfrac{1}{\omega_0^2 C}$ (5-11) $C = \dfrac{1}{\omega_0^2 L}$ (5-12)
Type II		$C_a = \dfrac{1}{\omega_0^2 L_a}$ (5-13) $L_b = \dfrac{1}{\omega_0^2 C_b}$ (5-14)
Type III		$C_1 = \dfrac{1}{\omega_0^2 L_1}$ (5-15) $L_2 = \dfrac{1}{\omega_0^2 C_2}$ (5-16)
Type IV		

Transformation of All-Pole Low-Pass Filters   The LC bandpass filters discussed in this section are probably the most important type of filter. These networks are directly obtained by the bandpass transformation from the LC low-pass values tabulated in chapter 12. Each normalized low-pass filter defines an infinitely large family of bandpass filters having a geometrically symmetrical response predetermined by the low-pass characteristics.

A low-pass transfer function can be transformed to a bandpass type by substitution of the frequency variable using equation (5-1). This transformation can also be made directly to the circuit elements by first scaling the low-pass filter to the required bandwidth and impedance level. Each coil is then resonated with a series capacitor to the center frequency f_0 and an inductor is introduced across each capacitor to form a parallel tuned circuit also resonant at f_0. Every low-pass branch is replaced by the associated bandpass branch, as illustrated by table 5-2.

Some of the effects of dissipation in low-pass filters were discussed in section 3.1. These effects are even more severe in bandpass filters. The minimum Q requirement for the low-pass elements can be obtained from figure 3-8 for a variety of filter types. These minimum values are based on the assumption that the filter elements are predistorted so that the theoretical response is obtained. Since this is not always the case, the branch Q's should be several times higher than the values indicated. When the network undergoes a low-pass to bandpass transformation, the Q requirement is increased by the bandpass Q of the filter. This can be stated as

$$Q_{min} \text{ (bandpass)} = Q_{min} \text{ (low-pass)} \times Q_{bp} \qquad (5\text{-}17)$$

where $Q_{bp} = f_0/BW_{3\,dB}$. As in the low-pass case the branch Q's should be several times higher than Q_{min}. Since capacitor losses are usually negligible, the branch Q is determined strictly by the inductor losses.

The spread of values in bandpass filters is usually wider than with low-pass filters. For some combinations of impedance and bandwidth the element values may be impossible or impractical to realize because of their magnitude or the effects of parasitics. When this situation occurs, the designer can use a variety of circuit transformations to obtain a more practical circuit. These techniques are covered in chapter 8.

The design method can be summarized as follows:

1. Convert the response requirement into a geometrically symmetrical specification.
2. Compute the bandpass steepness factor A_s. Select a normalized low-pass filter from the frequency-response curves of chapter 2 that makes the passband to stopband transition within a frequency ratio of A_s.
3. Scale the corresponding normalized low-pass filter from the tables of chapter 12 to the required bandwidth and impedance level of the bandpass filter.
4. Resonate each L and C to f_0 in accordance with table 5-2.
5. The final design may require manipulation by various transformations so that the values are more practical. In addition the branch Q's must be well in excess of Q_{min} (bandpass) as given by equation (5-17) to obtain near theoretical results.

Example 5-2

REQUIRED: Bandpass filter
            Center frequency of 1000 Hz
            3-dB points at 950 Hz and 1050 Hz
            25 dB minimum at 800 Hz and 1150 Hz
            $R_s = R_L = 600\ \Omega$
            Available inductor Q of 100

RESULT: (a) Convert to geometrically symmetrical bandpass requirement:
            First calculate the geometric center frequency.

$$f_0 = \sqrt{f_L f_u} = \sqrt{950 \times 1050} = 998.8 \text{ Hz} \qquad (2\text{-}14)$$

Compute the corresponding geometric frequency for each stop-band frequency given using equation (2-18).

$$f_1 f_2 = f_0^2 \qquad (2\text{-}18)$$

f_1	f_2	$f_2 - f_1$
800 Hz	1247 Hz	447 Hz
867 Hz	1150 Hz	283 Hz

The second pair of frequencies will be retained, since they represent the more severe requirement. The resulting geometrically symmetrical requirement can be summarized as

$$f_0 = 998.8 \text{ Hz}$$
$$BW_{3\text{ dB}} = 100 \text{ Hz}$$
$$BW_{25\text{ dB}} = 283 \text{ Hz}$$

(b) Compute the bandpass steepness factor.

$$A_s = \frac{\text{stopband bandwidth}}{\text{passband bandwidth}} = \frac{283 \text{ Hz}}{100 \text{ Hz}} = 2.83 \qquad (2\text{-}19)$$

(c) Select a normalized low-pass filter that makes the transition from 3 dB to more than 25 dB within a frequency ratio of 2.83:1. Figure 2-34 indicates that an $n = 3$ Butterworth type will satisfy the response requirement. The normalized low-pass filter is found in table 12-2 and is shown in figure 5-3a.

(d) Denormalize the low-pass filter using a Z of 600 and a frequency-scaling factor (FSF) of $2\pi f_c$ or 628 where $f_c = 100$ Hz.

$$C_1' = C_3' = \frac{C}{\text{FSF} \times Z} = \frac{1}{628 \times 600} = 2.653 \ \mu\text{F} \qquad (2\text{-}10)$$

$$L_2' = \frac{L \times Z}{\text{FSF}} = \frac{2 \times 600}{628} = 1.91 \text{ H} \qquad (2\text{-}9)$$

The denormalized low-pass filter is illustrated in figure 5-3b.

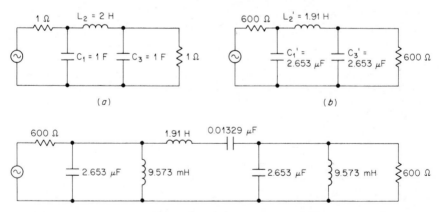

(a)

(b)

(c)

Fig. 5-3 Bandpass filter of example 5-2: (a) normalized $N = 3$ Butterworth low-pass filter; (b) low-pass filter scaled to 600 Ω and f_c of 100 Hz; (c) transformed bandpass filter.

(e) To make the low-pass to bandpass transformation resonate each capacitor with a parallel inductor and each inductor with a series capacitor using a resonate frequency of $f_0 = 998.8$ Hz.

$$L_1' = \frac{1}{\omega_0^2 C_1'} = \frac{1}{(6275)^2 \times 2.653 \times 10^{-6}} = 9.573 \text{ mH} \quad (5\text{-}11)$$

$$L_3' = L_1' = 9.573 \text{ mH}$$

$$C_2' = \frac{1}{\omega_0^2 L_2'} = \frac{1}{(6275)^2 \times 1.91} = 0.01329 \ \mu\text{F} \quad (5\text{-}12)$$

where $\omega_0 = 2\pi f_0$. The resulting bandpass filter is given in figure 5-3c.

(f) Estimate if the available inductor Q of 100 is sufficient.

$$Q_{\min} \text{ (bandpass)} = Q_{\min} \text{ (low-pass)} \times Q_{\text{bp}} = 2 \times 10 = 20 \quad (5\text{-}17)$$

where $Q_{\min}$ (low-pass) was obtained from figure 3-8 and Q_{bp} is $f_0/\text{BW}_{3 \text{ dB}}$. Since the available Q is well in excess of $Q_{\min}$ (bandpass), the filter response will closely agree with the theoretical predictions.

The response requirement of example 5-2 was converted to a geometrically symmetrical specification by calculating the corresponding frequency for each stopband frequency specified at a particular attenuation level using the relationship $f_1 f_2 = f_0^2$. The pair of frequencies having the lesser separation was chosen, since this would represent the steeper filter requirement. This technique represents a general method for obtaining the geometrically related frequencies that determine the response requirements of the normalized low-pass filter.

Stopband requirements are frequently specified in an arithmetically symmetrical manner where the deviation on both sides of the center frequency is the same for a given attenuation. Because of the geometric symmetry of bandpass filters the attenuation for a particular deviation below the center frequency will be greater than for the same deviation above the center frequency. The response curve would then appear compressed on the low side of the passband if plotted on a linear frequency axis. On a logarithmic scale the curve would be symmetrical.

When the specification is stated in arithmetic terms the stopband bandwidth on a geometric basis can be computed directly by

$$\text{BW} = f_2 - \frac{f_0^2}{f_2} \quad (5\text{-}18)$$

where f_2 is the upper stopband frequency and f_0 is the geometric center frequency as determined from the passband limits. This approach is demonstrated in the following example.

Example 5-3

REQUIRED: Bandpass filter
Center frequency of 50 kHz
3-dB points at $\pm$ 3 kHz (47 kHz, 53 kHz)
30 dB minimum at $\pm$ 7.5 kHz (42.5 kHz, 57.5 kHz)
40 dB minimum at $\pm$ 10.5 kHz (39.5 kHz, 60.5 kHz)
$R_s = 150 \ \Omega$ $R_L = 300 \ \Omega$

RESULT: (a) Convert to the geometrically symmetrical bandpass requirement.

$$f_0 = \sqrt{f_L f_u} = \sqrt{47 \times 53 \times 10^6} = 49.91 \text{ kHz} \quad (2\text{-}14)$$

Since the stopband requirement is arithmetically symmetrical, compute the stopband bandwidth using equation (5-18).

$$BW_{30 \text{ dB}} = f_2 - \frac{f_0^2}{f_2} = 57.5 \times 10^3 - \frac{(49.91 \times 10^3)^2}{57.5 \times 10^3} = 14.18 \text{ kHz}$$

$$BW_{40 \text{ dB}} = 19.33 \text{ kHz}$$

Requirement:

$$f_0 = 49.91 \text{ kHz}$$
$$BW_{3 \text{ dB}} = 6 \text{ kHz}$$
$$BW_{30 \text{ dB}} = 14.18 \text{ kHz}$$
$$BW_{40 \text{ dB}} = 19.33 \text{ kHz}$$

(b)   Since two stopband bandwidth requirements are given, they must both be converted into bandpass steepness factors.

$$A_s(30 \text{ dB}) = \frac{\text{stopband bandwidth}}{\text{passband bandwidth}} = \frac{14.18 \text{ kHz}}{6 \text{ kHz}} = 2.36 \qquad (2\text{-}19)$$

$$A_s(40 \text{ dB}) = 3.22$$

(c)   A normalized low-pass filter must be chosen that provides over 30 dB of rejection at 2.36 rad/s and more than 40 dB at 3.22 rad/s. Figure 2-41 indicates that a fourth-order 0.01-dB Chebyshev filter will meet this requirement. The corresponding low-pass filter can be found in table 12-27. Since a 2:1 ratio of R_L to R_s is required, the design for a normalized R_s of 2 Ω is chosen and is turned end for end. The circuit is shown in figure 5-4a.

(d)   The circuit is now scaled to an impedance level Z of 150 and a cutoff of $f_c = 6$ kHz. All inductors are multiplied by Z/FSF, and

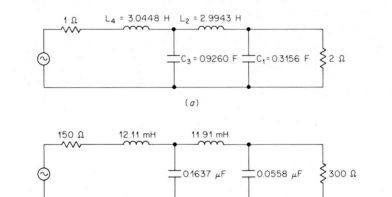

(a)

(b)

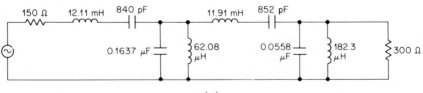

(c)

Fig. 5-4   Bandpass filter of example 5-3: (a) normalized low-pass filter; (b) scaled low-pass filter; (c) transformed bandpass filter.

the capacitors are divided by $Z \times FSF$, where FSF is $2\pi f_c$. The 1-Ω source and 2-Ω load become 150 and 300 Ω, respectively. The denormalized network is illustrated in figure 5-4b.

(e) The scaled low-pass filter is transformed to a bandpass filter at $f_0 = 49.91$ kHz by resonating each capacitor with a parallel inductor and each inductor with a series capacitor using the general relationship $\omega_0^2 LC = 1$. The resulting bandpass filter is shown in figure 5-4c.

Design of Parallel Tuned Circuits The simple RC low-pass circuit of figure 5-5a has a 3-dB cutoff corresponding to

$$f_c = \frac{1}{2\pi RC} \tag{5-19}$$

If a bandpass transformation is performed, the circuit of figure 5-5b results, where

$$L = \frac{1}{\omega_0^2 C} \tag{5-11}$$

The center frequency is f_0 and the 3-dB bandwidth is equal to f_c. The bandpass Q is given by

$$Q_{bp} = \frac{f_0}{BW_{3\,dB}} = \frac{f_0}{f_c} = \omega_0 RC \tag{5-20}$$

Since the magnitudes of the capacitive and inductive susceptances are equal at resonance by definition, we can substitute $1/\omega_0 L$ for $\omega_0 C$ in equation (5-20) and obtain

$$Q_{bp} = \frac{R}{\omega_0 L} \tag{5-21}$$

The element R may be a single resistor as in figure 5-5b or the parallel combination of both the input and output terminations if an output load resistor is also present.

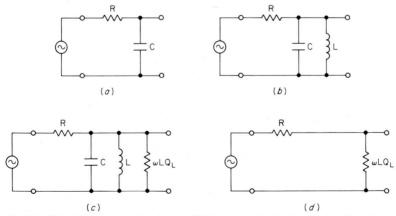

(a)   (b)

(c)   (d)

Fig. 5-5 The single tuned circuit: (a) RC low-pass circuit; (b) result of bandpass transformation; (c) representation of coil losses; (d) equivalent circuit at resonance.

The circuit of figure 5-5*b* is somewhat ideal, since inductor losses are usually unavoidable. (The slight losses usually associated with the capacitor will be neglected.) If the inductor Q is given as Q_L, the inductor losses can be represented as a parallel resistor of $\omega L Q_L$ as shown in figure 5-5*c*. The effective Q of the circuit becomes

$$Q_{eff} = \frac{\dfrac{R}{\omega_0 L} Q_L}{\dfrac{R}{\omega_0 L} + Q_L} \qquad (5\text{-}22)$$

As a result, the effective circuit Q is somewhat less than the values computed by equations (5-20) or (5-21). To compensate for the effect of finite inductor Q, the design Q should be somewhat higher. This value can be found from

$$Q_d = \frac{Q_{eff} Q_L}{Q_L - Q_{eff}} \qquad (5\text{-}23)$$

At resonance the equivalent circuit is represented by the resistive voltage divider of figure 5-5*d*, since the reactive elements cancel. The insertion loss at f_0 can be determined by the expression

$$IL_{dB} = 20 \log \left(1 + \frac{1}{k-1}\right) \qquad (5\text{-}24)$$

where $k = Q_L / Q_{eff}$ and can be obtained directly from the curve of figure 5-6. Clearly the insertion loss increases dramatically as the inductor Q approaches the required effective Q of the circuit.

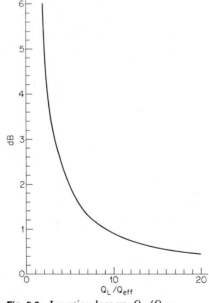

Fig. 5-6   Insertion loss vs. Q_L / Q_{eff}.

The frequency response of a single tuned circuit is expressed by

$$A_{dB} = 10 \log \left[1 + \left(\frac{BW_x}{BW_{3\,dB}} \right)^2 \right] \qquad (5\text{-}25)$$

where BW_x is the bandwidth of interest and $BW_{3\,dB}$ is the 3-dB bandwidth. The response characteristics are identical to an $n = 1$ Butterworth; so the attenuation curves of figure 2-34 can be applied using $BW_x/BW_{3\,dB}$ as the normalized frequency in radians per second.

The phase shift is given by

$$\theta = \tan^{-1} \left(\frac{2\Delta f}{BW_{3\,dB}} \right) \qquad (5\text{-}26)$$

where Δf is the frequency deviation from f_0. The output phase shift lags by 45° at the upper 3-dB frequency and leads by 45° at the lower 3-dB frequency. At DC and infinity the phase shift reaches +90° and −90°, respectively. Equation (5-26) is plotted in figure 5-7.

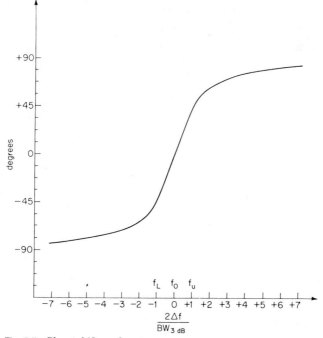

Fig. 5-7 Phase shift vs. frequency.

The group delay can be estimated by the slope of the phase shift at f_0 and results in the approximation

$$T_{gd} \approx \frac{318}{BW_{3\,dB}} \qquad (5\text{-}27)$$

where $BW_{3\,dB}$ is the 3-dB bandwidth in hertz and T_{gd} is the resulting group delay in milliseconds.

Example 5-4

REQUIRED: *LC* bandpass filter
Center frequency of 10 kHz
3 dB at $\pm$100 Hz (9.9 kHz, 10.1 kHz)
15 dB minimum at $\pm$1 kHz (9 kHz, 11 kHz)
Inductor $Q_L = 200$
$R_s = R_L = 6$ kΩ

RESULT: (a) Convert to the geometrically symmetrical bandpass specification. Since the bandpass Q is much greater than 10, the specified arithmetically symmetrical frequencies are used to determine the following design requirements:

$$f_0 = 10 \text{ kHz}$$
$$\text{BW}_{3\text{ dB}} = 200 \text{ Hz}$$
$$\text{BW}_{15\text{ dB}} = 2000 \text{ Hz}$$

(b) Compute the bandpass steepness factor.

$$A_s = \frac{\text{stopband bandwidth}}{\text{passband bandwidth}} = \frac{2000}{200} = 10 \qquad (2\text{-}19)$$

Figure 2-34 indicates that a single tuned circuit ($n = 1$) provides more than 15 dB of attenuation within a bandwidth ratio of 10:1.

(c) Calculate the design Q to obtain a Q_{eff} equal to $f_0/\text{BW}_{3\text{ dB}} = 50$ considering the inductor Q_L of 200.

$$Q_d = \frac{Q_{\text{eff}}Q_L}{Q_L - Q_{\text{eff}}} = \frac{50 \times 200}{200 - 50} = 66.7 \qquad (5\text{-}23)$$

(d) Since the source and load are both 6 kΩ, the total resistive loading on the tuned circuit is the parallel combination of both terminations; so $R = 3$ kΩ. The design Q can now be used in equation (5-20) to compute C.

$$C = \frac{Q_{\text{bp}}}{\omega_0 R} = \frac{66.7}{6.28 \times 10 \times 10^3 \times 3000} = 0.354 \ \mu\text{F} \qquad (5\text{-}20)$$

The inductance is given by equation (5-11).

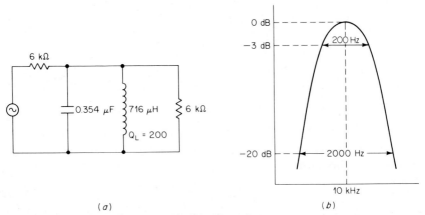

(a)   (b)

Fig. 5-8 Tuned circuit of example 5-4: (a) circuit; (b) frequency response.

$$L = \frac{1}{\omega_0^2 C} = \frac{1}{(2\pi \times 10 \text{ kHz})^2 \times 3.54 \times 10^{-7}} = 716 \ \mu\text{H} \qquad (5\text{-}11)$$

The resulting circuit is shown in figure 5-8a, which has the frequency response of figure 5-8b. See section 8.1 for a more practical implementation using a tapped inductor.

(e) The circuit insertion loss can be calculated from

$$IL_{\text{dB}} = 20 \log \left(1 + \frac{1}{k-1} \right) = 20 \log 1.333 = 2.5 \text{ dB} \qquad (5\text{-}24)$$

where $\quad k = \dfrac{Q_L}{Q_{\text{eff}}} = \dfrac{200}{50} = 4$

The low-pass to bandpass transformation illustrated in figure 5-5 can also be examined from a pole-zero perspective. The RC low-pass filter has a single real pole at $1/RC$ as shown in figure 5-9a and a zero at infinity. The bandpass transformation results in a pair of complex poles and zeros at the origin and infinity as illustrated in figure 5-9b. The radial distance from the origin to the

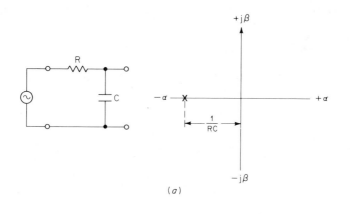

(a)

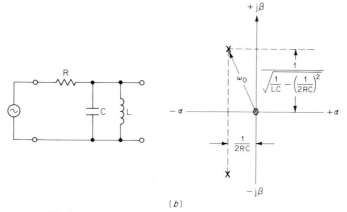

(b)

Fig. 5-9 The bandpass transformation: (a) low-pass circuit; (b) bandpass circuit.

pole is $1/(LC)^{1/2}$ corresponding to ω_0, the resonant frequency. The Q can be expressed by

$$Q = \frac{\omega_0}{2\alpha} \tag{5-28}$$

where α, the real part, is $1/2RC$. The transfer function of the circuit of figure 5-9b becomes

$$T(s) = \frac{s}{s^2 + \dfrac{\omega_0}{Q}s + \omega_0^2} \tag{5-29}$$

At ω_0 the impedance of the parallel resonant circuit is a maximum and is purely resistive, resulting in zero phase shift. If the Q is much less than 10, these effects do not both occur at precisely the same frequency. Series losses of the inductor will also displace the zero from the origin onto the negative real axis.

The Series Tuned Circuit   The losses of an inductor can be conveniently represented by a series resistor determined by

$$R_{coil} = \frac{\omega L}{Q_L} \tag{5-30}$$

If we form a series resonant circuit and include the source and load resistors, we obtain the circuit of figure 5-10. Equations (5-24) through (5-27) for insertion

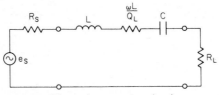

Fig. 5-10   The series resonant circuit.

loss, frequency response, phase shift, and group delay of the parallel tuned circuit apply, since the two circuits are duals of each other. The inductance is calculated from

$$L = \frac{R_s + R_L}{\omega_0 \left(\dfrac{1}{Q_{bp}} - \dfrac{1}{Q_L} \right)} \tag{5-31}$$

where Q_{bp} is the required Q and Q_L is the inductor Q. The capacitance is given by

$$C = \frac{1}{\omega_0^2 L} \tag{5-12}$$

Example 5-5

REQUIRED: Series tuned circuit
Center frequency of 100 kHz

3-dB bandwidth of 2 kHz
$R_s = R_L = 100 \ \Omega$
Inductor Q of 400

RESULT: (a) Compute the bandpass Q.

$$Q_{bp} = \frac{f_0}{BW_{3 \ dB}} = \frac{100 \ kHz}{2 \ kHz} = 50 \qquad (2\text{-}16)$$

(b) Calculate the element values.

$$L = \frac{R_s + R_L}{\omega_0 \left(\dfrac{1}{Q_{bp}} - \dfrac{1}{Q_L} \right)} = \frac{200}{2 \ \pi \times 10^5 \left(\dfrac{1}{50} - \dfrac{1}{400} \right)} = 18.2 \ mH \qquad (5\text{-}31)$$

$$C = \frac{1}{\omega_0^2 \ L} = 139 \ pF \qquad (5\text{-}12)$$

The circuit is shown in figure 5-11.

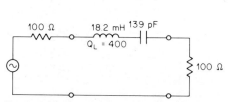

Fig. 5-11 Series tuned circuit of example 5-5.

The reader may recall from AC circuit theory that one of the effects of series resonance is a buildup of voltage across both reactive elements. The voltage across either reactive element at resonance is equal to Q times the input voltage and may be excessively high, causing inductor saturation or capacitor breakdown. In addition, the L/C ratio becomes large as the bandwidth is reduced and will result in impractical element values where high Q's are required. As a result, series resonant circuits are less desirable than parallel tuned circuits.

Synchronously Tuned Filters Tuned circuits can be cascaded to obtain bandpass filters of a higher complexity. Each stage must be isolated from the previous section. If all circuits are tuned to the same frequency, a synchronously tuned filter is obtained. The characteristics of synchronously tuned bandpass filters are discussed in section 2.8, and the normalized frequency response is illustrated by the curves of figure 2-77. The design Q of each section was given by

$$Q_{section} = Q_{overall} \sqrt{2^{1/n} - 1} \qquad (2\text{-}45)$$

where $Q_{overall}$ is defined by the ratio $f_0/BW_{3 \ dB}$ of the composite filter. The individual circuits may be of either the series or the parallel resonant type.

Synchronously tuned filters are the simplest approximation to a bandpass response. Since all stages are identical and tuned to the same frequency, they are simple to construct and easy to align. The Q requirement of each individual section is less than the overall Q, whereas the opposite is true for conventional bandpass filters. The transient behavior exhibits no overshoot or ringing. On the other hand, the selectivity is extremely poor. To obtain a particular attenuation for a given steepness factor A_s, many more stages are required than for the other filter types. In addition, each section must be isolated from the previous section; so interstage amplifiers are required. The disadvantages generally out-weigh the advantages; so synchronously tuned filters are usually restricted to special applications such as IF and RF amplifiers.

Example 5-6

REQUIRED: Synchronously tuned bandpass filter
               Center frequency of 455 kHz
               3 dB at ± 5 kHz
               30 dB minimum at ± 35 kHz
               Inductor Q of 400

RESULT: *(a)*   Compute the bandpass steepness factor.

$$A_s = \frac{\text{stopband bandwidth}}{\text{passband bandwidth}} = \frac{70 \text{ kHz}}{10 \text{ kHz}} = 7 \qquad (2\text{-}19)$$

The curves of figure 2-77 indicate that a third-order ($n = 3$) synchronously tuned filter satisfies the attenuation requirement.

(b)   Three sections are required which are all tuned to 455 kHz and have identical Q. To compute the Q of the individual sections, first calculate the overall Q, which is given by

$$Q_{bp} = \frac{f_0}{\text{BW}_{3 \text{ dB}}} = \frac{455 \text{ kHz}}{10 \text{ kHz}} = 45.5 \qquad (2\text{-}16)$$

The section Q's can be found from

$$Q_{\text{section}} = Q_{\text{overall}} \sqrt{2^{1/n} - 1} = 45.5 \sqrt{2^{1/3} - 1} = 23.2 \qquad (2\text{-}45)$$

(c)   The tuned circuits can now be designed using either a series or parallel realization. Let us choose a parallel tuned circuit configuration using a single source resistor of 10 kΩ and a high-impedance termination. Since an effective circuit Q of 23.2 is desired and the inductor Q is 400, the design Q is calculated from

$$Q_d = \frac{Q_{\text{eff}} Q_L}{Q_L - Q_{\text{eff}}} = \frac{23.2 \times 400}{400 - 23.2} = 24.6 \qquad (5\text{-}23)$$

The inductance is then given by

$$L = \frac{R}{\omega_0 Q_{bp}} = \frac{10 \times 10^3}{2 \pi 455 \times 10^3 \times 24.6} = 142 \text{ } \mu\text{H} \qquad (5\text{-}21)$$

The resonating capacitor can be obtained from

$$C = \frac{1}{\omega_0^2 L} = 862 \text{ pF} \qquad (5\text{-}12)$$

The final circuit is shown in figure 5-12 utilizing buffer amplifiers to isolate the three sections.

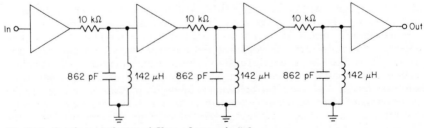

Fig. 5-12   Synchronously tuned filter of example 5-6.

Narrow-Band Coupled Resonators

Narrow-band bandpass filters can be designed by using coupling techniques where parallel tuned circuits are interconnected by coupling elements such as inductors or capacitors. Figure 5-13 illustrates some typical configurations.

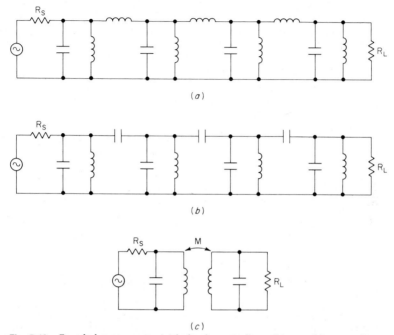

Fig. 5-13 Coupled resonators: *(a)* inductive coupling; *(b)* capacitive coupling; *(c)* magnetic coupling.

Coupled resonator configurations are desirable for narrow-band filters having bandpass Q's of 10 or more. The values are generally more practical than the elements obtained by the low-pass to bandpass transformation, especially for very high Q's. The tuning is also simpler, since it turns out that all nodes are resonated to the same frequency. Of the three configurations shown in figure 5-13, the capacitive coupled configuration is the most desirable from the standpoint of economy and ease of manufacture.

The theoretical justification for the design method is based on the assumption that the coupling elements have a constant impedance with frequency. This assumption is approximately accurate over narrow bandwidths. At DC the coupling capacitors will introduce additional response zeros. This causes the frequency response to be increasingly unsymmetrical both geometrically and arithmetically as we deviate from the center frequency. The response shape will be somewhat steeper on the low-frequency side of the passband.

The general form of a capacitive coupled resonator filter is shown in figure 5-14. An nth-order filter requires n parallel tuned circuits and contains n nodes. Tables 5-3 through 5-12 present in tabular form q and k parameters for all-

TABLE 5-3   Butterworth Capacitive Coupled Resonators

n	q_1	q_n	k_{12}	k_{23}	k_{34}	k_{45}	k_{56}	k_{67}	k_{78}
2	1.414	1.414	0.707						
3	1.000	1.000	0.707	0.707					
4	0.765	0.765	0.841	0.541	0.841				
5	0.618	0.618	1.000	0.556	0.556	1.000			
6	0.518	0.518	1.169	0.605	0.518	0.605	1.169		
7	0.445	0.445	1.342	0.667	0.527	0.527	0.667	1.342	
8	0.390	0.390	1.519	0.736	0.554	0.510	0.554	0.736	1.519

TABLE 5-4   0.01-dB Chebyshev Capacitive Coupled Resonators

n	q_1	q_n	k_{12}	k_{23}	k_{34}	k_{45}	k_{56}	k_{67}	k_{78}
2	1.483	1.483	0.708						
3	1.181	1.181	0.682	0.682					
4	1.046	1.046	0.737	0.541	0.737				
5	0.977	0.977	0.780	0.540	0.540	0.780			
6	0.937	0.937	0.809	0.550	0.518	0.550	0.809		
7	0.913	0.913	0.829	0.560	0.517	0.517	0.560	0.829	
8	0.897	0.897	0.843	0.567	0.520	0.510	0.520	0.567	0.843

TABLE 5-5   0.1-dB Chebyshev Capacitive Coupled Resonators

n	q_1	q_n	k_{12}	k_{23}	k_{34}	k_{45}	k_{56}	k_{67}	k_{78}
2	1.638	1.638	0.711						
3	1.433	1.433	0.662	0.662					
4	1.345	1.345	0.685	0.542	0.685				
5	1.301	1.301	0.703	0.536	0.536	0.703			
6	1.277	1.277	0.715	0.539	0.518	0.539	0.715		
7	1.262	1.262	0.722	0.542	0.516	0.516	0.542	0.722	
8	1.251	1.251	0.728	0.545	0.516	0.510	0.516	0.545	0.728

TABLE 5-6   0.5-dB Chebyshev Capacitive Coupled Resonators

n	q_1	q_n	k_{12}	k_{23}	k_{34}	k_{45}	k_{56}	k_{67}	k_{78}
2	1.950	1.950	0.723						
3	1.864	1.864	0.647	0.647					
4	1.826	1.826	0.648	0.545	0.648				
5	1.807	1.807	0.652	0.534	0.534	0.652			
6	1.796	1.796	0.655	0.533	0.519	0.533	0.655		
7	1.790	1.790	0.657	0.533	0.516	0.516	0.533	0.657	
8	1.785	1.785	0.658	0.533	0.515	0.511	0.515	0.533	0.658

TABLE 5-7   1-dB Chebyshev Capacitive Coupled Resonators

n	q_1	q_n	k_{12}	k_{23}	k_{34}	k_{45}	k_{56}	k_{67}
2	2.210	2.210	0.739					
3	2.210	2.210	0.645	0.645				
4	2.210	2.210	0.638	0.546	0.638			
5	2.210	2.210	0.633	0.535	0.538	0.633		
6	2.250	2.250	0.631	0.531	0.510	0.531	0.631	
7	2.250	2.250	0.631	0.530	0.517	0.517	0.530	0.631

TABLE 5-8 Bessel Capacitive Coupled Resonators

n	q_1	q_n	k_{12}	k_{23}	k_{34}	k_{45}	k_{56}	k_{67}	k_{78}
2	0.5755	2.148	0.900						
3	0.337	2.203	1.748	0.684					
4	0.233	2.240	2.530	1.175	0.644				
5	0.394	0.275	1.910	0.750	0.650	1.987			
6	0.415	0.187	2.000	0.811	0.601	1.253	3.038		
7	0.187	0.242	3.325	1.660	1.293	0.695	0.674	2.203	
8	0.139	0.242	4.284	2.079	1.484	1.246	0.678	0.697	2.286

TABLE 5-9 Linear Phase with Equiripple Error of 0.05° Capacitive Coupled Resonators

n	q_1	q_n	k_{12}	k_{23}	k_{34}	k_{45}	k_{56}	k_{67}	k_{78}
2	0.648	2.109	0.856						
3	0.433	2.254	1.489	0.652					
4	0.493	0.718	1.632	0.718	0.739				
5	0.547	0.446	1.800	0.848	0.584	1.372			
6	0.397	0.468	1.993	1.379	0.683	0.661	1.553		
7	0.316	0.484	2.490	1.442	1.446	0.927	0.579	1.260	
8	0.335	0.363	2.585	1.484	1.602	1.160	0.596	0.868	1.733

TABLE 5-10 Linear Phase with Equiripple Error of 0.5° Capacitive Coupled Resonators

n	q_1	q_n	k_{12}	k_{23}	k_{34}	k_{45}	k_{56}	k_{67}	k_{78}
2	0.825	1.980	0.783						
3	0.553	2.425	1.330	0.635					
4	0.581	1.026	1.575	0.797	0.656				
5	0.664	0.611	1.779	0.919	0.576	1.162			
6	0.552	0.586	1.874	1.355	0.641	0.721	1.429		
7	0.401	0.688	2.324	1.394	1.500	1.079	0.590	1.045	
8	0.415	0.563	2.410	1.470	1.527	1.409	0.659	0.755	1.335

TABLE 5-11 Transitional Gaussian to 6-dB Capacitive Coupled Resonators

n	q_1	q_n	k_{12}	k_{23}	k_{34}	k_{45}	k_{56}	k_{67}	k_{78}
3	0.404	2.338	1.662	0.691					
4	0.570	0.914	1.623	0.798	0.682				
5	0.891	0.670	1.418	0.864	0.553	1.046			
6	0.883	0.752	1.172	1.029	0.595	0.605	1.094		
7	0.736	0.930	1.130	0.955	0.884	0.534	0.633	1.104	
8	0.738	0.948	1.124	0.866	0.922	0.708	0.501	0.752	1.089

TABLE 5-12 Transitional Gaussian to 12-dB Capacitive Coupled Resonators

n	q_1	q_n	k_{12}	k_{23}	k_{34}	k_{45}	k_{56}	k_{67}	k_{78}
3	0.415	2.345	1.631	0.686					
4	0.419	0.766	1.989	0.833	0.740				
5	0.534	0.503	2.085	0.976	0.605	1.333			
6	0.543	0.558	1.839	1.442	0.686	0.707	1.468		
7	0.492	0.665	1.708	1.440	1.181	0.611	0.781	1.541	
8	0.549	0.640	1.586	1.262	1.296	0.808	0.569	1.023	1.504

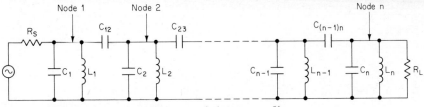

Fig. 5-14   General form of capacitive coupled resonator filter.

pole filters. These parameters are used to generate the component values for filters having the form shown in figure 5-14. For each network a q_1 and q_n is given corresponding to the first and last resonant circuit. The k parameters are given in terms of k_{12}, k_{23}, etc., and are related to the coupling capacitors shown in figure 5-14. The design method proceeds as follows:

1. Compute the desired filter's passband Q which was given by

$$Q_{bp} = \frac{f_0}{BW_{3\ dB}} \qquad (2\text{-}16)$$

2. Determine the q's and k's from the tables corresponding to the chosen filter type and the order of complexity n. Denormalize these coefficients as follows:

$$Q_1 = Q_{bp} \times q_1 \qquad (5\text{-}32)$$
$$Q_n = Q_{bp} \times q_n \qquad (5\text{-}33)$$
$$K_{xy} = \frac{k_{xy}}{Q_{bp}} \qquad (5\text{-}34)$$

3. Choose a convenient inductance value L. The source and load terminations are found from

$$R_s = \omega_0 L\, Q_1 \qquad (5\text{-}35)$$

and

$$R_L = \omega_0 L\, Q_n \qquad (5\text{-}36)$$

4. The total nodal capacitance is determined by

$$C_{node} = \frac{1}{\omega_0^2 L} \qquad (5\text{-}37)$$

The coupling capacitors are then computed from

$$C_{xy} = K_{xy}\, C_{node} \qquad (5\text{-}38)$$

5. The total capacity connected to each node must be equal to C_{node}. Therefore, the shunt capacitors of the parallel tuned circuits are equal to the total nodal capacitance C_{node} less the values of the coupling capacitors connected to that node. For example,

$$C_1 = C_{node} - C_{12}$$
$$C_2 = C_{node} - C_{12} - C_{23}$$
$$C_7 = C_{node} - C_{67} - C_{78}$$

Each node is tuned to f_0 with the adjacent nodes shorted to ground so that the coupling capacitors connected to that node are placed in parallel across the tuned circuit.

The completed filter may require impedance scaling so that the source and load terminating requirements are met. In addition some of the impedance transformations discussed in chapter 8 may have to be applied.

The k and q values tabulated in tables 5-3 through 5-12 are based on infinite inductor Q. In reality, satisfactory results will be obtained for inductor Q's several times higher than Q_{min} (bandpass) determined by equation (5-17) in conjunction with figure 3-8 which shows the minimum theoretical low-pass Q's.

Example 5-7

REQUIRED: Bandpass filter
Center frequency of 100 kHz
3 dB at ±2.5 kHz
35 dB minimum at ±12.5 kHz
Constant delay over the passband

RESULT: *(a)* Since a constant delay is required, a Bessel filter type will be chosen. The low-pass constant delay properties will undergo a minimum of distortion for the bandpass case since the bandwidth is relatively narrow, i.e., the bandpass Q is high. Because the bandwidth is narrow, we can treat the requirements on an arithmetically symmetrical basis.

The bandpass steepness factor is given by

$$A_s = \frac{\text{stopband bandwidth}}{\text{passband bandwidth}} = \frac{25\text{ kHz}}{5\text{ kHz}} = 5 \qquad (2\text{-}19)$$

The frequency-response curves of figure 2-56 indicate that an $n = 4$ Bessel filter provides over 35 dB of attenuation at 5 rad/s. A capacitive coupled resonator configuration will be used for the implementation.

(b) The q and k parameters for a Bessel filter corresponding to $n = 4$ are found in table 5-8 and are as follows:

$$q_1 = 0.233$$
$$q_4 = 2.240$$
$$k_{12} = 2.530$$
$$k_{23} = 1.175$$
$$k_{34} = 0.644$$

To denormalize these values, divide each k by the bandpass Q and multiply each q by the same factor as follows:

$$Q_{bp} = \frac{f_0}{BW_{3\text{ dB}}} = \frac{100\text{ kHz}}{5\text{ kHz}} = 20 \qquad (2\text{-}16)$$

The resulting values are

$$Q_1 = Q_{bp} \times q_1 = 20 \times 0.233 = 4.66 \qquad (5\text{-}32)$$
$$Q_4 = 44.8$$

$$K_{12} = \frac{k_{12}}{Q_{bp}} = \frac{2.530}{20} = 0.1265 \qquad (5\text{-}34)$$

$$K_{23} = 0.05875$$
$$K_{34} = 0.0322$$

(c) Let us choose an inductance of $L = 2.5$ mH. The source and load terminations are

$$R_s = \omega_0 L Q_1 = 6.28 \times 10^5 \times 2.5 \times 10^{-3} \times 4.66 = 7.32\text{ k}\Omega \qquad (5\text{-}35)$$

and

$$R_L = \omega_0 L Q_4 = 70.37\text{ k}\Omega \qquad (5\text{-}36)$$

where

$$\omega_0 = 2\pi f_0$$

(d)   The total nodal capacitance is determined by

$$C_{\text{node}} = \frac{1}{\omega_0^2 L} = 1013 \text{ pF} \qquad (5\text{-}37)$$

The coupling capacitors can now be calculated.

$$
\begin{aligned}
C_{12} &= K_{12}\, C_{\text{node}} = 0.1265 \times 1.013 \times 10^{-9} = 128.1 \text{ pF} \qquad (5\text{-}38) \\
C_{23} &= K_{23}\, C_{\text{node}} = 59.5 \text{ pF} \\
C_{34} &= K_{34}\, C_{\text{node}} = 32.6 \text{ pF}
\end{aligned}
$$

The shunt capacitors are determined from

$$
\begin{aligned}
C_1 &= C_{\text{node}} - C_{12} = 884.9 \text{ pF} \\
C_2 &= C_{\text{node}} - C_{12} - C_{23} = 825.4 \text{ pF} \\
C_3 &= C_{\text{node}} - C_{23} - C_{34} = 920.9 \text{ pF} \\
C_4 &= C_{\text{node}} - C_{34} = 980.4 \text{ pF}
\end{aligned}
$$

The final circuit is shown in figure 5-15.

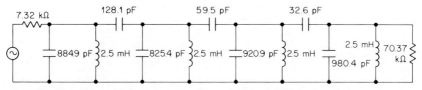

Fig. 5-15   Capacitive coupled resonator filter of example 5-7.

Predistorted Bandpass Filters

The inductor Q requirements of bandpass filters are higher than those of low-pass filters, since the minimum theoretical branch Q is given by

$$Q_{\min} \text{ (bandpass)} = Q_{\min} \text{ (low-pass)} \times Q_{\text{bp}} \qquad (5\text{-}17)$$

where $Q_{\text{bp}} = f_0/\text{BW}_{3 \text{ dB}}$. In the cases where the filter required is extremely narrow, a branch Q many times higher than the minimum theoretical Q may be difficult to obtain. Predistorted bandpass filters can then be used so that exact theoretical results can be obtained with reasonable branch Q's.

Predistorted bandpass filters can be obtained from the normalized predistorted low-pass filters given in chapter 12 by the conventional bandpass transformation. The low-pass filters must be of the uniform dissipation type, since the lossy-L networks would be transformed to a bandpass filter having losses in the series branches only.

The uniform dissipation networks are tabulated for different values of dissipation factor d. These values can be related to the required inductor Q by the relationship

$$Q_L = \frac{Q_{\text{bp}}}{d} \qquad (5\text{-}39)$$

where $Q_{\text{bp}} = f_0/\text{BW}_{3 \text{ dB}}$.

The losses of a predistorted low-pass filter having uniform dissipation are evenly distributed and occur as series losses in the inductors and shunt losses

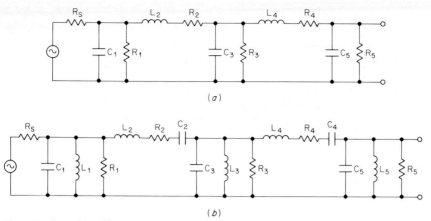

Fig. 5-16 Location of losses in uniformly predistorted filters: *(a)* low-pass filter; *(b)* band-pass filter.

across the capacitors. The equivalent circuit of the filter is shown in figure 5-16*a*. The inductor losses were previously given by

$$R_L = \frac{\omega L}{Q} \tag{3-2}$$

and the capacitor losses were defined by

$$R_c = \frac{Q}{\omega C} \tag{3-3}$$

When the circuit is transformed to a bandpass filter, the losses are still required to be distributed in series with the series branches and in parallel with the shunt branches, as shown in figure 5-16*b*. In reality, the capacitor losses are minimal and the inductor losses occur in series with the inductive elements in both the series and shunt branches. Therefore, as a narrow-band approximation, the losses may be distributed between the capacitors and inductors in an arbitrary manner. The only restriction is that the combination of inductor and capacitor losses in each branch results in a total branch Q equal to the value computed by equation (5-39). The combined Q of a lossy inductor and a lossy capacitor in a resonant circuit is given by

$$Q_T = \frac{Q_L\,Q_c}{Q_L + Q_c} \tag{5-40}$$

where Q_T is the total branch Q, Q_L is the inductor Q, and Q_c is the Q of the capacitor.

The predistorted networks tabulated in chapter 12 require an infinite termination on one side. In practice if the resistance used to approximate the infinite termination is large compared with the source termination, satisfactory results will be obtained. If the dual configuration is used, which ideally requires a zero impedance source, the source impedance should be much less than the load termination.

It is usually difficult to obtain inductor Q's precisely equal to the values computed from equation (5-39). A Q accuracy within 5 or 10% at f_0 is usually sufficient. If greater accuracy is required, an inductor Q higher than the calculated value is used. The Q is then degraded to the exact required value by adding resistors.

Example 5-8

REQUIRED: Bandpass filter
Center frequency of 10 kHz
3 dB at ±250 Hz
60 dB minimum at ±750 Hz
$R_s = 100\ \Omega$ $R_L = 10\ k\Omega$ minimum
Available inductor Q of 225

RESULT: (a)   Since the filter is narrow in bandwidth, the requirement is treated in its arithmetically symmetrical form. The bandpass steepness factor is obtained from

$$A_s = \frac{\text{stopband bandwidth}}{\text{passband bandwidth}} = \frac{1500\ \text{Hz}}{500\ \text{Hz}} = 3 \qquad (2\text{-}19)$$

The curves of figure 2-43 indicate that a fifth-order ($n = 5$) 0.25-dB Chebyshev filter will meet these requirements. A predistorted design will be used. The corresponding normalized low-pass filters are found in table 12-33.

(b)   The specified inductor Q can be used to compute the required d of the low-pass filter as follows:

$$d = \frac{Q_{bp}}{Q_L} = \frac{20}{225} = 0.0889 \qquad (5\text{-}39)$$

where $Q_{bp} = f_0/BW_{3\ dB}$. The circuit corresponding to $n = 5$ and $d = 0.0919$ will be selected, since this d is sufficiently close to the computed value. The schematic is shown in figure 5-17a.

(c)   Denormalize the low-pass filter using a frequency-scaling factor (FSF) of $2\pi f_c$ or 3140 where $f_c = 500$ Hz, the required bandwidth of the bandpass filter, and an impedance-scaling factor Z of 100.

$$C_1' = \frac{C}{\text{FSF} \times Z} = \frac{1.0397}{3140 \times 100} = 3.309\ \mu F \qquad (2\text{-}10)$$

$$C_3' = 7.014\ \mu F$$
$$C_5' = 3.660\ \mu F$$

and $$L_2' = \frac{L \times Z}{\text{FSF}} = \frac{1.8181 \times 100}{3140} = 57.87\ \text{mH} \qquad (2\text{-}9)$$

$$L_4' = 55.79\ \text{mH}$$

The denormalized low-pass filter is shown in figure 5-17b, where the termination has been scaled to 100 Ω. The filter has also been turned end for end, since the high-impedance termination is required at the output and the 100-Ω source at the input.

(d)   To transform the circuit into a bandpass filter, resonate each capacitor with an inductor in parallel and each inductor with a series capacitor using a resonant frequency of $f_0 = 10$ kHz. The parallel inductor is computed from

$$L = \frac{1}{\omega_0^2 C} \qquad (5\text{-}11)$$

and the series capacitor is calculated by

$$C = \frac{1}{\omega_0^2 L} \qquad (5\text{-}12)$$

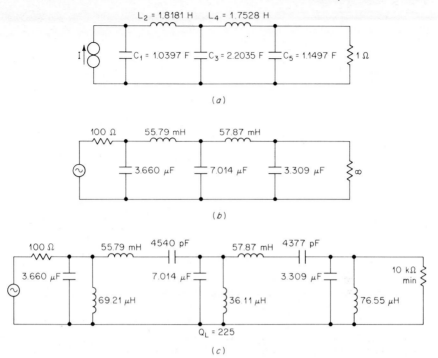

Fig. 5-17 Predistorted bandpass filter of example 5-8: *(a)* normalized low-pass filter; *(b)* frequency- and impedance-scaled network; *(c)* resulting bandpass filter.

where both formulas are forms of the general relationship for resonance $\omega_0^2 LC = 1$. The resulting bandpass filter is given in figure 5-17c. The large spread of values can be reduced by applying some of the techniques discussed in chapter 8.

Elliptic-Function Bandpass Filters

Elliptic-function low-pass filters were clearly shown to be far superior to the other filter types in terms of achieving a required attenuation within a given frequency ratio. This superiority is mainly the result of the presence of transmission zeros beginning just outside the passband.

Elliptic-function LC low-pass filters have been extensively tabulated by Saal and Ulbrich and by Zverev (see references). Some of these tables are reproduced in section 12. These networks can be transformed into bandpass filters in the same manner as the all-pole filter types. The elliptic-function bandpass filters will then exhibit the same superiority over the all-pole types as their low-pass counterparts.

When an elliptic-function low-pass filter is transformed into a bandpass filter, each low-pass transmission zero is converted in a pair of zeros, one above and one below the passband and geometrically related to the center frequency. (For the purposes of this discussion, negative zeros will be disregarded.) The low-pass zeros are directly determined by the resonances of the parallel tuned circuits in the series branches. When each series branch containing a parallel tuned

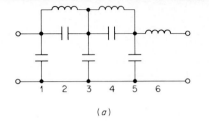

(a)

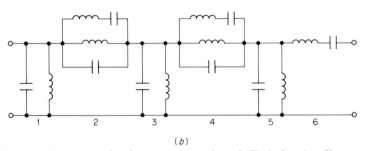

(b)

Fig. 5-18 Low-pass to bandpass transformation of elliptic-function filter:
(a) $N = 6$ low-pass filter; (b) transformed bandpass configuration.

circuit is modified by the bandpass transformation, two parallel branch resonances are introduced corresponding to the upper and lower zeros.

A sixth-order elliptic-function low-pass filter structure is shown in figure
5-18a. After frequency- and impedance-scaling the low-pass values, we can make
a bandpass transformation by resonating each inductor with a series capacitor
and each capacitor with a parallel inductor, where the resonant frequency is
f_0, the filter center frequency. The circuit of figure 5-18b results. The configuration obtained in branches 2 and 4 corresponds to a type III network from table
5-2.

The type III network realizes two parallel resonances corresponding to a
geometrically related pair of transmission zeros above and below the passband.
The circuit configuration itself is not very desirable. The elements corresponding
to both parallel resonances are not distinctly isolated. Each resonance is determined by the interaction of a number of elements; so tuning is made difficult.
Also for very narrow filters the values may become unreasonable. Fortunately
an alternate circuit exists that provides a more practical relationship between
the coils and capacitors. The two equivalent configurations are shown in figure
5-19. The alternate configuration utilizes two parallel tuned circuits where each
condition of parallel resonance directly corresponds to a transmission zero.

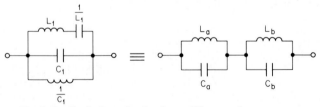

Fig. 5-19 Equivalent circuit of type III network.

The type III network of figure 5-19 is shown with reciprocal element values. These result when we normalize the bandpass filter to a center frequency of 1 rad/s. Since the general equation for resonance $\omega_0^2 LC = 1$ reduces to $LC = 1$ at $\omega_0 = 1$, the resonant elements become reciprocals of each other.

The reason for this normalization is to greatly simplify the equations for the transformation of figure 5-19. Otherwise the equations relating the two circuits would be extremely complex and unsuitable for computation by hand.

To obtain the normalized bandpass filter, first multiply all L and C values of the normalized low-pass filter by Q_{bp}, which is equal to f_0/BW, where BW is the passband bandwidth. The network can then be transformed directly into a normalized bandpass filter by resonating each inductor with a series capacitor and each capacitor with a parallel inductor. The resonant elements are merely reciprocals of each other, since $\omega_0 = 1$.

The transformation of figure 5-19 can now be performed. First calculate

$$\beta = 1 + \frac{1}{2 L_1 C_1} + \sqrt{\frac{1}{4 L_1^2 C_1^2} + \frac{1}{L_1 C_1}} \tag{5-41}$$

The values are then obtained from

$$L_a = \frac{1}{C_1(\beta + 1)} \tag{5-42}$$

$$L_b = \beta \, L_a \tag{5-43}$$

$$C_a = \frac{1}{L_b} \tag{5-44}$$

$$C_b = \frac{1}{L_a} \tag{5-45}$$

The resonant frequencies are given by

$$\Omega_{\infty,a} = \sqrt{\beta} \tag{5-46}$$

and

$$\Omega_{\infty,b} = \frac{1}{\Omega_{\infty,a}} \tag{5-47}$$

After the transformation of figure 5-19 is made wherever applicable, the normalized bandpass filter is scaled to the required center frequency and impedance level by multiplying all inductors by Z/FSF and dividing all capacitors by Z × FSF. The frequency-scaling factor in this case is equal to ω_0 ($\omega_0 = 2\pi f_0$, where f_0 is the desired center frequency of the filter). The resonant frequencies in hertz can be found by multiplying all normalized radian resonant frequencies by f_0.

The design of an elliptic-function bandpass filter is demonstrated by the following example.

Example 5-9

REQUIRED: Bandpass filter
1 dB maximum variation from 15 to 20 kHz
50 dB minimum below 14 kHz and above 23 kHz
$R_s = R_L = 10$ kΩ

RESULT: (a) Convert to geometrically symmetrical bandpass requirement: First calculate the geometric center frequency.

$$f_0 = \sqrt{f_L f_u} = \sqrt{15 \times 20 \times 10^6} = 17.32 \text{ kHz} \quad (2\text{-}14)$$

Compute the corresponding geometric frequency for each stop-band frequency given using the relationship

$$f_1 f_2 = f_0^2 \quad (2\text{-}18)$$

f_1	f_2	$f_2 - f_1$
14.00 kHz	21.43 kHz	7.43 kHz
13.04 kHz	23.00 kHz	9.96 kHz

The first pair of frequencies has the lesser separation and therefore represents the more severe requirement and will be retained. The geometrically symmetrical requirements can be summarized as

$$f_0 = 17.32 \text{ kHz}$$
$$\text{BW}_{1 \text{ dB}} = 5 \text{ kHz}$$
$$\text{BW}_{50 \text{ dB}} = 7.43 \text{ kHz}$$

(b) Compute the bandpass steepness factor.

$$A_s = \frac{\text{stopband bandwidth}}{\text{passband bandwidth}} = \frac{7.43 \text{ kHz}}{5 \text{ kHz}} = 1.486 \quad (2\text{-}19)$$

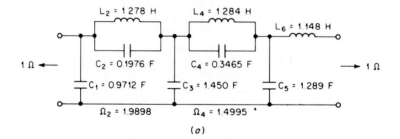

(a)

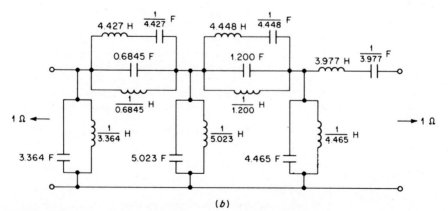

(b)

Fig. 5-20 Elliptic-function bandpass filter: (a) normalized low-pass filter; (b) bandpass filter normalized to $\omega_0 = 1$; (c) transformed type III branches; (d) final scaled circuit; (e) frequency response.

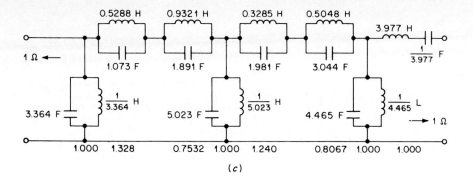

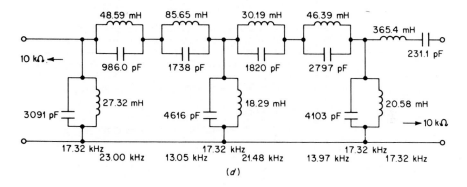

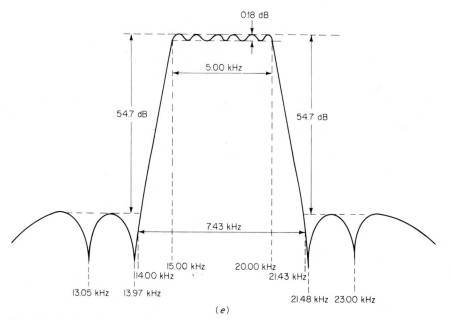

Fig. 5-20 *Continued*

(c) Select a normalized low-pass filter that makes the transition from less than 1 dB to more than 50 dB within a frequency ratio of 1.486. Since the requirement is rather steep, an elliptic-function type will be chosen.

Figure 2-86 indicates that for a ρ of 20% (0.18-dB ripple) a filter complexity of $n = 6$ is required. Using the tables of section 12-56, a design corresponding to C06 20C $\theta = 46°$ has the parameters $\Omega_s = 1.458$ and $A_{min} = 54.7$ dB, which meets these requirements. The normalized circuit is shown in figure 5-20a.

(d) The filter must now be converted to a normalized bandpass filter having a center frequency of $\omega_0 = 1$. The bandpass Q is first computed from

$$Q_{bp} = \frac{f_0}{BW} = \frac{17.32 \text{ kHz}}{5 \text{kHz}} = 3.464$$

Multiply all inductance and capacitance values by Q_{bp}. Then transform the network into a bandpass filter centered at $\omega_0 = 1$ by resonating each capacitor with a parallel inductor and each inductor with a series capacitor. The resonating elements introduced are simply the reciprocal values as shown in figure 5-20b.

(e) The type III branches will now be transformed in accordance with figure 5-19.

For the first branch:

$$L_1 = 4.427 \text{ H} \qquad C_1 = 0.6845 \text{ F}$$

First compute

$$\beta = 1 + \frac{1}{2 L_1 C_1} + \sqrt{\frac{1}{4 L_1^2 C_1^2} + \frac{1}{L_1 C_1}} = 1.763 \qquad (5\text{-}41)$$

then

$$L_a = \frac{1}{C_1(\beta + 1)} = 0.5288 \text{ H} \qquad (5\text{-}42)$$

$$L_b = \beta\, L_a = 0.9321 \text{ H} \qquad (5\text{-}43)$$

$$C_a = \frac{1}{L_b} = 1.073 \text{ F} \qquad (5\text{-}44)$$

$$C_b = \frac{1}{L_a} = 1.891 \text{ F} \qquad (5\text{-}45)$$

The resonant frequencies are

$$\Omega_{\infty,a} = \sqrt{\beta} = 1.328 \qquad (5\text{-}46)$$

$$\Omega_{\infty,b} = \frac{1}{\Omega_{\infty,a}} = 0.7532 \qquad (5\text{-}47)$$

For the second branch:

$$L_1 = 4.448 \text{ H} \qquad C_1 = 1.200 \text{ F}$$

then

$$\beta = 1.5365$$
$$L_a = 0.3285 \text{ H}$$
$$L_b = 0.5048 \text{ H}$$
$$C_a = 1.981 \text{ F}$$
$$C_b = 3.044 \text{ F}$$
$$\Omega_{\infty,a} = 1.240$$
$$\Omega_{\infty,b} = 0.8067$$

The transformed filter is shown in figure 5-20c. The resonant frequencies in radians per second are indicated below the schematic.

(f) To complete the design, denormalize the filter to a center frequency (f_0) of 17.32 kHz and an impedance level of 10 kΩ. Multiply all inductors by Z/FSF and divide all capacitors by $Z \times FSF$ where $Z = 10^4$ and $FSF = 2\pi f_0$ or 1.0882×10^5. The final filter is shown in figure 5-20d. The resonant frequencies were obtained by direct multiplication of the normalized resonant frequencies of figure 5-20c by the geometric center frequency f_0. The frequency response is shown in figure 5-20e.

5.2 ACTIVE BANDPASS FILTERS

When the separation between the upper and lower cutoff frequencies exceeds a ratio of approximately 2, the bandpass filter is considered a wide-band type. The specifications are then separated into individual low-pass and high-pass requirements and met by a cascade of active low-pass and high-pass filters.

Narrow-band LC bandpass filters are usually designed by transforming a low-pass configuration directly into the bandpass circuit. Unfortunately, no such circuit transformation exists for active networks. The general approach involves transforming the low-pass transfer function into a bandpass type. The bandpass poles and zeros are then implemented by a cascade of bandpass filter sections.

Wide-Band Filters

When LC low-pass and high-pass filters were cascaded, care had to be taken to minimize terminal impedance variations so that each filter maintained its individual response in the cascaded form. Active filters can be interconnected with no interaction because of the inherent buffering of the operational amplifiers. The only exception occurs in the case of the elliptic-function VCVS filters of sections 3.2 and 4.2 where the last sections are followed by an RC network to provide the real poles. An amplifier must then be introduced for isolation.

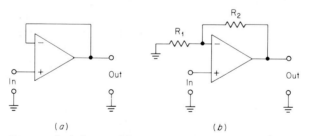

Fig. 5-21 Isolation amplifiers: (a) voltage follower; (b) noninverting amplifier.

Figure 5-21 shows two simple amplifier configurations which can be used after an active elliptic filter. The gain of the voltage follower is unity. The noninverting amplifier has a gain equal to $R_2/R_1 + 1$. The resistors R_1 and R_2 can have any convenient values, since only their ratio is of significance.

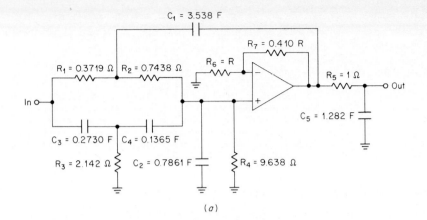

(a)

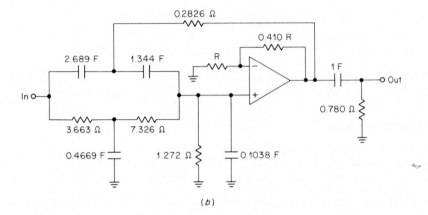

(b)

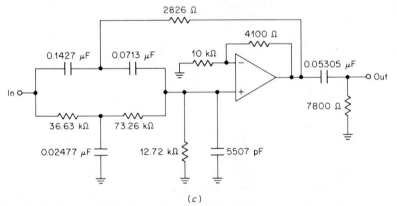

(c)

Fig. 5-22 Wide-band bandpass filter of example 5-10: (a) normalized elliptic-function low-pass filter; (b) transformed high-pass filter; (c) scaled high-pass filter; (d) normalized low-pass filter; (e) denormalized low-pass filter; (f) bandpass filter configuration.

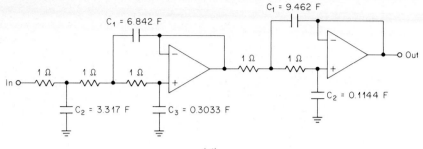

(d)

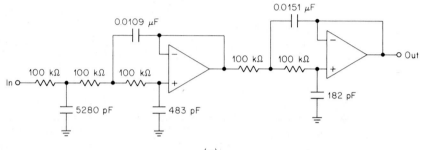

(e)

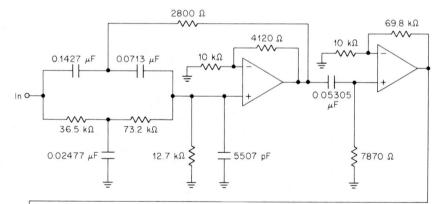
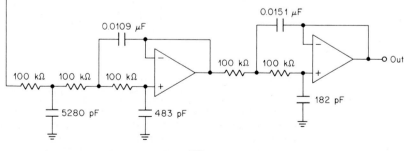

(f)

Fig. 5-22 *Continued*

Example 5-10

REQUIRED: Active bandpass filter
1 dB maximum variation from 300 to 900 Hz
35 dB minimum below 100 Hz and above 1800 Hz
Gain of +20 dB

RESULT: (a)   Since the ratio of upper cutoff frequency to lower cutoff frequency is well in excess of an octave, the design will be treated as a cascade of low-pass and high-pass filters. The frequency-response requirement can be restated as the following set of individual low-pass and high-pass specifications:

High-pass filter:	Low-pass filter:
1 dB maximum at 300 Hz	1 dB maximum at 900 Hz
35 dB minimum below 100 Hz	35 dB minimum above 1800 Hz

(b)   To design the high-pass filter, first compute the high-pass steepness factor.

$$A_s = \frac{f_c}{f_s} = \frac{300 \text{ Hz}}{100 \text{ Hz}} = 3 \qquad (2\text{-}13)$$

A normalized low-pass filter must now be chosen that makes the transition from less than 1 dB to more than 35 dB within a frequency ratio of $3:1$. An elliptic-function type will be selected. The curves of figure 2-86 indicate that for a ρ of 25% ($R_{dB} = 0.28$ dB) a third-order filter will provide over 35 dB of attenuation. The following parameters are found in table 12-57 for a filter corresponding to CO3 25 $\theta = 20°$:

$$n = 3$$
$$R_{dB} = 0.28 \text{ dB}$$
$$\Omega_s = 2.924$$
$$A_{min} = 39.48 \text{ dB}$$

The normalized low-pass filter from table 12-57 is shown in figure 5.22a.

To transform the normalized low-pass filter into a high-pass structure, the filter elements are converted to reciprocal values. Capacitors are replaced by resistors and vice versa. The resulting normalized high-pass filter is given in figure 5-22b. To denormalize the filter network, all capacitors are divided by $Z \times$ FSF, where Z is arbitrarily chosen at 10^4 and FSF is given by $2\pi f_c$ or 1885, where f_c is 300 Hz. The resistors are multiplied by Z. The denormalized high-pass filter is given in figure 5-22c.

(c)   The low-pass filter is now designed. The low-pass steepness factor is computed by

$$A_s = \frac{f_s}{f_c} = \frac{1800 \text{ Hz}}{900 \text{ Hz}} = 2 \qquad (2\text{-}11)$$

A low-pass filter must be selected that makes the transition from less than 1 dB to more than 35 dB within a frequency ratio of $2:1$. The curves of figure 2-44 indicate that the attenuation of a normalized 0.5-dB Chebyshev filter of a complexity of $n = 5$ is less than 1 dB at 0.9 rad/s and more than 35 dB at 1.8 rad/s, which satisfies the requirements. The corresponding active filter is found in table 12-39 and is shown in figure 5-22d.

To denormalize the low-pass circuit, first compute the FSF, which is given by

$$\text{FSF} = \frac{\text{desired reference frequency}}{\text{existing reference frequency}}$$
$$= \frac{2\pi \times 900 \text{ rad/s}}{0.9 \text{ rad/s}} = 6283 \tag{2-1}$$

The filter is then denormalized by dividing all capacitors by $Z \times \text{FSF}$ and multiplying all resistors by Z, where Z is arbitrarily chosen at 10^5. The denormalized circuit is shown in figure 5-22e.

(d) To complete the design, the low-pass and high-pass filters are cascaded. Since the real-pole RC network of the elliptic low-pass filter must be buffered and since a gain of $+20$ dB is required, the noninverting amplifier of figure 5-21b will be used.

A gain of 1.263 or $+2$ dB is tabulated in table 12-57 alongside the element values of the active elliptic-function filter chosen. An additional gain of $+18$ dB is then needed. The ratio of $R_2/R_1 + 1$ is then equal to 7.94 corresponding to 18 dB. If we choose R_1 to be 10 kΩ, the value of R_2 is determined as 69.4 kΩ.

The finalized design is shown in figure 5-22f, where the resistors have been rounded off to standard 1% values.

Bandpass Transformation of Low-Pass Poles and Zeros

Active bandpass filters are designed directly from a bandpass transfer function. To obtain the bandpass poles and zeros from the low-pass transfer function, a low-pass to bandpass transformation must be performed. It was shown in section 5.1 how this transformation can be accomplished by replacing the frequency variable by a new variable which was given by

$$f_{bp} = f_0 \left(\frac{f}{f_0} - \frac{f_0}{f} \right) \tag{5-1}$$

This substitution maps the low-pass frequency response into a bandpass magnitude characteristic.

Two sets of bandpass poles are obtained from each low-pass complex pole pair. If the low-pass pole is real, a single pair of complex poles result for the bandpass case. Also, each pair of imaginary axis zeros is transformed into two pairs of conjugate zeros. This is shown in figure 5-23.

Clearly, the total number of poles and zeros is doubled when the bandpass transformation is performed. However, it is conventional to disregard the conjugate bandpass poles and zeros below the real axis. An n-pole low-pass filter is said to result in an nth-order bandpass filter even though the bandpass transfer function is of the order $2n$. An nth-order active bandpass filter will then consist of n bandpass sections.

Each all-pole bandpass section has a second-order transfer function given by

$$T(s) = \frac{Hs}{s^2 + \dfrac{\omega_r}{Q} s + \omega_r^2} \tag{5-48}$$

where ω_r is equal to $2\pi f_r$, the pole resonant frequency in radians per second, Q is the bandpass section Q, and H is a gain constant.

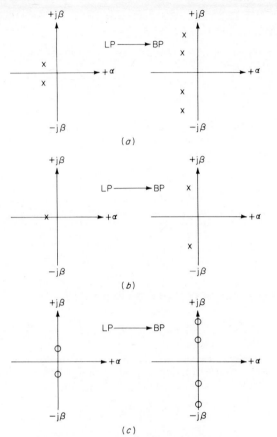

Fig. 5-23 Low-pass to bandpass transformation: (a) low-pass complex pole pair; (b) low-pass real pole; (c) low-pass pair of imaginary zeros.

If transmission zeros are required, the section transfer function will then take the form

$$T(s) = \frac{H(s^2 + \omega_\infty^2)}{s^2 + \frac{\omega_r}{Q} s + \omega_r^2} \tag{5-49}$$

where ω_∞ is equal to $2\pi f_\infty$, the frequency of the transmission zero in radians per second.

Active bandpass filters are designed by the following sequence of operations:

1. Convert the bandpass specification to a geometrically symmetrical requirement as described in section 2.1.
2. Calculate the bandpass steepness factor A_s using equation (2-19) and select a normalized filter type from chapter 2.
3. Look up the corresponding normalized poles (and zeros) from the tables of chapter 12 and transform these coordinates into bandpass parameters.

4. Select the appropriate bandpass circuit configuration from the types pre-
    sented in this chapter and cascade the required number of sections.

It is convenient to specify each bandpass filter section in terms of its center
frequency and Q. Elliptic-function filters will also require zeros. These parameters
can be directly transformed from the poles (and zeros) of the normalized low-
pass transfer function. A numerical procedure will be described for making
this transformation.

First make the preliminary calculation

$$Q_{bp} = \frac{f_0}{BW} \qquad (2\text{-}16)$$

where f_0 is the geometric bandpass center frequency and BW is the passband
bandwidth. The bandpass transformation is made as follows:

Complex Poles   Complex poles occur in the tables of chapter 12 having
the form

$$-\alpha \pm j\beta$$

where α is the real coordinate and β is the imaginary part. Given α, β, Q_{bp},
and f_0, the following series of calculations results in two sets of values for Q
and center frequency which define a pair of bandpass filter sections:

$$C = \alpha^2 + \beta^2 \qquad (5\text{-}50)$$

$$D = \frac{2\alpha}{Q_{bp}} \qquad (5\text{-}51)$$

$$E = \frac{C}{Q_{bp}^2} + 4 \qquad (5\text{-}52)$$

$$G = \sqrt{E^2 - 4D^2} \qquad (5\text{-}53)$$

$$Q = \sqrt{\frac{E+G}{2D^2}} \qquad (5\text{-}54)$$

$$M = \frac{\alpha Q}{Q_{bp}} \qquad (5\text{-}55)$$

$$W = M + \sqrt{M^2 - 1} \qquad (5\text{-}56)$$

$$f_{ra} = \frac{f_0}{W} \qquad (5\text{-}57)$$

$$f_{rb} = W f_0 \qquad (5\text{-}58)$$

The two bandpass sections have resonant frequencies of f_{ra} and f_{rb} (in hertz)
and identical Q's as given by equation (5-54).

Real Poles   A normalized low-pass real pole having a real coordinate of
magnitude α_0 is transformed into a single bandpass section having a Q defined
by

$$Q = \frac{Q_{bp}}{\alpha_0} \qquad (5\text{-}59)$$

The section is tuned to f_0, the geometric center frequency of the filter.

Imaginary Zeros   Elliptic-function low-pass filters contain transmission zeros as well as poles. These zeros must be transformed along with the poles when a bandpass filter is required. Imaginary-axis zeros are tabulated alongside the low-pass active-elements values in table 12-57 and are of the form $\pm j\omega_\infty$. The bandpass zeros can be obtained as follows:

$$H = \frac{\omega_\infty^2}{2\,Q_{bp}^2} + 1 \tag{5-60}$$

$$Z = \sqrt{H + \sqrt{H^2 - 1}} \tag{5-61}$$

$$f_{\infty,a} = \frac{f_0}{Z} \tag{5-62}$$

$$f_{\infty,b} = Z \times f_0 \tag{5-63}$$

A pair of imaginary bandpass zeros are obtained occurring at $f_{\infty,a}$ and $f_{\infty,b}$ (in hertz) from each low-pass zero.

Determining Section Gain   The gain of a single bandpass section at the filter geometric center frequency f_0 is given by

$$A_0 = \frac{A_r}{\sqrt{1 + Q^2\left(\dfrac{f_0}{f_r} - \dfrac{f_r}{f_0}\right)^2}} \tag{5-64}$$

where A_r is the section gain at its resonant frequency f_r. The section gain will always be less at f_0 than at f_r, since the circuit is peaked to f_r, except for transformed real poles where $f_r = f_0$. Equation (5-64) will then simplify to $A_0 = A_r$. The composite filter gain is determined by the product of the A_0 values of all the sections.

If the section Q is relatively high ($Q > 10$), equation (5-64) can be simplified to

$$A_0 = \frac{A_r}{\sqrt{1 + \left(\dfrac{2\,Q\Delta f}{f_r}\right)^2}} \tag{5-65}$$

where Δf is the frequency separation between f_0 and f_r.

Example 5-11

REQUIRED: Determine the pole locations and section gains for a third-order Butterworth bandpass filter having a geometric center frequency of 1000 Hz, a 3-dB bandwidth of 100 Hz, and a midband gain of +30 dB.

RESULT: (a) The normalized pole locations for an $n = 3$ Butterworth low-pass filter are obtained from table 12-1 and are
$$-0.500 \pm j0.8660$$
$$-1.000$$
To obtain the bandpass poles first compute

$$Q_{bp} = \frac{f_0}{BW_{3\,dB}} = \frac{1000\ \text{Hz}}{100\ \text{Hz}} = 10 \tag{2-16}$$

The low-pass to bandpass pole transformation is performed as follows:

Complex Pole:

$$\alpha = 0.5000 \qquad \beta = 0.8660$$
$$C = \alpha^2 + \beta^2 = 1.000000 \tag{5-50}$$

$$D = \frac{2\alpha}{Q_{bp}} = 0.100000 \tag{5-51}$$

$$E = \frac{C}{Q_{bp}^2} + 4 = 4.010000 \tag{5-52}$$

$$G = \sqrt{E^2 - 4D^2} = 4.005010 \tag{5-53}$$

$$Q = \sqrt{\frac{E+G}{2D^2}} = 20.018754 \tag{5-54}$$

$$M = \frac{\alpha Q}{Q_{bp}} = 1.000938 \tag{5-55}$$

$$W = M + \sqrt{M^2 - 1} = 1.044261 \tag{5-56}$$

$$f_{ra} = \frac{f_0}{W} = 957.6 \text{ Hz} \tag{5-57}$$

$$f_{rb} = Wf_0 = 1044.3 \text{ Hz} \tag{5-58}$$

Real pole:

$$\alpha_0 = 1.0000$$

$$Q = \frac{Q_{bp}}{\alpha_0} = 10 \tag{5-59}$$

$$f_r = f_0 = 1000 \text{ Hz}$$

(b) Since a composite midband gain of +30 dB is required, let us distribute the gain uniformly among the three sections. Therefore, $A_0 = 3.162$ for each section corresponding to +10 dB.

The gain at section resonant frequency f_r is obtained from the following form of equation (5-64):

$$A_r = A_0 \sqrt{1 + Q^2 \left(\frac{f_0}{f_r} - \frac{f_r}{f_0}\right)^2}$$

The resulting values are

Section 1: $f_r = 957.6 \text{ Hz}$

         $Q = 20.02$
         $A_r = 6.333$

Section 2: $f_r = 1044.3 \text{ Hz}$

         $Q = 20.02$
         $A_r = 6.335$

Section 3: $f_r = 1000.0 \text{ Hz}$

         $Q = 10.00$
         $A_r = 3.162$

The block diagram of the realization is shown in figure 5-24.

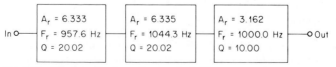

Fig. 5-24 Block realization of example 5-11.

The calculations required for the bandpass pole transformation should be maintained to more than four significant figures after the decimal point to obtain accurate results, since differences of close numbers are involved. Equations (5-55) and (5-56) are especially critical; so the value of M should be computed to five or six places after the decimal point.

Sensitivity in Active Bandpass Circuits

Sensitivity defines the amount of change of a dependent variable resulting from the variation of an independent variable. Mathematically, the sensitivity of y with respect to x is expressed as

$$S_x^y = \frac{dy/y}{dx/x} \qquad (5\text{-}66)$$

Sensitivity is used as a figure of merit to measure the change in a particular filter parameter such as Q or resonant frequency for a given change in a component value.

Deviations of components from their nominal values occur because of the effects of temperature, aging, humidity, and other environmental conditions in addition to the errors because of tolerances. These variations cause changes in parameters such as Q and center frequency from their design values.

As an example let us assume we are given the parameter $S_{R_1}^Q = -3$ for a particular circuit. This means that for a 1% increment of R_1, the circuit Q will change 3% in the opposite direction.

In addition to component value sensitivity, the operation of a filter is dependent on the active elements as well. The Q and resonant frequency can be a function of amplifier open-loop gain and phase shift; so the sensitivity to these active parameters is useful in determining an amplifier's suitability for a particular design.

The Q sensitivity of a circuit is a good measure of its stability. With some circuits the Q can increase to infinity, which implies a self-oscillation. Low Q sensitivity of a circuit usually indicates that the configuration will be practical from a stability point of view.

Sometimes the sensitivity is expressed as an equation instead of a numerical value such as $S_A^Q = 2Q^2$. This expression implies that the sensitivity of Q with respect to amplifier gain A increases with Q^2; so the circuit is not suitable for high Q realizations.

The frequency-sensitivity parameters of a circuit are useful in determining whether the circuit will require resistive trimming and indicate which element should be made variable. It should be mentioned that in general, only resonant frequency is made adjustable when the bandpass filter is sufficiently narrow. Q variations of 5 or 10% are usually tolerable, whereas a comparable frequency error would be disastrous in narrow filters. However, in the case of a state-variable realization, a Q-enhancement effect occurs caused by amplifier phase shift. The Q may increase very dramatically; so Q adjustment is usually required in addition to resonant frequency.

All-Pole Bandpass Configurations

Multiple-Feedback Bandpass (MFBP)   The circuit of figure 5-25a realizes a bandpass pole pair and is commonly referred to as a multiple-feedback bandpass

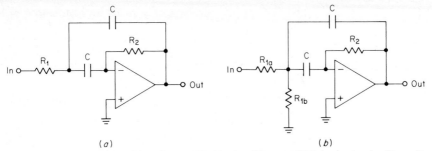

Fig. 5-25 Multiple-feedback bandpass (MFBP) ($Q < 20$): (a) MFBP basic circuit; (b) modified configuration.

(MFBP) configuration. This circuit features a minimum number of components and low sensitivity to component tolerances. The transfer function is given by

$$T(s) = -\frac{sC/R_1}{s^2C^2 + s2C/R_2 + 1/R_1R_2} \tag{5-67}$$

If we equate the coefficients of this transfer function with the general bandpass transfer function of equation (5-48), we can derive the following expressions for the element values:

$$R_2 = \frac{Q}{\pi f_r C} \tag{5-68}$$

and

$$R_1 = \frac{R_2}{4Q^2} \tag{5-69}$$

where C is arbitrary.

The circuit gain at resonant frequency f_r is given by

$$A_r = 2Q^2 \tag{5-70}$$

The open-loop gain of the operational amplifier at f_r should be well in excess of $2Q^2$, so that the circuit performance is controlled mainly by the passive elements. This requirement places a restriction on realizable Q's to values typically below 20, depending upon the amplifier type and frequency range.

Extremely high gains occur for moderate Q values because of the Q^2 gain proportionality; so there will be a tendency for clipping at the amplifier output for moderate input levels. Also the circuit gain is fixed by the Q, which limits flexibility.

An alternate and preferred form of the circuit is shown in figure 5-25 b. The input resistor R_1 has been split into two resistors, R_{1a} and R_{1b}, to form a voltage divider so that the circuit gain can be controlled. The parallel combination of the two resistors is equal to R_1 to retain the resonant frequency. The transfer function of the modified circuit is given by

$$T(s) = -\frac{sR_2C}{s^2R_{1a}R_2C^2 + s2R_{1a}C + (1 + R_{1a}/R_{1b})} \tag{5-71}$$

The values of R_{1a} and R_{1b} are computed from

$$R_{1a} = \frac{R_2}{2A_r} \tag{5-72}$$

and
$$R_{1b} = \frac{R_2/2}{2\,Q^2 - A_r} \tag{5-73}$$

where A_r is the desired gain at resonant frequency f_r and cannot exceed $2\,Q^2$. The value of R_2 is still computed from equation (5-68).

The circuit sensitivities can be determined as follows:

$$S_{R_{1a}}^{Q} = S_{R_{1a}}^{f_r} = \frac{A_r}{4\,Q^2} \tag{5-74}$$

$$S_{R_{1b}}^{Q} = S_{R_{1b}}^{f_r} = \frac{1}{2}\,(1 + A_r/2\,Q^2) \tag{5-75}$$

$$S_{R_2}^{f_r} = S_{C}^{f_r} = -\frac{1}{2} \tag{5-76}$$

$$S_{R_2}^{Q} = \frac{1}{2} \tag{5-77}$$

For $Q^2/A_r \gg 1$, the resonant frequency can be directly controlled by R_{1b}, since $S_{R_{1b}}^{f_r}$ approaches ½. To use this result, let us assume that the capacitors have 2% tolerances and the resistors have a tolerance of 1%, which could result in a possible 3% frequency error. If frequency adjustment is desired, R_{1b} should be made variable over a minimum resistance range of ±6%. This would then permit a frequency adjustment of ±3%, since $S_{R_{1b}}^{f_r}$ is equal to ½. Resistor R_{1b} should be composed of a fixed resistor in series with a single-turn potentiometer to provide good resolution.

Adjustment of Q can be accomplished by making R_2 adjustable. However, this will affect resonant frequency and in any event is not necessary for most filters if 1 or 2% tolerance parts are used. The section gain can be varied by making R_{1a} adjustable, but again resonant frequency may be affected.

In conclusion, this circuit is highly recommended for low Q requirements. Although a large spread in resistance values can occur and the Q is limited by amplifier gain, the circuit simplicity, low element sensitivity, and ease of frequency adjustment make it highly desirable.

The following example demonstrates the design of a bandpass filter using the MFBP configuration.

Example 5-12

REQUIRED: Design an active bandpass filter having the following specifications:
Center frequency of 300 Hz
3 dB at ±10 Hz
25 dB minimum at ±40 Hz
Essentially zero overshoot to a 300-Hz carrier pulse step
Gain of +12 dB at 300 Hz

RESULT: (a) Since the bandwidth is narrow, the requirement can be treated on an arithmetically symmetrical basis. The bandpass steepness factor is given by

$$A_s = \frac{\text{stopband bandwidth}}{\text{passband bandwidth}} = \frac{80\text{ Hz}}{20\text{ Hz}} = 4 \tag{2-19}$$

The curves of figures 2-69 and 2-74 indicate that an $n = 3$ transitional gaussian to 6-dB filter will meet the frequency- and step-response requirements.

(b) The pole locations for the corresponding normalized low-pass filter are found in table 12-50 and are as follows:

$$-0.9622 \pm j1.2214$$
$$-0.9776$$

First compute the bandpass Q:

$$Q_{bp} = \frac{f_0}{BW_{3\,dB}} = \frac{300 \text{ Hz}}{20 \text{ Hz}} = 15 \qquad (2\text{-}16)$$

The low-pass poles are transformed to the bandpass form in the following manner:

Complex pole:

$$\alpha = 0.9622 \qquad \beta = 1.2214$$

$C = 2.417647$	(5-50)
$D = 0.128293$	(5-51)
$E = 4.010745$	(5-52)
$G = 4.002529$	(5-53)
$Q = 15.602243$	(5-54)
$M = 1.000832$	(5-55)
$W = 1.041630$	(5-56)
$f_{ra} = 288.0 \text{ Hz}$	(5-57)
$f_{rb} = 312.5 \text{ Hz}$	(5-58)

Real pole:

$\alpha_0 = 0.9776$	
$Q = 15.34$	(5-59)
$f_r = 300.0 \text{ Hz}$	

(c) A midband gain of +12 dB is required. Let us allocate a gain of +4 dB to each section corresponding to $A_0 = 1.585$. The value of A_r, the resonant frequency gain for each section, is obtained from equation (5-64) and is listed in the following table, which summarizes the design parameters of the filter sections:

	f_r	Q	A_r
Section 1	288.0 Hz	15.60	2.567
Section 2	312.5 Hz	15.60	2.567
Section 3	300.0 Hz	15.34	1.585

(d) Three MFBP bandpass sections will be connected in tandem. The following element values are computed where C is set equal to 0.1 μF:

Section 1:

$$R_2 = \frac{Q}{\pi f_r C} = \frac{15.6}{\pi \times 288 \times 10^{-7}} = 172.4 \text{ k}\Omega \qquad (5\text{-}68)$$

$$R_{1a} = \frac{R_2}{2A_r} = \frac{172.4 \times 10^3}{2 \times 2.567} = 33.6 \text{ k}\Omega \qquad (5\text{-}72)$$

$$R_{1b} = \frac{R_2/2}{2Q^2 - A_r} = \frac{86.2 \times 10^3}{2 \times 15.6^2 - 2.567} = 178 \text{ }\Omega \quad (5\text{-}73)$$

Section 2:

$R_2 = 158.9 \text{ k}\Omega$
$R_{1a} = 30.9 \text{ k}\Omega$
$R_{1b} = 164 \text{ }\Omega$

Section 3:

$R_2 = 162.8 \text{ k}\Omega$
$R_{1a} = 51.3 \text{ k}\Omega$
$R_{1b} = 174 \text{ }\Omega$

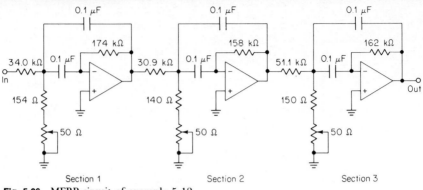

Fig. 5-26   MFBP circuit of example 5-12.

The final circuit is shown in figure 5-26. Resistor values have been rounded off to standard 1% values, and resistor R_{1b} has been made variable in each section for tuning purposes.

Each filter section can be adjusted by applying a sine wave at the section f_r to the filter input. The phase shift of the section being adjusted is monitored by connecting one channel of an oscilloscope to the section input and the other channel to the section output. A Lissajous pattern is obtained. Resistor R_{1b} is adjusted until the ellipse closes to a straight line.

Dual-Amplifier Bandpass (DABP) Structure   The bandpass circuit of figure 5-27 was first introduced by Sedra and Espinoza (see references). Truly remarkable performance in terms of available Q, low sensitivity, and flexibility can be obtained in comparison with alternate schemes involving two amplifiers.
The transfer function is given by

$$T(s) = \frac{s2/R_1C}{s^2 + s1/R_1C + 1/R_2R_3C^2}$$ (5-78)

If we compare this expression with the general bandpass transfer function of equation (5-48) and let $R_2R_3 = R^2$, the following design equations for the element values can be obtained.
First compute

$$R = \frac{1}{2\pi f_r C}$$ (5-79)

then $$R_1 = QR$$ (5-80)

$$R_2 = R_3 = R$$ (5-81)

where C is arbitrary. The value of R' in figure 5-27 can also be chosen at any convenient value. Circuit gain at f_r is equal to 2.
The following sensitivities can be derived:

$$S_{R_1}^Q = 1$$ (5-82)

$$S_{R_2}^{f_r} = S_{R_3}^{f_r} = S_{R_4}^{f_r} = S_C^{f_r} = -\tfrac{1}{2}$$ (5-83)

$$S_{R_5}^{f_r} = \tfrac{1}{2}$$ (5-84)

An interesting result of sensitivity studies is that if the bandwidths of both amplifiers are nearly equivalent, extremely small deviations of Q from the design values will occur. This is especially advantageous at higher frequencies where the amplifier poles have to be taken into account. It is then suggested that a dual-type amplifier be used for each filter section, since both amplifier halves will be closely matched to each other.

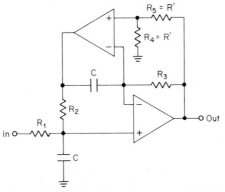

Fig. 5-27 Dual-amplifier bandpass (DABP) configuration ($Q < 150$).

A useful feature of this circuit is that resonant frequency and Q can be independently adjusted. Alignment can be accomplished by first adjusting R_2 for resonance at f_r. Resistor R_1 can then be adjusted for the desired Q without affecting the resonant frequency.

Since each section provides a fixed gain of 2 at f_r, a composite filter may require an additional amplification stage if higher gains are required. If a gain reduction is desired, resistor R_1 can be split into two resistors to form a voltage divider in the same manner as in figure 5-25b. The resulting values are

$$R_{1a} = \frac{2R_1}{A_r} \tag{5-85}$$

and

$$R_{1b} = \frac{R_{1a} A_r}{2 - A_r} \tag{5-86}$$

where A_r is the desired gain at resonance.

The spread of element values of the MFBP section previously discussed is equal to $4Q^2$. In comparison this circuit has a ratio of resistances determined by Q; so the spread is much less.

The DABP configuration has been found very useful for designs covering a wide range of Q's and frequencies. Component sensitivity is small, resonant frequency and Q are easily adjustable, and the element spread is low. The following example illustrates the use of this circuit.

Example 5-13

REQUIRED: Design an active bandpass filter to meet the following specifications:
Center frequency of 3000 Hz
3 dB at ±30 Hz
20 dB minimum at ±120 Hz

RESULT: (a)   If we consider the requirement as being arithmetically symmetrical, the bandpass steepness factor becomes

$$A_s = \frac{\text{stopband bandwidth}}{\text{passband bandwidth}} = \frac{240 \text{ Hz}}{60 \text{ Hz}} = 4 \qquad (2\text{-}19)$$

We can determine from the curve of figure 2-34 that a second-order Butterworth low-pass filter provides over 20 dB of rejection within a frequency ratio of 4:1. The corresponding poles of the normalized low-pass filter are found in table 12-1 and are as follows:

$$-0.7071 \pm j0.7071$$

(b)   To convert these poles to the bandpass form first compute:

$$Q_{bp} = \frac{f_0}{BW_{3 \text{ dB}}} = \frac{3000 \text{ Hz}}{60 \text{ Hz}} = 50 \qquad (2\text{-}16)$$

The bandpass pole transformation is performed in the following manner:

$$\alpha = 0.7071 \qquad \beta = 0.7071$$

$$C = 1.000000 \qquad\qquad\qquad\qquad (5\text{-}50)$$

$$D = 0.028284 \qquad\qquad\qquad\qquad (5\text{-}51)$$

$$E = 4.000400 \qquad\qquad\qquad\qquad (5\text{-}52)$$

$$G = 4.000000 \qquad\qquad\qquad\qquad (5\text{-}53)$$

$$Q = 70.713124 \qquad\qquad\qquad\qquad (5\text{-}54)$$

$$M = 1.000025 \qquad\qquad\qquad\qquad (5\text{-}55)$$

$$W = 1.007096 \qquad\qquad\qquad\qquad (5\text{-}56)$$

$$f_{ra} = 2978.9 \text{ Hz} \qquad\qquad\qquad\qquad (5\text{-}57)$$

$$f_{rb} = 3021.3 \text{ Hz} \qquad\qquad\qquad\qquad (5\text{-}58)$$

(c)   Two DABP sections will be used. The element values are now computed, where C is set equal to 0.01 μF and R' is 10 kΩ.

Section 1:

$$f_r = 2978.9 \text{ Hz}$$

$$Q = 70.7$$

$$R = \frac{1}{2\pi f_r C} = \frac{1}{2\pi \times 2978.9 \times 10^{-8}} = 5343 \text{ } \Omega \qquad (5\text{-}79)$$

$$R_1 = QR = 70.7 \times 5343 = 377.7 \text{ k}\Omega \qquad (5\text{-}80)$$

$$R_2 = R_3 = R = 5343 \text{ } \Omega \qquad (5\text{-}81)$$

Section 2:

$$f_r = 3021.3 \text{ Hz}$$

$$Q = 70.7$$

$$R = 5268 \text{ } \Omega \qquad\qquad\qquad (5\text{-}79)$$

$$R_1 = 372.4 \text{ k}\Omega \qquad\qquad\qquad (5\text{-}80)$$

$$R_2 = R_3 = 5268 \text{ } \Omega \qquad\qquad\qquad (5\text{-}81)$$

The circuit is illustrated in figure 5-28, where resistors have been rounded off to standard 1% values and R_2 is made adjustable for tuning.

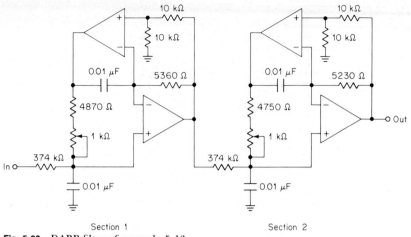

Section 1   Section 2

Fig. 5-28 DABP filter of example 5-13.

State-Variable (Biquad) All-Pole Circuit The state-variable or biquad configuration was first introduced in section 3.2 for use as a low-pass filter section. A bandpass output is available as well. The biquad approach features excellent sensitivity properties and the capability to control resonant frequency and Q independently. It is especially suited for constructing precision-active filters in a standard form. A variety of manufacturers offer hybrid biquad filter sections where frequency and Q are determined by a few external resistors.

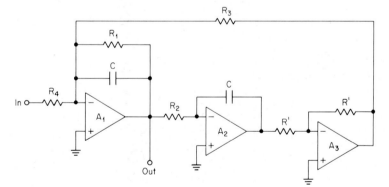

Fig. 5-29 Biquad all-pole circuit ($Q < 200$).

The circuit of figure 5-29 is the all-pole bandpass form of the general biquadratic configuration. The transfer function is given by

$$T(s) = -\frac{s/CR_4}{s^2 + s/CR_1 + 1/R_2R_3C^2} \qquad (5\text{-}87)$$

If we equate this expression to the general bandpass transfer function of equation (5-48), the circuit resonant frequency and 3-dB bandwidth can be expressed as

$$f_r = \frac{1}{2\pi \, C\sqrt{R_2 R_3}} \tag{5-88}$$

and

$$BW_{3\,dB} = \frac{1}{2\pi \, R_1 C} \tag{5-89}$$

where $BW_{3\,dB}$ is equal to f_r/Q.

Equations (5-88) and (5-89) indicate that the resonant frequency and 3-dB bandwidth can be independently controlled. This feature is highly desirable and can lead to many useful applications such as variable filters.

If we substitute f_r/Q for $BW_{3\,dB}$ and set $R_2 = R_3$, the following design equations can be derived for the section:

$$R_1 = \frac{Q}{2\pi f_r C} \tag{5-90}$$

$$R_2 = R_3 = \frac{R_1}{Q} \tag{5-91}$$

and

$$R_4 = \frac{R_1}{A_r} \tag{5-92}$$

where A_r is the desired gain at resonant frequency f_r. The values of C and R' in figure 5-29 can be conveniently selected. By making R_3 and R_1 adjustable, the resonant frequency and Q respectively can be adjusted.

The sensitivity factors are

$$S_{R_2}^{f_r} = S_{R_3}^{f_r} = S_C^{f_r} = -\tfrac{1}{2} \tag{5-93}$$

$$S_{R_1}^Q = 1 \tag{5-94}$$

and

$$S_\mu^Q = \frac{2Q}{\mu} \tag{5-95}$$

where μ is the open-loop gain of amplifiers A_1 and A_2. The section Q is then limited by the finite gain of the operational amplifier.

Another serious limitation occurs because of finite amplifier bandwidth. Thomas (see references) has shown that as the resonant frequency increases for a fixed design Q, the actual Q remains constant over a broad band and then begins to increase, eventually becoming infinite (oscillatory). This effect is called "Q enhancement."

If we assume that the open-loop transfer function of the amplifier has a single pole, the effective Q can be approximated by

$$Q_{eff} = \frac{Q_{design}}{1 - \dfrac{2\,Q_{design}}{\mu_0 \omega_c}(2\omega_r - \omega_c)} \tag{5-96}$$

where ω_r is the resonant frequency, ω_c is the 3-dB breakpoint of the open-loop amplifier gain, and μ_0 is the open-loop gain at DC. As ω_r is increased, the denominator approaches zero.

The Q-enhancement effect can be minimized by having a high gain-bandwidth product. If the amplifier requires external frequency compensation, the compensation can be made lighter than the recommended values. The state-variable circuit is well suited for light compensation, since the structure contains two integrators which have a stabilizing effect.

A solution suggested by Thomas is to introduce a leading phase component in the feedback loop which compensates for the lagging phase caused by finite amplifier bandwidth. This can be achieved by introducing a capacitor in parallel with resistor R_3 having the value

$$C_p = \frac{4}{\mu_0 \omega_c R_3} \tag{5-97}$$

Probably the most practical solution is to make resistor R_1 variable. The Q may be determined by measuring the 3-dB bandwidth. R_1 is adjusted until the ratio $f_r/BW_{3\,dB}$ is equal to the required Q.

As the Q is enhanced, the section gain is increased also. Empirically it has been found that correcting for the gain enhancement compensates for the Q enhancement as well. R_1 can be adjusted until the measured gain at f_r is equal to the design value of A_r used in equation (5-92). Although this technique is not as accurate as determining the actual Q from the 3-dB bandwidth, it certainly is much more convenient and will usually be sufficient.

The biquad is a low-sensitivity filter configuration suitable for precision applications. Circuit Q's of up to 200 are realizable over a broad frequency range. The following example demonstrates the use of this structure.

Example 5-14

REQUIRED: Design an active bandpass filter which satisfies the following specifications:

Center frequency of 2500 Hz
3 dB at ±15 Hz
15 dB minimum at ±45 Hz
Gain of +12 dB at 2500 Hz

RESULT: (a)   The bandpass steepness factor is determined from

$$A_s = \frac{\text{stopband bandwidth}}{\text{passband bandwidth}} = \frac{90 \text{ Hz}}{30 \text{ Hz}} = 3 \tag{2-19}$$

Using figure 2-42, we find that a second-order 0.1-dB Chebyshev normalized low-pass filter will meet the attenuation requirements. The corresponding poles are found in table 12-23 and are as follows:

$$-0.6125 \pm j0.7124$$

(b)   To transform these low-pass poles to the bandpass form first compute

$$Q_{bp} = \frac{f_0}{BW_{3\,dB}} = \frac{2500 \text{ Hz}}{30 \text{ Hz}} = 83.33 \tag{2-16}$$

The bandpass poles are determined from the following series of computations. Since the filter is very narrow, an extended number of significant figures will be used in equations (5-50) through (5-58) to maintain accuracy.

$$\alpha = 0.6125 \qquad \beta = 0.7124$$

$$C = 0.882670010 \tag{5-50}$$

$$D = 0.014700588 \tag{5-51}$$

$$E = 4.000127115 \tag{5-52}$$

$$G = 4.000019064 \tag{5-53}$$
$$Q = 136.0502228 \tag{5-54}$$
$$M = 1.000009138 \tag{5-55}$$
$$W = 1.004284182 \tag{5-56}$$
$$f_{ra} = 2489.3 \text{ Hz} \tag{5-57}$$
$$f_{rb} = 2510.7 \text{ Hz} \tag{5-58}$$

(c) Since a midband gain of +12 dB is required, each section will be allocated a midband gain of +6 dB corresponding to $A_0 = 2.000$. The gain A_r at the resonant frequency of each section is determined from equation (5-65) and is listed in the following table:

	f_r	Q	A_r
Section 1	2489.3 Hz	136	3.069
Section 2	2510.7 Hz	136	3.069

(d) Two biquad sections in tandem will be used. C is chosen to be 0.1 μF and R' is 10 kΩ. The element values are computed as follows:

Section 1:

$$R_1 = \frac{Q}{2\pi f_r C} = \frac{136}{2\pi \times 2489.3 \times 10^{-7}} = 86.9 \text{ k}\Omega \tag{5-90}$$

$$R_2 = R_3 = \frac{R_1}{Q} = \frac{86.9 \times 10^3}{136} = 639 \ \Omega \tag{5-91}$$

$$R_4 = \frac{R_1}{A_r} = \frac{86.9 \times 10^3}{3.069} = 28.3 \text{ k}\Omega \tag{5-92}$$

Section 2:

$$R_1 = 86.2 \text{ k}\Omega \tag{5-90}$$
$$R_2 = R_3 = 634 \ \Omega \tag{5-91}$$
$$R_4 = 28.1 \text{ k}\Omega \tag{5-92}$$

The final circuit is shown in figure 5-30. Resistors R_3 and R_1 are made variable so that resonant frequency and Q can be adjusted. Standard values of 1% resistors have been used.

The Q-Multiplier Approach   Certain active bandpass structures such as the MFBP configuration of section 5.2 are severely Q-limited because of insufficient amplifier gain or other inadequacies. The technique outlined in this section uses a low-Q-type bandpass circuit within a Q-multiplier structure which increases the circuit Q to the desired value.

A bandpass transfer function having unity gain at resonance can be expressed as

$$T(s) = \frac{\dfrac{\omega_r}{Q} s}{s^2 + \dfrac{\omega_r}{Q} s + \omega_r^2} \tag{5-98}$$

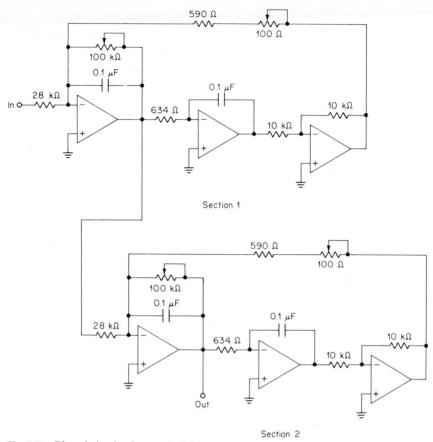

Fig. 5-30 Biquad circuit of example 5-14.

If the corresponding circuit is combined with a summing amplifier in the manner shown in figure 5-31a, where β is an attenuation factor, the following overall transfer function can be derived:

$$T(s)=\frac{\dfrac{\omega_r}{Q}s}{s^2+\dfrac{\dfrac{\omega_r}{Q}}{1-\beta}s+\omega_r^2} \tag{5-99}$$

The middle term of the denominator has been modified so that the circuit Q is given by $Q/(1-\beta)$, where $0 < \beta < 1$. By selecting a β sufficiently close to unity, the Q can be increased by the factor $1/(1-\beta)$. The circuit gain is also increased by the same factor.

If we use the MFBP section for the bandpass circuit, the Q-multiplier configuration will take the form of figure 5-31b. Since the MFBP circuit is inverting, an inverting amplifier can also be used for summing.

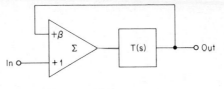

(a)

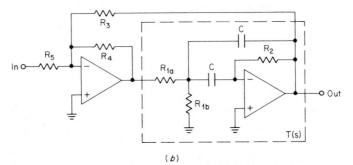

(b)

Fig. 5-31 Q-multiplier circuit: (a) block diagram; (b) realization using MFBP section.

The value of β can be found from

$$\beta = 1 - \frac{Q_r}{Q_{\text{eff}}} \tag{5-100}$$

where Q_{eff} is the effective overall Q and Q_r is the design Q of the bandpass section. The component values are determined by the following equations:

$$R_3 = \frac{R}{\beta} \tag{5-101}$$

$$R_4 = R \tag{5-102}$$

and

$$R_5 = \frac{R}{(1 - \beta) A_r} \tag{5-103}$$

where R can be conveniently chosen and A_r is the desired gain at resonance.

Design equations for the MFBP section were derived in section 5.2 and are repeated here corresponding to unity gain.

$$R_2 = \frac{Q_r}{\pi f_r C} \tag{5-68}$$

$$R_{1a} = \frac{R_2}{2} \tag{5-72}$$

and

$$R_{1b} = \frac{R_{1a}}{2 Q_r^2 - 1} \tag{5-73}$$

The value of C can be freely chosen.

The configuration of figure 5-31b is not restricted to the MFBP section. The state-variable all-pole bandpass circuit may be used instead. The only requirements are that the filter section be of an inverting type and that the gain is unity at resonance. This last requirement is especially critical because of the positive feedback nature of the circuit. Small gain errors could result in large overall Q variations when β is close to 1. It may then be desirable to adjust section gain precisely to unity.

Example 5-15

REQUIRED: Design a single bandpass filter section having the following characteristics:

  Center frequency of 3600 Hz
  3-dB bandwidth of 60 Hz
  Gain of 3

RESULT: (a) The bandpass Q is given by

$$Q=\frac{f_0}{\text{BW}_{3\,\text{dB}}}=\frac{3600\ \text{Hz}}{60\ \text{Hz}}=60 \qquad (2\text{-}16)$$

A Q-multiplier implementation using the MFBP section will be employed.

(b) Let us use a Q_r of 10 for the MFBP circuit. The following component values are computed where C is set equal to 0.01 μF.

$$R_2=\frac{Q_r}{\pi f_r C}=\frac{10}{\pi 3600\times 10^{-8}}=88.4\ \text{k}\Omega \qquad (5\text{-}68)$$

$$R_{1a}=\frac{R_2}{2}=\frac{88.4\times 10^3}{2}=44.2\ \text{k}\Omega \qquad (5\text{-}72)$$

$$R_{1b}=\frac{R_{1a}}{2\,Q_r^2-1}=\frac{44.2\times 10^3}{2\times 10^2-1}=222\ \Omega \qquad (5\text{-}73)$$

The remaining values are given by the following design equations where R is chosen at 10 kΩ and gain A_r is equal to 3:

$$\beta=1-\frac{Q_r}{Q_{\text{eff}}}=1-\frac{10}{60}=0.8333 \qquad (5\text{-}100)$$

$$R_3=\frac{R}{\beta}=\frac{10^4}{0.8333}=12.0\ \text{k}\Omega \qquad (5\text{-}101)$$

$$R_4=R=10\ \text{k}\Omega \qquad (5\text{-}102)$$

$$R_5=\frac{R}{(1-\beta)\,A_r}=\frac{10^4}{(1-0.8333)3}=20\ \text{k}\Omega \qquad (5\text{-}103)$$

The resulting circuit is shown in figure 5-32 using standard resistor values. R_{1b} has been made adjustable for tuning.

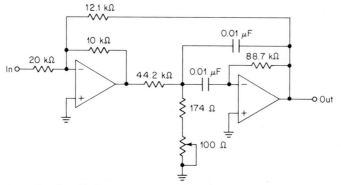

Fig. 5-32 *Q-multiplier section of example 5-15.*

Elliptic-Function Bandpass Filters

An active elliptic-function bandpass filter is designed by first transforming the low-pass poles and zeros to the bandpass form using the formulas of section 5.2. The bandpass poles and zeros are then implemented using active structures.

Low-pass poles and zeros for elliptic-function low-pass filters have been extensively tabulated by Zverev (see references) for complexities ranging from $n = 3$ through $n = 7$. Table 12-57 contains some of these values for odd-order n's alongside the active low-pass filter elements.

The general form of a bandpass transfer function containing zeros was given in section 5.2 as

$$T(s) = \frac{H(s^2 + \omega_\infty^2)}{s^2 + \dfrac{\omega_r}{Q} s + \omega_r^2} \tag{5-49}$$

Elliptic-function bandpass filters are comprised of cascaded first-order bandpass sections. When n is odd, $n - 1$ zero producing sections are required along with a single all-pole section. When n is even, $n - 2$ zero producing sections are used along with two all-pole networks.

This section discusses the VCVS and biquad configurations which have a transfer function in the form of equation (5-49) and their use in the design of active elliptic-function bandpass filters.

VCVS Network Section 3.2 discussed the design of active elliptic-function low-pass filters using an RC section containing a voltage-controlled voltage source (VCVS). The circuit is repeated in figure 5-33a. This structure is not restricted to the design of low-pass filters exclusively. Transmission zeros can be obtained at frequencies either above or below the pole locations as required by the bandpass transfer function.

The element values are computed from the following equations:

$$R_1 = \frac{f_\infty^2 + F_f^2}{3 f_\infty^2} R' \tag{5-104}$$

$$R_2 = 2 R_1 \tag{5-105}$$

$$R_3 = \frac{f_\infty^2 + f_f^2}{4.5 f_f^2} R' \tag{5-106}$$

$$R_4 = 4.5 R_3 \tag{5-107}$$

$$C_1 = \frac{1.5}{2\pi f_r R_1} \tag{5-108}$$

$$C_2 = \frac{C_1}{4.5} \tag{5-109}$$

$$C_3 = \frac{f_r}{2\pi R_1 f_\infty^2} \tag{5-110}$$

$$C_4 = \frac{C_3}{2} \tag{5-111}$$

$$K = \frac{\left(2.5 - \dfrac{1}{Q}\right)\left(\dfrac{f_f^2}{f_\infty^2} + 1\right)}{1.5} \tag{5-112}$$

$$R_6 = R \tag{5-113}$$

$$R_7 = (K - 1)\,R \tag{5-114}$$

where R' and R can be arbitrarily chosen.

In the event K is less than 1, the circuit of figure 5-33b is used. Resistor R_4 is split into two resistors, R_{4a} and R_{4b}, which are given by

$$R_{4a} = (1 - K)\,R_4 \tag{5-115}$$

and

$$R_{4b} = KR_4 \tag{5-116}$$

The section Q can be controlled independent of resonant frequency by making R_6 or R_7 adjustable when $K > 1$. The resonant frequency, however, is not easily adjusted. Experience has shown that with 1% resistors and capacitors, section Q's of up to 10 can be realized with little degradation of overall filter response due to component tolerances.

The actual circuit Q cannot be measured directly, since the section's 3-dB bandwidth is determined not only by the design Q but by the transmission zero as well. Another undesirable characteristic is that the capacitor values can

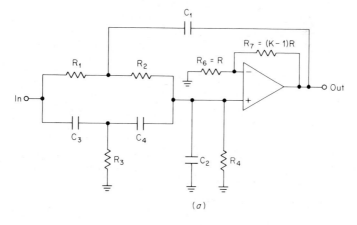

(a)

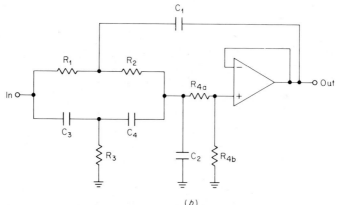

(b)

Fig. 5-33 VCVS elliptic-function bandpass section: (a) circuit for $K > 1$; (b) circuit for $K < 1$.

have a large spread[1] and cannot be forced to standard values easily. Nevertheless the VCVS configuration uses a minimum number of amplifiers and is widely used for low-Q elliptic-function realizations. The design technique is demonstrated in the following example.

Example 5-16

REQUIRED: Active bandpass filter
Center frequency of 500 Hz
1 dB maximum at ±100 Hz (400 Hz, 600 Hz)
35 dB minimum at ±375 Hz (125 Hz, 875 Hz)

RESULT: (a)   Convert to geometrically symmetrical bandpass requirement: First calculate geometric center frequency

$$f_0 = \sqrt{f_L f_u} = \sqrt{400 \times 600} = 490.0 \text{ Hz} \qquad (2\text{-}14)$$

Since the stopband requirement is arithmetically symmetrical compute stopband bandwidth using equation (5-18).

$$\text{BW}_{35 \text{ dB}} = f_2 - \frac{f_0^2}{f_2} = 875 - \frac{490^2}{875} = 600.6 \text{ Hz}$$

The bandpass steepness factor is given by

$$A_s = \frac{\text{stopband bandwidth}}{\text{passband bandwidth}} = \frac{600.6 \text{ Hz}}{200 \text{ Hz}} = 3.003 \quad (2\text{-}19)$$

(b)   An elliptic-function filter type will be used. A normalized low-pass filter is required that makes a transition from less than 1 dB to more than 35 dB within a frequency ratio of 3.003:1. A satisfactory filter can be found in table 12-57 corresponding to CO3 25 $\theta = 20°$; it has the following parameters:

$$n = 3$$
$$\rho = 25\% \text{ (0.28-dB ripple)}$$
$$\Omega_s = 2.924$$
$$A_{\min} = 39.48 \text{ dB}$$

The pole-zero coordinates can be found alongside the element values and are as follows:

Poles:

$-0.3449 \pm j1.086$
-0.7801

Zero:

$\pm j3.350$

(c)   To determine the bandpass parameters, first compute

$$Q_{bp} = \frac{f_0}{\text{BW}_{1 \text{ dB}}} = \frac{490 \text{ Hz}}{200 \text{ Hz}} = 2.45 \qquad (2\text{-}16)$$

The poles and zeros are transformed as follows:

Complex pole:

$$\alpha = 0.3449 \qquad \beta = 1.086$$
$$C = 1.298352 \qquad (5\text{-}50)$$
$$D = 0.281551 \qquad (5\text{-}51)$$

---

[1] Section 3.2 also discusses a VCVS uniform capacitor structure for elliptic-function filters which allows an additional degree of freedom in capacitor selection at the expense of sensitivity.

$$E = 4.216302 \qquad (5\text{-}52)$$

$$G = 4.178531 \qquad (5\text{-}53)$$

$$Q = 7.276691 \qquad (5\text{-}54)$$

$$M = 1.024380 \qquad (5\text{-}55)$$

$$W = 1.246538 \qquad (5\text{-}56)$$

$$f_{ra} = 393 \text{ Hz} \qquad (5\text{-}57)$$

$$f_{rb} = 611 \text{ Hz} \qquad (5\text{-}58)$$

Real pole:

$$\alpha_0 = 0.7801$$

$$Q = 3.14 \qquad (5\text{-}59)$$

$$f_r = 490 \text{ Hz}$$

Zero:

$$\omega_\infty = 3.350$$

$$H = 1.934819 \qquad (5\text{-}60)$$

$$Z = 1.895040 \qquad (5\text{-}61)$$

$$f_{\infty,a} = 259 \text{ Hz} \qquad (5\text{-}62)$$

$$f_{\infty,b} = 929 \text{ Hz} \qquad (5\text{-}63)$$

The bandpass parameters are summarized in the following table, where the zeros are arbitrarily assigned to the first two sections:

Section	f_r	Q	f_∞
1	393 Hz	7.28	259 Hz
2	611 Hz	7.28	929 Hz
3	490 Hz	3.14	

(d) Sections 1 and 2 are realized using the VCVS configuration of figure 5-33. The element values are computed as follows, where R' and R are both 10 kΩ.

Section 1:

$$f_r = 393 \text{ Hz}$$
$$Q = 7.28$$
$$f_\infty = 259 \text{ Hz}$$

$$R_1 = \frac{f_\infty^2 + f_r^2}{3 f_\infty^2} R' = \frac{259^2 + 393^2}{3 \times 259^2} \, 10^4 = 11.0 \text{ k}\Omega \qquad (5\text{-}104)$$

$$R_2 = 2 R_1 = 2 \times 11 \times 10^3 = 22 \text{ k}\Omega \qquad (5\text{-}105)$$

$$R_3 = \frac{f_\infty^2 + f_r^2}{4.5 f_r^2} R' = \frac{259^2 + 393^2}{4.5 \times 393^2} \, 10^4 = 3.19 \text{ k}\Omega \qquad (5\text{-}106)$$

$$R_4 = 4.5 R_3 = 4.5 \times 3190 = 14.3 \text{ k}\Omega \qquad (5\text{-}107)$$

$$C_1 = \frac{1.5}{2\pi f_r R_1} = \frac{1.5}{2\pi \times 393 \times 11 \times 10^3} = 0.0552 \ \mu\text{F} \qquad (5\text{-}108)$$

$$C_2 = \frac{C_1}{4.5} = 0.0123 \ \mu\text{F} \qquad (5\text{-}109)$$

$$C_3 = \frac{f_r}{2\pi R_1 f_\infty^2} = \frac{393}{2\pi \times 11 \times 10^3 \times 259^2} = 0.0848 \ \mu\text{F} \qquad (5\text{-}110)$$

$$C_4 = \frac{C_3}{2} = \frac{0.0848 \times 10^{-6}}{2} = 0.0424 \ \mu\text{F} \qquad (5\text{-}111)$$

$$K = \frac{\left(2.5 - \dfrac{1}{Q}\right)\left(\dfrac{f_r^2}{f_\infty^2} + 1\right)}{1.5} = \frac{\left(2.5 - \dfrac{1}{7.28}\right)\left(\dfrac{393^2}{259^2} + 1\right)}{1.5} \qquad (5\text{-}112)$$

$$= 5.202$$

$$R_6 = R = 10 \ \text{k}\Omega \qquad (5\text{-}113)$$

$$R_7 = (K-1)\,R = 4.202 \times 10^4 = 42.0 \ \text{k}\Omega \qquad (5\text{-}114)$$

Section 2:

$$f_r = 611 \ \text{Hz}$$
$$Q = 7.28$$
$$f_\infty = 929 \ \text{Hz}$$

$$R_1 = 4.78 \ \text{k}\Omega \qquad (5\text{-}104)$$

$$R_2 = 9.55 \ \text{k}\Omega \qquad (5\text{-}105)$$

$$R_3 = 7.36 \ \text{k}\Omega \qquad (5\text{-}106)$$

$$R_4 = 33.1 \ \text{k}\Omega \qquad (5\text{-}107)$$

$$C_1 = 0.0817 \ \mu\text{F} \qquad (5\text{-}108)$$

$$C_2 = 0.0182 \ \mu\text{F} \qquad (5\text{-}109)$$

$$C_3 = 0.0236 \ \mu\text{F} \qquad (5\text{-}110)$$

$$C_4 = 0.0118 \ \mu\text{F} \qquad (5\text{-}111)$$

$$K = 2.257 \qquad (5\text{-}112)$$

$$R_6 = 10 \ \text{k}\Omega \qquad (5\text{-}113)$$

$$R_7 = 12.6 \ \text{k}\Omega \qquad (5\text{-}114)$$

(e) Section 3 is required to be of the all-pole type; so the MFBP configuration of figure 5-25b will be used, where C is chosen at 0.01 μF and section gain A_r is set to unity.

Section 3:

$$f_r = 490 \ \text{Hz}$$
$$Q = 3.14$$

$$R_2 = 204 \ \text{k}\Omega \qquad (5\text{-}68)$$

$$R_{1a} = 102 \ \text{k}\Omega \qquad (5\text{-}72)$$

$$R_{1b} = 5.45 \ \text{k}\Omega \qquad (5\text{-}73)$$

The complete circuit is shown in figure 5-34 using standard 1% resistor values.

State-Variable (Biquad) Circuit The all-pole bandpass form of the state-variable or biquad section was discussed in section 5.2. With the addition of an operational amplifier the circuit can be used to realize transmission zeros as well as poles. The configuration is shown in figure 5-35. This circuit is identical to the elliptic-function low-pass and high-pass filter configurations of sections 3.2 and 4.2. By connecting R_5 either to node 1 or to node 2, the zero can be located above or below the resonant frequency.

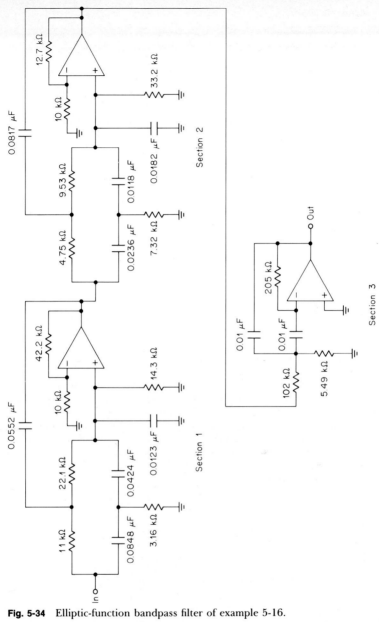

Fig. 5-34 Elliptic-function bandpass filter of example 5-16.

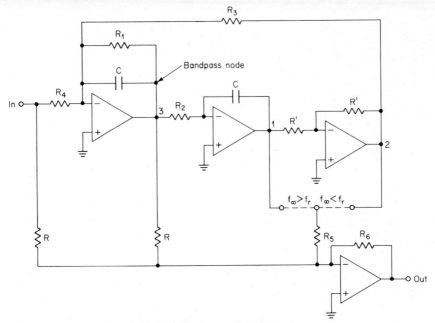

Fig. 5-35 Biquad elliptic-function bandpass configuration.

On the basis of sensitivity and flexibility the biquad configuration has been found to be the optimum method of constructing precision active elliptic-function bandpass filters. Section Q's of up to 200 can be obtained, whereas the VCVS section is limited to Q's below 10. Resonant frequency f_r, Q and notch frequency f_∞ can be independently monitored and adjusted.

For the case where $f_\infty < f_r$, the transfer function is given by

$$T(s) = -\frac{R_6}{R}\frac{s^2 + \dfrac{1}{R_2 R_3 C^2}\left(1 - \dfrac{R_3 R}{R_4 R_5}\right)}{s^2 + \dfrac{1}{R_1 C}s + \dfrac{1}{R_2 R_3 C^2}} \tag{5-117}$$

and when $f_\infty > f_r$, the corresponding transfer function is

$$T(s) = -\frac{R_6}{R}\frac{s^2 + \dfrac{1}{R_2 R_3 C^2}\left(1 + \dfrac{R_3 R}{R_4 R_5}\right)}{s^2 + \dfrac{1}{R_1 C}s + \dfrac{1}{R_2 R_3 C^2}} \tag{5-118}$$

If we equate the transfer-function coefficients to those of the general bandpass transfer function (with zeros) of equation (5-49), the following series of design equations can be derived:

$$R_1 = R_4 = \frac{Q}{2\pi f_r C} \tag{5-119}$$

$$R_2 = R_3 = \frac{R_1}{Q} \tag{5-120}$$

$$R_5 = \frac{f_r^2 R}{Q|f_r^2 - f_\infty^2|} \qquad (5\text{-}121)$$

for $f_\infty > f_r$:
$$R_6 = \frac{f_r^2 R}{f_\infty^2} \qquad (5\text{-}122)$$

and when $f_\infty < f_r$:
$$R_6 = R \qquad (5\text{-}123)$$

where C and R can be conveniently selected. The value of R_6 is based on unity section gain. The gain can be raised or lowered by proportionally changing R_6.

The section can be tuned by implementing the following steps in the indicated sequence. Both resonant frequency and Q are monitored at the bandpass output occurring at node 3, whereas the notch frequency f_∞ is observed at the section output.

1. *Resonant frequency f_r:* If R_3 is made variable, the section resonant frequency can be adjusted. Resonance is monitored at node 3 (see figure 5-35) and can be determined by the 180° phase shift method.
2. *Q adjustment:* The section Q is controlled by R_1 and can be directly measured at node 3. The configuration is subject to the Q-enhancement effect discussed in section 5.2 under All-Pole Bandpass Configurations; so a Q adjustment is normally required. The Q can be monitored in terms of the 3-dB bandwidth at node 3 or R_1 can be adjusted until unity gain occurs between the section input and node 3 with f_r applied.
3. *Notch frequency f_∞:* Adjustment of the notch frequency (transmission zero) usually is not required if the circuit is previously tuned to f_r, since f_∞ will usually then fall in. If an adjustment is desired, the notch frequency can be controlled by making R_5 variable.

The biquad approach is a highly stable and flexible implementation for precision active elliptic-function filters. Independent adjustment capability for resonant frequency, Q and notch frequency preclude its use when Q's in excess of 10 are required. Stable Q's of up to 200 are obtainable.

Example 5-17

REQUIRED: Active bandpass filter
Center frequency 500 Hz
0.2 dB maximum at ±50 Hz (450 Hz, 550 Hz)
30 dB minimum at ±150 Hz (350 Hz, 650 Hz)

RESULT: (a) Convert to geometrically symmetrical requirement

$$f_0 = \sqrt{f_L f_u} = \sqrt{450 \times 550} = 497.5 \text{ Hz} \qquad (2\text{-}14)$$

$$\text{BW}_{30 \text{ dB}} = f_2 - \frac{f_0^2}{f_2} = 650 - \frac{497.5^2}{650} = 269 \text{ Hz} \qquad (5\text{-}18)$$

$$A_s = \frac{\text{stopband bandwidth}}{\text{passband bandwidth}} = \frac{269 \text{ Hz}}{100 \text{ Hz}} = 2.69 \quad (2\text{-}19)$$

(b) A normalized low-pass filter must be chosen that makes the transition from less than 0.2 dB to more than 30 dB within a frequency ratio of 2.69 : 1. An elliptic-function filter will be chosen from table 12-57. The design corresponding to C03 15 $\theta = 23°$ has the following characteristics:

$$n = 3$$
$$\rho = 15\% \ (0.098\text{-dB ripple})$$
$$\Omega_s = 2.559$$
$$A_{\min} = 31.13 \text{ dB}$$

The pole-zero coordinates are tabulated alongside the low-pass element values and are as follows:

Poles:

$$-0.4249 \pm j1.2169$$
$$-1.0544$$

Zero:

$$\pm j2.9256$$

(c)   The bandpass pole-zero transformation is now performed. First compute

$$Q_{bp} = \frac{f_0}{BW_{0.2\ dB}} = \frac{497.5\ \text{Hz}}{100\ \text{Hz}} = 4.975 \qquad (2\text{-}16)$$

The transformation proceeds as follows:

Complex pole:

$$\alpha = 0.4249 \qquad \beta = 1.2169$$

$C = 1.661386$	(5-50)
$D = 0.170814$	(5-51)
$E = 4.067125$	(5-52)
$G = 4.052752$	(5-53)
$Q = 11.796041$	(5-54)
$M = 1.007465$	(5-55)
$W = 1.129880$	(5-56)
$f_{ra} = 440.3\ \text{Hz}$	(5-57)
$f_{rb} = 562.1\ \text{Hz}$	(5-58)

Real pole:

$$\alpha_0 = 1.0544$$
$$Q = 4.718 \qquad (5\text{-}59)$$
$$f_r = 497.5\ \text{Hz}$$

Zero:

$$\omega_\infty = 2.9256$$

$H = 1.172907$	(5-60)
$Z = 1.336361$	(5-61)
$f_{\infty,a} = 372.3\ \text{Hz}$	(5-62)
$f_{\infty,b} = 664.8\ \text{Hz}$	(5-63)

The computed bandpass parameters are summarized in the following table. The zeros are assigned to the first two sections.

Section	f_r	Q	f_∞
1	440.3 Hz	11.8	372.3 Hz
2	562.1 Hz	11.8	664.8 Hz
3	497.5 Hz	4.72	

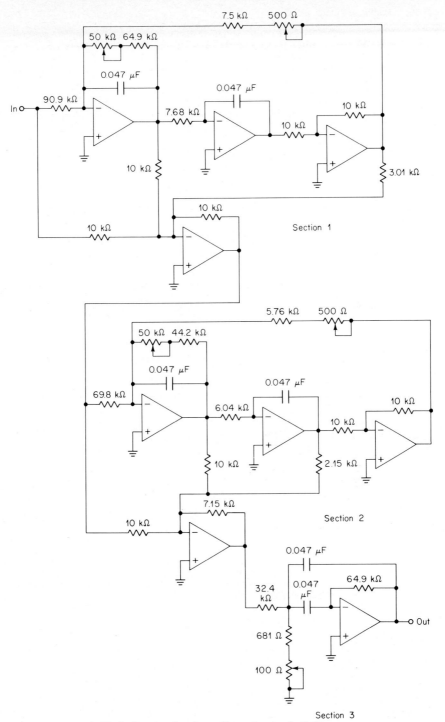

Fig. 5-36 Biquad elliptic-function bandpass filter of example 5-17.

(d) Sections 1 and 2 will be realized in the form of the biquad configuration of figure 5-35 where R' and R are both 10 kΩ and $C = 0.047$ μF.

Section 1:

$f_r = 440.3$ Hz
$Q = 11.8$
$f_\infty = 372.3$ Hz

$$R_1 = R_4 = \frac{Q}{2\pi f_r C} \tag{5-119}$$

$$= \frac{11.8}{2\pi \times 440.3 \times 4.7 \times 10^{-8}} = 90.8 \text{ k}\Omega$$

$$R_2 = R_3 = \frac{R_1}{Q} = \frac{90.8 \times 10^3}{11.8} = 7.69 \text{ k}\Omega \tag{5-120}$$

$$R_5 = \frac{f_r^2 R}{Q \, |f_r^2 - f_\infty^2|} = \frac{440.3^2 \times 10^4}{11.8|440.3^2 - 372.3^2|} = 2.97 \text{ k}\Omega \tag{5-121}$$

$$R_6 = R = 10 \text{ k}\Omega \tag{5-123}$$

Section 2:

$f_r = 562.1$ Hz
$Q = 11.8$
$f_\infty = 664.8$ Hz

$$R_1 = R_4 = 71.1 \text{ k}\Omega \tag{5-119}$$

$$R_2 = R_3 = 6.02 \text{ k}\Omega \tag{5-120}$$

$$R_5 = 2.13 \text{ k}\Omega \tag{5-121}$$

$$R_6 = 7.15 \text{ k}\Omega \tag{5-122}$$

(e) The MFBP configuration of figure 5-25b will be used for the all-pole circuit of section 3. The value of C is 0.047 μF, and A_r is unity.

Section 3:

$f_r = 497.5$ Hz
$Q = 4.72$

$$R_2 = 64.3 \text{ k}\Omega \tag{5-68}$$

$$R_{1a} = 32.1 \text{ k}\Omega \tag{5-72}$$

$$R_{1b} = 738 \text{ }\Omega \tag{5-73}$$

The resulting filter is shown in figure 5-36, where standard 1% resistors are used. The resonant frequency and Q of each section have been made adjustable.

REFERENCES

Huelsman, L. P., "Theory and Design of Active RC Circuits," McGraw-Hill Book Company, New York, 1968.

Sedra, A. S., and Espinoza, J. L., "Sensitivity and Frequency Limitations of Biquadratic Active Filters," IEEE Transactions on Circuits and Systems, Vol. CAS–22, No. 2, February 1975.

Thomas, L. C., "The Biquad: Part I—Some Practical Design Considerations," IEEE Transactions on Circuit Theory, Vol. CT–18, May 1971.

Tow, J., "A Step-by-Step Active Filter Design," IEEE Spectrum, Vol. 6, December 1969.

Williams, A. B., "Active Filter Design," Artech House, Dedham, Mass., 1975.

Williams, A. B., "Q-Multiplier Techniques Increase Filter Selectivity," EDN, pp. 74–76, October 5, 1975.

Zverev, A. I., "Handbook of Filter Synthesis," John Wiley and Sons, New York, 1967.

6

Band-Reject Filters

6.1 LC BAND-REJECT FILTERS

Normalization of a band-reject requirement and the definitions of the response shape parameters were discussed in section 2.1. Like bandpass filters, band-reject networks can also be derived from a normalized low-pass filter by a suitable transformation.

In section 5.1 we discussed the design of wide-band bandpass filters by cascading a low-pass filter and a high-pass filter. In a similar manner, wide-band band-reject filters could also be obtained by combining low-pass and high-pass filters. Both the input and output terminals are paralleled, and each filter must have a high input and output impedance in the band of the other filter to prevent interaction. Therefore, the order n must be odd and the first and last branches should consist of series elements. These restrictions make the design of band-reject filters by combining low-pass and high-pass filters undesirable. The impedance interaction between filters is a serious problem unless the separation between cutoffs is many octaves; so the design of band-reject filters is best approached by transformation techniques.

The Band-Reject Circuit Transformation

Bandpass filters were obtained by first designing a low-pass filter having a cutoff frequency equivalent to the required bandwidth and then resonating each element to the desired center frequency. The response of the low-pass filter at DC will correspond to the response of the bandpass filter at center frequency.

Band-reject filters are designed by initially transforming the normalized low-pass filter into a high-pass network having a cutoff frequency equal to the required bandwidth and at the desired impedance level. Every high-pass element is then resonated to the center frequency in the same manner as bandpass filters.

This corresponds to replacing the frequency variable in the high-pass transfer function by a new variable which is given by

$$f_{br} = f_0 \left(\frac{f}{f_0} - \frac{f_0}{f} \right) \tag{6-1}$$

As a result, the response of the high-pass filter at DC is transformed to the band-reject network at the center frequency. The bandwidth response of the band-reject filter is identical to the frequency response of the high-pass filter. The high-pass to band-reject transformation is shown in figure 6-1. Negative frequencies of course are strictly of theoretical interest; so only the response shape corresponding to positive frequencies is applicable. As in the case of bandpass filters, the response curve exhibits geometric symmetry.

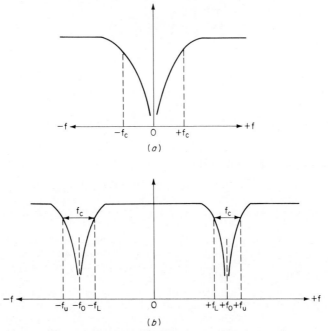

Fig. 6-1 The band-reject transformation: *(a)* high-pass filter response; *(b)* transformed band-reject filter response.

The design procedure can be summarized as follows:

1. Normalize the band-reject filter specification and select a normalized low-pass filter that provides the required attenuation within the computed steepness factor.
2. Transform the normalized low-pass filter to a normalized high-pass filter. Then scale the high-pass filter to a cutoff frequency equal to the desired bandwidth and to the preferred impedance level.
3. Resonate each element to the center frequency by introducing a capacitor in series with each inductor and an inductor in parallel with each capacitor to complete the design. The transformed circuit branches are summarized in table 6-1.

All-Pole Band-Reject Filters

Band-reject filters can be derived from any all-pole or elliptic-function *LC* low-pass network. Although not as efficient as elliptic-function filters, the all-pole

TABLE 6-1   The High-Pass to Band-Reject Transformation

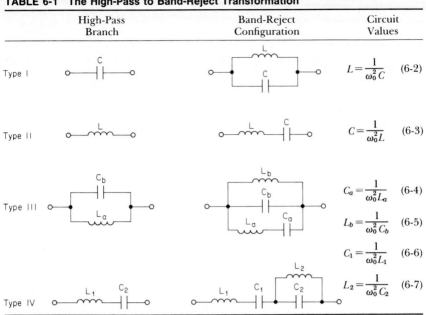

High-Pass Branch	Band-Reject Configuration	Circuit Values
Type I		$L=\dfrac{1}{\omega_0^2 C}$ (6-2)
Type II		$C=\dfrac{1}{\omega_0^2 L}$ (6-3)
Type III		$C_a=\dfrac{1}{\omega_0^2 L_a}$ (6-4) $L_b=\dfrac{1}{\omega_0^2 C_b}$ (6-5)
Type IV		$C_1=\dfrac{1}{\omega_0^2 L_1}$ (6-6) $L_2=\dfrac{1}{\omega_0^2 C_2}$ (6-7)

approach results in a simpler band-reject structure where all sections are tuned to the center frequency.

The following example demonstrates the design of an all-pole band-reject filter.

Example 6-1

REQUIRED: Band-reject filter
Center frequency of 10 kHz
3 dB at ±250 Hz (9.75 kHz, 10.25 kHz)
30 dB minimum at ±100 Hz (9.9 kHz, 10.1 kHz)
Source and load impedance of 600 Ω

RESULT: (a)   Convert to geometrically symmetrical requirement. Since the bandwidth is relatively narrow, the specified arithmetically symmetrical frequencies will determine the following design parameters:

$$f_0 = 10 \text{ kHz}$$
$$\text{BW}_{3 \text{ dB}} = 500 \text{ Hz}$$
$$\text{BW}_{30 \text{ dB}} = 200 \text{ Hz}$$

(b)   Compute the band-reject steepness factor.

$$A_s = \frac{\text{passband bandwidth}}{\text{stopband bandwidth}} = \frac{500 \text{ Hz}}{200 \text{ Hz}} = 2.5 \qquad (2\text{-}20)$$

The response curves of figure 2-45 indicate that an $n = 3$ Chebyshev normalized low-pass filter having a 1-dB ripple provides over 30 dB of attenuation within a frequency ratio of 2.5:1. The corresponding circuit is found in table 12-31 and is shown in figure 6-2a.

(c) To transform the normalized low-pass circuit into a normalized high-pass filter, replace inductors with capacitors and vice versa using reciprocal element values. The transformed structure is shown in figure 6-2b.

(d) The normalized high-pass filter is scaled to a cutoff frequency of 500 Hz corresponding to the desired bandwidth and to an

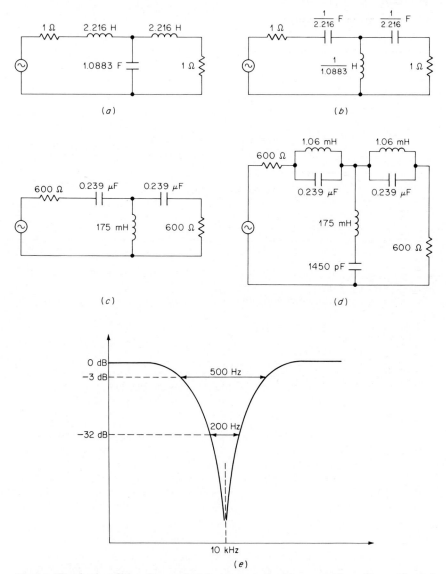

(a)

(b)

(c)

(d)

(e)

Fig. 6-2 Band-reject filter of example 6-1: (a) normalized low-pass filter; (b) transformed normalized high-pass filter; (c) frequency- and impedance-scaled high-pass filter; (d) transformed band-reject filter; (e) frequency response.

impedance level of 600 Ω. The capacitors are divided by $Z \times$ FSF and the inductors are multiplied by Z/FSF, where Z is 600 and the FSF (frequency-scaling factor) is given by $2\pi f_c$, where f_c is 500 Hz. The scaled high-pass filter is illustrated in figure 6-2c.

(e) To make the high-pass to band-reject transformation, resonate each capacitor with a parallel inductor and each inductor with a series capacitor. The resonating inductors for the series branches are both given by

$$L = \frac{1}{\omega_0^2 C} = \frac{1}{(2\pi 10 \times 10^3)^2 \times 0.239 \times 10^{-6}} = 1.06 \text{ mH} \quad (6\text{-}2)$$

The tuning capacitor for the shunt inductor is determined from

$$C = \frac{1}{\omega_0^2 L} = \frac{1}{(2\pi 10 \times 10^3)^2 \times 0.175} = 1450 \text{ pF} \quad (6\text{-}3)$$

The final filter is shown in figure 6-2d, where all peaks are tuned to the center frequency of 10 kHz. The theoretical frequency response is illustrated in figure 6-2e.

When a low-pass filter undergoes a high-pass transformation followed by a band-reject transformation, the minimum Q requirement is increased by a factor equal to the Q of the band-reject filter. This can be expressed as

$$Q_{min} \text{ (band-reject)} = Q_{min} \text{ (low-pass)} \times Q_{br} \quad (6\text{-}8)$$

where values for Q_{min} (low-pass) are given in figure 3-8 and $Q_{br} = f_0/BW_{3 \text{ dB}}$. The branch Q should be several times larger than Q_{min} (band-reject) to obtain near theoretical results.

The equivalent circuit of a band-reject filter at center frequency can be determined by replacing each parallel tuned circuit by a resistor of $\omega_0 L Q_L$ and each series tuned circuit by a resistor of $\omega_0 L/Q_L$. These resistors correspond to the branch impedances at resonance, where ω_0 is $2\pi f_0$, L is the branch inductance, and Q_L is the branch Q which is normally determined only by the inductor losses.

It is then apparent that at center frequency, the circuit can be replaced by a resistive voltage divider. The amount of attenuation that can be obtained is then directly controlled by the branch Q's. Let us determine the attenuation of the circuit of example 6-1 for a finite value of inductor Q.

Example 6-2

REQUIRED: Estimate the amount of rejection obtainable at the center frequency of 10 kHz for the band-reject filter of example 6-1. An inductor Q of 100 is available and the capacitors are assumed to be lossless. Also determine if the Q is sufficient to retain the theoretical passband characteristics.

RESULT: (a) Compute the equivalent resistances at resonance for all tuned circuits.

Parallel tuned circuits:

$$R = \omega_0 L Q_L = 2\pi \times 10^4 \times 1.06 \times 10^{-3} \times 100 = 6660 \text{ Ω} \quad (5\text{-}21)$$

Series tuned circuit:

$$R = \frac{\omega_0 L}{Q_L} = \frac{2\pi \times 10^4 \times 0.175}{100} = 110 \text{ Ω} \quad (5\text{-}30)$$

(b) The equivalent circuit at 10 kHz is shown in figure 6-3. Using conventional circuit analysis methods such as mesh equations or approximation techniques, the overall loss is found to be 58 dB. Since the flat loss due to the 600-Ω terminations is 6 dB, the relative attenuation at 10 kHz will be 52 dB.

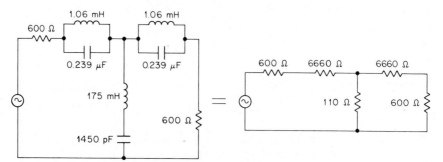

Fig. 6-3   Equivalent circuit at center frequency for filter of figure 6-2.

(c) The curves of figure 3-8 indicate that an $n = 3$ Chebyshev filter with a 1-dB ripple has a minimum theoretical Q requirement of 4.5. The minimum Q of the band-reject filter is given by

$$Q_{min} \text{ (band-reject)} = Q_{min} \text{ (low-pass)} \times Q_{br}$$
$$= 4.5 \times \frac{10,000}{500} = 90 \qquad (6\text{-}8)$$

Therefore, the available Q of 100 is barely adequate and some passband rounding will occur in addition to the reduced stopband attenuation. The resulting effect on frequency response is shown in figure 6-4.

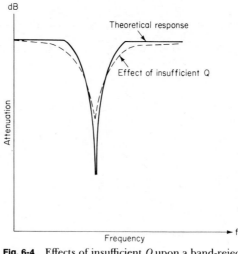

Fig. 6-4   Effects of insufficient Q upon a band-reject filter.

Elliptic-Function Band-Reject Filters

The superior properties of the elliptic-function family of filters can also be applied to band-reject requirements. Extremely steep characteristics in the transition region between passband and stopband can be achieved much more efficiently than with all-pole filters.

Saal and Ulbrich as well as Zverev (see references) have extensively tabulated the LC values for normalized elliptic-function low-pass networks. These circuits can be transformed to high-pass filters and subsequently to a band-reject filter in the same manner as the all-pole filters.

Since each tabulated low-pass filter can be realized in dual forms as defined by the upper and lower schematics shown in the tables, the resulting band-reject filters can also take on two different configurations, as illustrated in figure 6-5.

Branch 2 of the standard band-reject filter circuit corresponds to the type III network shown in table 6-1. This branch provides a pair of geometrically related zeros, one above and one below the center frequency. These zeros result from two conditions of parallel resonance. However, the circuit configuration

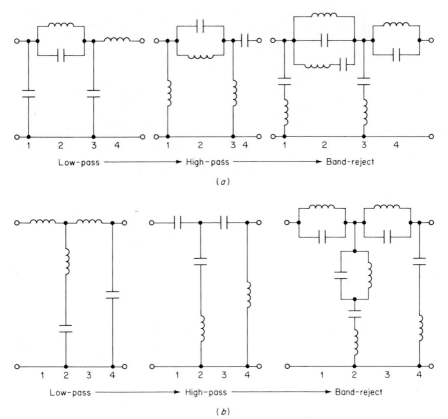

Fig. 6-5 The band-reject transformation of elliptic-function filters: (*a*) standard configuration; (*b*) dual configuration.

itself is not very desirable. The elements corresponding to the individual parallel resonances are not distinctly isolated, since each resonance is determined by the interaction of a number of elements. This makes tuning somewhat difficult. For very narrow filters the element values also become somewhat unreasonable.

An identical situation occurred during the bandpass transformation of an elliptic-function low-pass filter discussed in section 5.1. An equivalent configuration was presented as an alternate and is repeated in figure 6-6.

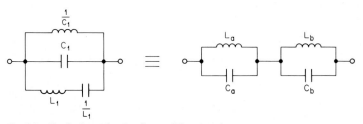

Fig. 6-6   Equivalent circuit of type III network.

The type III network of figure 6-6 has reciprocal element values which occur when the band-reject filter has been normalized to a 1-rad/s center frequency, since the equation for resonance, $\omega_0^2 LC = 1$, then reduces to $LC = 1$. The reason for this normalization is to greatly simplify the transformation equations.

To normalize the band-reject filter circuit, first transform the normalized low-pass filter to a normalized high-pass configuration in the conventional manner by replacing inductors with capacitors and vice versa using reciprocal element values. The high-pass elements are then multiplied by the factor Q_{br}, which is equal to f_0 /BW, where f_0 is the geometric center frequency of the band-reject filter and BW is the bandwidth. The normalized band-reject filter can be directly obtained by resonating each inductor with a series capacitor and each capacitor with a parallel inductor using reciprocal values.

To make the transformation of figure 6-6 first compute

$$\beta = 1 + \frac{1}{2L_1 C_1} + \sqrt{\frac{1}{4L_1^2 C_1^2} + \frac{1}{L_1 C_1}} \tag{6-9}$$

The values are then found from

$$L_a = \frac{1}{C_1(\beta + 1)} \tag{6-10}$$

$$L_b = \beta L_a \tag{6-11}$$

$$C_a = \frac{1}{L_b} \tag{6-12}$$

$$C_b = \frac{1}{L_a} \tag{6-13}$$

The resonant frequencies for each tuned circuit are given by

$$\Omega_{\infty,a} = \sqrt{\beta} \tag{6-14}$$

and
$$\Omega_{\infty,b} = \frac{1}{\Omega_{\infty,a}} \tag{6-15}$$

After the normalized band-reject filter has undergone the transformation of figure 6-6 wherever applicable, the circuit can be scaled to the desired impedance level and frequency. The inductors are multiplied by Z/FSF and capacitors are divided by $Z \times FSF$. The value of Z is the desired impedance level, and the frequency-scaling factor (FSF) in this case is equal to ω_0 ($\omega_0 = 2\pi f_0$). The resulting resonant frequencies in hertz are determined by multiplying the normalized radian resonant frequencies by f_0.

Branch 2 of the band-reject filter derived from the dual low-pass structure of figure 6-5b corresponds to the type IV network of table 6-1. This configuration realizes a pair of finite zeros resulting from two conditions of series resonance. However, as in the case of the type III network, the individual resonances are determined by the interaction of all the elements, which makes tuning difficult and can result in unreasonable values for narrow filters. An alternate configuration is shown in figure 6-7 consisting of two series resonant circuits in parallel.

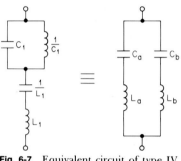

Fig. 6-7   Equivalent circuit of type IV network.

To simplify the transformation equations, the type IV network requires reciprocal values; so the band-reject filter must be normalized to a 1-rad/s center frequency. This is accomplished as previously described, and the filter is subsequently denormalized after the transformations have been made.

The transformation is accomplished as follows:

First compute

$$\beta = 1 + \frac{1}{2L_1 C_1} + \sqrt{\frac{1}{4L_1^2 C_1^2} + \frac{1}{L_1 C_1}} \tag{6-16}$$

then

$$L_a = \frac{(\beta + 1)L_1}{\beta} \tag{6-17}$$

$$C_a = \frac{1}{(\beta + 1)L_1} \tag{6-18}$$

$$L_b = \frac{1}{C_a} \tag{6-19}$$

$$C_b = \frac{1}{L_a} \tag{6-20}$$

$$\Omega_{\infty,a} = \sqrt{\beta} \tag{6-21}$$

$$\Omega_{\infty,b} = \frac{1}{\Omega_{\infty,a}} \tag{6-22}$$

The standard configuration of the elliptic-function filter is usually preferred over the dual circuit so that the transformed low-pass zeros can be realized using the structure of figure 6-6. Parallel tuned circuits are generally more desirable than series tuned circuits since they can be transformed to alternate L/C ratios to optimize Q and reduce capacitor values (see section 8.2 on tapped inductors).

Example 6-3

REQUIRED: Design a band-reject filter to satisfy the following requirements:
1 dB maximum at 2200 Hz and 2800 Hz
50 dB minimum at 2300 Hz and 2700 Hz
Source and load impedance of 600 Ω

RESULT: (a)   Convert to geometrically symmetrical requirement. First calculate the geometric center frequency.

$$f_0 = \sqrt{f_L f_u} = \sqrt{2200 \times 2800} = 2482 \text{ Hz} \tag{2-14}$$

Compute the corresponding geometric frequency for each stopband frequency given using equation (2-18).

$$f_1 f_2 = f_0^2 \tag{2-18}$$

f_1	f_2	$f_2 - f_1$
2300 Hz	2678 Hz	378 Hz
2282 Hz	2700 Hz	418 Hz

The second pair of frequencies are retained, since they represent the steeper requirement. The complete geometrically symmetrical specification can be stated as

$$f_0 = 2482 \text{ Hz}$$
$$BW_{1 \text{ dB}} = 600 \text{ Hz}$$
$$BW_{50 \text{ dB}} = 418 \text{ Hz}$$

(b)   Compute the band-reject steepness factor.

$$A_s = \frac{\text{passband bandwidth}}{\text{stopband bandwidth}} = \frac{600 \text{ Hz}}{418 \text{ Hz}} = 1.435 \tag{2-20}$$

A normalized low-pass filter must be chosen that makes the transition from less than 1 dB to more than 50 dB within a frequency ratio of 1.435. An elliptic-function filter will be selected.

Figure 2-86 indicates that with $\rho = 20\%$ (0.18-dB ripple) a complexity of $n = 6$ is required. Examination of table 12-56 results in the selection of a design corresponding to C06 20C $\theta = 47°$ which has the parameters $\Omega_s = 1.433$ and $A_{\min} = 53.4$ dB. This filter can then provide over 50 dB of attenuation within a frequency ratio of 1.435 as required by the specification. The normalized circuit is shown in figure 6-8a.

(c)   The normalized low-pass filter is now transformed into a normalized high-pass structure by replacing all inductors with capacitors and vice versa using reciprocal values. The resulting filter is given in figure 6-8b.

(d) To obtain a normalized band-reject filter so that the transformation of figure 6-6 can be performed, first multiply all the high-pass elements by Q_{br}, which is given by

$$Q_{br} = \frac{f_0}{BW_{1\ dB}} = \frac{2482\ Hz}{600\ Hz} = 4.137$$

The modified high-pass filter is shown in figure 6-8c.

(e) Each high-pass inductor is resonated with a series capacitor and each capacitor is resonated with a parallel inductor to obtain the normalized band-reject filter. Since the center frequency is 1 rad/s, the resonant elements are simply the reciprocals of each other as illustrated in figure 6-8d.

(f) The type III networks of the second and fourth branches are now transformed to the equivalent circuit of figure 6-6 as follows:

Type III network of second branch:

$$L_1 = 19.90\ H \qquad C_1 = 3.268\ F$$

$$\beta = 1 + \frac{1}{2L_1 C_1} + \sqrt{\frac{1}{4L_1^2 C_1^2} + \frac{1}{L_1 C_1}} = 1.1319 \qquad (6\text{-}9)$$

$$L_a = \frac{1}{C_1(\beta + 1)} = 0.1435\ H \qquad (6\text{-}10)$$

$$L_b = \beta L_a = 0.1625\ H \qquad (6\text{-}11)$$

$$C_a = \frac{1}{L_b} = 6.155\ F \qquad (6\text{-}12)$$

$$C_b = \frac{1}{L_a} = 6.969\ F \qquad (6\text{-}13)$$

$$\Omega_{\infty,a} = \sqrt{\beta} = 1.0639 \qquad (6\text{-}14)$$

$$\Omega_{\infty,b} = \frac{1}{\Omega_{\infty,a}} = 0.9399 \qquad (6\text{-}15)$$

Type III network of fourth branch:

$$L_1 = 11.31\ H \qquad C_1 = 3.283\ F$$
$$\beta = 1.1781$$
$$L_a = 0.1398\ H$$
$$L_b = 0.1648\ H$$
$$C_a = 6.068\ F$$
$$C_b = 7.153\ F$$
$$\Omega_{\infty,a} = 1.0854$$
$$\Omega_{\infty,b} = 0.9213$$

The resulting normalized band-reject filter is shown in figure 6-8e.

(g) The final filter can now be obtained by frequency- and impedance-scaling the normalized band-reject filter to a center frequency of 2482 Hz and 600 Ω. The inductors are multiplied by Z/FSF, and the capacitors are divided by Z × FSF, where Z is 600 and the FSF is $2\pi f_0$, where f_0 is 2482 Hz. The circuit is given in figure 6-8f, where the resonant frequencies of each section were obtained by multiplying the normalized frequencies by f_0. The frequency response is illustrated in figure 6-8g.

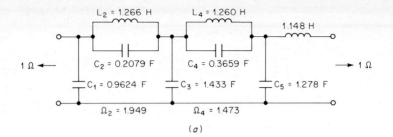

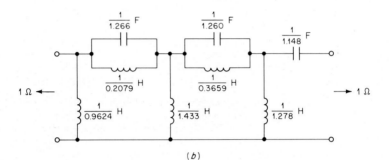

(a)

(b)

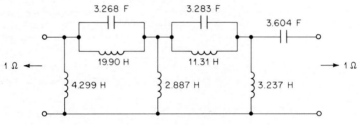

(c)

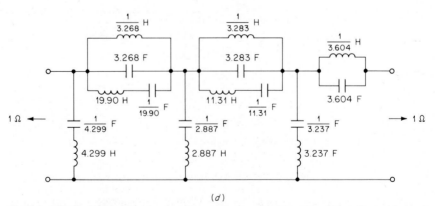

(d)

Fig. 6-8 Elliptic-function band-reject filter: (a) normalized low-pass filter; (b) transformed high-pass filter; (c) high-pass filter with elements multiplied by Q_{br}; (d) normalized band-

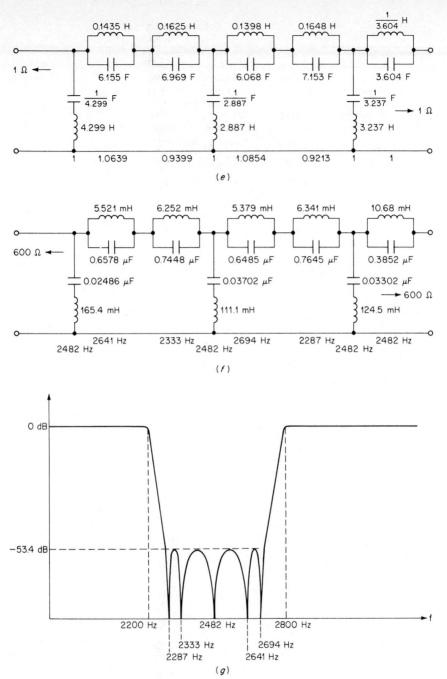

(e)

(f)

(g)

reject filter; (e) transformed type III networks; (f) frequency- and impedance-scaled circuit; (g) frequency response.

Null Networks

A null network can be loosely defined as a circuit intended to reject a single frequency or a very narrow band of frequencies and is frequently referred to as a trap. Notch depth rather than rate of roll-off is the prime consideration, and the circuit is restricted to a single section.

Parallel Resonant Trap

The RC high-pass circuit of figure 6-9a has a 3-dB cutoff given by

$$f_c = \frac{1}{2\pi RC} \tag{6-23}$$

A band-reject transformation will result in the circuit of figure 6-9b. The value of L is computed from

$$L = \frac{1}{\omega_0^2 C} \tag{6-24}$$

where $\omega_0 = 2\pi f_0$. The center frequency is f_0 and the 3-dB bandwidth is f_c.

The frequency response of a first-order band-reject filter can be expressed as

$$A_{dB} = 10 \log \left[1 + \left(\frac{BW_{3\,dB}}{BW_{x\,dB}} \right)^2 \right] \tag{6-25}$$

where $BW_{3\,dB}$ is the 3-dB bandwidth corresponding to f_c in equation (6-23) and $BW_{x\,dB}$ is the bandwidth of interest. The response can also be determined from the normalized Butterworth attenuation curves of figure 2-34 corresponding to $n = 1$, where $BW_{3\,dB}/BW_{x\,dB}$ is the normalized bandwidth.

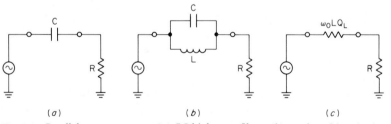

Fig. 6-9   Parallel resonant trap: (a) RC high-pass filter; (b) results of band-reject transformation; (c) equivalent circuit at f_0.

The impedance of a parallel tuned circuit at resonance is equal to $\omega_0 L Q_L$, where Q_L is the inductor Q and the capacitor is assumed to be lossless. We can then represent the band-reject filter at f_0 by the equivalent circuit of figure 6-9c. After some algebraic manipulation involving equations (6-23) and (6-24) and the circuit of figure 6-9c, we can derive the following expression for the attenuation at resonance of the $n = 1$ band-reject filter of figure 6-9:

$$A_{dB} = 20 \log \left(\frac{Q_L}{Q_{br}} + 1 \right) \qquad (6\text{-}26)$$

where $Q_{br} = f_0/BW_{3\,dB}$. Equation (6-26) is plotted in figure 6-10. When Q_{br} is high, the required inductor Q may become prohibitively large in order to attain sufficient attenuation at f_0.

The effect of insufficient inductor Q will not only reduce relative attenuation but will also cause some rounding of the response near the cutoff frequencies. Therefore, the ratio Q_L/Q_{br} should be as high as possible.

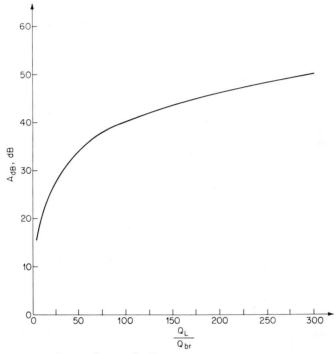

Fig. 6-10   Attenuation vs. Q_L/Q_{br}.

Example 6-4

REQUIRED: Design a parallel resonant circuit which has a 3-dB bandwidth of 500 Hz and a center frequency of 7500 Hz. The source resistance is zero and the load is 1 kΩ. Also determine the minimum inductor Q for a relative attenuation of at least 30 dB at 7500 Hz.

RESULT: (a)   Compute the value of the capacitor from

$$C = \frac{1}{2\pi f_c R} = \frac{1}{2\pi 500 \times 1000} = 0.3183\ \mu F \qquad (6\text{-}23)$$

The inductance is given by

$$L = \frac{1}{\omega_0^2 C} = \frac{1}{(2\pi 7500)^2 \times 3.183 \times 10^{-7}} = 1.415\ mH \qquad (6\text{-}24)$$

The resulting circuit is shown in figure 6-11.

(b)   The required ratio of Q_L/Q_{br} for 30-dB attenuation at f_0 can be determined from figure 6-10 or equation (6-26) and is approximately 30. Therefore, the inductor Q should exceed 30 Q_{br} or 450 where $Q_{br} = f_0/BW_{3\text{ dB}}$.

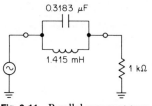

0.3183 μF

1.415 mH

1 kΩ

Fig. 6-11   Parallel resonant trap of example 6-4.

Frequently it is desirable to operate the band-reject network between equal source and load terminations instead of a voltage source as in figure 6-9. If a source and load resistor are specified where both are equal to R, equation (6-23) is modified to

$$f_c = \frac{1}{4\pi RC} \tag{6-27}$$

When the source and load are unequal, the cutoff frequency is given by

$$f_c = \frac{1}{2\pi(R_s + R_L)C} \tag{6-28}$$

Series Resonant Trap   An $n = 1$ band-reject filter can also be derived from the RL high-pass filter of figure 6-12a. The 3-dB cutoff is determined from

$$f_c = \frac{R}{2\pi L} \tag{6-29}$$

The band-reject filter of figure 6-12b is obtained by resonating the coil with a series capacitor where

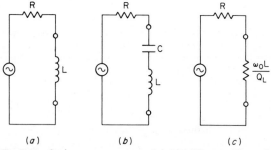

R

R

R

C

L

L

$\dfrac{\omega_0 L}{Q_L}$

(a)   (b)   (c)

Fig. 6-12   Series resonant trap: (a) RL high-pass filter; (b) result of band-reject transformation; (c) equivalent circuit at f_0.

$$C = \frac{1}{\omega_0^2 L} \qquad (6\text{-}24)$$

The center frequency is f_0 and the 3-dB bandwidth is equal to f_c. The series losses of an inductor can be represented by a resistor of $\omega_0 L/Q_L$. The equivalent circuit of the band-reject network at resonance is given by the circuit of figure 6-12c and the attenuation computed from equation (6-26) or figure 6-10.

Example 6-5

REQUIRED: Design a series resonant circuit having a 3-dB bandwidth of 500 Hz and a center frequency of 7500 Hz as in the previous example. The source impedance is 1 kΩ and the load is assumed infinite.

RESULT: Compute the element values from the following relationships:

$$L = \frac{R}{2\pi f_c} = \frac{1000}{2\pi 500} = 0.318 \text{ H} \qquad (6\text{-}29)$$

and $$C = \frac{1}{\omega_0^2 L} = \frac{1}{(2\pi 7500)^2 0.318} = 1420 \text{ pF} \qquad (6\text{-}24)$$

The circuit is given in figure 6-13.

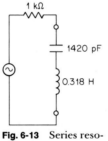

Fig. 6-13   Series resonant trap of example
6-5.

When a series resonant trap is to be terminated with a load resistance equal to the source, the high-pass 3-dB cutoff and resulting 3-dB bandwidth of the band-reject filter are given by

$$f_c = \frac{R}{4\pi L} \qquad (6\text{-}30)$$

For the more general case where source and load are unequal, the cutoff frequency is determined from

$$f_c = \frac{R_{eq}}{2\pi L} \qquad (6\text{-}31)$$

where R_{eq} is the equivalent value of the source and load resistors in parallel.

The Bridged-T Configuration   The resonant traps previously discussed suffer severe degradation of notch depth unless the inductor Q is many magnitudes greater than Q_{br}. The bridged-T band-reject structure can easily provide rejection of 60 dB or more with practical values of inductor Q. The configuration is shown in figure 6-14a.

To understand the operation of the circuit, let us first consider the equivalent circuit of a center-tapped inductor having a coefficient of magnetic coupling equal to unity which is shown in figure 6-14b. The inductance between terminals A and C corresponds to L of figure 6-14a. The inductance between A and B or B and C is equal to $L/4$ since, as the reader may recall, the impedance across one-half of a center-tapped autotransformer is one-fourth the overall impedance. This occurs because the impedance is proportional to the turns ratio squared.

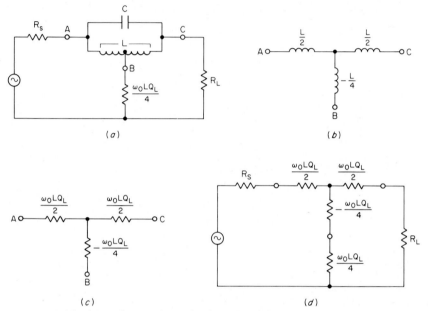

Fig. 6-14 Bridged-T null network: (*a*) circuit configuration; (*b*) equivalent circuit of center-tapped inductor; (*c*) tuned circuit equivalent at resonance; (*d*) bridged-T equivalent circuit at resonance.

The impedance of a parallel tuned circuit at resonance was previously determined to be equivalent to a resistor of $\omega_0 LQ_L$. Since the circuit of figure 6-14a is center-tapped, the equivalent three-terminal network is shown in figure 6-14c. The impedance between A and C is still $\omega_0 LQ_L$. A negative resistor must then exist in the middle shunt branch so that the impedance across one-half the tuned circuit is one-fourth the overall impedance or $\omega_0 LQ_L/4$. Of course negative resistors or inductors are physically impossible as individual passive two-terminal elements but can be embedded within an equivalent circuit.

If we combine the equivalent circuit of figure 6-14c with the bridged-T network of figure 6-14a, we obtain the circuit of figure 6-14d. The positive and negative resistors in the center branch will cancel, resulting in infinite rejection at center frequency. The degree of rejection actually obtained is dependent upon a variety of factors such as center-tap accuracy, coefficient of coupling, and magnitude of Q_L. When the bridged-T configuration is implemented by modifying a parallel trap design of figure 6-9b by adding a center tap and a resistor of $\omega_0 LQ_L/4$, a dramatic improvement in notch depth will usually occur.

A center-tapped inductor is not always available or practical. An alternate form of a bridged-T is given in figure 6-15. The parallel resonant trap design of figure 6-9 is modified by splitting the capacitor into two capacitors of twice the value, and a resistor of $\omega_0 LQ_L/4$ is introduced. The two capacitors should be closely matched.

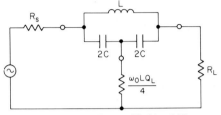

Fig. 6-15   Alternate form of bridged-T.

In conclusion, the bridged-T structure is an economical and effective means of increasing the available notch rejection of a parallel resonant trap without increasing the inductor Q. However, as a final general comment, a single null section can provide high rejection only at a single frequency or relatively narrow band of frequencies for a given 3-dB bandwidth, since $n = 1$. The stability of the circuit then becomes a significant factor. A higher-order band-reject filter design can have a wider stopband and yet maintain the same 3-dB bandwidth.

6.2   ACTIVE BAND-REJECT FILTERS

This section considers the design of active band-reject filters for both wide-band and narrow-band applications. Active null networks are covered and the popular twin-T circuit is discussed in detail.

Wide-Band Active Band-Reject Filters

Wide-band filters can be designed by first separating the specification into individual low-pass and high-pass requirements. Low-pass and high-pass filters are then independently designed and combined by paralleling the inputs and summing both outputs to form the band-reject filter.

A wide-band approach is valid when the separation between cutoffs is an octave or more for all-pole filters so that minimum interaction occurs in the stopband when the outputs are summed (see section 2.1 and figure 2-13). Elliptic-function networks will require less separation, since their characteristics are steeper.

An inverting amplifier is used for summing and can also provide gain. Filters can be combined using the configuration of figure 6-16a, where R is arbitrary and A is the desired gain. The individual filters should have a low output impedance to avoid loading by the summing resistors.

The VCVS elliptic-function low-pass and high-pass filters of sections 3.2 and 4.2 each require an RC termination on the last stage to provide the real pole. These elements can be combined with the summing resistors, resulting in the circuit of figure 6-16b. R_a and C_a correspond to the denormalized values of R_5 and C_5 for the low-pass filter of figure 3-20. The denormalized high-pass filter real-pole values are R_b and C_b. If only one filter is of the VCVS type, the summing network of the filter having the low output impedance can be replaced by a single resistor having a value of R.

When one or both filters are of the elliptic-function type, the ultimate attenuation obtainable is determined by the filter having the lesser value of A_{min}, since the stopband output is the summation of the contributions of both filters.

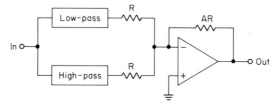

(a)

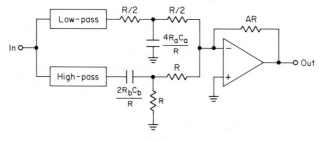

(b)

Fig. 6-16   Wide-band band-reject filters: (a) combining of filters having low output impedance; (b) combined filters requiring RC real poles.

Example 6-6

REQUIRED: Design an active band-reject filter having 3-dB points at 100 and 400 Hz and greater than 35 dB of attenuation between 175 and 225 Hz.

RESULT: (a) Since the ratio of upper cutoff to lower cutoff is well in excess of an octave, a wide-band approach can be used. First separate the specification into individual low-pass and high-pass requirements.

Low-pass:	High-pass:
3 dB at 100 Hz	3 dB at 400 Hz
35 dB minimum at 175 Hz	35 dB minimum at 225 Hz

(b) The low-pass and high-pass filters can now be independently designed as follows:

Low-pass filter:

Compute the steepness factor.

$$A_s = \frac{f_s}{f_c} = \frac{175 \text{ Hz}}{100 \text{ Hz}} = 1.75 \qquad (2\text{-}11)$$

An $n = 5$ Chebyshev filter having a 0.5-dB ripple is chosen using figure 2-44. The normalized active low-pass filter values are given in table 12-39 and the circuit is shown in figure 6-17a.

To denormalize the filter, multiply all resistors by Z and divide all capacitors by $Z \times$ FSF, where Z is conveniently selected at 10^5 and the FSF is $2\pi f_c$, where f_c is 100 Hz. The denormalized low-pass filter is given in figure 6-17b.

High-pass filter:

Compute the steepness factor.

$$A_s = \frac{f_c}{f_s} = \frac{400 \text{ Hz}}{225 \text{ Hz}} = 1.78 \qquad (2\text{-}13)$$

An $n = 5$ Chebyshev filter with a 0.5-dB ripple will also satisfy the high-pass requirement. A high-pass transformation can be performed on the normalized low-pass filter of figure 6-17a to obtain the circuit of figure 6-17c. All resistors have been replaced with capacitors and vice versa using reciprocal element values.

The normalized high-pass filter is then frequency- and impedance-scaled by multiplying all resistors by Z and dividing all capacitors by $Z \times$ FSF, where Z is chosen at 10^5 and FSF is $2\pi f_c$, using an f_c of 400 Hz. The denormalized high-pass filter is shown in figure 6-17d using standard 1% resistor values.

(c) The individual low-pass and high-pass filters can now be combined using the configuration of figure 6-16a. Since no gain is required, A is set equal to unity. The value of R is conveniently selected at 10 kΩ, resulting in the circuit of figure 6-17e.

Band-Reject Transformation of Low-Pass Poles

The wide-band approach to the design of band-reject filters using combined low-pass and high-pass networks is applicable to bandwidths of typically an octave or more. If the separation between cutoffs is insufficient, interaction in the stopband will occur, resulting in inadequate stopband rejection (see figure 2-13).

A more general approach involves normalizing the band-reject requirement and selecting a normalized low-pass filter type that meets these specifications. The corresponding normalized low-pass poles are then directly transformed to the band-reject form and realized using active sections.

A band-reject transfer function can be derived from a low-pass transfer function by substituting the frequency variable f by a new variable given by

$$f_{br} = \frac{1}{f_0 \left(\dfrac{f}{f_0} - \dfrac{f_0}{f} \right)} \qquad (6\text{-}32)$$

This transformation combines the low-pass to high-pass and subsequent band-reject transformation discussed in section 6.1 so that a band-reject filter can be obtained directly from the low-pass transfer function.

The band-reject transformation results in two pairs of complex poles and a pair of second-order imaginary zeros from each low-pass complex pole pair. A single low-pass real pole is transformed into a complex pole pair and a pair of first-order imaginary zeros. These relationships are illustrated in figure 6-18. The zeros occur at center frequency and result from the transformed low-pass zeros at infinity.

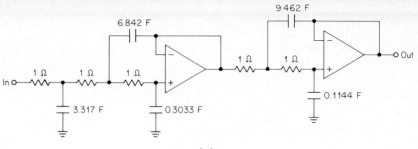

(a)

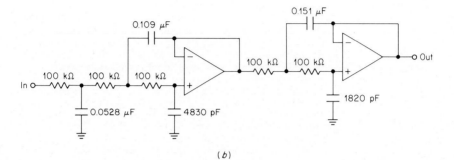

(b)

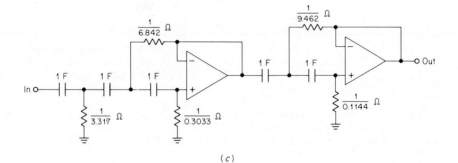

(c)

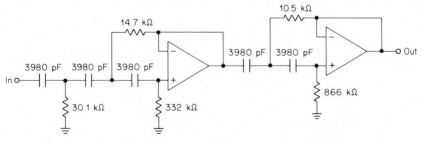

(d)

6-22

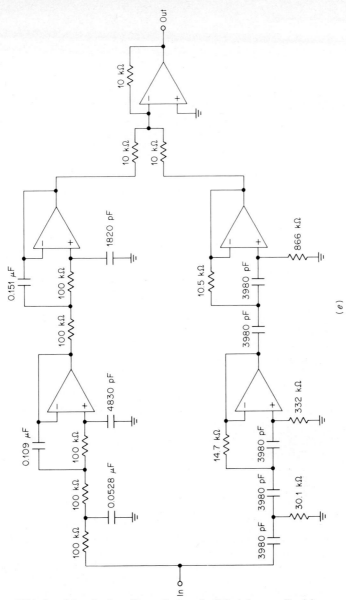

Fig. 6-17 Wide-band band-reject filter of example 6-6: *(a)* normalized low-pass filter; *(b)* denormalized low-pass filter; *(c)* transformed normalized high-pass filter; *(d)* denormalized high-pass filter; *(e)* combining filters to obtain band-reject response.

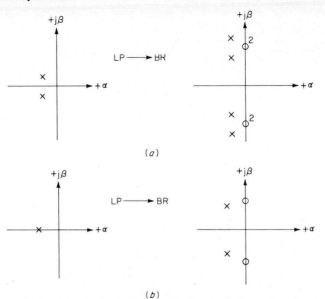

(a)

(b)

Fig. 6-18 Band-reject transformation of low-pass poles: (a) low-pass complex pole pair; (b) low-pass real pole.

The band-reject pole-zero pattern of figure 6-18a corresponds to two band-reject sections where each section provides a zero at center frequency and provides one of the pole pairs. The pattern of figure 6-18b is realized by a single band-reject section where the zero also occurs at center frequency.

To make the low-pass to band-reject transformation, first compute

$$Q_{br} = \frac{f_0}{BW} \tag{6-33}$$

where f_0 is the geometric center frequency and BW is the passband bandwidth. The transformation then proceeds as follows:

Complex Poles   The tables of chapter 12 contain tabulated poles corresponding to the all-pole low-pass filter families discussed in chapter 2. Complex poles are given in the form: $-\alpha \pm j\beta$ where α is the real coordinate and β is the imaginary part. Given α, β, Q_{br}, and f_0, the following computations result in two sets of values for Q and frequency which defines two band-reject filter sections. Each section also has a zero at f_0.

$$C = \alpha^2 + \beta^2 \tag{6-34}$$

$$D = \frac{\alpha}{Q_{br}C} \tag{6-35}$$

$$E = \frac{\beta}{Q_{br}C} \tag{6-36}$$

$$F = E^2 - D^2 + 4 \tag{6-37}$$

$$G = \sqrt{\frac{F}{2} + \sqrt{\frac{F^2}{4} + D^2 E^2}} \tag{6-38}$$

$$H = \frac{DE}{G} \tag{6-39}$$

$$K = \frac{1}{2} \sqrt{(D+H)^2 + (E+G)^2} \tag{6-40}$$

$$Q = \frac{K}{D+H} \tag{6-41}$$

$$f_{ra} = \frac{f_0}{K} \tag{6-42}$$

$$f_{rb} = K f_0 \tag{6-43}$$
$$f_\infty = f_0 \tag{6-44}$$

The two band-reject sections have resonant frequencies of f_{ra} and f_{rb} (in hertz) and identical Q's given by equation (6-41). In addition each section has a zero at f_0, the filter geometric center frequency.

Real Poles   A normalized low-pass real pole having a real coordinate of α_0 is transformed into a single band-reject section having a Q given by

$$Q = Q_{br} \, \alpha_0 \tag{6-45}$$

The section resonant frequency is equal to f_0, and the section must also have a transmission zero at f_0.

Example 6-7

REQUIRED: Determine the pole and zero locations for a band-reject filter having the following specifications:
Center frequency of 3600 Hz
3 dB at ±150 Hz
40 dB minimum at ±30 Hz

RESULT: (a) Since the filter is narrow, the requirement can be treated directly in its arithmetically symmetrical form:

$$f_0 = 3600 \text{ Hz}$$
$$BW_{3 \text{ dB}} = 300 \text{ Hz}$$
$$BW_{40 \text{ dB}} = 60 \text{ Hz}$$

The band-reject steepness factor is given by

$$A_s = \frac{\text{passband bandwidth}}{\text{stopband bandwidth}} = \frac{300 \text{ Hz}}{60 \text{ Hz}} = 5 \tag{2-20}$$

(b) An $n = 3$ Chebyshev normalized low-pass filter having a 0.1-dB ripple is selected using figure 2-42. The corresponding pole locations are found in table 12-23 and are
$-0.3500 \pm j0.8695$
-0.6999
First make the preliminary computation:

$$Q_{br} = \frac{f_0}{BW_{3 \text{ dB}}} = \frac{3600 \text{ Hz}}{300 \text{ Hz}} = 12 \tag{6-33}$$

The low-pass to band-reject pole transformation is performed as follows:

Complex-pole transformation:

$$\alpha = 0.3500 \qquad \beta = 0.8695$$

$$C = \alpha^2 + \beta^2 = 0.878530 \tag{6-34}$$

$$D = \frac{\alpha}{Q_{br}C} = 0.033199 \tag{6-35}$$

$$E = \frac{\beta}{Q_{br}C} = 0.082477 \tag{6-36}$$

$$F = E^2 - D^2 + 4 = 4.005700 \tag{6-37}$$

$$G = \sqrt{\frac{F}{2} + \sqrt{\frac{F^2}{4} + D^2 E^2}} = 2.001425 \tag{6-38}$$

$$H = \frac{DE}{G} = 0.001368 \tag{6-39}$$

$$K = \frac{1}{2}\sqrt{(D+H)^2 + (E+G)^2} = 1.042094 \tag{6-40}$$

$$Q = \frac{K}{D+H} = 30.15 \tag{6-41}$$

$$f_{ra} = \frac{f_0}{K} = 3455 \text{ Hz} \tag{6-42}$$

$$f_{rb} = Kf_0 = 3752 \text{ Hz} \tag{6-43}$$

$$f_\infty = f_0 = 3600 \text{ Hz} \tag{6-44}$$

Real-pole transformation:

$$\alpha_0 = 0.6999$$

$$Q = Q_{br}\,\alpha_0 = 8.40 \tag{6-45}$$

$$f_r = f_\infty = f_0 = 3600 \text{ Hz}$$

The block diagram is shown in figure 6-19.

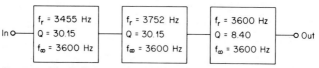

In o— | f_r = 3455 Hz | f_r = 3752 Hz | f_r = 3600 Hz | —o Out

Fig. 6-19 Block diagram of example 6-7.

Narrow-Band Active Band-Reject Filters

Narrow-band active band-reject filters are designed by first transforming a set of normalized low-pass poles to the band-reject form. The band-reject poles are computed in terms of resonant frequency f_r, Q, and f_∞ using the results of section 6.2 and are then realized with active band-reject sections.

The VCVS Band-Reject Section   Complex low-pass poles result in a set of band-reject parameters where f_r and f_∞ do not occur at the same frequency. Band-reject sections are then required that permit independent selection of f_r

and f_∞ in their design procedure. Both the VCVS and biquad circuits covered in section 5.2 under Elliptic-Function Bandpass Filters have this degree of freedom.

The VCVS realization is shown in figure 6-20. The design equations were given in section 5.2 under Elliptic-Function Bandpass Filters and are repeated

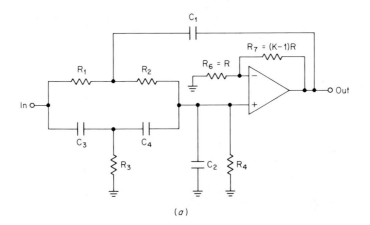

(a)

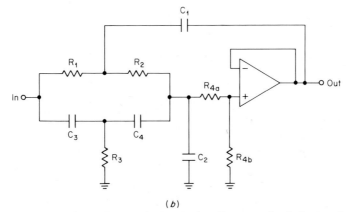

(b)

Fig. 6-20 VCVS realization for band-reject filters: (a) circuit for $K > 1$; (b) circuit for $K < 1$.

here for convenience, where f_r, Q, and f_∞ are obtained by the band-reject transformation procedure of section 6.2. The values are computed from

$$R_1 = \frac{f_\infty^2 + f_r^2}{3f_\infty^2} R'$$
(6-46)

$$R_2 = 2R_1$$
(6-47)

$$R_3 = \frac{f_\infty^2 + f_r^2}{4.5f_r^2} R'$$
(6-48)

$$R_4 = 4.5R_3 \tag{6-49}$$

$$C_1 = \frac{1.5}{2\pi f_r R_1} \tag{6-50}$$

$$C_2 = \frac{C_1}{4.5} \tag{6-51}$$

$$C_3 = \frac{f_r}{2\pi R_1 f_\infty^2} \tag{6-52}$$

$$C_4 = \frac{C_3}{2} \tag{6-53}$$

$$K = \frac{\left(2.5 - \dfrac{1}{Q}\right)\left(\dfrac{f_r^2}{f_\infty^2} + 1\right)}{1.5} \tag{6-54}$$

$$R_6 = R \tag{6-55}$$
$$R_7 = (K-1)R \tag{6-56}$$

where R' and R can be arbitrarily chosen.

The circuit of figure 6-20a is used when $K > 1$. In the cases where $K < 1$, the configuration of figure 6-20b is utilized, where

$$R_{4a} = (1 - K)R_4 \tag{6-57}$$
and
$$R_{4b} = KR_4 \tag{6-58}$$

The section gain at DC is given by

$$A_{dc} = \frac{f_\infty^2}{f_r^2 + f_\infty^2} \tag{6-59}$$

The gain of the composite filter in the passband is the product of the DC gains of all the sections.

The VCVS structure has a number of undesirable characteristics. Although the circuit Q can be adjusted by making R_6 or R_7 variable when $K > 1$, the Q cannot be independently measured since the 3-dB bandwidth at the output is affected by the transmission zero. Resonant frequency f_r or the notch frequency f_∞ cannot be easily adjusted, since these parameters are determined by the inter-action of a number of elements. Also the section gain is fixed by the design parameters. Another disadvantage of the circuit is that a large spread in capacitor values[1] may occur so that standard values cannot be easily used. Nevertheless the VCVS realization makes effective use of a minimum number of operational amplifiers in comparison with other implementations and is widely used. However, because of its lack of adjustment capability, its application is generally restricted to Q's below 10 and with 1% component tolerances.

The State-Variable Band-Reject Section The biquad or state-variable ellip-tic-function bandpass filter section discussed in section 5.2 is highly suitable for implementing band-reject transfer functions. The circuit is given in figure 6-21. By connecting resistor R_5 to either node 1 or to node 2, the notch frequency f_∞ will be located above or below the pole resonant frequency f_r.

---

[1] The elliptic-function configuration of the VCVS uniform capacitor structure given in section 3.2 can be used at the expense of additional sensitivity.

Section Q's of up to 200 can be obtained. The design parameters f_r, Q, and f_∞ as well as the section gain can be independently chosen, monitored, and adjusted. From the point of view of low sensitivity and maximum flexibility the biquad approach is the most desirable method of realization.

The design equations were stated in section 5.2 under Elliptic-Function Band-pass Filters and are repeated here for convenience, where f_r, Q, and f_∞ are given and the values of C, R, and R' can be arbitrarily chosen.

$$R_1 = R_4 = \frac{Q}{2\pi f_r C} \tag{6-60}$$

$$R_2 = R_3 = \frac{R_1}{Q} \tag{6-61}$$

$$R_5 = \frac{f_r^2 R}{Q|f_r^2 - f_\infty^2|} \tag{6-62}$$

for $f_\infty > f_r$:
$$R_6 = \frac{f_r^2 R}{f_\infty^2} \tag{6-63}$$

and when $f_\infty < f_r$:
$$R_6 = R \tag{6-64}$$

The value of R_6 is based on unity section gain at DC. The gain can be raised or lowered by proportionally increasing or decreasing R_6.

Resonance is adjusted by monitoring the phase shift between the section input and node 3 using a Lissajous pattern and adjusting R_3 for 180° phase shift with an input frequency of f_r.

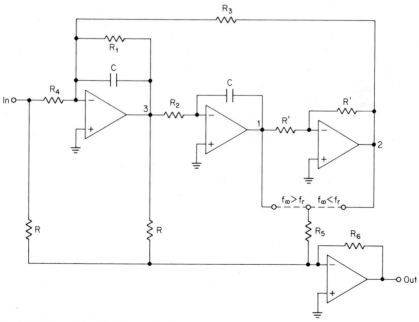

Fig. 6-21   Biquad band-reject realization.

The Q is controlled by R_1 and can be measured at node 3 in terms of section 3-dB bandwidth, or R_1 can be adjusted until unity gain occurs between the input and node 3 with f_r applied. Because of the Q enhancement effect discussed in section 5.2 under All-Pole Bandpass Configurations a Q adjustment is usually necessary.

The notch frequency is then determined by monitoring the section output for a null. Adjustment is normally not required, since tuning of f_r will usually bring in f_∞ with acceptable accuracy. If an adjustment is desired, R_5 can be made variable.

Sections for Transformed Real Poles   When a real pole undergoes a band-reject transformation, the result is a single pole pair and a single set of imaginary zeros. Complex poles resulted in two sets of pole pairs and two sets of zeros. The resonant frequency f_r of the transformed real pole is exactly equal to the notch frequency f_∞; so the design flexibility of the VCVS and biquad structures is not required.

A general second-order bandpass transfer function can be expressed as

$$T(s) = \frac{\dfrac{\omega_r}{Q} s}{s^2 + \dfrac{\omega_r}{Q} s + \omega_r^2} \tag{6-65}$$

where the gain is unity at ω_r. If we realize the circuit of figure 6-22 where $T(s)$ corresponds to the above transfer function, the composite transfer function at the output is given by

$$T(s) = \frac{s^2 + \omega_r^2}{s^2 + \dfrac{\omega_r}{Q} s + \omega_r^2} \tag{6-66}$$

This corresponds to a band-reject transfer function having a transmission zero at f_r (i.e., $f_\infty = f_r$). The occurrence of this zero can also be explained intuitively from the structure of figure 6-22. Since $T(s)$ is unity only at f_r, both input signals to the summing amplifier will then cancel, resulting in no output signal.

These results indicate that band-reject sections for transformed real poles can be obtained by combining any of the all-pole bandpass circuits of section 5.2 in the configuration of figure 6-22. The basic design parameters are the required f_r and Q of the band-reject section which are directly used in the design equations for the bandpass circuits.

By combining these bandpass sections with summing amplifiers, the three band-reject structures of figure 6-23 can be derived. The design equations for the bandpass sections were given in section 5.2 and are repeated here where C, R, and R' can be arbitrarily chosen.

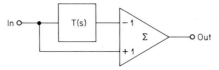

Fig. 6-22   Band-reject configuration for $f_r = f_\infty$.

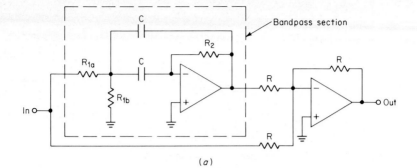

(a)

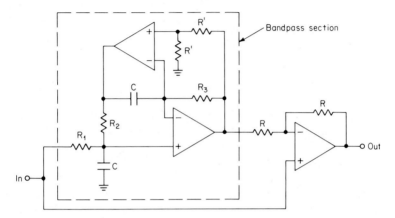

(b)

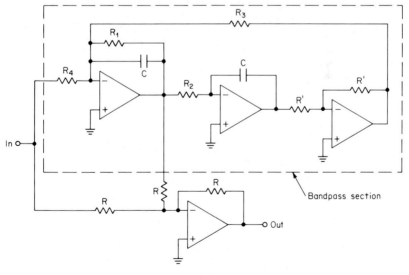

(c)

Fig. 6-23 Band-reject circuits for $f_r = f_\infty$ (a) MFBP band-reject section ($Q < 20$);
(b) DABP band-reject section ($Q < 150$); (c) biquad band-reject section ($Q < 200$).

The MFBP band-reject section $(f_r = f_\infty)$ is given by

$$R_2 = \frac{Q}{\pi f_r C} \qquad (6\text{-}67)$$

$$R_{1a} = \frac{R_2}{2} \qquad (6\text{-}68)$$

$$R_{1b} = \frac{R_{1a}}{2Q^2 - 1} \qquad (6\text{-}69)$$

The DABP band-reject section $(f_r = f_\infty)$ is given by

$$R_1 = \frac{Q}{2\pi f_r C} \qquad (6\text{-}70)$$

$$R_2 = R_3 = \frac{R_1}{Q} \qquad (6\text{-}71)$$

The biquad band-reject section $(f_r = f_\infty)$ is given by

$$R_1 = R_4 = \frac{Q}{2\pi f_r C} \qquad (6\text{-}72)$$

$$R_2 = R_3 = \frac{R_1}{Q} \qquad (6\text{-}73)$$

These equations correspond to unity bandpass gain for the MFBP and biquad circuits so that cancellation at f_r will occur when the section input and bandpass output signals are equally combined by the summing amplifiers. Since the DABP section has a gain of 2 and has a noninverting output, the circuit of figure 6-23 b has been modified accordingly so that cancellation occurs.

Tuning can be accomplished by making R_{1b}, R_2, and R_3 variable in the MFBP, DABP, and biquad circuits, respectively. In addition the biquad circuit will usually require R_1 to be made adjustable to compensate for the Q-enhancement effect (see section 5.2 under All-Pole Bandpass Configurations). The circuit can be tuned by adjusting the indicated elements for either a null at f_r measured at the circuit output or for 0° or 180° phase shift at f_r observed between the input and the output of the bandpass section. If the bandpass section gain is not sufficiently close to unity for the MFBP and biquad case and 2 for the DABP circuit, the null depth may be inadequate.

Example 6-8

REQUIRED: Design an active band-reject filter from the band-reject parameters determined in example 6-7 having a gain of +6 dB.

RESULT: (a)   The band-reject transformation in example 6-7 resulted in the following set of requirements for a three-section filter:

Section	f_r	Q	f_∞
1	3455 Hz	30.15	3600 Hz
2	3752 Hz	30.15	3600 Hz
3	3600 Hz	8.40	3600 Hz

(b)   Two biquad circuits in tandem will be used for sections 1 and 2 followed by a DABP band-reject circuit for section 3. The value of C is chosen at 0.01 μF and R as well as R' at 10 kΩ. Since the DABP section has a gain of 2 at DC, which satisfies the 6-dB gain requirement, both biquad sections should then have unity gain. The element values are determined as follows:

Section 1 (biquad of figure 6-21):

$$f_r = 3455 \text{ Hz} \qquad Q = 30.15 \qquad f_\infty = 3600 \text{ Hz}$$

$$R_1 = R_4 = \frac{Q}{2\pi f_r C} = \frac{30.15}{2\pi \times 3455 \times 10^{-8}} = 138.9 \text{ k}\Omega \quad (6\text{-}60)$$

$$R_2 = R_3 = \frac{R_1}{Q} = \frac{138.9 \times 10^3}{30.15} = 4610 \ \Omega \qquad (6\text{-}61)$$

$$R_5 = \frac{f_r^2 R}{Q|f_r^2 - f_\infty^2|} = \frac{3455^2 \times 10^4}{30.15|3455^2 - 3600^2|} = 3870 \ \Omega \quad (6\text{-}62)$$

$$R_6 = \frac{f_r^2 R}{f_\infty^2} = \frac{3455^2 \times 10^4}{3600^2} = 9210 \ \Omega \qquad (6\text{-}63)$$

Section 2 (biquad of figure 6-21):

$$f_r = 3752 \text{ Hz} \qquad Q = 30.15 \qquad f_\infty = 3600 \text{ Hz}$$

$$R_1 = R_4 = 127.9 \text{ k}\Omega \qquad\qquad\qquad (6\text{-}60)$$
$$R_2 = R_3 = 4240 \ \Omega \qquad\qquad\qquad (6\text{-}61)$$
$$R_5 = 4180 \ \Omega \qquad\qquad\qquad (6\text{-}62)$$
$$R_6 = 10 \text{ k}\Omega \qquad\qquad\qquad (6\text{-}64)$$

Section 3 (DABP of figure 6-23):

$$f_r = f_\infty = 3600 \text{ Hz} \qquad Q = 8.40$$

$$R_1 = \frac{Q}{2\pi f_r C} = \frac{8.40}{2\pi \times 3600 \times 10^{-8}} = 37.1 \text{ k}\Omega \quad (6\text{-}70)$$

$$R_2 = R_3 = \frac{R_1}{Q} = \frac{37.1 \times 10^3}{8.40} = 4420 \ \Omega \qquad (6\text{-}71)$$

The final circuit is shown in figure 6-24 with standard 1% resistor values. The required resistors have been made variable so that the resonant frequencies can be adjusted for all sections, and in addition the Q is variable for the biquad circuits.

Active Null Networks

Active null networks are single sections used to provide attenuation at a single frequency or over a narrow band of frequencies. The most popular sections are of the twin-T form; so this circuit will be discussed in detail along with some other structures.

The Twin-T The twin-T was first discovered by H. W. Augustadt in 1934. Although this circuit is passive by nature, it is also used in many active configurations to obtain a variety of different characteristics.

The circuit of figure 6-25a is an RC bridge structure where balance or an output null occurs at 1 rad/s when all arms have an equal impedance (0.5 − j0.5 Ω). The circuit is redrawn in the form of a symmetrical lattice in figure 6-25b (refer to Guillemin and Stewart in references for detailed discussions of the lattice). The lattice of figure 6-25b can be redrawn again in the form of two parallel lattices as shown in figure 6-25c.

If identical series elements are present in both the series and shunt branches of a lattice, the element may be extracted and symmetrically placed outside

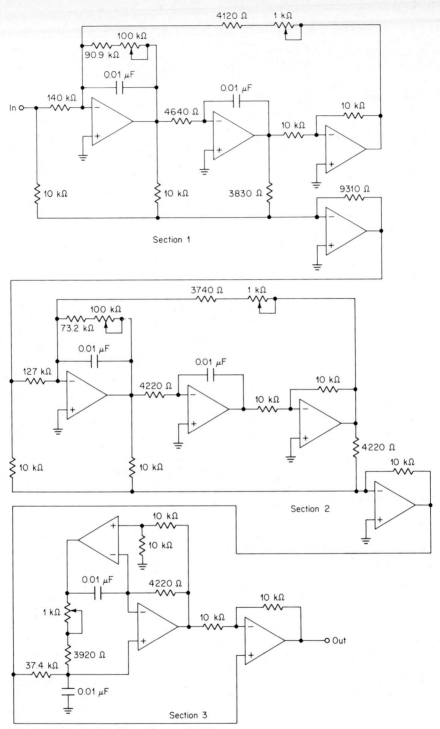

Fig. 6-24 Band-reject filter of example 6-8.

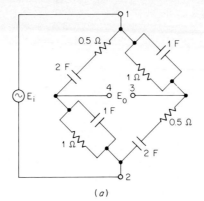

(a)

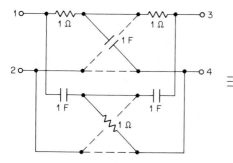

(b)

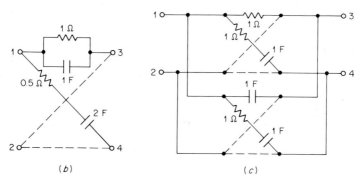

(c)

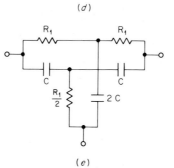

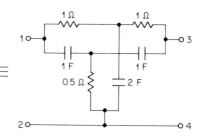

(d)

(e)

Fig. 6-25 Derivation of the twin-T: *(a)* RC bridge; *(b)* lattice circuit; *(c)* parallel lattice; *(d)* twin-T equivalent; *(e)* general form of twin-T.

the lattice structure. A 1-Ω resistor satisfies this requirement for the upper lattice and a 1-F capacitor for the lower lattice. Removal of these components to outside the lattice results in the twin-T of figure 6-25d.

The general form of a twin-T is shown in figure 6-25e. The value of R_1 is computed from

$$R_1 = \frac{1}{2\pi f_0 C} \tag{6-74}$$

where C is arbitrary. This denormalizes the circuit of figure 6-25d so that the null now occurs at f_0 instead of at 1 rad/s.

When a twin-T is driven from a voltage source and terminated in an infinite load,[2] the transfer function is given by

$$T(s) = \frac{s^2 + \omega_0^2}{s^2 + 4\omega_0 s + \omega_0^2} \tag{6-75}$$

If we compare this expression with the general transfer function of a second-order pole-zero section as given by equation (6-66), we can determine that a twin-T provides a notch at f_0 with a Q of $\frac{1}{4}$. The attenuation at any bandwidth can be computed by

$$A_{dB} = 10 \, \log \left[1 + \left(\frac{4f_0}{BW_{x\,dB}} \right)^2 \right] \tag{6-76}$$

The frequency response is shown in figure 6-26. The requirement for geometric symmetry applies.

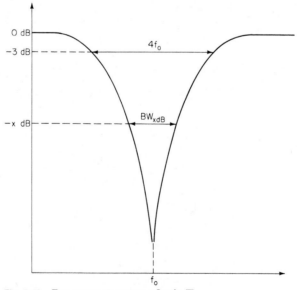

Fig. 6-26   Frequency response of twin-T.

[2] Since the source and load are always finite, the value of R_1 should be in the vicinity of $\sqrt{R_s R_L}$, provided that the ratio R_L / R_s is in excess of 10.

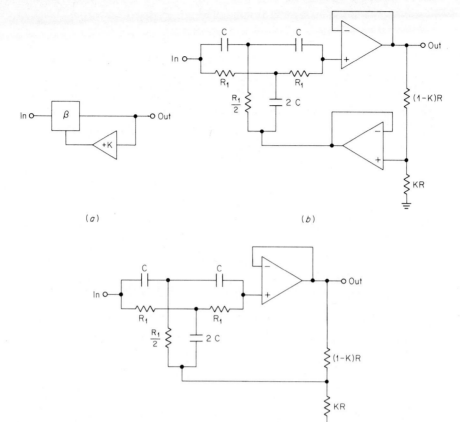

Fig. 6-27 Twin-T with positive feedback: (a) block diagram; (b) circuit realization; (c) simplified configuration $R_1 \gg (1 - K)\ R$.

Twin-T with Positive Feedback The twin-T has gained widespread usage as a general-purpose null network. However, a major shortcoming is a fixed Q of $\frac{1}{4}$. This limitation can be overcome by introducing positive feedback.

The transfer function of the circuit of figure 6-27a can be derived as

$$T(s) = \frac{\beta}{1 + K(\beta - 1)} \tag{6-77}$$

If β is replaced by equation (6-75), the transfer function of a twin-T, the resulting circuit transfer function expression becomes

$$T(s) = \frac{s^2 + \omega_0^2}{s^2 + 4\omega_0 (1 - K)s + \omega_0^2} \tag{6-78}$$

The corresponding Q is

$$Q = \frac{1}{4(1 - K)} \tag{6-79}$$

By selecting a positive K of < 1 and sufficiently close to unity, the circuit Q can be dramatically increased. The required value of K can be determined by

$$K = 1 - \frac{1}{4Q} \tag{6-80}$$

The block diagram of figure 6-27a can be implemented using the circuit of figure 6-27b, where R is arbitrary. By choosing C and R so that $R_1 \gg (1 - K)R$, the circuit may be simplified to the configuration of figure 6-27c, which uses only one amplifier.

The attenuation at any bandwidth is given by

$$A_{dB} = 10 \log \left[1 + \left(\frac{f_0}{Q \times BW_{x\ dB}} \right)^2 \right] \tag{6-81}$$

Equation (6-81) is the general expression for the attentuation of a single band-reject section where the resonant frequency and notch frequency are identical (i.e., $f_r = f_\infty$). The attenuation formula can be expressed in terms of the 3-dB bandwidth as follows:

$$A_{dB} = 10 \log \left[1 + \left(\frac{BW_{3\ dB}}{BW_{x\ dB}} \right)^2 \right] \tag{6-82}$$

The attenuation characteristics can also be determined from the frequency-response curve of a normalized $n = 1$ Butterworth low-pass filter (see figure 2-34) by using the ratio $BW_{3\ dB}/BW_{x\ dB}$ for the normalized frequency.

The twin-T in its basic form or in the positive-feedback configuration is widely used for single-section band-reject sections. However, it suffers from the fact that tuning cannot be easily accomplished. Tight component tolerances may then be required to ensure sufficient accuracy of tuning and adequate notch depth. About 40- to 60-dB rejection at the notch could be expected using 1% components.

Example 6-9

REQUIRED: Design a single null network having a center frequency of 1000 Hz and a 3-dB bandwidth of 100 Hz. Also determine the attenuation at the 30-Hz bandwidth.

RESULT: (a) A twin-T structure with positive feedback will be used. To design the twin-T, first choose a capacitance C of 0.01 μF. The value of R_1 is given by

$$R_1 = \frac{1}{2\pi f_0 C} = \frac{1}{2\pi \times 10^3 \times 10^{-8}} = 15.9\ \text{k}\Omega \tag{6-74}$$

(b) The required value of K for the feedback network is calculated from

$$K = 1 - \frac{1}{4Q} = 1 - \frac{1}{4 \times 10} = 0.975 \tag{6-80}$$

where $Q = f_0/BW_{3\ dB}$

(c) The single amplifier circuit of figure 6-27c will be used. If R is chosen at 1 kΩ, the circuit requirement for $R_1 \gg (1 - K)R$ is satisfied. The resulting section is shown in figure 6-28.

(d) To determine the attenuation at a bandwidth of 30 Hz, calculate

$$A_{dB} = 10 \log \left[1 + \left(\frac{BW_{3\ dB}}{BW_{x\ dB}} \right)^2 \right] = 10 \log \left[1 + \left(\frac{100\ \text{Hz}}{30\ \text{Hz}} \right)^2 \right]$$
$$= 10.8\ \text{dB} \tag{6-82}$$

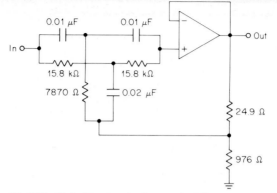

Fig. 6-28   Twin-T network of example 6-9.

Bandpass Structure Null Networks   Section 6.2 under Narrow-Band Active Band-Reject Filters showed how a first-order bandpass section can be combined with a summing amplifier to obtain a band-reject circuit for transformed real poles where $f_r = f_\infty$. Three types of sections were illustrated in figure 6-23 corresponding to different Q ranges of operation. These same sections can be used as null networks. They offer more flexibility than the twin-T, since the null frequency can be adjusted to compensate for component tolerances. In addition the DABP and biquad circuits permit Q adjustment as well.

The design formulas were given by equations (6-67) through (6-73). The values of f_r and Q in the equations correspond to the section center frequency and Q, respectively.

Frequently a bandpass and band-reject output are simultaneously required. A typical application might involve separation of signals for comparison of in-band and out-of band spectral energy. The band-reject sections of figure 6-23 can each provide a bandpass output from the bandpass section along with the null output signal. An additional feature of this technique is that the bandpass and band-reject outputs will track.

REFERENCES

Guillemin, E. A., "Communication Networks," Vol. 2, John Wiley and Sons, New York, 1935.

Saal, R., and Ulbrich, E., "On the Design of Filters by Synthesis," IRE Transactions on Circuit Theory, December 1958.

Stewart, J. L., "Circuit Theory and Design," John Wiley and Sons, New York, 1956.

Tow, J., "A Step-by-Step Active Filter Design," IEEE Spectrum, Vol. 6, pp. 64–68, December 1969.

Williams, A. B., "Active Filter Design," Artech House, Dedham, Mass., 1975.

Zverev, A. I., "Handbook of Filter Synthesis," John Wiley and Sons, New York, 1967.

7

Networks for the Time Domain

7.1 ALL-PASS TRANSFER FUNCTIONS

Up until now, the networks we discussed were used to obtain a desired amplitude versus frequency characteristic. No less important is the all-pass family of filters. This class of networks exhibits a flat frequency response but introduces a pre-scribed phase shift versus frequency. All-pass filters are frequently called "delay equalizers."

If a network is to be of an all-pass type, the absolute magnitudes of the numerator and denominator of the transfer function must be related by a fixed constant at all frequencies. This condition will be satisfied if the zeros are the images of the poles. Since poles are restricted to the left half quadrants of the complex frequency plane to maintain stability, the zeros must occur in the right half plane as the mirror image of the poles about the $j\omega$ axis. Figure 7-1 illustrates the all-pass pole-zero representations in the complex frequency plane for first-order and second-order all-pass transfer functions.

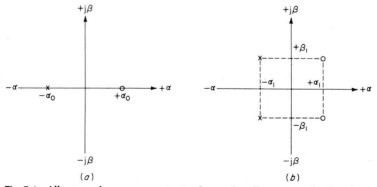

Fig. 7-1 All-pass pole-zero patterns: (a) first-order all-pass transfer function; (b) second-order all-pass transfer function.

First-Order All-Pass Transfer Functions

The real pole-zero pair of figure 7-1a has a separation of $2\alpha_0$ between the pole and zero and corresponds to the following first-order all-pass transfer function:

$$T(s) = \frac{s - \alpha_0}{s + \alpha_0} \qquad (7\text{-}1)$$

To determine the absolute magnitude of $T(s)$, compute

$$|T(s)| = \frac{|s - \alpha_0|}{|s + \alpha_0|} = \frac{\sqrt{\alpha_0^2 + \omega^2}}{\sqrt{\alpha_0^2 + \omega^2}} = 1 \qquad (7\text{-}2)$$

where $s = j\omega$. For any value of frequency the numerator and denominator of equation (7-2) are equal; so the transfer function is clearly all-pass and has an absolute magnitude of unity at all frequencies.

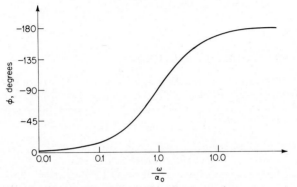

Fig. 7-2   Phase shift of first-order all-pass section.

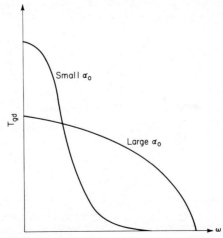

Fig. 7-3   Group delay of first-order all-pass transfer functions.

The phase shift is given by

$$\beta(\omega) = -2 \tan^{-1} \frac{\omega}{\alpha_0} \tag{7-3}$$

where $\beta(\omega)$ is in radians.

The phase shift versus the ratio ω/α_0 as defined by equation (7-3) is plotted in figure 7-2. The phase angle is $0°$ at DC and $-90°$ at $\omega = \alpha_0$. The phase shift asymptotically approaches $-180°$ with increasing frequency.

The group delay was defined in section 2.2, under Effect of Nonuniform Time Delay, as the derivative of the phase shift which results in

$$T_{gd} = -\frac{d\beta(\omega)}{d\omega} = \frac{2\alpha_0}{\alpha_0^2 + \omega^2} \tag{7-4}$$

If equation (7-4) is plotted with respect to ω for different values of α_0, a family of curves are obtained as shown in figure 7-3. First-order all-pass sections exhibit maximum delay at DC and decreasing delay with increasing frequency. For small values of α_0 the delay becomes large at low frequencies and decreases quite rapidly above this range. The delay at DC is found by setting ω equal to zero in equation (7-4), which results in

$$T_{gd} \text{ (DC)} = \frac{2}{\alpha_0} \tag{7-5}$$

Second-Order All-Pass Transfer Functions

The second-order all-pass transfer function represented by the pole-zero pattern of figure 7-1b is given by

$$T(s) = \frac{s^2 - \dfrac{\omega_r}{Q} s + \omega_r^2}{s^2 + \dfrac{\omega_r}{Q} s + \omega_r^2} \tag{7-6}$$

where ω_r and Q are the pole resonant frequency (in radians per second) and the pole Q. These terms may also be computed from the real and imaginary pole-zero coordinates of figure 7-1b by

$$\omega_r = \sqrt{\alpha_1^2 + \beta_1^2} \tag{7-7}$$

and

$$Q = \frac{\omega_r}{2\alpha_1} \tag{7-8}$$

The absolute magnitude of $T(s)$ is found to be

$$|T(s)| = \frac{\sqrt{(\omega_r^2 - \omega^2)^2 + \dfrac{\omega^2 \omega_r^2}{Q^2}}}{\sqrt{(\omega_r^2 - \omega^2)^2 + \dfrac{\omega^2 \omega_r^2}{Q^2}}} = 1 \tag{7-9}$$

which is all-pass.

The phase shift in radians is

$$\beta(\omega) = -2 \tan^{-1} \frac{\dfrac{\omega \omega_r}{Q}}{\omega_r^2 - \omega^2} \tag{7-10}$$

and the group delay is given by

$$T_{gd} = \frac{2Q\,\omega_r(\omega^2 + \omega_r^2)}{Q^2(\omega^2 - \omega_r^2)^2 + \omega^2\omega_r^2} \tag{7-11}$$

The phase and delay parameters of first-order transfer functions are relatively simple to manipulate, since they are a function of a single design parameter α_0. A second-order type, however, has two design parameters, Q and ω_r.

The phase shift of a second-order transfer function is $-180°$ at $\omega = \omega_r$. At DC the phase shift is zero and at frequencies well above ω_r the phase asymptotically approaches $-360°$.

The group delay reaches a peak which occurs very close to ω_r. As the Q is made larger, the peak delay increases, the delay response becomes sharper, and the delay at DC decreases as shown in figure 7-4.

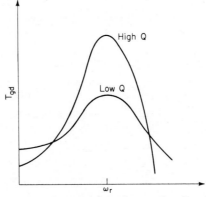

Fig. 7-4 Group delay of second-order all-pass transfer functions.

The frequency of maximum delay is slightly below ω_r and is expressed in radians per second by

$$\omega(T_{gd,\max}) = \omega_r \sqrt{\sqrt{4 - \frac{1}{Q^2}} - 1} \tag{7-12}$$

For all practical purposes the maximum delay occurs at ω_r for Q's in excess of 2.

By setting $\omega = \omega_r$ in equation (7-11), the delay at ω_r is given by

$$T_{gd,\max} = \frac{4Q}{\omega_r} = \frac{2Q}{\pi f_r} \tag{7-13}$$

If we set $\omega = 0$, the delay at DC is found from

$$T_{gd}\,(DC) = \frac{2}{Q\omega_r} = \frac{1}{Q\pi f_r} \tag{7-14}$$

7.2 DELAY EQUALIZER SECTIONS

Passive or active networks that realize first- or second-order all-pass transfer functions are called "delay equalizers," since they are normally used to provide a required delay characteristic without disturbing the amplitude response. All-pass networks can be realized in a variety of configurations both passive and active. Equalizers with adjustable characteristics can also be designed and are discussed in section 7.6.

LC All-Pass Structures

First-Order Constant-Resistance Circuit The lattice of figure 7-5a realizes a first-order all-pass transfer function. The network is also a constant-resistance type, which means that the input impedance has a constant value of R over

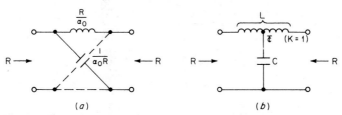

Fig. 7-5   First-order *LC* equalizer section: (*a*) lattice form; (*b*) unbalanced form.

the entire frequency range. Constant-resistance networks can be cascaded with no interaction so that composite delay curves can be built up by accumulating the individual delay contributions. The lattice has an equivalent unbalanced form shown in figure 7-5b. The design formulas are given by

$$L = \frac{2R}{\alpha_0} \tag{7-15}$$

$$C = \frac{2}{\alpha_0 R} \tag{7-16}$$

where R is the desired impedance level and α_0 is the real pole-zero coordinate. The phase shift and delay properties were defined by equations (7-3) through (7-5).

The circuit of figure 7-5b requires a center-tapped inductor having a coefficient of magnetic coupling K equal to unity.

Second-Order Constant-Resistance Sections A second-order all-pass lattice with constant-resistance properties is shown in figure 7-6a. The circuit may be transformed into the unbalanced bridged-T form of figure 7-6b. The elements are given by

$$L_a = \frac{2R}{\omega_r Q} \tag{7-17}$$

$$C_a = \frac{Q}{\omega_r R} \tag{7-18}$$

$$L_b = \frac{QR}{2\omega_r} \tag{7-19}$$

$$C_b = \frac{2Q}{\omega_r (Q^2 - 1)R} \tag{7-20}$$

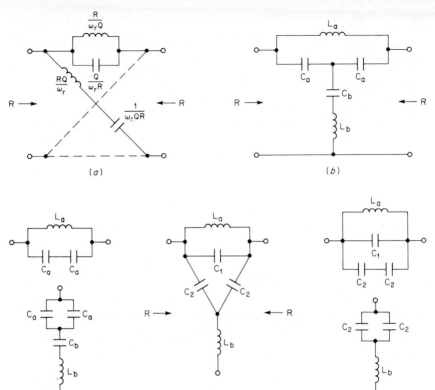

Fig. 7-6   Second-order section $Q > 1$: *(a)* lattice form; *(b)* unbalanced form; *(c)* circuit for measuring branch resonances; *(d)* circuit modified by T to pi transformation; *(e)* resonant branches of modified circuit.

For tuning and test purposes the section can be split into parallel and series resonant branches by opening the shunt branch and shorting the bridging or series branch as shown in figure 7-6c. Both circuits will resonate at ω_r.

The T to pi transformation was first introduced in section 4.1. This transformation may be applied to the T of capacitors that are embedded in the section of figure 7-6b to reduce capacitor values if desired. The resulting circuit is given in figure 7-6d. Capacitors C_1 and C_2 are computed as follows:

$$C_1 = \frac{C_a^2}{2C_a + C_b} \tag{7-21}$$

$$C_2 = \frac{C_a C_b}{2C_a + C_b} \tag{7-22}$$

The branch resonances are obtained by opening the shunt branch and then shorting the bridging branch, which results in the parallel and series resonant circuits of figure 7-6e. Both resonances occur at ω_r.

Close examination of equation (7-20) indicates that C_b will be negative if the Q is less than 1. (If $Q = 1$, C_b can be replaced by a short.) This restricts the circuits of figure 7-6 to those cases where the Q is in excess of unity. Fortunately this is true in most instances.

In those cases where the Q is below 1, the configurations of figure 7-7 are

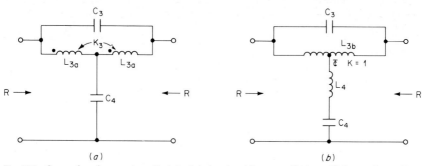

(a)   (b)

Fig. 7-7 Second-order section $Q < 1$: (a) circuit with controlled coefficient of coupling; (b) circuit with unity coefficient of coupling.

used. The circuit of figure 7-7a uses a single inductor with a controlled coefficient of coupling given by

$$K_3 = \frac{1 - Q^2}{1 + Q^2} \tag{7-23}$$

The element values are given by

$$L_{3a} = \frac{(Q^2 + 1)R}{2\,Q\,\omega_r} \tag{7-24}$$

$$C_3 = \frac{Q}{2\omega_r R} \tag{7-25}$$

$$C_4 = \frac{2}{Q\,\omega_r\,R} \tag{7-26}$$

It is not always convenient to control the coefficient of coupling of a coil to obtain the specific value required by equation (7-23). A more practical approach uses the circuit of figure 7-7b. The inductor L_{3b} is center-tapped and requires a unity coefficient of coupling (typical values of 0.95 or greater can usually be obtained and are acceptable). The values of L_{3b} and L_4 are computed from

$$L_{3b} = 2(1 + K_3)L_{3a} \tag{7-27}$$

$$L_4 = \frac{(1 - K_3)L_{3a}}{2} \tag{7-28}$$

The sections of figure 7-7 may be tuned to ω_r in the same manner as in the equalizers in figure 7-6. A parallel resonant circuit is obtained by opening C_4 and a series resonant circuit will result by placing a short across C_3.

The second-order section of figures 7-6 and 7-7 may not always be all-pass.

If the inductors have insufficient Q, a notch will occur at the resonances and will have a notch depth that can be approximated by

$$A_{\text{dB}} = 20 \log \frac{Q_L + 4Q}{Q_L - 4Q} \qquad (7\text{-}29)$$

where Q_L is the inductor Q. If the notch is unacceptable, adequate coil Q must be provided or amplitude-equalization techniques are used as discussed in section 8.4.

Minimum Inductor All-Pass Sections   The bridged-T circuit of figure 7-8 realizes a second-order all-pass transfer function with a single inductor. The section is not a constant-resistance type and operates between a zero impedance

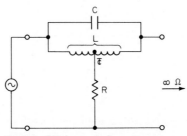

Fig. 7-8   Minimum inductor type, second-order section.

source and an infinite load. If the ratio of load to source is well in excess of 10, satisfactory results will be obtained. The elements are computed by

$$C = \frac{Q}{4\omega_r R} \qquad (7\text{-}30)$$

and

$$L = \frac{1}{\omega_r^2 C} \qquad (7\text{-}31)$$

The value of R can be chosen as the geometric mean of the source and load impedance ($\sqrt{R_s R_L}$). The LC circuit is parallel resonant at ω_r.

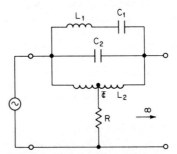

Fig. 7-9   Fourth-order minimum inductance, all-pass structure.

The reader is reminded that the design parameters Q and ω_r are determined from the delay parameters defined in section 7.1, which covers all-pass transfer functions.

A notch will occur at resonance due to finite inductor Q and can be calculated from

$$A_{\text{dB}} = 20 \log \frac{4R + \omega_r L Q_L}{4R - \omega_r L Q_L} \tag{7-32}$$

If R is set equal to $\omega_r L Q_L / 4$, the notch attenuation becomes infinite and the circuit is then identical to the bridged-T null network of section 6.1.

Two sets of all-pass poles and zeros corresponding to a fourth-order transfer function can also be obtained by using a minimum-inductance-type structure. The circuit configuration is shown in figure 7-9.

Upon being given two sets of equalizer parameters Q_1, ω_{r1} and Q_2, ω_{r2} as defined in section 7.1, the following equations are used to determine the element values:

First compute

$$A = \omega_1^2 \omega_{r2}^2 \tag{7-33}$$

$$B = A\left(\frac{1}{\omega_{r2} Q_2} + \frac{1}{\omega_{r1} Q_1}\right) \tag{7-34}$$

$$C = \frac{Q_1 Q_2}{A(Q_2 \omega_{r1} + Q_1 \omega_{r2})} \tag{7-35}$$

$$D = \frac{Q_1 Q_2 (\omega_{r1}^2 + \omega_{r2}^2) + \omega_{r1} \omega_{r2}}{ABQ_1 Q_2} - C - \frac{1}{AB^2 C} \tag{7-36}$$

$$E = \frac{1}{ABCD} \tag{7-37}$$

The element values are then given by

$$L_1 = \frac{4ER}{A} \tag{7-38}$$

$$C_1 = \frac{AD}{4R} \tag{7-39}$$

$$L_2 = \frac{4BR}{A} \tag{7-40}$$

$$C_2 = \frac{AC}{4R} \tag{7-41}$$

The value of R is generally chosen as the geometric mean of the source and load terminations as with the second-order minimum-inductance section. The series and parallel branch resonant frequencies are found from

$$\omega_{L1C1} = \frac{1}{\sqrt{ED}} \tag{7-42}$$

and

$$\omega_{L2C2} = \frac{1}{\sqrt{BC}} \tag{7-43}$$

Active All-Pass Structures

First- and second-order all-pass transfer functions can be obtained by using an active approach. The general form of the active all-pass section is represented by the block diagram of figure 7-10, where $T(s)$ is a first- or second-order transfer function having a gain of unity.

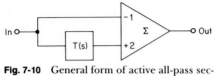

Fig. 7-10   General form of active all-pass section.

First-Order Sections   The transfer function of the circuit of figure 7-10 is given by

$$\frac{E_{out}}{E_{in}} = 2\,T(s) - 1 \tag{7-44}$$

If $T(s)$ is a first-order RC high-pass network having the transfer function $sCR/(sCR + 1)$, the composite transfer function becomes

$$\frac{E_{out}}{E_{in}} = \frac{s - 1/RC}{s + 1/RC} \tag{7-45}$$

This expression corresponds to the first-order all-pass transfer function of equation (7-1), where

$$\alpha_0 = \frac{1}{RC} \tag{7-46}$$

The circuit can be directly implemented by the configuration of figure 7-11a, where R' is arbitrary. The phase shift is then given by

$$\beta(\omega) = -2\,\tan^{-1}\omega RC \tag{7-47}$$

and the delay is found from

$$T_{gd} = \frac{2RC}{(\omega RC)^2 + 1} \tag{7-48}$$

At DC the delay is a maximum and is computed from

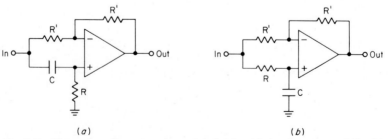

(a)   (b)

Fig. 7-11   First-order all-pass sections: (a) circuit with lagging phase shift; (b) circuit with leading phase shift.

$$T_{gd}(\text{DC}) = 2RC \tag{7-49}$$

The corresponding phase shift is shown in figure 7-2. A phase shift of $-90°$ occurs at $\omega = 1/RC$ and approaches $-180°$ and $0°$ at DC and infinity, respectively. By making the element R variable, an all-pass network can be obtained having a phase shift adjustable between 0 and $-180°$.

A sign inversion of the phase will occur if the circuit of figure 7-11b is used. The circuit will remain all-pass and first-order, and the group delay is still defined by equations (7-48) and (7-49).

Second-Order Sections If $T(s)$ in figure 7-10 is a second-order bandpass network having the general bandpass transfer function

$$T(s) = \frac{\dfrac{\omega_r}{Q}s}{s^2 + \dfrac{\omega_r}{Q}s + \omega_r^2} \tag{7-50}$$

the composite transfer function then becomes

$$\frac{E_{\text{out}}}{E_{\text{in}}} = 2T(s) - 1 = -\frac{s^2 - \dfrac{\omega_r}{Q}s + \omega_r^2}{s^2 + \dfrac{\omega_r}{Q}s + \omega_r^2} \tag{7-51}$$

which corresponds to the second-order all-pass expression given by equation (7-6) (except for a simple sign inversion). Therefore, a second-order all-pass equalizer can be obtained by implementing the structure of figure 7-10 using a single active bandpass section for $T(s)$.

Section 5-2 discussed the MFBP, DABP, and biquad all-pole bandpass sections. Each circuit can be combined with a summing amplifier to generate a delay equalizer.

The MFBP equalizer section is shown in figure 7-12a. The element values are given by

$$R_2 = \frac{2Q}{\omega_r C} = \frac{Q}{\pi f_r C} \tag{7-52}$$

$$R_{1a} = \frac{R_2}{2} \tag{7-53}$$

$$R_{1b} = \frac{R_{1a}}{2Q^2 - 1} \tag{7-54}$$

The values of C and R can be arbitrarily chosen, and A in figure 7-12a corresponds to the desired gain.

The maximum delay which occurs at f_r was given by equation (7-13). This expression can be combined with equations (7-52) and (7-54); so the element values can alternately be expressed in terms of $T_{gd,\text{max}}$ as follows for $Q > 2$:

$$R_2 = \frac{T_{gd,\text{max}}}{2C} \tag{7-55}$$

$$R_{1b} = \frac{R_2}{(\pi f_r T_{gd,\text{max}})^2 - 2} \tag{7-56}$$

where R_{1a} remains $R_2/2$.

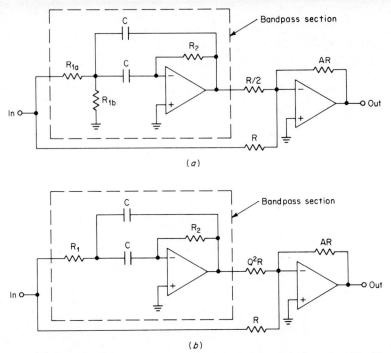

Fig. 7-12   MFBP delay equalizer $Q < 20$: (a) circuit for $0.707 < Q < 20$; (b) circuit for $Q < 0.707$.

The MFBP section can be tuned by making R_{1b} variable. R_{1b} can then be adjusted until 180° of phase shift occurs between the input and output of the bandpass section at f_r. In order for the response to be all-pass, the bandpass section gain must be exactly unity at resonance. Otherwise an amplitude ripple will occur in the frequency-response characteristic in the vicinity of f_r.

The section Q is limited to values below 20 or is expressed in terms of delay

$$T_{gd,\max} < \frac{40}{\pi f_r} \tag{7-57}$$

Experience has indicated that required Q's are usually well under 20; so this circuit will suffice in most cases. However, if the Q is below 0.707, the value of R_{1b} as given by equation (7-54) is negative; so the circuit of figure 7-12b is used. The value of R_1 is given by

$$R_1 = \frac{R_2}{4Q^2} \tag{7-58}$$

In the event higher Q's are required, the DABP section can be applied to the block diagram of figure 7-10. Since the DABP circuit has a gain of 2 and is noninverting, the implementation shown in figure 7-13 is used. The element values are given by

$$R_1 = \frac{Q}{\omega_r C} = \frac{Q}{2\pi f_r C} \tag{7-59}$$

and
$$R_2 = R_3 = \frac{R_1}{Q}$$

(7-60)

where C, R, and R' can be conveniently chosen. Resistor R_2 may be made variable if tuning is desired. The Q and therefore the delay can also be trimmed by making R_1 adjustable.

The biquad structure can be configured in the form of a delay equalizer. The circuit is shown in figure 7-14. The element values are computed from

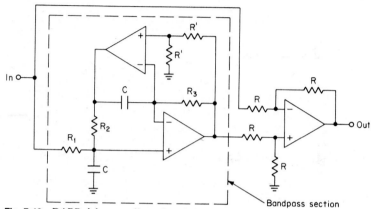

Fig. 7-13 DABP delay equalizer $Q < 150$.

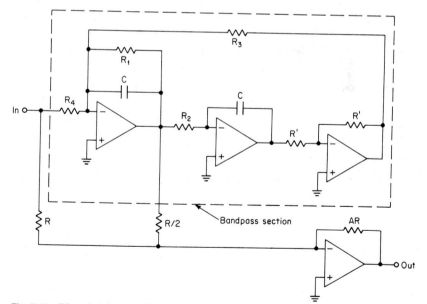

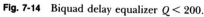

Fig. 7-14 Biquad delay equalizer $Q < 200$.

$$R_1 = R_4 = \frac{Q}{\omega_r C} = \frac{Q}{2\pi f_r C} \tag{7-61}$$

and

$$R_2 = R_3 = \frac{R_1}{Q} \tag{7-62}$$

where C, R, and R' are arbitrary.

Resistor R_3 can be made variable for tuning. The Q is adjusted for the nominal value by making R_1 variable and monitoring the 3-dB bandwidth at the output of the bandpass section or adjusting for unity bandpass gain at f_r. The biquad is subject to the Q-enchancement effect discussed in section 5.2, under All-Pole Bandpass Configurations; so a Q adjustment is usually required.

7.3   DESIGN OF DELAY LINES

The classical approach to the design of delay lines involves a cascade of identical LC sections (except for the end sections) and uses image-parameter theory (see Wallis in references). This technique is an approximation at best.

Modern network theory permits us to predict the delay of networks accurately and to obtain a required delay in a much more efficient manner than with the classical approach. The Bessel, linear phase with equiripple error and transitional filters all feature a constant delay. The curves in chapter 2 indicate that for n greater then 3, the flat delay region is extended well into the stopband. If a delay line is desired, a low-pass filter implementation is not a very desirable approach from a delay-bandwidth perspective. A significant portion of the constant delay region would be attenuated.

All the low-pass transfer functions covered can be implemented by using an all-pass realization to overcome the bandwidth limitations. This results in a very precise and efficient means of designing delay lines.

The Low-Pass to All-Pass Transformation

A low-pass transfer function can be transformed to an all-pass transfer function by simply introducing zeros in the right half plane of the $j\omega$ axis corresponding to each pole. If the real and complex poles tabulated in chapter 12 are realized using the first- and second-order all-pass structures of section 7.2, complementary zeros will also occur. When a low-pass to all-pass transformation is made, the low-pass delay is increased by a factor of exactly 2 because of the additional phase-shift contributions of the zeros.

TABLE 7-1   All-Pass Bessel Delay Characteristics

N	T_{gd} (DC)	1% Deviation		10% Deviation	
		ω_u	TU	ω_u	TU
2	2.72	0.412	1.121	0.801	2.179
3	3.50	0.691	2.419	1.109	3.882
4	4.26	0.906	3.860	1.333	5.679
5	4.84	1.120	5.421	1.554	7.521
6	5.40	1.304	7.042	1.737	9.380
7	5.90	1.478	8.720	1.912	11.280
8	6.34	1.647	10.440	2.079	13.180
9	6.78	1.794	12.160	2.227	15.100

An all-pass delay-bandwidth factor can be derived from the delay curves of chapter 2 which is given by

$$TU = \omega_u T_{gd} \, (\text{DC}) \tag{7-63}$$

The value of T_{gd} (DC) is the delay at DC, which is twice the delay shown in the curves because of the all-pass transformation, and ω_u is the upper limit radian frequency where the delay deviates a specified amount from the DC value.

Table 7-1 lists the delay at DC, ω_u, and the delay-bandwidth product TU for an all-pass realization of the Bessel maximally flat delay family. Values are provided for both 1 and 10% deviations of delay at ω_u.

To choose a transfer-function type and determine the complexity required, first compute

$$TU_{\text{req}} = 2\pi f_{gd} T_{gd} \tag{7-64}$$

where f_{gd} is the maximum desired frequency of operation and T_{gd} is the nominal delay needed. A network is then selected that has a delay-bandwidth factor TU that exceeds TU_{req}.

Compute the delay-scaling factor (DSF), which is the ratio of the normalized delay at DC to the required nominal delay, i.e.,

$$\text{DSF} = \frac{T_{gd}(\text{DC})}{T_{gd}} \tag{7-65}$$

The corresponding poles of the filter selected are denormalized by the DSF and can then be realized by the all-pass circuits of section 7.2.

A real pole α_0 is denormalized by the formula

$$\alpha_0' = \alpha_0 \times \text{DSF} \tag{7-66}$$

Complex poles tabulated in the form $\alpha + j\beta$ are denormalized and transformed into the all-pass section design parameters ω_r and Q by the relationships

$$\omega_r = \text{DSF}\sqrt{\alpha^2 + \beta^2} \tag{7-67}$$

$$Q = \frac{\omega_r}{2\alpha \, \text{DSF}} \tag{7-68}$$

The parameters α_0', ω_r, and Q are then directly used in the design equations for the circuits of section 7.2.

Sometimes the required delay-bandwidth factor TU_{req} as computed by equation (7-64) is in excess of the TU factors available from the standard filter families tabulated. The total delay required can then be subdivided into N smaller increments and realized by N delay lines in cascade, since the delays will add algebraically.

LC Delay Lines

LC delay lines are designed by first selecting a normalized filter type and then denormalizing the corresponding real and complex poles, all in accordance with section 7.3, under the Low-Pass to All-Pass Transformation.

The resulting poles and associated zeros are then realized using the LC all-pass circuits of section 7.2. This procedure is best illustrated by the following design example.

Example 7-1

REQUIRED: Design a passive delay line to provide 1 ms of delay constant within 10% from DC to 3200 Hz. The source and load impedances are both 10 kΩ.

RESULT: (a) Compute the required delay-bandwidth factor.

$$TU_{req} = 2\pi f_{gd} T_{gd} = 2\pi 3200 \times 0.001 = 20.1 \quad (7\text{-}64)$$

A linear phase design with an equiripple error of 0.5° will be chosen. The delay characteristics for the corresponding low-pass filters are shown in figure 2-64. The delay at DC of a normalized all-pass network for $n = 9$ is equal to 7.5 s, which is twice the value obtained from the curves. Since the delay remains relatively flat to 3 rad/s, the delay-bandwidth factor is given by

$$TU = \omega_u T_{gd} \text{ (DC)} = 3 \times 7.5 = 22.5 \quad (7\text{-}63)$$

Since TU is in excess of TU_{req}, the $n = 9$ design will be satisfactory.

(b) The low-pass poles are found in table 12-45 and are as follows:

$-0.5688 \pm j0.7595$
$-0.5545 \pm j1.5089$
$-0.5179 \pm j2.2329$
$-0.4080 \pm j2.9028$
-0.5728

Four second-order all-pass sections and a single first-order section will be required. The delay-scaling factor is given by

$$\text{DSF} = \frac{T_{gd} \text{ (DC)}}{T_{gd}} = \frac{7.5}{10^{-3}} = 7500 \quad (7\text{-}65)$$

The denormalized design parameters ω_r and Q for the second-order sections are computed by equations (7-67) and (7-68), respectively, and are tabulated as follows:

Section	α	β	ω_r	Q
1	0.5688	0.7595	7117	0.8341
2	0.5545	1.5089	12057	1.450
3	0.5179	2.2329	17191	2.213
4	0.4080	2.9028	21985	3.592

The design parameter α'_0 for section 5 corresponding to the real pole is found from

$$\alpha'_0 = \alpha_0 \times \text{DSF} = 4296 \quad (7\text{-}66)$$

where α_0 is 0.5728.

(c) The element values can now be computed as follows:

Section 1:

Since the Q is less than unity, the circuit of figure 7-7b will be used. The element values are found from

$$K_3 = \frac{1 - Q^2}{1 + Q^2} = \frac{1 - 0.8341^2}{1 + 0.8341^2} = 0.1794 \quad (7\text{-}23)$$

$$L_{3a} = \frac{(Q^2 + 1) R}{2 Q \omega_r} = \frac{(0.8341^2 + 1)10^4}{2 \times 0.8341 \times 7117} = 1.428 \text{ H} \quad (7\text{-}24)$$

$$C_3 = \frac{Q}{2\omega_r R} = \frac{0.8341}{2 \times 7117 \times 10^4} = 5860 \text{ pF} \quad (7\text{-}25)$$

$$C_4 = \frac{2}{Q\,\omega_r\,R} = \frac{2}{0.8341 \times 7117 \times 10^4} = 0.0337 \ \mu\mathrm{F} \quad (7\text{-}26)$$

$$L_{3b} = 2(1 + K_3)\,L_{3a} = 3.368 \ \mathrm{H} \quad (7\text{-}27)$$

$$L_4 = \frac{(1 - K_3)\,L_{3a}}{2} = 0.586 \ \mathrm{H} \quad (7\text{-}28)$$

Sections 2 through 4:

Since the Q's are in excess of unity, the circuit of figure 7-6b will be used. The values for section 2 are found from

$$L_a = \frac{2\,R}{\omega_r\,Q} = \frac{2 \times 10^4}{12{,}057 \times 1.450} = 1.144 \ \mathrm{H} \quad (7\text{-}17)$$

$$C_a = \frac{Q}{\omega_r\,R} = \frac{1.450}{12{,}057 \times 10^4} = 0.012 \ \mu\mathrm{F} \quad (7\text{-}18)$$

$$L_b = \frac{QR}{2\omega_r} = \frac{1.450 \times 10^4}{2 \times 12{,}057} = 0.601 \ \mathrm{H} \quad (7\text{-}19)$$

$$C_b = \frac{2\,Q}{\omega_r\,(Q^2 - 1)\,R} = \frac{2 \times 1.450}{12{,}057(1.45^2 - 1)10^4}$$
$$= 0.0218 \ \mu\mathrm{F} \quad (7\text{-}20)$$

In the same manner the remaining element values can be computed, which results in:

Section 3:

$$L_a = 0.526 \ \mathrm{H}$$
$$C_a = 0.0129 \ \mu\mathrm{F}$$
$$L_b = 0.644 \ \mathrm{H}$$
$$C_b = 6606 \ \mathrm{pF}$$

Section 4:

$$L_a = 0.253 \ \mathrm{H}$$
$$C_a = 0.0163 \ \mu\mathrm{F}$$
$$L_b = 0.817 \ \mathrm{H}$$
$$C_b = 2745 \ \mathrm{pF}$$

Section 5:

The remaining first-order all-pass section is realized using the circuit of figure 7-5b. The element values are given by

$$L = \frac{2\,R}{\alpha_0'} = \frac{2 \times 10^4}{4296} = 4.655 \ \mathrm{H} \quad (7\text{-}15)$$

$$C = \frac{2}{\alpha_0'\,R} = \frac{2}{4296 \times 10^4} = 0.0466 \ \mu\mathrm{F} \quad (7\text{-}16)$$

(d) The resulting delay line is illustrated in figure 7-15a. The resonant frequencies shown are in hertz corresponding to $\omega_r/2\pi$ for each section. The center-tapped inductors require a unity coefficient of coupling. The delay characteristics as a function of frequency are also shown in figure 7-15b.

The delay line of example 7-1 requires a total of nine inductors. If the classical design approach (see Wallis in references), which is based on image-parameter theory, were used, the resulting delay line would use about twice as many coils. Although the inductors would all be uniform in value (except for the end sections), this feature is certainly not justified by the added cost and complexity.

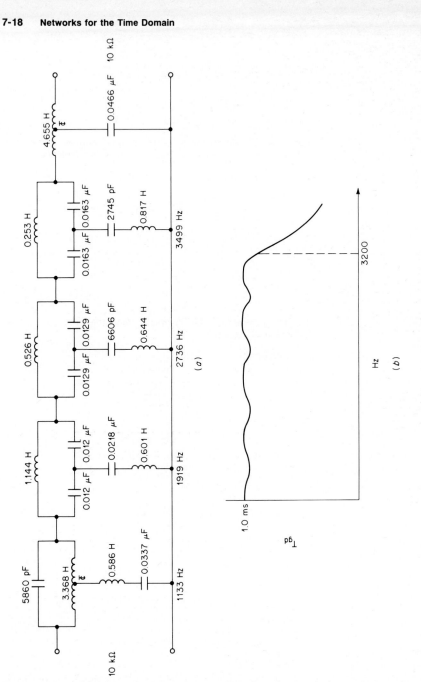

Fig. 7-15   1-ms delay line of example 7-1: (a) delay-line circuit; (b) frequency response.

Active Delay Lines

An active delay line is designed by initially choosing a normalized filter and then denormalizing the associated poles in the same manner as in the case of LC delay lines. The resulting all-pass design parameters are implemented using the first and second-order active structures of section 7.2.

Active delay lines do not suffer from the Q limitations of LC delay lines and are especially suited for low-frequency applications where inductor values may become impractical. The following example illustrates the design of an active delay line.

Example 7-2

REQUIRED: Design an active delay line having a delay of 100 μs constant within 3% to 3 kHz. A gain of 10 is also required.

RESULT: (a) Compute the required delay-bandwidth factor.

$$TU_{\text{req}} = 2\pi f_{gd} T_{gd} = 2\pi 3000 \times 10^{-4} = 1.885 \quad (7\text{-}64)$$

A Bessel-type all-pass network will be chosen. Table 7-1 indicates that for a delay deviation of 1%, a complexity of $n = 3$ has a delay-bandwidth factor of 2.419, which is in excess of the required value.

(b) The Bessel low-pass poles are given in table 12-41 and the corresponding values for $n = 3$ are

$$-1.0509 \pm j1.0025$$
$$-1.3270$$

Two sections are required consisting of a first-order and second-order type. The delay-scaling factor is computed to be

$$\text{DSF} = \frac{T_{gd}(\text{DC})}{T_{gd}} = \frac{3.5}{10^{-4}} = 3.5 \times 10^4 \quad (7\text{-}65)$$

where $T_{gd}(\text{DC})$ is obtained from table 7-1 and T_{gd} is 100 μs, the design value.

The second-order section design parameters are

$$\omega_r = \text{DSF}\sqrt{\alpha^2 + \beta^2} = 3.5 \times 10^4 \sqrt{1.0509^2 + 1.0025^2}$$
$$= 50{,}833 \quad (7\text{-}67)$$

and

$$Q = \frac{\omega_r}{2\,\alpha\,\text{DSF}} = \frac{50{,}833}{2 \times 1.0509 \times 3.5 \times 10^4} = 0.691 \quad (7\text{-}68)$$

The first-order section design parameter is given by

$$\alpha'_0 = \alpha_0 \times \text{DSF} = 1.327 \times 3.5 \times 10^4 = 46{,}450 \quad (7\text{-}66)$$

(c) The element values are computed as follows:

Second-order section:

The MFBP equalizer section of figure 7-12b will be used corresponding to $Q < 0.707$, where $R = 10$ kΩ, $C = 0.01$ μF, and $A = 10$. The element values are found from

$$R_2 = \frac{2Q}{\omega_r C} = \frac{2 \times 0.691}{50{,}833 \times 10^{-8}} = 2719 \ \Omega \quad (7\text{-}52)$$

$$R_1 = \frac{R_2}{4 Q^2} = \frac{2719}{4 \times 0.691^2} = 1424 \ \Omega \quad (7\text{-}58)$$

First-order section:

The first-order section of figure 7-11a will be used, where R' is chosen at 10 kΩ, C at 0.01 μF, and α_0' is 46,450. The value of R is given by

$$R = \frac{1}{\alpha_0'C} = \frac{1}{46,450 \times 10^{-8}} = 2153 \ \Omega \qquad (7\text{-}46)$$

(d)   The resulting 100-μs active delay line is shown in figure 7-16 using standard 1% resistor values.

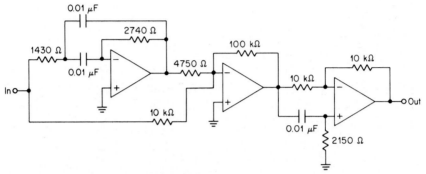

Fig. 7-16   100-μs delay line of example 7-2.

7.4   DELAY EQUALIZATION OF FILTERS

The primary emphasis in previous chapters has been the attenuation characteristics of filters. However, if the signal consists of a modulated waveform, the delay characteristics are also of importance. To minimize distortion of the input signal, a constant delay over the frequency range of interest is desirable.

The Bessel, linear phase with equiripple error, and transitional filter families all exhibit a flat delay. However, the amplitude response is less selective than that of other families. Frequently the only solution to an attenuation requirement is a Butterworth, Chebyshev, or elliptic-function filter type. To also maintain the delay constant, delay equalizers are required.

Delay equalizer networks are frequently at least as complex as the filter being equalized. The number of sections required is dependent on the initial delay curve, the portion of the curve to be equalized, and the degree of equalization necessary. A very crude approximation to the number of equalizer sections required is given by

$$n = 2\Delta_{\text{BW}}\Delta_T + 1 \qquad (7\text{-}69)$$

where Δ_{BW} is the bandwidth of interest in hertz and Δ_T is the delay distortion over Δ_{BW} in seconds.

The approach to delay equalization discussed in this section is graphical rather than analytical. A closed-form solution to the delay equalization of filters is not available. However, computer programs can be obtained that achieve a least-squares approximation to the required delay specifications and are preferred to trial-and-error techniques.

Simply stated, delay equalization of a filter involves designing a network that has a delay shape which complements the delay curve of the filter being equalized.

The composite delay will then exhibit the required flatness. Although the absolute delay increases as well, this result is usually of little significance, since it is the delay variation over the band of interest that disperses the spectral components of the signal. Typical delay curves of a bandpass filter, the delay equalizer network, and the composite characteristics are shown in figure 7-17.

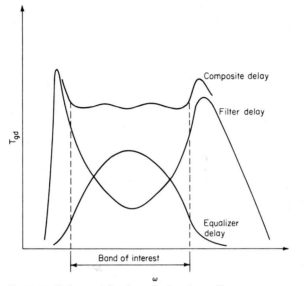

Fig. 7-17 Delay equalization of a bandpass filter.

To equalize the delay of a low-pass filter graphically, the highest frequency of interest and corresponding delay should be scaled to 1 rad/s so that the lower portion of the curve falls within the frequency region between DC and $\omega = 1$. This is accomplished by multiplying the delay axis by $2\pi f_h$, where f_h is the highest frequency to be equalized. The frequency axis is also divided by f_h and interpreted in radians per second so that f_h is transformed to 1 rad/s and all other frequencies are normalized to this point.

The normalized low-pass filter delay curves shown in section 2 for the various filter families may also be used directly. In either case the required equalizer delay characteristic is obtained by subtracting the delay curve from a constant equal to the maximum delay that occurs over the band. The resulting curve is then approximated by adding the delay contributions of equalizer sections. A sufficient number of sections are used to obtain the required composite delay flatness.

When a suitable match to the required curve is found, the equalizer parameters may be directly used to design the equalizer, and the circuit is then denormalized to the required frequency range and impedance level. Alternately the equalizer parameters can first be denormalized and the equalizer designed directly.

Bandpass filters are equalized in a manner similar to low-pass filters. The delay curve is first normalized by multiplying the delay axis by $2\pi f_0$, where f_0 is the filter center frequency. The frequency axis is divided by f_0 and interpreted in radians per second so that the center frequency is 1 rad/s and all other frequencies are normalized to the center frequency. A complementary curve is found and appropriate equalizer sections are used until a suitable fit occurs. The equalizer is then denormalized.

First-Order Equalizers

First-order all-pass transfer functions were first introduced in section 7.1. The delay of a first-order all-pass section is characterized by a maximum delay at low frequencies and decreasing delay with increasing frequency. As the value of α_0 is reduced, the delay tends to peak at DC and will roll off more rapidly with increasing frequencies.

The delay of a first-order all-pass section was given in section 7.1 by

$$T_{gd} = \frac{2\alpha_0}{\alpha_0^2 + \omega^2} \tag{7-4}$$

Working directly with equation (7-4) is somewhat tedious; so a table of delay values for α_0 ranging between 0.05 and 2.00 at frequencies from $\omega = 0$ to $\omega = 1$ is provided in table 7-2. This table can be directly used to determine the approximate α_0 necessary to equalize the normalized filter delay. A more exact value of α_0 can then be determined from equation (7-4) if desired.

TABLE 7-2   First-Order Equalizer Delay, Seconds

	ω, rad/s										
α_0	0	0.1	0.2	0.3	0.4	0.5	0.6	0.7	0.8	0.9	1.0
0.05	40.00	8.00	2.35	1.08	0.62	0.40	0.28	0.20	0.16	0.12	0.10
0.10	20.00	10.00	4.00	2.00	1.18	0.77	0.54	0.40	0.31	0.24	0.20
0.15	13.33	9.23	4.80	2.67	1.64	1.10	0.78	0.59	0.45	0.36	0.29
0.20	10.00	8.00	5.00	3.08	2.00	1.38	1.00	0.75	0.59	0.47	0.38
0.25	8.00	6.90	4.88	3.28	2.25	1.60	1.18	0.91	0.71	0.57	0.47
0.30	6.67	6.00	4.62	3.33	2.40	1.76	1.33	1.03	0.82	0.67	0.55
0.35	5.71	5.28	4.31	3.29	2.48	1.88	1.45	1.14	0.92	0.75	0.62
0.40	5.00	4.71	4.00	3.20	2.50	1.95	1.54	1.23	1.00	0.82	0.69
0.45	4.44	4.24	3.71	3.08	2.48	1.99	1.60	1.30	1.07	0.89	0.75
0.50	4.00	3.85	3.45	2.94	2.44	2.00	1.64	1.35	1.12	0.94	0.80
0.55	3.64	3.52	3.21	2.80	2.38	1.99	1.66	1.39	1.17	0.99	0.84
0.60	3.33	3.24	3.00	2.67	2.31	1.97	1.67	1.41	1.20	1.03	0.88
0.65	3.08	3.01	2.81	2.54	2.23	1.93	1.66	1.42	1.22	1.05	0.91
0.70	2.86	2.80	2.64	2.41	2.15	1.89	1.65	1.43	1.24	1.08	0.94
0.75	2.67	2.62	2.49	2.30	2.08	1.85	1.63	1.43	1.25	1.09	0.96
0.80	2.50	2.46	2.35	2.19	2.00	1.80	1.60	1.42	1.25	1.10	0.98
0.85	2.35	2.32	2.23	2.09	1.93	1.75	1.57	1.40	1.25	1.11	0.99
0.90	2.22	2.20	2.12	2.00	1.86	1.70	1.54	1.38	1.24	1.11	0.99
0.95	2.11	2.08	2.02	1.91	1.79	1.65	1.50	1.36	1.23	1.11	1.00
1.00	2.00	1.98	1.92	1.83	1.72	1.60	1.47	1.34	1.22	1.10	1.00
1.25	1.60	1.59	1.56	1.51	1.45	1.38	1.30	1.22	1.14	1.05	0.98
1.50	1.33	1.33	1.31	1.28	1.24	1.20	1.15	1.09	1.04	0.98	0.92
1.75	1.14	1.14	1.13	1.11	1.09	1.06	1.02	0.99	0.95	0.90	0.86
2.00	1.00	1.00	0.99	0.98	0.97	0.94	0.92	0.89	0.86	0.83	0.80

Use of table 7-2 is best illustrated by an example as follows.

Example 7-3

REQUIRED: Design a delay equalizer for an $n = 5$ Butterworth low-pass filter having a 3-dB cutoff of 1600 Hz. The delay variation should not exceed 75 μs from DC to 1600 Hz.

RESULT: (a)   The Butterworth normalized delay curves of figure 2-35 can be used directly, since the region between DC and 1 rad/s corresponds to the frequency range of interest. The curve for $n = 5$ indicates that the peak delay occurs near 0.9 rad/s and is approximately 1.9 s greater than the value at DC. This corresponds to a denormalized variation of $1.9/2\pi f_h$ or 190 μs where f_h is 1600 Hz; so an equalizer is required.

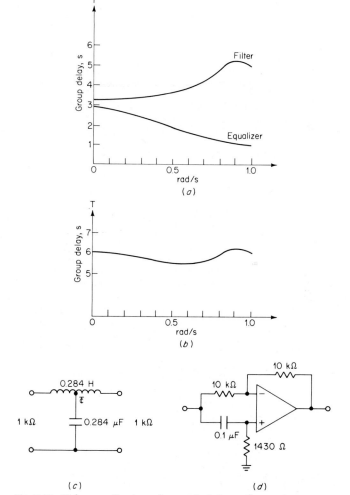

Fig. 7-18   Delay equalization of example 7-3: (a) filter and equalizer delay curves; (b) composite delay curve; (c) LC equalizer; (d) active equalizer.

(b) Examination of table 7-2 indicates that a first-order equalizer with an α_0 of 0.7 has a delay at DC that is approximately 1.8 s greater than the delay at 0.9 rad/s; so a reasonable fit to the required shape should occur.

The delay of the normalized filter and the first-order equalizer for $\alpha_0 = 0.7$ is shown in figure 7.18a. The combined delay is given in figure 7-18b. The peak-to-peak delay variation is about 0.7 s, which corresponds to a denormalized delay variation of $0.7/2\pi f_h$ or 70 μs.

(c) The first-order equalizer parameter $\alpha_0 = 0.7$ is denormalized by the factor $2\pi f_h$, resulting in $\alpha_0' = 7037$. The corresponding passive equalizer is designed as follows, where the impedance level R is chosen to be 1 kΩ:

$$L = \frac{2R}{\alpha_0'} = \frac{2 \times 10^3}{7037} = 0.284 \text{ H} \qquad (7\text{-}15)$$

$$C = \frac{2}{\alpha_0' R} = \frac{2}{7037 \times 10^3} = 0.284 \text{ } \mu\text{F} \qquad (7\text{-}16)$$

The first-order LC equalizer section is shown in figure 7-18c.

(d) An active first-order equalizer section can also be designed using the circuit of figure 7-11a. If we select a C of 0.1 μF and $R' = 10$ kΩ, the value of R is given by

$$R = \frac{1}{\alpha_0' C} = \frac{1}{7037 \times 10^{-7}} = 1421 \text{ } \Omega \qquad (7\text{-}46)$$

The active equalizer circuit is illustrated in figure 7-18d.

Highly selective low-pass filters such as the elliptic-function type have a corresponding delay characteristic that increases very dramatically near cutoff. First-order all-pass sections cannot then provide a complementary delay shape; so they are limited to applications involving low-pass filters of moderate selectivity.

Second-Order Equalizers

First-order equalizers have a maximum delay at DC and a single design parameter α_0 which limits their use. Second-order sections have two design parameters, ω_r and Q. The delay shape is bandpass in nature and can be made broad or sharp by changing the Q. The peak delay frequency is determined by the design parameter ω_r. As a result of this flexibility, second-order sections can be used to equalize virtually any type of delay curve. The only limitation is in the number of sections the designer is willing to use and the effort required to achieve a given degree of equalization.

The group delay of a second-order all-pass section was given by

$$T_{gd} = \frac{2 Q \omega_r (\omega^2 + \omega_r^2)}{Q^2 (\omega^2 - \omega_r^2)^2 + \omega^2 \omega_r^2} \qquad (7\text{-}11)$$

If we normalize this expression by setting ω_r equal to 1, we obtain

$$T_{gd} = \frac{2 Q (\omega^2 + 1)}{Q^2 (\omega^2 - 1)^2 + \omega^2} \qquad (7\text{-}70)$$

To determine the delay at DC, we can set ω equal to zero, which results in

$$T_{gd}(\text{DC}) = \frac{2}{Q} \qquad (7\text{-}71)$$

For Q's below 0.577, the maximum delay occurs at DC. As the Q is increased, the frequency of maximum delay approaches 1 rad/s and is given by

$$\omega(T_{gd,\text{max}}) = \sqrt{\sqrt{4 - \frac{1}{Q^2}} - 1} \tag{7-72}$$

For Q's of 2 or more, the maximum delay can be assumed to occur at 1 rad/s and may be determined from

$$T_{gd,\text{max}} = 4\,Q \tag{7-73}$$

Equations (7-70) through (7-72) are evaluated in table 7-3 for Q's ranging from 0.25 to 10.

To use table 7-3 directly, first normalize the curve to be equalized so that the minimum delay occurs at 1 rad/s. Then select an equalizer from the table that provides the best fit for a complementary curve.

A composite curve is then plotted. If the delay ripple is excessive, additional equalizer sections are required to fill in the delay gaps. The data of table 7-3 can again be used by scaling the region to be equalized to a 1-rad/s center and selecting a complementary equalizer shape from the table. The equalizer parameters can then be shifted to the region of interest by scaling.

The procedure described is an oversimplification of the design process. The equalizer responses will interact with each other; so each delay region to be filled in cannot be treated independently. Every time a section is added, the previous sections may require an adjustment of their design parameters.

Delay equalization generally requires considerably more skill than the actual design of filters. Standard pole-zero patterns are defined for the different filter families, whereas the design of equalizers involves the approximation problem where a pole-zero pattern must be determined for a suitable fit to a curve. The following example illustrates the use of second-order equalizer sections to equalize delay.

Example 7-4

REQUIRED: A bandpass filter with the delay measurements of table I must be equalized to within 700 μs.

TABLE I   Specified Delay

Frequency, Hz	Delay, μs
500	1600
600	960
700	640
800	320
900	50
1000	0
1100	160
1200	480
1300	800
1400	1120
1500	1500

TABLE 7-3 Delay of Normalized Second-Order Section ($\omega_r = 1$ rad/s)

Q	$T_{gd}(\text{DC})$	$\omega(T_{yd,\max})$	$T_{gd,\max}$	0.1	0.2	0.3	0.4	0.5	0.6	0.7	0.8	0.9
0.25	8.00	DC	8.00	7.09	5.33	3.85	2.84	2.19	1.76	1.42	1.27	1.11
0.50	4.00	DC	4.00	3.96	3.85	3.67	3.45	3.20	2.94	2.69	2.44	2.21
0.75	2.67	0.700	3.51	2.70	2.79	2.94	3.12	3.31	3.46	3.51	3.45	3.27
1.00	2.00	0.856	4.31	2.04	2.16	2.37	2.68	3.08	3.53	3.97	4.26	4.28
1.25	1.60	0.913	5.23	1.64	1.76	1.97	2.30	2.77	3.40	4.16	4.87	5.22
1.50	1.33	0.941	6.18	1.37	1.47	1.67	1.99	2.47	3.18	4.16	5.28	6.09
1.75	1.14	0.957	7.15	1.17	1.27	1.45	1.75	2.22	2.95	4.05	5.54	6.88
2.00	1.00	0.968	8.13	1.03	1.12	1.28	1.56	2.00	2.72	3.89	5.66	7.59
2.25	0.89	0.975	9.12	0.91	0.99	1.15	1.40	1.82	2.52	3.71	5.69	8.20
2.50	0.80	0.980	10.1	0.82	0.90	1.04	1.27	1.66	2.33	3.52	5.66	8.74
2.75	0.73	0.983	11.1	0.75	0.82	0.94	1.16	1.53	2.16	3.34	5.57	9.19
3.00	0.67	0.986	12.1	0.69	0.75	0.87	1.07	1.41	2.02	3.16	5.45	9.57
3.25	0.61	0.988	13.1	0.63	0.69	0.80	0.99	1.31	1.89	2.99	5.31	9.88
3.50	0.57	0.990	14.1	0.59	0.64	0.75	0.92	1.23	1.77	2.84	5.15	10.1
3.75	0.53	0.991	15.1	0.55	0.60	0.70	0.86	1.15	1.67	2.69	5.00	10.3
4.00	0.50	0.992	16.1	0.51	0.56	0.65	0.81	1.08	1.57	2.56	4.84	10.4
4.25	0.47	0.993	17.1	0.48	0.53	0.62	0.76	1.02	1.49	2.44	4.68	10.5
4.50	0.44	0.994	18.1	0.46	0.50	0.58	0.72	0.97	1.41	2.33	4.52	10.6
4.75	0.42	0.994	19.1	0.43	0.47	0.55	0.69	0.92	1.35	2.22	4.37	10.6
5.00	0.40	0.995	20.1	0.41	0.45	0.52	0.65	0.87	1.28	2.13	4.23	10.6
6.00	0.33	0.997	24.0	0.34	0.38	0.44	0.54	0.73	1.08	1.82	3.71	10.3
7.00	0.29	0.997	28.0	0.29	0.32	0.38	0.47	0.63	0.93	1.58	3.28	9.83
8.00	0.25	0.998	32.0	0.26	0.28	0.33	0.41	0.55	0.82	1.39	2.94	9.28
9.00	0.22	0.999	36.0	0.23	0.25	0.29	0.36	0.49	0.73	1.24	2.65	8.73
10.00	0.20	0.999	40.0	0.21	0.23	0.26	0.33	0.44	0.66	1.13	2.41	8.19

The corresponding delay curve is plotted in figure 7-19a.

RESULT: (*a*) Since the minimum delay occurs at 1000 Hz, normalize the curve by dividing the frequency axis by 1000 Hz and multiplying the delay axis by 2π 1000. The results are tabulated below and plotted in figure 7-19b.

TABLE II Normalized Delay

Frequency, rad/s	Delay, s
0.5	10.1
0.6	6.03
0.7	4.02
0.8	2.01
0.9	0.31
1.0	0
1.1	1.01
1.2	3.02
1.3	5.03
1.4	7.04
1.5	9.42

TABLE 7-3 (Continued)

ω, rad/s

1.0	1.1	1.2	1.3	1.4	1.5	1.6	1.7	1.8	1.9	2.0	3.0	4.0	5.0
1.00	0.91	0.84	0.78	0.73	0.69	0.66	0.62	0.60	0.57	0.55	0.38	0.28	0.21
2.00	1.81	1.64	1.49	1.35	1.23	1.12	1.03	0.94	0.87	0.80	0.40	0.24	0.15
3.00	2.69	2.36	2.06	1.79	1.56	1.36	1.19	1.05	0.93	0.83	0.33	0.18	0.11
4.00	3.52	2.99	2.48	2.05	1.71	1.43	1.20	1.03	0.88	0.77	0.27	0.14	0.09
5.00	4.32	3.50	2.76	2.18	1.73	1.40	1.15	0.96	0.81	0.69	0.23	0.11	0.07
6.00	5.06	3.90	2.92	2.20	1.69	1.33	1.07	0.88	0.73	0.62	0.20	0.10	0.06
7.00	5.75	4.20	2.99	2.17	1.62	1.24	0.98	0.80	0.66	0.55	0.17	0.08	0.05
8.00	6.38	4.41	2.99	1.10	1.53	1.16	0.91	0.73	0.60	0.50	0.15	0.07	0.04
9.00	6.94	4.54	2.95	2.01	1.44	1.08	0.83	0.66	0.54	0.45	0.14	0.06	0.04
10.0	7.44	4.60	2.88	1.92	1.35	1.00	0.77	0.61	0.50	0.41	0.12	0.06	0.04
11.0	7.88	4.62	2.80	1.82	1.27	0.93	0.72	0.57	0.46	0.38	0.11	0.05	0.03
12.0	8.25	4.60	2.70	1.73	1.20	0.87	0.67	0.53	0.43	0.35	0.10	0.05	0.03
13.0	8.57	4.55	2.60	1.65	1.13	0.82	0.62	0.49	0.40	0.33	0.09	0.05	0.03
14.0	8.84	4.48	2.50	1.56	1.06	0.77	0.58	0.46	0.37	0.31	0.09	0.04	0.03
15.0	9.06	4.40	2.41	1.49	1.01	0.73	0.55	0.43	0.35	0.29	0.08	0.04	0.02
16.0	9.23	4.30	2.31	1.42	0.95	0.69	0.52	0.41	0.33	0.27	0.08	0.04	0.02
17.0	9.36	4.20	2.22	1.35	0.91	0.65	0.49	0.38	0.31	0.26	0.07	0.04	0.02
18.0	9.46	4.10	2.14	1.29	0.86	0.62	0.47	0.36	0.29	0.24	0.07	0.03	0.02
19.0	9.52	3.99	2.06	1.24	0.82	0.59	0.44	0.35	0.29	0.23	0.07	0.03	0.02
20.0	9.56	3.89	1.98	1.18	0.79	0.56	0.42	0.33	0.27	0.22	0.06	0.03	0.02
24.0	9.48	3.48	1.71	1.01	0.67	0.47	0.36	0.28	0.22	0.18	0.05	0.03	0.02
28.0	9.18	3.13	1.51	0.88	0.58	0.41	0.31	0.24	0.19	0.16	0.04	0.02	0.01
32.0	8.77	2.82	1.34	0.78	0.51	0.36	0.27	0.21	0.17	0.14	0.04	0.02	0.01
36.0	8.32	2.57	1.20	0.70	0.45	0.32	0.24	0.19	0.15	0.12	0.03	0.02	0.01
40.0	7.87	2.35	1.09	0.69	0.41	0.30	0.22	0.17	0.13	0.11	0.03	0.02	0.01

(b) An equalizer is required that has a nominal delay peak of 10 s at 1 rad/s relative to the delay at 0.5 and 1.5 rad/s. Examination of table 7-3 indicates that the delay corresponding to a Q of 2.75 will meet this requirement.

   If we add this delay, point by point, to the normalized delay of table II, the following table of values is obtained.

TABLE III   Equalized Delay

Frequency, rad/s	Delay, s
0.5	11.6
0.6	8.19
0.7	7.36
0.8	7.58
0.9	9.50
1.0	11.0
1.1	8.89
1.2	7.64
1.3	7.83
1.4	8.86
1.5	10.7

The corresponding curve is plotted in figure 7-19c. This curve can be denormalized by dividing the delay by 2π 1000 and multiplying the frequency axis by 1000, resulting in the final curve of figure 7-19d. The differential delay variation over the band is about 675 μs.

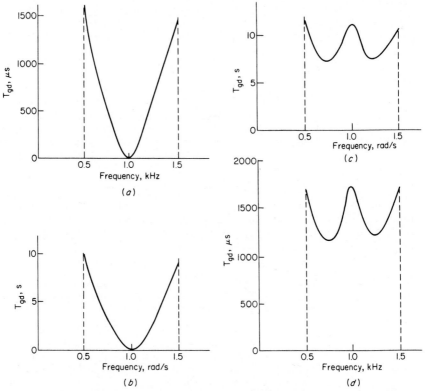

Fig. 7-19 Delay equalization of example 7-4: (a) unequalized delay; (b) normalized delay; (c) normalized equalized delay; (d) denormalized equalized delay.

The equalizer of example 7-4 provides over a 2:1 reduction in the differential delay. Further equalization can be obtained with two additional equalizers to fill in the concave regions around 750 and 1250 Hz.

7.5 WIDE-BAND 90° PHASE-SHIFT NETWORKS

Wide-band 90° phase-shift networks have a single input and two output ports. Both outputs maintain a constant phase difference of 90° within a prescribed error over a wide range of frequencies. The overall transfer function is all-pass. These networks are widely used in the design of single-sideband systems and in other applications requiring 90° phase splitting.

Bedrosian (see references) solved the approximation problem for this family of networks on a computer. The general structure is shown in figure 7-20a and consists of N and P networks. Each network provides real-axis pole-zero pairs and is all-pass. The transfer function is of the form

$$T(s) = \frac{(s - \alpha_1)(s - \alpha_2) \cdots (s - \alpha_{n/2})}{(s + \alpha_1)(s + \alpha_2) \cdots (s + \alpha_{n/2})} \qquad (7\text{-}74)$$

where $n/2$ is the order of the numerator and denominator polynomials. The total complexity of both networks is then n.

Real-axis all-pass transfer functions can be realized using a cascade of passive or active first-order sections. Both versions are shown in figure 7-20b and c.

The transfer functions tabulated in table 7-4 approximate a 90° phase differ-

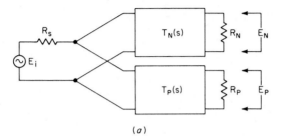

(a)

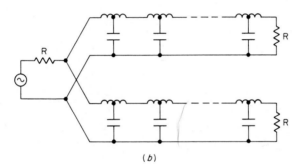

(b)

(c)

Fig. 7-20 Wide-band 90° phase-shift networks: (a) general structure; (b) passive realization; (c) active realization.

TABLE 7-4 Pole-Zero Locations for 90° Phase-Shift Networks*

n	$\Delta\phi$	α_N	α_P
		$\omega_u/\omega_L = 1146$	
6	6.84°	43.3862	8.3350
		2.0264	0.4935
		0.1200	0.0231
8	2.12°	59.7833	14.4159
		4.8947	1.6986
		0.5887	0.2043
		0.0694	0.0167
10	0.66°	75.8845	20.4679
		8.3350	3.5631
		1.5279	0.6545
		0.2807	0.1200
		0.0489	0.0132
		$\omega_u/\omega_L = 573.0$	
6	4.99°	34.3132	7.0607
		1.9111	0.5233
		0.1416	0.0291
8	1.39°	47.0857	11.8249
		4.3052	1.6253
		0.6153	0.2323
		0.0846	0.0212
10	0.39°	59.6517	16.5238
		7.0607	3.2112
		1.4749	0.6780
		0.3114	0.1416
		0.0605	0.0168
		$\omega_u/\omega_L = 286.5$	
4	13.9°	16.8937	2.4258
		0.4122	0.0592
6	3.43°	27.1337	5.9933
		1.8043	0.5542
		0.1669	0.0369
8	0.84°	37.0697	9.7136
		3.7944	1.5566
		0.6424	0.2636
		0.1030	0.0270
10	0.21°	46.8657	13.3518
		5.9933	2.8993
		1.4247	0.7019
		0.3449	0.1669
		0.0749	0.0213
		$\omega_u/\omega_L = 143.2$	
4	10.2°	13.5875	2.2308
		0.4483	0.0736

* Numerical values for table 7-4 obtained from S. D. Bedrosian, "Normalized Design of 90° Phase-Difference Networks," IRE Transactions on Circuit Theory, June 1960.

		$\omega_u/\omega_L = 143.2$	*Continued*
8	0.46°	29.3327	8.0126
		3.3531	1.4921
		0.6702	0.2982
		0.1248	0.0341
10	0.10°	37.0091	10.8375
		5.1050	2.6233
		1.3772	0.7261
		0.3812	0.1959
		0.0923	0.0270

		$\omega_u/\omega_L = 81.85$	
4	7.58°	11.4648	2.0883
		0.4789	0.0918
6	1.38°	18.0294	4.5017
		1.6316	0.6129
		0.2221	0.0555
8	0.25°	24.4451	6.8929
		3.0427	1.4432
		0.6929	0.3287
		0.1451	0.0409
10	0.046°	30.7953	9.2085
		4.5017	2.4248
		1.3409	0.7458
		0.4124	0.2221
		0.1086	0.0325

		$\omega_u/\omega_L = 57.30$	
4	6.06°	10.3270	2.0044
		0.4989	0.0968
6	0.99°	16.1516	4.1648
		1.5873	0.6300
		0.2401	0.0619
8	0.16°	21.8562	6.2817
		2.8648	1.4136
		0.7074	0.3491
		0.1592	0.0458
10	0.026°	27.5087	8.3296
		4.1648	2.3092
		1.3189	0.7582
		0.4331	0.2401
		0.1201	0.0364

		$\omega_u/\omega_L = 28.65$	
4	3.57°	8.5203	1.6157
		0.5387	0.1177
6	0.44°	13.1967	3.6059
		1.5077	0.6633
		0.2773	0.0758
8	0.056°	17.7957	5.2924
		2.5614	1.3599
		0.7354	0.3904
		0.1890	0.0562

$\omega_u/\omega_L = 28.65$	*Continued*		
10	0.0069°	22.3618	6.9242
		3.6059	2.1085
		1.2786	0.7821
		0.4743	0.2773
		0.1444	0.0447
$\omega_u/\omega_L = 11.47$			
4	1.31°	5.9339	1.5027
		0.5055	0.1280
6	0.10°	10.4285	3.0425
		1.4180	0.7052
		0.3287	0.0959
8	0.0075°	14.0087	4.3286
		2.2432	1.2985
		0.7701	0.4458
		0.2310	0.0714

ence in an equiripple manner. This approximation occurs within the bandwidth limits ω_L and ω_u as shown in figure 7-21. These frequencies are normalized so that $\sqrt{\omega_L\omega_u} = 1$. For a specified bandwidth ratio ω_u/ω_L, the individual band limits can be found from

$$\omega_L = \sqrt{\frac{\omega_L}{\omega_u}} \tag{7-75}$$

and

$$\omega_u = \sqrt{\frac{\omega_u}{\omega_L}} \tag{7-76}$$

As the total complexity n is made larger, the phase error decreases for a fixed bandwidth ratio, or for a fixed phase error, the bandwidth ratio will increase.

To use table 7-4, first determine the required bandwidth ratio from the frequencies given. A network is then selected that has a bandwidth ratio ω_u/ω_L that exceeds the requirement and a phase error $\pm\Delta\phi$ that is acceptable.

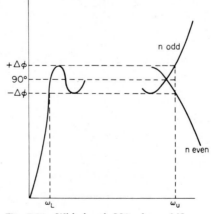

Fig. 7-21 Wide-band 90° phase-shift approximation.

A frequency-scaling factor (FSF) is determined from

$$FSF = 2\pi f_0 \qquad (7\text{-}77)$$

where f_0 is the geometric mean of the specified band limits or $\sqrt{f_L f_u}$. The tabulated α's are then multiplied by the FSF for denormalization. The resulting pole-zero pairs can be realized by a cascade of active or passive first-order sections for each network.

The following example illustrates the design of a 90° phase-shift network:

Example 7-5

REQUIRED: Design a network having dual outputs which maintain a phase difference of 90° within ±0.2° over the frequency range of 300 to 3000 Hz. The circuit should be all-pass and active.

RESULT: (a) Since a 10:1 bandwidth ratio is required (3000 Hz/300 Hz), the design corresponding to $n = 6$ and $\omega_u/\omega_L = 11.47$ is chosen. The phase-shift error will be ±0.1°.

(b) The normalized real pole-zero coordinates for both networks are given as follows:

P Network	N Network
$\alpha_1 = 10.4285$	$\alpha_4 = 3.0425$
$\alpha_2 = 1.4180$	$\alpha_5 = 0.7052$
$\alpha_3 = 0.3287$	$\alpha_6 = 0.0959$

The frequency-scaling factor is

$$FSF = 2\pi f_0 = 2\pi \times 948.7 = 5961 \qquad (7\text{-}77)$$

where f_0 is $\sqrt{300 \times 3000}$. The pole-zero coordinates are multiplied by the FSF, resulting in the following set of denormalized values for α:

P Network	N Network
$\alpha_1' = 62164$	$\alpha_4' = 18136$
$\alpha_2' = 8453$	$\alpha_5' = 4204$
$\alpha_3' = 1959$	$\alpha_6' = 571.7$

(c) The P and N networks can now be realized using the active first-order all-pass circuit of section 7.2 and figure 7-11a.

If we let $R' = 10$ kΩ and $C = 1000$ pF, the value of R is given by

$$R = \frac{1}{\alpha_0 C} \qquad (7\text{-}46)$$

Using the denormalized α's for the P and N networks, the following values are obtained:

Section	P Network	N Network
1	$R = 16.09$ kΩ	$R = 55.14$ kΩ
2	$R = 118.3$ kΩ	$R = 237.9$ kΩ
3	$R = 510.5$ kΩ	$R = 1.749$ MΩ

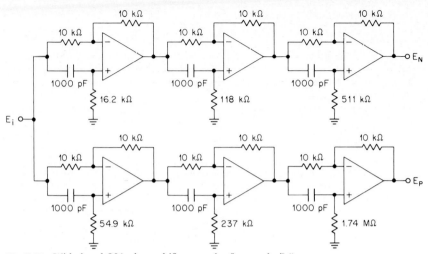

Fig. 7-22 Wide-band 90° phase-shift network of example 7-5.

The final circuit is shown in figure 7-22 using standard 1% resistor values.

7.6 ADJUSTABLE DELAY AND AMPLITUDE EQUALIZERS

Delay equalizers were discussed in section 7.2 and applied to the delay equalization of filters in section 7.4. Frequently a telephone channel must be equalized to reduce the delay and amplitude variation. This process is called "line conditioning." Since the initial parameters of lines vary and the line characteristics may change from time to time, the equalizer will consist of multiple sections where each stage is required to be adjustable.

LC Delay Equalizers

The circuit of figure 7-23a illustrates a simplified adjustable LC delay equalizer section. The emitter and collector load resistors R_e and R_c are equal; so Q_1 serves as a phase splitter. Transistor Q_2 is an emitter follower output stage.

The equivalent circuit is shown in figure 7-23b. The transfer function can be determined by superposition as

$$T(s) = \frac{s^2 - \dfrac{1}{RC}s + \dfrac{1}{LC}}{s^2 + \dfrac{1}{RC}s + \dfrac{1}{LC}} \tag{7-78}$$

This expression is of the same form as the general second-order all-pass transfer function of equation (7-6). By equating coefficients we obtain

$$\omega_r = \frac{1}{\sqrt{LC}} \tag{7-79}$$

and

$$Q = \omega_r RC \tag{7-80}$$

Equation (7-80) can be substituted in equation (7-13) for the maximum delay of a second-order section resulting in

$$T_{gd,\max} = 4RC \tag{7-81}$$

By making R variable, the delay can be directly controlled while retaining the all-pass properties. The peak delay will occur at or near the LC resonant frequency.

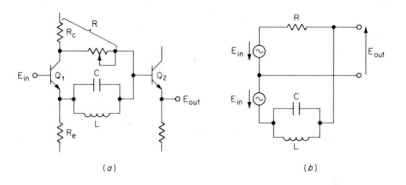

(a)   (b)

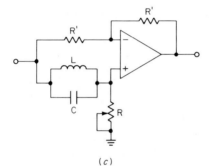

(c)

Fig. 7-23   Adjustable LC delay equalizer: (a) adjustable equalizer; (b) equivalent circuit; (c) operational-amplifier realization.

The all-pass transfer function of equation (7-78) can also be implemented using an operational amplifier. This configuration is shown in figure 7-23c, where R' is arbitrary. Design equations (7-79) through (7-81) still apply.

Example 7-6

REQUIRED: Design an adjustable LC delay equalizer using the two-transistor circuit of figure 7-23a. The delay should be variable from 0.5 to 2.5 ms with a center frequency of 1700 Hz.

RESULT: Using a capacitor C of 0.05 μF, the range of resistance R is given by

$$R_{\min} = \frac{T_{gd,\max}}{4C} = \frac{0.5 \times 10^{-3}}{4 \times 0.05 \times 10^{-6}} = 2500 \ \Omega \tag{7-81}$$

$$R_{\max} = \frac{2.5 \times 10^{-3}}{4 \times 0.05 \times 10^{-6}} = 12.5 \ \text{k}\Omega$$

The inductor is computed by the general formula for resonance $\omega^2 LC = 1$, resulting in an inductance of 175 mH. The circuit is shown in figure 7-24a. The emitter resistor R_e is composed of two resistors for proper biasing of phase splitter Q_1, and electrolytic capacitors are used for DC blocking. The delay extremes are shown in the curves of figure 7-24b.

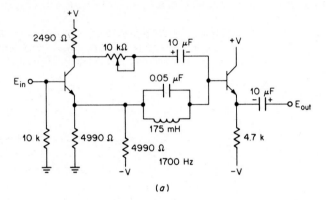

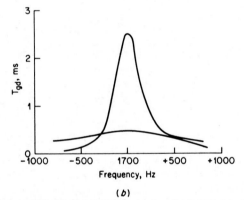

Fig. 7-24 Adjustable delay equalizer of example 7-6: (a) equalizer circuit; (b) delay adjustment range.

LC Delay and Amplitude Equalizers

Frequently the magnitude response of a transmission channel must be equalized along with the delay. An equalizer circuit featuring both adjustable amplitude and delay is shown in figure 7-25a. Transistor Q_1 serves as a phase splitter where the signal applied to emitter follower Q_2 is K times the input signal. The equivalent circuit is illustrated in figure 7-25b. The transfer function can be determined by superposition as

$$T(s) = \frac{s^2 - \dfrac{K}{RC}s + \dfrac{1}{LC}}{s^2 + \dfrac{K}{RC}s + \dfrac{1}{LC}} \tag{7-82}$$

If K is set equal to unity, this expression is then equivalent to equation (7-78) corresponding to a second-order all-pass transfer function. As K increases or decreases from unity, a boost or null occurs at midfrequency with an asymptotic return to unity gain at DC and infinity.

The amount of amplitude equalization at midfrequency in decibels is given by

$$A_{dB} = 20 \log K \tag{7-83}$$

The maximum delay occurs at the LC resonant frequency and can be derived as

$$T_{gd,\max} = \frac{2\,RC}{K} + 2\,RC \tag{7-84}$$

If K is unity, equation (7-84) reduces to $4\,RC$, which is equivalent to equation (7-81) for the all-pass circuit of figure 7-23.

An operational-amplifier implementation is also shown in figure 7-25c. The value of R' is arbitrary, and equations (7-83) and (7-84) are still applicable.

The following conclusions may be reached based on evaluation of equations (7-82) through (7-84):

1. The maximum delay is equal to $4\,RC$ for $K = 1$; so R is a delay magnitude control.

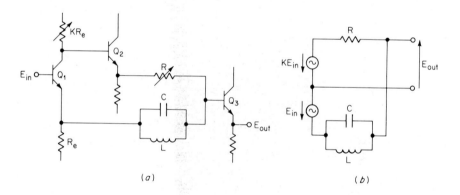

(a)   (b)

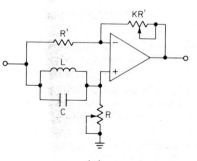

(c)

Fig. 7-25 Adjustable LC delay and amplitude equalizer: (a) adjustable delay and amplitude equalizer; (b) equivalent circuit; (c) operational-amplifier realization.

2. The maximum delay will be minimally affected by a nonunity K, as is evident from equation (7-84).
3. The amount of amplitude equalization at the LC resonant frequency is independent of the delay setting and is strictly a function of K. However, the selectivity of the amplitude response is a function of the delay setting and becomes more selective with increased delay.

The curves of figure 7-26 show some typical delay and amplitude characteristics. The interaction between delay and amplitude is not restricted to LC equalizers and will occur whenever the same resonant element, either passive or active, is used to provide both the amplitude and delay equalization. However, for small amounts of amplitude correction such as ± 3 dB, the effect on the delay is minimal.

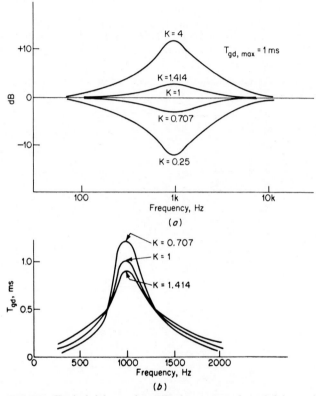

Fig. 7-26 Typical delay and amplitude response for LC delay and amplitude equalizer: (a) amplitude characteristics for a fixed delay; (b) delay variation for $\pm$ 3 dB of amplitude equalization.

Active Delay and Amplitude Equalizers

The dual-amplifier bandpass (DABP) delay equalizer structure of section 7.2, under Active All-Pass Structures, and figure 7-13 has a fixed gain and remains

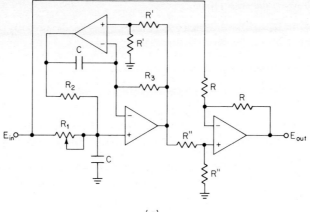

(a)

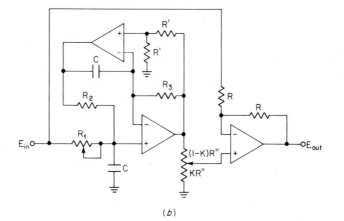

(b)

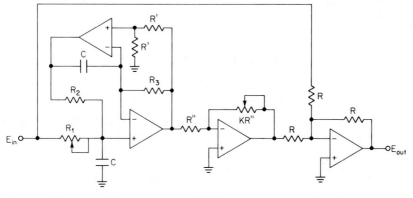

(c)

Fig. 7-27 DABP delay and amplitude equalizer: (a) adjustable delay equalizer; (b) adjustable delay and amplitude equalizer; (b) adjustable delay and amplitude equalizer with extended amplitude range.

all-pass regardless of the design Q. If resistor R_1 is made variable, the Q and therefore the delay can be directly adjusted with no effect on resonant frequency or the all-pass behavior. The adjustable delay equalizer is shown in figure 7-27a. The design equations are

$$T_{gd,\max} = 4\,R_1 C \qquad (7\text{-}85)$$

and
$$R_2 = R_3 = \frac{1}{\omega_r C} \qquad (7\text{-}86)$$

where C, R, R', and R'' can be conveniently chosen. Resistor R_2 can be made variable for frequency trimming.

If amplitude equalization capability is also desired, a potentiometer can be introduced, resulting in the circuit of figure 7-27b. The amplitude equalization at ω_r is given by

$$A_{\text{dB}} = 20 \log{(4K - 1)} \qquad (7\text{-}87)$$

where a K variation of 0.25 to 1 covers an amplitude equalization range of $-\infty$ to $+9.5$ dB.

To extend the equalization range above $+9.5$ dB, an additional amplifier can be introduced, as illustrated in figure 7-27c. The amplitude equalization at ω_r is then obtained from

$$A_{\text{dB}} = 20 \log{(2K - 1)} \qquad (7\text{-}88)$$

where a K variation of 0.5 to ∞ results in an infinite range of equalization capability. In reality a ± 15 dB maximum range has been found to be more than adequate for most equalization requirements.

Example 7-7

REQUIRED: Design an adjustable active delay and amplitude equalizer that has a delay adjustment range of 0.5 to 3 ms, an amplitude range of ± 12 dB, and a center frequency of 1000 Hz.

RESULT: The circuit of figure 7-27c will provide the required delay and amplitude adjustment capability.

If we choose $C = 0.01$ μF and $R = R' = R'' = 10$ kΩ, the element values are computed as follows:

$$R_{1,\min} = \frac{T_{gd,\max}}{4C} = \frac{0.5 \times 10^{-3}}{4 \times 10^{-8}} = 12.5 \text{ k}\Omega \qquad (7\text{-}85)$$

$$R_{1,\max} = \frac{3 \times 10^{-3}}{4 \times 10^{-8}} = 75 \text{ k}\Omega$$

$$R_2 = R_3 = \frac{1}{\omega_r C} = \frac{1}{2\pi \times 1000 \times 10^{-8}} = 15.9 \text{ k}\Omega \qquad (7\text{-}86)$$

The extreme values of K for ± 12 dB of amplitude equalization are found from

$$K = \frac{1}{2}\left[\log^{-1}\left(\frac{A_{\text{dB}}}{20}\right) + 1\right] \qquad (7\text{-}88)$$

The range of K is then 0.626 to 2.49. The equalizer section is shown in figure 7-28. Resistor R_2 has also been made adjustable for frequency trimming.

An active delay equalizer having adjustable delay was implemented by combining a second-order bandpass section with a summing amplifier. The bandpass section was required to have a fixed gain and a resonant frequency which were both independent of the Q setting. If amplitude equalization alone is needed,

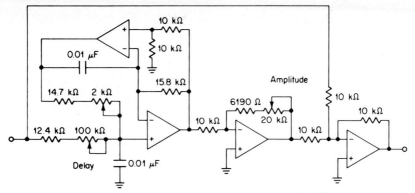

Fig. 7-28 Adjustable delay and amplitude equalizer of example 7-7.

the bandpass section can operate with a fixed design Q. The low-complexity MFBP delay equalizer section of figure 7-12a can then be used as an adjustable amplitude equalizer by making one of the summing resistors variable.

This circuit is shown in figure 7-29a. The design equations are given by

$$R_2 = \frac{2Q}{\omega_r C} \tag{7-89}$$

$$R_{1a} = \frac{R_2}{2} \tag{7-90}$$

$$R_{1b} = \frac{R_{1a}}{2Q^2 - 1} \tag{7-91}$$

The amount of amplitude equalization at ω_r is computed from

$$A_{dB} = 20 \log \left(\frac{1}{K} - 1 \right) \tag{7-92}$$

where K will range from 0 to 1 for an infinite range of amplitude equalization.

If the Q is below 0.707, the value of R_{1b} becomes negative; so the circuit of figure 7-29b is used. R_2 is given by equation (7-89) and R_1 is found from

$$R_1 = \frac{R_2}{4Q^2} \tag{7-93}$$

The attenuation or boost at resonance is computed from

$$A_{dB} = 20 \log \left(\frac{2Q^2}{K} - 1 \right) \tag{7-94}$$

The magnitude of Q determines the selectivity of the response in the region of resonance and is limited to values typically below 20 because of amplifier limitations.

If higher Q's are required or if a circuit featuring independently adjustable Q and amplitude equalization is desired, the DABP circuits of figure 7-27 may be used, where R_1 becomes the Q adjustment and is given by

$$R_1 = QR_2 \tag{7-95}$$

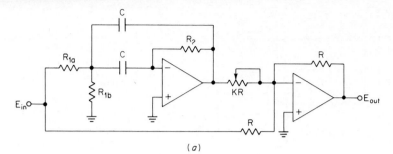

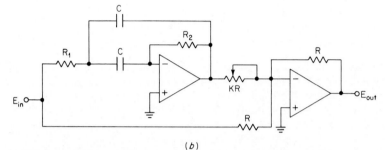

Fig. 7-29 MFBP amplitude equalizer: (a) amplitude equalizer $0.707 < Q < 20$; (b) amplitude equalizer $0 < Q < 20$.

To compute the required Q of an amplitude equalizer, first define f_b, which is the frequency corresponding to one-half the pad loss (in decibels). The Q is then given by

$$Q = \frac{f_b\, b^2 \sqrt{K_r}}{f_r(b^2 - 1)} \tag{7-96}$$

where

$$K_r = \log^{-1}\left(\frac{A_{dB}}{20}\right) = 10^{A_{dB}/20} \tag{7-97}$$

and

$$b = \frac{f_b}{f_r} \tag{7-98}$$

or

$$b = \frac{f_r}{f_b} \tag{7-99}$$

whichever b is greater than unity.

Example 7-8

REQUIRED: Design a fixed active amplitude equalizer that provides a +12-dB boost at 3200 Hz and has a boost of +6 dB at 2500 Hz.

RESULT: (a) First compute

$$K_r = 10^{A_{dB}/20} = 10^{12/20} = 3.98 \tag{7-97}$$

and

$$b = \frac{f_r}{f_b} = \frac{3200\ \text{Hz}}{2500\ \text{Hz}} = 1.28 \tag{7-99}$$

The Q is then found from

$$Q = \frac{f_b b^2 \sqrt{K_r}}{f_r(b^2 - 1)} = \frac{2500 \times 1.28^2 \sqrt{3.98}}{3200 (1.28^2 - 1)} = 4.00 \quad (7\text{-}96)$$

(b) The MFBP amplitude equalizer circuit of figure 7-29a will be used. Using a C of 0.0047 μF and a R of 10 kΩ, the element values are given by

$$R_2 = \frac{2Q}{\omega_r C} = \frac{2 \times 4}{2\pi 3200 \times 4.7 \times 10^{-9}} = 84.6 \text{ k}\Omega \quad (7\text{-}89)$$

$$R_{1a} = \frac{R_2}{2} = 42.3 \text{ k}\Omega \quad (7\text{-}90)$$

$$R_{1b} = \frac{R_{1a}}{2Q^2 - 1} = \frac{42.3 \times 10^3}{2 \times 4^2 - 1} = 1365 \text{ }\Omega \quad (7\text{-}91)$$

$$K = \frac{1}{1 + 10^{A_{dB}/20}} = \frac{1}{1 + 10^{12/20}} = 0.200 \quad (7\text{-}92)$$

The equalizer circuit and corresponding frequency response are shown in figure 7-30.

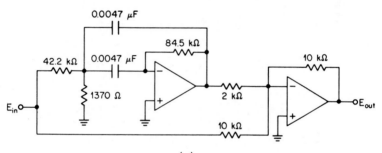

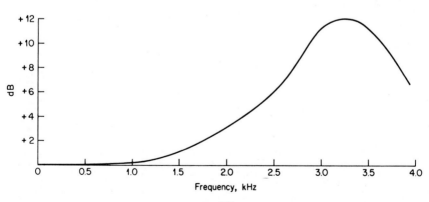

Fig. 7-30 Amplitude equalizer of example 7-8: (a) amplitude equalizer circuit; (b) frequency response.

REFERENCES

Bedrosian, S. D., "Normalized Design of 90° Phase-Difference Networks," IRE Transactions on Circuit Theory, Vol. CT-7, June 1960.

Geffe, P. R., "Simplified Modern Filter Design," John F. Rider, New York, 1963.

Lindquist, C. S., "Active Network Design," Steward and Sons, Long Beach, Calif., 1977.

Wallis, C. M., "Design of Low-Frequency Constant Time Delay Lines," AIEE Proceedings, Vol. 71, 1952.

Williams, A. B., "An Active Equalizer with Adjustable Amplitude and Delay," IEEE Transactions on Circuit Theory, Vol. CT-16, November 1969.

Williams, A. B., "Active Filter Design," Artech House, Dedham, Mass., 1975.

8

Refinements in LC Filter Design

8.0 REFINEMENTS IN LC FILTER DESIGN

The straightforward application of the design techniques outlined for LC filters will not always result in practical element values or desirable circuit configurations. Extreme cases of impedance or bandwidth can produce designs which may be extremely difficult or even impossible to realize. This chapter is concerned mainly with circuit transformations so that impractical designs can be transformed into alternate configurations having the identical response and using more practical elements.

8.1 TAPPED INDUCTORS

An extremely useful tool for eliminating impractical element values is the transformer. As the reader may recall from introductory AC circuit analysis, a transformer having a turns ratio N will transform an impedance by a factor of N^2. A parallel element can be shifted between the primary and secondary at will provided that its impedance is modified by N^2.

Figure 8-1 illustrates how a tapped inductor is used to reduce the value of a resonating capacitor. The tuned circuit of figure 8-1a is first modified by introducing an impedance step-up transformer as shown in figure 8-1b so that

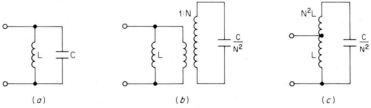

(a) (b) (c)

Fig. 8-1 The tapped inductor: (a) basic tuned circuit; (b) introduction of transformer; (c) absorbed transformer.

capacitor C can be moved to the secondary and reduced by a factor of N^2. This can be carried a step further, resulting in the circuit of figure 8-1c. The transformer has been absorbed as a continuation of the inductor, resulting in an autotransformer. The ratio of the overall inductance to the tap inductance becomes N^2.

As an example, let us modify the tuned circuit of example 5-4 shown in figure 8-2a. To reduce the capacitor from 0.354 to 0.027 μF, the overall inductance is increased by the impedance ratio 0.354 μF/0.027 μF, resulting in the circuit of figure 8-2b. The resonant frequency remains unchanged, since the overall LC product is still the same.

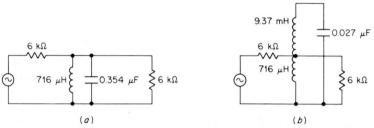

(a)   (b)

Fig. 8-2   Reducing resonant capacitor value: (a) tuned circuit; (b) modified circuit.

As a further example, let us consider LC elliptic-function low-pass filters. The parallel resonant circuits may also contain high capacity values which can be reduced by this method. Figure 8-3 shows a section of the low-pass filter of example 3-2. To reduce the resonating capacitor to 0.1 μF, the overall inductance is increased by the factor 1.055 μF/0.1 μF and a tap is provided at the original inductance value.

The tapped coil is useful not only for reducing resonating capacitors but also for transforming entire sections of a filter including terminations. The usefulness of the tapped inductor is limited only by the ingenuity and resourcefulness of the designer. Figure 8-4 illustrates some applications of this technique using designs from previous examples. In the case of figure 8-4a, where a tapped coil enables operation from unequal terminations, the same result could have been achieved using Bartlett's bisection theorem or other methods (see section 3.1). However, the transformer approach results in maximum power transfer (minimum insertion loss). The circuits of figure 8-4b and c demonstrate how

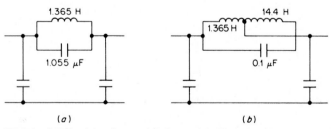

(a)   (b)

Fig. 8-3   Application of tapped inductor in elliptic-function low-pass filters: (a) filter section; (b) tapped inductor.

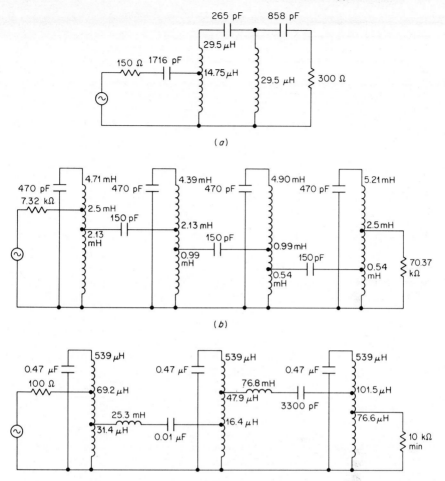

Fig. 8-4 Applications of tapped inductors: *(a)* high-pass filter of example 4-1 modified for unequal terminations; *(b)* filter of example 5-7 modified for standard capacitor values; *(c)* filter of example 5-8 modified for standard capacitor values.

element values can be manipulated by taps. The tapped inductance values shown are all measured from the grounded end of the shunt inductors. Series branches can be manipulated up or down in impedance level by multiplying the shunt inductance taps on both sides of the branch by the desired impedance-scaling factor.

Transformers or autotransformers are by no means ideal. Imperfect coupling within the magnetic structure will result in a leakage inductance which can cause spurious responses at higher frequencies, as shown in figure 8-5. These effects can be minimized by using near unity turns ratios. Another solution is to leave a portion of the original capacity at the tap for high-frequency bypassing. This method is shown in figure 8-6.

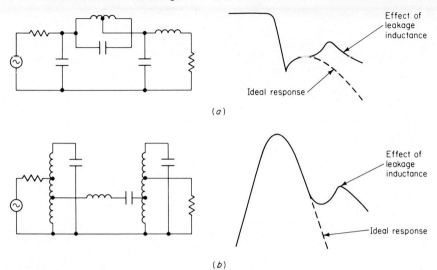

(a)

(b)

Fig. 8-5 Spurious responses from leakage inductance: *(a)* low-pass filter; *(b)* bandpass filter.

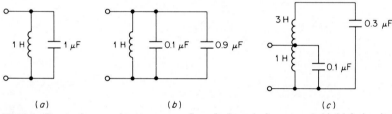

(a)   (b)   (c)

Fig. 8-6 Preventing spurious response from leakage inductance: *(a)* initial circuit; *(b)* split capacity; *(c)* transformed circuit.

8.2 CIRCUIT TRANSFORMATIONS

Circuit transformations fall into two categories, equivalent circuits or narrow-band approximations. The impedance of a circuit branch can be expressed as a ratio of two polynomials in s, similar to a transfer function. If two branches are equivalent, their impedance expressions are identical. A narrow-band approximation to a particular filter branch is valid only over a small frequency range. Outside of this region the impedances depart considerably; so the filter response is affected. As a result, narrow-band approximations are essentially limited to small percentage bandwidth bandpass filters.

Norton's Capacitance Transformer

Let us consider the circuit of figure 8-7*a* consisting of impedance Z interconnected between impedances Z_1 and Z_2. If it is desired to raise impedance Z_2 by a factor of N^2 without disturbing an overall transfer function (except for possibly a constant multiplier), a transformer can be introduced as shown in figure 8-7b.

Determinant manipulation can provide us with an alternate approach. The nodal determinant of a two-port network is given by

$$\begin{vmatrix} Y_{11} & -Y_{12} \\ -Y_{21} & Y_{22} \end{vmatrix}$$

where Y_{11} and Y_{22} are the input and output nodal admittances, respectively, and Y_{12} and Y_{21} are the transfer admittances, which are normally equal to each other.

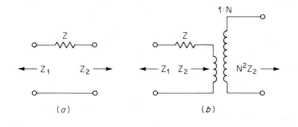

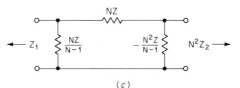

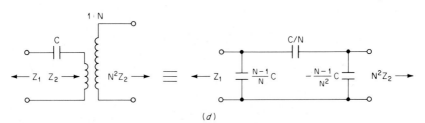

Fig. 8-7 Norton's capacitance transformer: (a) general two-port network; (b) transformer step-up of output impedance; (c) Norton impedance transformation; (d) Norton capacitance transformation.

If we consider the two-port network of figure 8-7a, the nodal determinant becomes

$$\begin{vmatrix} \dfrac{1}{Z_1} + \dfrac{1}{Z} & -\dfrac{1}{Z} \\ -\dfrac{1}{Z} & \dfrac{1}{Z_2} + \dfrac{1}{Z} \end{vmatrix}$$

To raise the impedance of the output or Y_{22} node by N^2, the second row and second column are multiplied by $1/N$, resulting in

$$
\begin{vmatrix}
\dfrac{1}{Z_1} + \dfrac{1}{Z} & -\dfrac{1}{NZ} \\[3mm]
-\dfrac{1}{NZ} & \dfrac{1}{N^2 Z_2} + \dfrac{1}{N^2 Z}
\end{vmatrix}
$$

This determinant corresponds to the circuit of figure 8-7c. The Y_{11} total nodal admittance is unchanged and the Y_{22} total nodal admittance has been reduced by N^2 or the impedance has been increased by N^2. This result was originated by Norton and is called "Norton's transformation."

If the element Z is a capacitor C, this transformation can be applied to obtain the equivalent circuit of figure 8-7d. This transformation is important, since it can be used to modify the impedance on one side of a capacitor by a factor of N^2 without a transformer. However, the output shunt capacitor introduced is negative. A positive capacitor must then be present external to the network so that the negative capacitance can be absorbed.

If an N^2 of less than unity is used, the impedance at the output node will be reduced. The shunt capacitor at the input node will then become negative and must be absorbed by an external positive capacitor across the input.

The following example illustrates the use of the capacitance transformer.

Example 8-1

REQUIRED: Using the capacitance transformation, modify the bandpass filter circuit of figure 5-3c so that the 1.91-H inductor is reduced to 100 mH. The source and load impedances should remain 600 Ω.

RESULT: (a)   The circuit to be transformed is shown in figure 8-8a. To facilitate the capacitance transformation, the 0.01329-μF series capacitor is split into two equal capacitors of twice the value and redrawn in figure 8-8b.

To reduce the 1.91-H inductor to 100 mH, let us first lower the impedance of the network to the right of the dashed line in figure 8-8b by factor of 100 mH/1.91 H or 0.05236. Using the capacitance transformation of figure 8-7d, where $N^2 = 0.05236$, the circuit of figure 8-8c is obtained where the input negative capacitor has been absorbed.

(b)   To complete the transformation, the output node must be transformed back up in impedance to restore the 600-Ω termination. Again using the capacitance transformation with an N^2 of 600 Ω/31.42 Ω or 19.1, the final circuit of figure 8-8d is obtained. Because of the symmetrical nature of the circuit of figure 8-8b both capacitor transformations are also symmetrical.

(c)   Each parallel resonant circuit is tuned by opening the inductors of the adjacent series resonant circuits, and each series resonant circuit is resonated by shorting the inductors of the adjacent parallel tuned circuits as shown in figure 8-8e.

Narrow-Band Approximations

A narrow-band approximation to a circuit branch consists of an alternate network which is theoretically equivalent only at a single frequency. Nevertheless good results can be obtained with bandpass filters having small percentage bandwidths typically up to 20%.

The series and parallel RL and RC circuits of table 8-1 are narrow-band approximations which are equivalent at ω_0. This frequency is generally set equal to the bandpass center frequency in equations (8-1) through (8-8). These equations

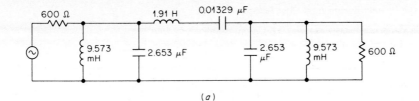

(a)

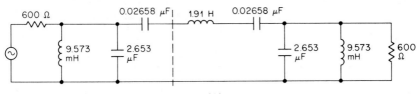

(b)

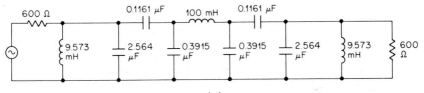

(c)

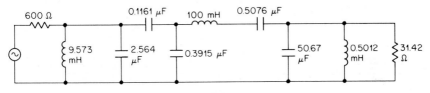

(d)

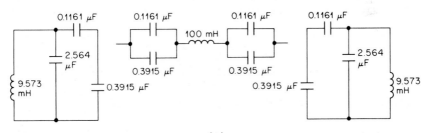

(e)

Fig. 8-8 Capacitance transformation applied to filter of example 5-2: (a) bandpass filter of example 5-2; (b) split series capacitors; (c) reduction of 1.91-H inductor using capacitance transformation; (d) restoration of 600-Ω output impedance using capacitance transformation; (e) equivalent circuits for tuning.

TABLE 8-1   Narrow-Band Approximations

Circuit	Design Equations
	$$L_1 = L_a + \frac{R_a^2}{\omega_0^2 L_a} \tag{8-1}$$
	$$R_1 = R_a + \frac{\omega_0^2 L_a^2}{R_a} \tag{8-2}$$
	$$L_a = \frac{L_1 R_1^2}{R_1^2 + \omega_0^2 L_1^2} \tag{8-3}$$
	$$R_a = \frac{\omega_0^2 L_1^2 R_1}{R_1^2 + \omega_0^2 L_1^2} \tag{8-4}$$
	$$C_2 = \frac{C_b}{1 + \omega_0^2 C_b^2 R_b^2} \tag{8-5}$$
	$$R_2 = R_b + \frac{1}{\omega_0^2 C_b^2 R_b} \tag{8-6}$$
	$$C_b = C_2 + \frac{1}{\omega_0^2 R_2^2 C_2} \tag{8-7}$$
	$$R_b = \frac{R_2}{1 + \omega_0^2 C_2^2 R_2^2} \tag{8-8}$$

were derived simply by determining the expressions for the network impedances and equating the real parts and the imaginary parts to solve for the resistive and reactive components, respectively.

Narrow-band approximations can be used to manipulate the source and load terminations of bandpass filters. If a parallel *RC* network is converted to a series *RC* circuit, it is apparent from equation (8-8) that the resistor value decreases. When we apply this approximation to a bandpass filter having a parallel resonant circuit as the terminating branch, the source or load resistor can be made smaller. To control the degree of reduction so that a desired termination can be obtained, the shunt capacitor is first subdivided into two capacitors where only one capacitor is associated with the termination.

These results are illustrated in figure 8-9. The element values are given by

$$C_2 = \frac{1}{\omega_0 \sqrt{R_1 R_2 - R_2^2}} \tag{8-9}$$

and

$$C_1 = C_T - \frac{1}{\omega_0} \sqrt{\frac{R_1 - R_2}{R_1^2 R_2}} \tag{8-10}$$

where the restrictions $R_2 < R_1$ and $(R_1 - R_2)/(R_1^2 R_2) < \omega_0^2 C_T^2$ apply.

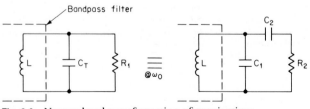

Fig. 8-9   Narrow-band transformation of terminations.

Example 8-2

REQUIRED: Modify the 100-kHz bandpass filter of example 5-7 for a source impedance of 600 Ω.

RESULT: The filter is shown in figure 8-10a. If we use the narrow-band source transformation of figure 8-9, the values are given by

$$C_2 = \frac{1}{\omega_0 \sqrt{R_1 R_2 - R_2^2}} = \frac{1}{2\pi \times 10^5 \sqrt{7.32 \times 6 \times 10^5 - 600^2}}$$
$$= 792.6 \text{ pF} \tag{8-9}$$

$$C_1 = C_T - \frac{1}{\omega_0} \sqrt{\frac{R_1 - R_2}{R_1^2 R_2}} = 884.9 \times 10^{-12}$$

$$- \frac{1}{2\pi \times 10^5} \sqrt{\frac{7.32 \times 10^3 - 600}{7320^2 \times 600}} = 157.3 \text{ pF} \tag{8-10}$$

The resulting filter is illustrated in figure 8-10b.

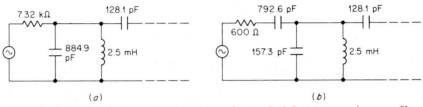

Fig. 8-10   Narrow-band source transformation of example 8-2: *(a)* source input to filter of example 5-7; *(b)* transformed source.

8.3   DESIGNING WITH PARASITIC CAPACITANCE

As a first approximation, inductors and capacitors are considered pure lumped reactive elements. Most physical capacitors are nearly perfect reactances. Inductors, on the other hand, have impurities which can be detrimental in many cases. In addition to the highly critical resistive losses, distributed capacity across the coil will occur because of interturn capacitance of the coil winding and other stray capacities involving the core, etc. The equivalent circuit of an inductor is shown in figure 8-11.

The result of this distributed capacitance is to create the effect of a parallel resonant circuit instead of an inductor. If the coil is to be located in shunt with an external capacitance, the external capacitor value can be decreased accordingly, thus absorbing the distributed capacitance.

The distributed capacity across the inductor in a series resonant circuit causes parallel resonances resulting in nulls in the frequency response. If the self-resonant frequency is too low, the null may even occur in the passband, thus severely distorting the expected response.

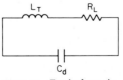

Fig. 8-11   Equivalent circuit of inductor.

To determine the effective inductance of a practical inductor, the coil is reso-nated to the frequency of interest with an external capacitor and the effective inductance is calculated using the standard formula for resonance. The effective inductance can also be found from

$$L_{\text{eff}} = \frac{L_T}{1 - \left(\dfrac{f}{f_r}\right)^2} \qquad (8\text{-}11)$$

where L_T is the true (low-frequency) inductance, f is the frequency of interest, and f_r is the inductor's self-resonant frequency. As f approaches f_r, the value of L_{eff} will increase quite dramatically and will become infinite at self-resonance. Equation (8-11) is plotted in figure 8-12.

To compensate for the effect of distributed capacity in a series resonant circuit, the true inductance L_T can be appropriately decreased so that the effective inductance given by equation (8-11) is the required value. However, the Q of a practical series resonant circuit is given by

$$Q_{\text{eff}} = Q_L \left[1 - \left(\frac{f}{f_r}\right)^2 \right] \qquad (8\text{-}12)$$

where Q_L is the Q of the inductor as determined by the series losses (i.e., $\omega L_T / R_L$). The effective Q is therefore reduced by the distributed capacity.

Distributed capacity is determined by the mechanical parameters of the core and winding and as a result is subject to change due to mechanical stresses, etc. Therefore, for maximum stability the distributed capacity should be kept as small as possible. Techniques for minimizing inductor capacity are discussed in chapter 9.

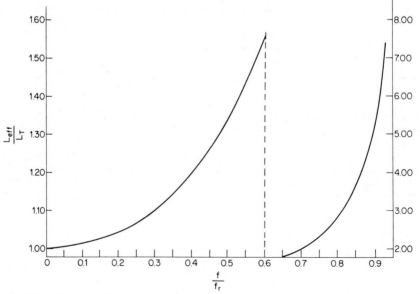

Fig. 8-12   Effective inductance with frequency.

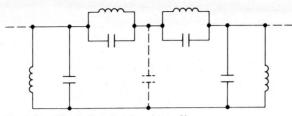

Fig. 8-13 Elliptic-function bandpass filter.

Another form of parasitic capacity is stray capacitance between the circuit nodes and ground. These strays may be especially harmful at high frequencies and with high-impedance nodes. In the case of low-pass filters where the circuit nodes already have shunt capacitors to ground, these strays can usually be neglected, especially when the impedance levels are low.

A portion of an elliptic-function bandpass filter is shown in figure 8-13. The stray capacity at nodes not connected to ground by a design capacitor may cause problems, since these nodes have high impedances.

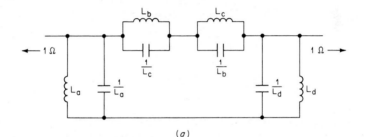

(*a*)

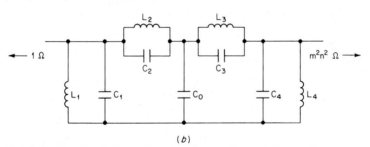

(*b*)

Fig. 8-14 Transformation to absorb stray capacitance: (*a*) normalized filter section; (*b*) transformed circuit.

Geffe (see references) has derived a transformation to introduce a design capacitor from the junction of the parallel tuned circuits to ground. The stray capacity can then be absorbed. The design of elliptic-function bandpass filters was discussed in section 5.1. This transformation is performed upon the filter while it is normalized to a 1-rad/s center frequency and a 1-Ω impedance level. A section of the normalized network is shown in figure 8-14a. The transformation proceeds as follows:

Choose an arbitrary value of $m < 1$. Then

$$n = 1 - \frac{L_b}{L_c}\frac{1-m}{m^2} \tag{8-13}$$

$$C_0 = \frac{1-n}{n^2 L_c} - \frac{1-m}{mn^2 L_b} \tag{8-14}$$

$$C_1 = \frac{1}{L_a} - \frac{1-n}{nL_c} \tag{8-15}$$

$$C_2 = \frac{1}{nL_c} \tag{8-16}$$

$$C_3 = \frac{1}{mn^2 L_b} \tag{8-17}$$

$$C_4 = \frac{1-m}{m^2 n^2 L_b} + \frac{1}{n^2 L_d} \tag{8-18}$$

$$L_1 = \frac{L_a}{1 - \dfrac{L_a}{L_b}\dfrac{1-n}{n}} \tag{8-19}$$

$$L_2 = nL_b \tag{8-20}$$

$$L_3 = mn^2 L_c \tag{8-21}$$

$$L_4 = \frac{m^2 n^2 L_c}{1 + \dfrac{L_c}{L_d}\dfrac{m^2}{1-m}} \tag{8-22}$$

The resulting network is given in figure 8-14b. The output node has been transformed to an impedance level of $m^2 n^2 \ \Omega$. Therefore, all the circuitry to the right of this node up to and including the termination must be impedance-scaled by this same factor. The filter is subsequently denormalized by scaling to the desired center frequency and impedance level.

8.4 AMPLITUDE EQUALIZATION FOR INADEQUATE Q

Insufficient element Q will cause a sagging or rounding of the frequency response in the region of cutoff. Some typical cases are shown in figure 8-15, where

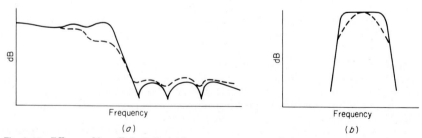

Fig. 8-15   Effects of insufficient Q: (a) low-pass response; (b) bandpass response.

the solid curve represents the theoretical response. Finite Q will also result in less rejection in the vicinity of any stopband zeros and increased filter insertion loss.

Amplitude-equalization techniques can be applied to compensate for the sagging response near cutoff. A passive amplitude equalizer will not actually "boost" the corner response, since a gain cannot be achieved as with active equalizer circuits. However, the equalizer will introduce attenuation except in the region of interest, therefore resulting in a boost in terms of the relative response.

Amplitude equalizers used for low Q compensation are of the bandpass type. They have either constant-impedance or nonconstant-impedance characteristics. The constant-impedance types can be cascaded with each other and the filter with no interaction. The nonconstant-impedance equalizer sections are less complex but will result in some interaction when cascaded with other networks. However, for a boost of 1 or 2 dB, these effects are usually minimal and can be neglected.

Both types of equalizers are shown in figure 8-16. The nonconstant-impedance type can be used in either the series or shunt form. In general, the shunt form is preferred, since the resonating capacitor may be reduced by tapping the inductor.

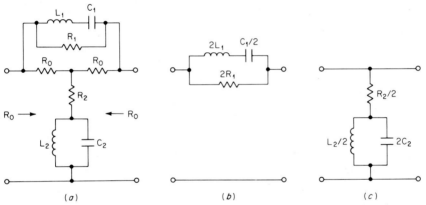

Fig. 8-16   Bandpass-type amplitude equalizers: (a) constant-impedance type; (b) series nonconstant-impedance type; (c) shunt nonconstant-impedance type.

To design a bandpass equalizer, the following characteristics must be determined from the curve to be equalized:

A_{dB} = total amount of equalization required in decibels
  f_r = frequency corresponding to A_{dB}
  f_b = frequency corresponding to $A_{dB}/2$

These parameters are illustrated in figure 8-17, where the corner response and corresponding equalizer are shown for both upper and lower cutoff frequencies.

To design the equalizer, first compute K from

$$A_{dB} = 20 \log K \tag{8-23}$$

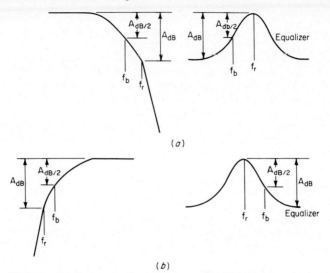

(a)

(b)

Fig. 8-17 Bandpass equalization of corner response: (a) equalization of upper cutoff; (b) equalization of lower cutoff.

Then calculate b where

$$b = \frac{f_r}{f_b} \qquad (8\text{-}24)$$

or

$$b = \frac{f_b}{f_r} \qquad (8\text{-}25)$$

selecting whichever b is greater than unity.

The element values corresponding to the sections of figure 8-16 are found as follows:

$$L_1 = \frac{R_0(K-1)}{2\pi f_b(b^2-1)\sqrt{K}} \qquad (8\text{-}26)$$

$$C_1 = \frac{1}{(2\pi f_r)^2 L_1} \qquad (8\text{-}27)$$

$$L_2 = \frac{R_0(b^2-1)\sqrt{K}}{2\pi f_b b^2(K-1)} \qquad (8\text{-}28)$$

$$C_2 = \frac{1}{(2\pi f_r)^2 L_2} \qquad (8\text{-}29)$$

$$R_1 = R_0(K-1) \qquad (8\text{-}30)$$

$$R_2 = \frac{R_0}{K-1} \qquad (8\text{-}31)$$

where R_0 is the terminating impedance of the filter.

To equalize a low-pass or high-pass filter, a single equalizer is required at the cutoff. For bandpass or band-reject filters a pair of equalizer sections are needed for the upper and lower cutoff frequencies.

The following example illustrates the design of an equalizer to compensate for low Q.

Example 8-3

REQUIRED: A low-pass filter should have a theoretical roll-off of 0.1 dB at 2975 Hz but has instead the following response in the vicinity of cutoff due to insufficient Q:

2850 Hz   −0.5 dB
2975 Hz   −1.0 dB

Design a shunt nonconstant-impedance equalizer to restore the sagging response. The filter impedance level is 1000 Ω.

RESULT: First make the following preliminary computations:

$$K = 10^{A_{dB}/20} = 10^{1/20} = 1.122 \tag{8-23}$$

$$b = \frac{f_r}{f_b} = \frac{2975 \text{ Hz}}{2850 \text{ Hz}} = 1.0439 \tag{8-24}$$

then

$$L_2 = \frac{R_0(b^2 - 1)\sqrt{K}}{2\pi f_b b^2 (K - 1)} = \frac{10^3(1.0439^2 - 1)\sqrt{1.122}}{2\pi 2850 \times 1.0439^2(1.122 - 1)}$$
$$= 40.0 \text{ mH} \tag{8-28}$$

$$C_2 = \frac{1}{(2\pi f_r)^2 L_2} = \frac{1}{(2\pi 2975)^2 \times 0.04} = 0.0715 \ \mu\text{F} \tag{8-29}$$

$$R_2 = \frac{R_0}{K - 1} = \frac{1000}{1.122 - 1} = 8197 \ \Omega \tag{8-31}$$

The resulting equalizer is shown in figure 8-18 using the circuit of figure 8-16c.

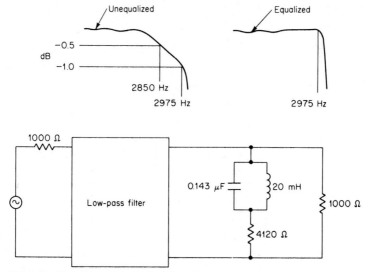

Fig. 8-18 Equalizer of example 8-3.

8.5 COIL-SAVING ELLIPTIC-FUNCTION BANDPASS FILTERS

If an even-order elliptic-function low-pass filter as shown in figure 8 19a is transformed into a bandpass filter using the methods of section 5.1, the bandpass circuit of figure 8-19b is obtained.

A method has been developed to transform the low-pass filter into the configuration of figure 8-19c. The transfer function is unchanged except for a constant multiplier, and $1/2 \, (n - 2)$ coils are saved in comparison with the conventional transformation. These structures are called minimum-inductance or zigzag bandpass filters. However, this transformation requires a very large number of calculations (see Saal and Ulbrich in references) and is therefore considered impractical without a computer.

Geffe (see references) has presented a series of formulas so that this transformation can be performed on an $n = 4$ low-pass network. A tabulation of low-pass element values for $n = 4$ can be found in either Zverev's "Handbook of Filter Synthesis" or Saal's "Der Entwurf von Filtern mit Hilfe des Kataloges Normierter Tiefpasse" (see references) as well as in table 12-56.

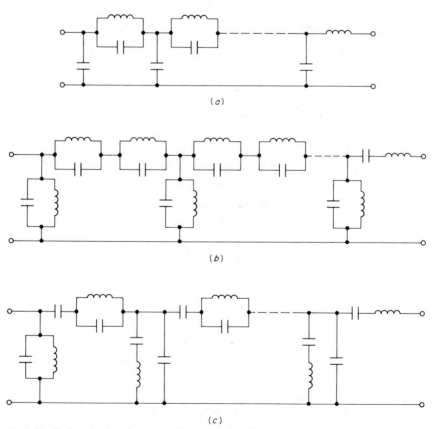

Fig. 8-19 Coil-saving bandpass transformation: *(a)* elliptic-function low-pass filter; *(b)* conventional bandpass transformation; *(c)* minimum-inductance bandpass transformation.

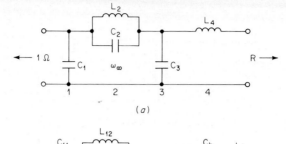

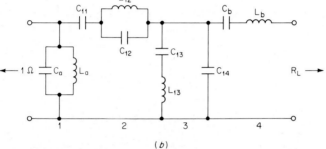

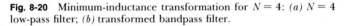

(b)

Fig. 8-20 Minimum-inductance transformation for $N = 4$: (a) $N = 4$ low-pass filter; (b) transformed bandpass filter.

The low-pass filter and corresponding bandpass network are shown in figure 8-20. The following preliminary computations are required:

$$Q_{bp} = \frac{f_0}{BW} \tag{8-32}$$

$$a = \frac{\omega_\infty}{2 Q_{bp}} \tag{8-33}$$

$$x = 1 + \sqrt{a^2 + 1} \tag{8-34}$$

$$t_1 = 1 + \frac{C_3}{C_2} \tag{8-35}$$

$$T = \frac{1 + t_1 x^2}{t_1 + x^2} \tag{8-36}$$

$$k = \frac{Q_{bp} T}{t_1} \tag{8-37}$$

$$t_2 = \frac{x^2}{x^2 + t_1} \tag{8-38}$$

$$t_3 = \frac{t_1 t_2}{T} \tag{8-39}$$

$$\alpha = 1 - \frac{1}{x^2} \tag{8-40}$$

$$\beta = x^2 - 1 \tag{8-41}$$

$$A = \frac{C_3 k \alpha}{T} \tag{8-42}$$

$$B = \frac{t_2 t_3}{C_3 k \beta}$$

(8-43)

The bandpass element values can now be computed as follows:

$$R_L = t_3^2 R$$

(8-44)

$$C_{11} = \frac{C_3 k \beta}{t_1 t_2}$$

(8-45)

$$C_{12} = \frac{C_3 k \alpha}{T - 1}$$

(8-46)

$$L_{12} = \frac{1}{x^2 C_{12}}$$

(8-47)

$$C_{13} = \frac{C_{11}(T - 1)}{t_2}$$

(8-48)

$$L_{13} = \frac{x^2}{C_{13}}$$

(8-49)

$$C_{14} = \frac{C_3 k \alpha}{t_2}$$

(8-50)

$$L_a = \frac{1}{Q_{bp}\left(C_1 + \dfrac{C_3}{t_1}\right)}$$

(8-51)

$$C_a = \frac{1}{L_a} - A$$

(8-52)

$$L_b = t_3^2 Q_{bp} L_4$$

(8-53)

$$C_b = \frac{1}{L_b - B}$$

(8-54)

The bandpass filter of figure 8-20b must be denormalized to the required impedance level and center frequency f_0. Since the source and load impedance levels are unequal, either the tapped inductor or the capacitance transformation can be used to obtain equal terminations if required.

The transmission zero above the passband is provided by the parallel resonance of $L_{12}C_{12}$ in branch 2 and the lower zero corresponds to the series resonance of $L_{13}C_{13}$ in branch 3. The circuits of branches 2 and 3 each have conditions of both series and parallel resonance and can be transformed from one form to the other. The following equations relate the type 1 and 2 networks shown in figure 8-21:

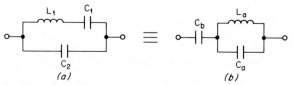

Fig. 8-21 Equivalent branches: (a) type 1 network; (b) type 2 network.

For a type 1 network:

$$L_1 = L_a \left(1 + \frac{C_a}{C_b}\right)^2 \qquad (8\text{-}55)$$

$$C_1 = C_b \frac{1}{1 + \dfrac{C_a}{C_b}} \qquad (8\text{-}56)$$

$$C_2 = C_a \frac{1}{1 + \dfrac{C_a}{C_b}} \qquad (8\text{-}57)$$

$$f_{\text{series}} = \frac{1}{2\pi\sqrt{L_1 C_1}} \qquad (8\text{-}58)$$

$$f_{\text{par}} = \frac{1}{2\pi\sqrt{\dfrac{L_1 C_1 C_2}{C_1 + C_2}}} \qquad (8\text{-}59)$$

For a type 2 network:

$$L_a = L_1 \frac{1}{\left(1 + \dfrac{C_2}{C_1}\right)^2} \qquad (8\text{-}60)$$

$$C_a = C_2 \left(1 + \frac{C_2}{C_1}\right) \qquad (8\text{-}61)$$

$$C_b = C_1 + C_2 \qquad (8\text{-}62)$$

$$f_{\text{series}} = \frac{1}{2\pi\sqrt{L_1(C_a + C_b)}} \qquad (8\text{-}63)$$

$$f_{\text{par}} = \frac{1}{2\pi\sqrt{L_a C_a}} \qquad (8\text{-}64)$$

In general, the bandpass series arms are of the type 2 form and the shunt branches are of the type 1 form as in figure 8-19c. The tuning usually consists of adjusting the parallel resonances of the series branches and the series resonances of the shunt branches, i.e., the transmission zeros.

8.6 FILTER TUNING METHODS

LC filters are typically assembled using elements with 1 or 2% tolerances. For many applications the deviation in the desired response caused by component variations may be unacceptable; so adjustment of elements will be required. It has been found that wherever resonances occur, the LC product of the resonant circuit is significantly more critical than the L/C ratio. As a result, filter adjustment normally involves adjusting each tuned circuit for resonance at the specified frequency.

Adjustment techniques are based on the impedance extremes that occur at resonance. In the circuit of figure 8-22a, an output null will occur at parallel resonance because of voltage-divider action. Series LC circuits are tuned using the circuit of figure 8-22b, where an output null will also occur at resonance.

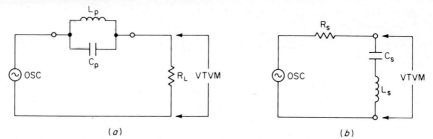

Fig. 8-22   Test circuits for adjusting resonant frequencies: (*a*) adjusting parallel resonance; (*b*) adjusting series resonance.

The adjustment method in both cases involves setting the oscillator for the required frequency and adjusting the variable element, usually the inductor, for an output null. Resistors R_L and R_s are chosen so that an approximately 20- to 30-dB drop occurs between the oscillator and the output at resonance. These values can be estimated from

$$R_L \approx \frac{2\pi f_r L_p Q_L}{20}$$
(8-65)

and
$$R_s \approx \frac{40\pi f_r L_s}{Q_L}$$
(8-66)

where Q_L is the inductor Q. A feature of this technique is that no tuning errors result from stray capacity across the VTVM. Care should be taken that the oscillator does not have excessive distortion, since a sharp null may then be difficult to obtain. Also, excessive levels should be avoided, as detuning can occur from inductor saturation effects.

When inductor Q's are below 10, sharp nulls cannot be obtained. A more desirable tuning method is to adjust for the condition of zero phase shift at resonance, which will be more distinct than the null. The circuits of figure 8-22 can still be used in conjunction with an oscilloscope having both vertical and horizontal inputs. One channel monitors the oscillator and the other channel is connected to the output instead of using the VTVM. A "Lissajous pattern" is obtained, and the tuned circuit is then adjusted for a closed ellipse.

Certain construction practices must be used so that the assembled filter can be tuned. There must be provision for access to each tuned circuit on an individual basis. This is usually accomplished by leaving all the grounds disconnected until after tuning so that each branch can be individually inserted into the tuning configurations of figure 8-22 with all the other branches present.

8.7   MEASUREMENT METHODS

This section discusses some major filter parameters and describes techniques for their measurement. Also some misconceptions associated with these characteristics are clarified. All measurements should be made using rated operating levels so that the results are meaningful. After fabrication, filters should be subjected to insertion-loss and frequency-response measurements as a minimum production test.

Insertion Loss and Frequency Response

The frequency response of filters is always considered as relative to the attenuation occurring at a particular reference frequency. The actual attenuation at this reference is called "insertion loss."

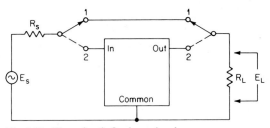

Fig. 8-23 Test circuit for insertion loss.

The classical definition of insertion loss is the decrease in power delivered to the load when a filter is inserted between the source and the load. Using figure 8-23, the insertion loss is given by

$$IL_{dB} = 10 \log \frac{P_{L1}}{P_{L2}} \tag{8-67}$$

where P_{L1} is the power delivered to the load with both switches in position 1 (filter bypassed) and P_{L2} is the output power with both switches in position 2. Equation (8-67) can also be expressed in terms of a voltage ratio as

$$IL_{dB} = 10 \log \frac{E_{L1}^2/R_L}{E_{L2}^2/R_L} = 20 \log \frac{E_{L1}}{E_{L2}} \tag{8-68}$$

so a decibel meter can be used at the output to measure insertion loss directly in terms of output voltage.

The classical definition of insertion loss may be somewhat inapplicable when the source and load terminations are unequal. In reality, if the filter were not used, the source and load would probably be connected through an impedance-matching transformer instead of a direct connection. Therefore, the comparison of figure 8-23 would be invalid.

An alternate definition is transducer loss, which is defined as the decrease in power delivered to the load when an ideal impedance-matching transformer is replaced by the filter. The test circuit of figure 8-23 can still be used if a correction factor is added to equation (8-68). The resulting expression becomes

$$IL_{dB} = 20 \log \frac{E_{L1}}{E_{L2}} + 20 \log \frac{R_s + R_L}{2\sqrt{R_s R_L}} \tag{8-69}$$

Frequency response or relative attenuation is measured using the test circuit of figure 8-24. The input source E_s is set to the reference frequency and the level is arbitrarily set for a 0-dB reference at the output. As the input frequency is changed, the variation in output level is the relative attenuation.

It must be understood that the variation of the ratio E_L/E_s is the frequency response. The ratio E_L/E_i is of no significance, since it reflects the frequency response of the filter when driven by a voltage source. As the source frequency is varied, the voltage E_s must be kept constant. Any attempt to keep E_i constant will distort the response shape, since voltage-divider action between R_s and the filter input impedance must normally occur to satisfy the transfer function.

The oscillator source itself, E_s, may contain some internal impedance. Nevertheless the value of R_s should correspond to the design source impedance, since the internal impedance of E_s is allowed for by maintaining the terminal voltage of E_s constant.

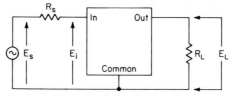

Fig. 8-24   Test circuit for frequency response.

Input Impedance of Filter Networks

The input or output impedance of filters must frequently be determined to ensure compatibility with external circuitry. The input impedance of a filter is the impedance measured at the input terminals with the output appropriately terminated. Conversely, the output impedance can be measured by terminating the input.

Let us first consider the test circuit of figure 8-25a. A common fallacy is to adjust the value of R until $|E_2|$ is equal to $\frac{1}{2}|E_1|$, i.e., a 6-dB drop.

The input impedance Z_{11} is then said to be equal to R. However, this will be true only if Z_{11} is purely resistive. As an example, if Z_{11} is purely reactive and its absolute magnitude is equal to R, the value of E_2 will be 0.707 or 3 dB below E_1 and not 6 dB.

Using the circuit of figure 8-25a, an alternate approach will result in greater accuracy. If R is adjusted until a 20-dB drop occurs between $|E_2|$ and $|E_1|$, then $|Z_{11}|$ is determined by $R/10$. The accuracy will be within 10%. For even more accurate results, the 40-dB method can be used where R is adjusted for a 40-dB drop. The magnitude of Z_{11} is then given by $R/100$.

If a more precise measurement is required, a floating meter can be used in the configuration of figure 8-25b. The input impedance is then directly given by

$$|Z_{11}| = \frac{|E_b|}{|E_a|} R_s \qquad (8\text{-}70)$$

Return Loss   Return loss is a figure of merit which indicates how closely a measured impedance matches a standard impedance, both in magnitude and in phase angle. Return loss is expressed as

$$A_\rho = 20 \log \left| \frac{Z_s + Z_x}{Z_s - Z_x} \right| \qquad (8\text{-}71)$$

where Z_s is the standard impedance and Z_x is the measured impedance. For a perfect match, the return loss would be infinite.

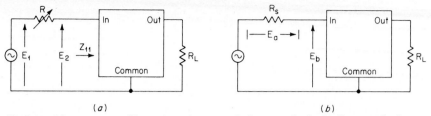

(a)   (b)

Fig. 8-25   Measurement of input impedance: *(a)* indirect method; *(b)* direct method.

Return loss can be directly measured using the bridge arrangement of figure 8-26. The return loss is given by

$$A_\rho = 20 \log \left| \frac{V_{01}}{V_{02}} \right| \qquad (8\text{-}72)$$

where V_{01} is the output voltage with the switch closed and V_{02} is the output voltage with the switch open. The return loss can then be read directly using a decibel meter. The value of R is arbitrary, but both resistors must be closely matched to each other.

The family of curves in figure 8-27 represents the return loss using a standard impedance of 600 Ω with phase angle of impedance as a parameter. Clearly the return loss is very sensitive to phase angle. If the impedance were 600 Ω at an angle of only 10°, the return loss would be 21 dB. If there was no phase shift, an impedance error of as much as 100 Ω would correspond to 21 dB of return loss.

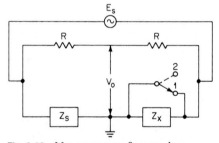

Fig. 8-26   Measurement of return loss.

Time-Domain Characteristics

Step Response   The step response of a filter network is a useful criterion since low transient distortion is a necessary requirement for good transmission of modulated signals. To determine the step response of a low-pass filter, an input DC step is applied. For bandpass filters a carrier step is used where the carrier frequency is equal to the filter center frequency f_0. Since it is difficult to view a single transient on an oscilloscope unless it is of the storage type, a square-wave generator is used instead of the DC step and a tone-burst generator is substituted for the carrier step. However, the repetition rate must be chosen slow enough so that the transient behavior has stabilized prior to the next pulse to obtain meaningful results.

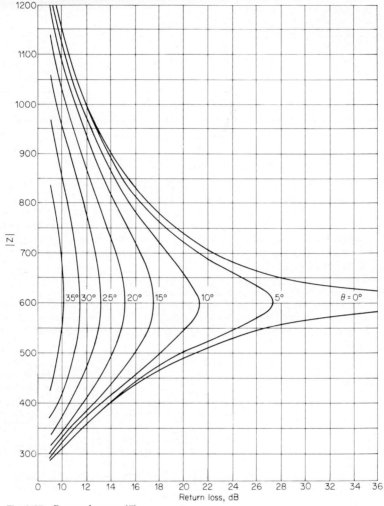

Fig. 8-27 Return loss vs. $|Z|$.

The test configuration is shown in figure 8-28*a*. The output waveforms are depicted in figures 8-28*b* and *c* for a DC step and tone burst, respectively. The following definitions are applicable:

Percent Overshoot P_T: The difference between the peak response and the final steady-state value expressed as a percentage.

Rise Time T_r: The interval between 10 and 90% of the final value.

Settling Time T_s: Time required for the response to settle within a specified percent of its final value.

The waveform definitions shown in figure 8-28*b* also apply to figure 8-28*c* if we consider the envelope of the carrier waveform instead of the instantaneous values.

Group Delay The phase shift of a filter can be measured by using the "Lissa-jous pattern" method. By connecting the vertical channel of an oscilloscope to the input source and the horizontal channel to the load, an ellipse is obtained as shown in figure 8-29. The phase angle in degrees is given by

$$\phi = \sin^{-1}\frac{Y_{int}}{Y_{max}} \tag{8-73}$$

Since group delay is the derivative with respect to frequency of the phase shift, we can measure the phase shift at two closely spaced frequencies and approximate the delay as follows:

$$T_{gd} \approx \frac{\Delta\phi}{360\Delta f} \tag{8-74}$$

where $\Delta\phi$ is $\phi_2 - \phi_1$ in degrees, Δf is $f_2 - f_1$ in hertz, and T_{gd} is the group delay at the midfrequency, i.e., $(f_1 + f_2)/2$.

A less accurate method involves determining the 180° phase shift points. As the frequency is varied throughout the passband of a high-order filter, the phase shift will go through many integer multiples of 180° where the Lissajous pattern adopts a straight line at either 45° or 225°. If we record the separation between adjacent 180° points, the nominal group delay at the midfrequency can be approximated by

$$T_{gd} \approx \frac{1}{2\Delta f} \tag{8-75}$$

The classical approach for the measurement of group delay directly is shown in figure 8-30. A sine-wave source, typically 25 Hz, is applied to an amplitude

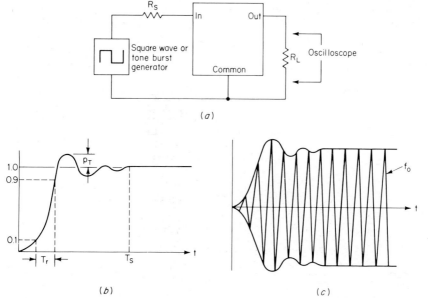

(a)

(b) *(c)*

Fig. 8-28 Step response of networks: *(a)* test circuit; *(b)* step response to DC step; *(c)* step response to tone burst.

modulator along with a carrier signal. The output consists of an amplitude-modulated signal comprised of the carrier and two sidebands at ±25 Hz on either side of the carrier. The signal is then applied to the network under test.

The output signal from the network is of the same form as the input, but the 25-Hz envelope has been shifted in time by an amount equal to the group delay at the carrier frequency. The output envelope is recovered by an AM detector and applied to a phase detector along with a reference 25-Hz signal.

The phase detector output is a DC signal proportional to the phase shift between the 25-Hz reference and the demodulated 25-Hz carrier envelope. As the carrier is varied in frequency, the DC signal will vary in accordance with the change in group delay (differential delay distortion) and the delay can be displayed on a DC meter having the proper calibration. If an adjustable phase-shift network is interposed between the AM detector and phase detector, the meter indication can be adjusted to establish a reference level at a desired reference frequency.

The theoretical justification for this scheme is based on the fact that the delay of the envelope is determined by the slope of a line segment interconnecting the phase shift of the two sidebands at ±25 Hz about the carrier, i.e., $(\phi_2 - \phi_1)/(\omega_2 - \omega_1)$. This definition is sometimes called the "envelope delay," for obvious reasons. As the separation between sidebands is decreased, the envelope delay approaches the theoretical group delay at the carrier frequency, since group delay is defined as the derivative of the phase shift. For most measurements a modulation rate of 25 Hz is adequate.

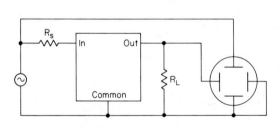
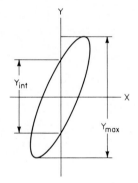

Fig. 8-29 Measurement of phase shift.

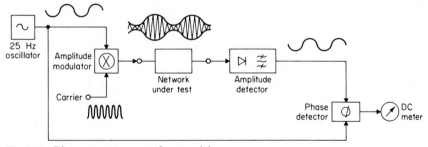

Fig. 8-30 Direct measurement of group delay.

Measuring the Q of Inductors

A device called a "Q meter" is frequently used to measure the Q of inductors at a specified frequency. The principle of the Q meter is based on the fact that in a series resonant circuit, the voltage across each reactive element is Q times the voltage applied to the resonant circuit. The Q can then be directly determined by the ratio of two voltages.

A test circuit is shown in figure 8-31a. Care should be taken that the applied voltage does not result in an excessive voltage developed across the inductor during the measurement. Also the resonating capacitor should have a much higher Q than the inductor, and the meter used for the voltage measurements should be a high-impedance type to avoid loading errors.

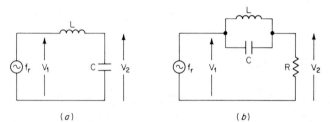

(a) (b)

Fig. 8-31 Measurement of coil Q: (a) Q meter method; (b) parallel resonant circuit method.

An alternate approach involves measuring the impedance of a parallel resonant circuit consisting of the inductor and the required resonating capacitor for the frequency of interest. Using the voltage divider of figure 8-31b, the Q at resonance is found from

$$Q = \frac{R}{2\pi f_r L}\left(\frac{V_1}{V_2} - 1\right)$$

(8-76)

A null will occur in V_2 at resonance.

For meaningful Q measurements it is important that the measurement frequency corresponds to the frequency of interest of the filter, since coil Q can decrease quite dramatically outside a particular range. In low-pass and high-pass filters the Q should be measured at the cutoff, and for bandpass and band-reject filters the center frequency is the frequency of interest.

REFERENCES

Geffe, P. R., "Simplified Modern Filter Design," John F. Rider, New York, 1963.

Saal, R., "Der Entwurf von Filtern mit Hilfe des Kataloges Normierter Tiefpasse," Telefunken GMBH, Backnang, West Germany, 1963.

Saal, R., and Ulbrich, E., "On the Design of Filters by Synthesis," IRE Transactions on Circuit Theory, Vol. CT–5, December 1958.

Zverev, A. I., "Handbook of Filter Synthesis," John Wiley and Sons, New York, 1967.

9

Design of Magnetic Components

9.1 BASIC PRINCIPLES OF MAGNETIC-CIRCUIT DESIGN

Units of Measurement

Magnetic permeability is represented by the symbol μ and is defined by

$$\mu = \frac{B}{H} \tag{9-1}$$

B is the magnetic flux density in lines per square centimeter and is measured in gauss and H is the magnetizing force in oersteds that produced the flux. Permeability is dimensionless and can be considered a figure of merit of a particular magnetic material, since it represents the ease of producing a magnetic flux for a given input. The permeability of air or that of a vacuum is 1.

Magnetizing force is caused by current flowing through turns of wire; so H can be determined from ampere-turns by

$$H = \frac{4\pi NI}{10 \text{ mL}} \tag{9-2}$$

where N is the number of turns, I is the current in amperes, and mL is the mean length of the magnetic path in centimeters.

The inductance of a coil is directly proportional to the number of flux linkages per unit current. The total flux is found from

$$\phi = BA = \mu HA = \frac{4\pi NI\mu A}{10 \text{ mL}} \tag{9-3}$$

where A is the cross-sectional area in square centimeters.

The inductance proportionality may then be expressed as

$$L \propto \frac{4\pi NI\mu A}{10 \text{ mL}} \quad \frac{N}{I} \tag{9-4}$$

or directly in henrys by

$$L = \frac{4\pi N^2 \mu A}{mL} \, 10^{-9} \qquad (9\text{-}5)$$

A number of things should be apparent from equation (9-5). First of all, the inductance of a coil is directly proportional to the permeability of the core material. If an iron core is inserted into an air-core inductor, the inductance will increase in direct proportion to the iron core's permeability. The inductance is also proportional to N^2.

All the previous design equations make the assumption that the magnetic path is uniform and closed with negligible leakage flux in the surrounding air such as would occur with a single-layer toroidal coil structure. However, this assumption is really never completely valid; so some deviations from the theory can be expected.

The induced voltage of an inductor can be related to the flux density by

$$E_{rms} = 4.44 \, BNfA \times 10^{-8} \qquad (9\text{-}6)$$

where B is the maximum flux density in gauss, N is the number of turns, f is the frequency in hertz, and A is the cross-sectional area of the core in square centimeters. This important equation is derived from Faraday's law.

Saturation and DC Polarization

A plot of B vs. H is shown in figure 9-1. Let us start at point A and increase the magnetizing force to point B. A decrease in magnetizing force will pass through point C and then D and E as the magnetizing force is made negative. An increasing magnetizing force, again in the positive direction, will travel to B through point F. The enclosed area formed by the curve is called a "hysteresis loop" and results from the energy required to reverse the magnetic molecules of the core. The magnitude of H between points D and A is called "coercive force" and is the amount of H necessary to reduce the residual magnetism in the core to zero.

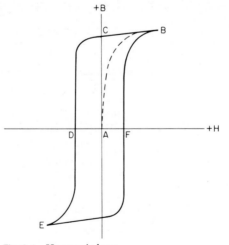

Fig. 9-1 Hysteresis loop.

Permeability was defined as the ratio B/H and can be obtained from the slope of the BH curve. Since we normally deal with low-level AC signals, the region of interest is restricted to a relatively narrow range. We can then assume that the permeability is determined by the derivative of the curve at the origin. The derivative of a B/H curve is sometimes called "incremental permeability."

If a DC bias is introduced, the quiescent point will move from the origin to a point farther out on the curve. Since the curve tends to flatten out with higher values of H, the incremental permeability will decrease, which reduces the inductance. This effect is known as "saturation" and can also occur without a DC bias for large AC signals. Severe waveform distortion usually accompanies saturation. The B/H curve for an air core is a straight line at a 45° angle through the origin. The permeability is unity, and no saturation can occur.

Inductor Losses

The Q of a coil can be found from

$$Q = \frac{\omega L}{R_{dc} + R_{ac} + R_d} \tag{9-7}$$

where R_{dc} is the copper loss, R_{ac} is the core loss, and R_d is the dielectric loss. Copper loss consists strictly of the DC winding resistance and is determined by the wire size and total length of wire required. The core loss is comprised mostly of losses due to eddy currents and hysteresis. Eddy currents are induced in the core material by changing magnetic fields. These circulating currents produce losses that are proportional to the square of the inducing frequency.

When a core is subjected to an AC or pulsating DC magnetic field, the B vs. H characteristics can be represented by the curve of figure 9-1. The enclosed area was called a hysteresis loop and resulted from the energy required to reverse the magnetic molecules of the core. These core losses increase in direct proportion to frequency, since each cycle traverses the hysteresis loop.

The dielectric losses are important at higher frequencies and are determined by the power factor of the distributed capacity. Keeping the distributed capacity small as well as using wire insulation with good dielectric properties will minimize dielectric losses.

Above approximately 50 kHz the current will tend to travel on the surface of a conductor rather than through the cross section. This phenomenon is called "skin effect." To reduce this effect, litz wire is commonly used. This wire consists of many braided strands of insulated conductors so that a larger surface area is available in comparison with a single solid conductor of the equivalent cross section. Above 1 or 2 MHz solid wire can again be used.

A figure of merit of the efficiency of a coil at low frequencies is the ratio of ohms per henry (Ω/H), where the ohms correspond to R_{dc}, i.e., the copper losses. For a given coil structure and permeability, the ratio Ω/H is a constant independent of the total number of turns, provided that the winding cross-sectional area is kept constant.

Effect of an Air Gap

If an ideal toroidal core has a narrow air gap introduced, the flux will decrease and the permeability is reduced. The resulting effective permeability can be found from

$$\mu_e = \frac{\mu_i}{1 + \mu_i \left(\dfrac{g}{mL}\right)}$$ (9-8)

where μ_i is the initial permeability of the core and g/mL is the ratio of gap to length of the magnetic path. Equation (9-8) applies to closed magnetic structures of any shape if the initial permeability is high and the gap ratio small.

The effect of an air gap is to reduce the permeability and make the coil's characteristics less dependent upon the initial permeability of the core material. However, lower permeability requires more turns and the associated copper losses; so a suitable compromise is required.

TABLE 9-1   Wire Chart—Round Heavy Film Insulated Solid Copper*

AWG	Diameter over Bare, in			Insulation Additions		Diameter over Insulation		Pounds per 1000 Feet
	Minimum	Nominal	Maximum	Minimum	Maximum	Minimum	Maximum	
4	0.2023	0.2043	0.2053	0.0037	0.0045	0.2060	0.2098	127.20
5	0.1801	0.1819	0.1828	0.0036	0.0044	0.1837	0.1872	100.84
6	0.1604	0.1620	0.1628	0.0035	0.0043	0.1639	0.1671	80.00
7	0.1429	0.1443	0.1450	0.0034	0.0041	0.1463	0.1491	63.51
8	0.1272	0.1285	0.1292	0.0033	0.0040	0.1305	0.1332	50.39
9	0.1133	0.1144	0.1150	0.0032	0.0039	0.1165	0.1189	39.98
10	0.1009	0.1019	0.1024	0.0031	0.0037	0.1040	0.1061	31.74
11	0.0898	0.0907	0.0912	0.0030	0.0036	0.0928	0.0948	25.16
12	0.0800	0.0808	0.0812	0.0029	0.0035	0.0829	0.0847	20.03
13	0.0713	0.0720	0.0724	0.0028	0.0033	0.0741	0.0757	15.89
14	0.0635	0.0641	0.0644	0.0032	0.0038	0.0667	0.0682	12.60
15	0.0565	0.0571	0.0574	0.0030	0.0035	0.0595	0.0609	10.04
16	0.0503	0.0508	0.0511	0.0029	0.0034	0.0532	0.0545	7.95
17	0.0448	0.0453	0.0455	0.0028	0.0033	0.0476	0.0488	6.33
18	0.0399	0.0403	0.0405	0.0026	0.0032	0.0425	0.0437	5.03
19	0.0355	0.0359	0.0361	0.0025	0.0030	0.0380	0.0391	3.99
20	0.0317	0.0320	0.0322	0.0023	0.0029	0.0340	0.0351	3.18
21	0.0282	0.0285	0.0286	0.0022	0.0028	0.0302	0.0314	2.53
22	0.0250	0.0253	0.0254	0.0021	0.0027	0.0271	0.0281	2.00
23	0.0224	0.0226	0.0227	0.0020	0.0026	0.0244	0.0253	1.60
24	0.0199	0.0201	0.0202	0.0019	0.0025	0.0218	0.0227	1.26
25	0.0177	0.0179	0.0180	0.0018	0.0023	0.0195	0.0203	1.00
26	0.0157	0.0159	0.0160	0.0017	0.0022	0.0174	0.0182	0.794
27	0.0141	0.0142	0.0143	0.0016	0.0021	0.0157	0.0164	0.634
28	0.0125	0.0126	0.0127	0.0016	0.0020	0.0141	0.0147	0.502
29	0.0112	0.0113	0.0114	0.0015	0.0019	0.0127	0.0133	0.405
30	0.0099	0.0100	0.0101	0.0014	0.0018	0.0113	0.0119	0.318
31	0.0088	0.0089	0.0090	0.0013	0.0018	0.0101	0.0108	0.253
32	0.0079	0.0080	0.0081	0.0012	0.0017	0.0091	0.0098	0.205
33	0.0070	0.0071	0.0072	0.0011	0.0016	0.0081	0.0088	0.162
34	0.0062	0.0063	0.0064	0.0010	0.0014	0.0072	0.0078	0.127
35	0.0055	0.0056	0.0057	0.0009	0.0013	0.0064	0.0070	0.101
36	0.0049	0.0050	0.0051	0.0008	0.0012	0.0057	0.0063	0.0805
37	0.0044	0.0045	0.0046	0.0008	0.0011	0.0052	0.0057	0.0655
38	0.0039	0.0040	0.0041	0.0007	0.0010	0.0046	0.0051	0.0518
39	0.0034	0.0035	0.0036	0.0006	0.0009	0.0040	0.0045	0.0397
40	0.0030	0.0031	0.0032	0.0006	0.0008	0.0036	0.0040	0.0312
41	0.0027	0.0028	0.0029	0.0005	0.0007	0.0032	0.0036	0.0254
42	0.0024	0.0025	0.0026	0.0004	0.0006	0.0028	0.0032	0.0203

* Courtesy Belden Corp.

Design of Coil Windings

Inductors are normally wound using insulated copper wire. The general method used to express wire size is the American Wire Gauge (AWG) system. As the wire size numerically decreases, the diameter increases, where the ratio of diameters of one size to the next larger size is 1.1229. The ratio of cross-sectional areas of adjacent wire sizes corresponds to the square of the diameter, or 1.261. Therefore, for an available cross-sectional winding area, reducing the wire by one size permits 1.261 times as many turns. Two wire sizes correspond to a factor of 1.261^2, or 1.59, and three wire sizes permits twice as many turns.

TABLE 9-1 *Continued*

Weight		Resistance at 20°C (68°F)			Turns		
Feet per Pound	Pounds per Cubic Inch	Ohms per 1000 Feet	Ohms per Pound	Ohms per Cubic Inch	Per Linear Inch	Per Square Inch	AWG
7.86	0.244	0.2485	0.001954	0.0004768	4.80	24.0	4
9.92	0.243	0.3134	0.003108	0.0007552	5.38	28.9	5
12.50	0.242	0.3952	0.004940	0.001195	6.03	36.4	6
15.75	0.241	0.4981	0.007843	0.001890	6.75	45.6	7
19.85	0.240	0.6281	0.01246	0.002791	7.57	57.3	8
25.0	0.239	0.7925	0.01982	0.004737	8.48	71.9	9
31.5	0.238	0.9988	0.03147	0.007490	9.50	90.3	10
39.8	0.237	1.26	0.0501	0.0119	10.6	112	11
49.9	0.236	1.59	0.0794	0.0187	11.9	142	12
62.9	0.235	2.00	0.126	0.0296	13.3	177	13
82.9	0.230	2.52	0.200	0.0460	14.8	219	14
99.6	0.229	3.18	0.317	0.0726	16.6	276	15
126	0.228	4.02	0.506	0.115	18.5	342	16
158	0.226	5.05	0.798	0.180	20.7	428	17
199	0.224	6.39	1.27	0.284	23.1	534	18
251	0.223	8.05	2.02	0.450	25.9	671	19
314	0.221	10.1	3.18	0.703	28.9	835	20
395	0.219	12.8	5.06	1.11	32.3	1,043	21
500	0.217	16.2	8.10	1.76	36.1	1,303	22
625	0.215	20.3	12.7	2.73	40.2	1,616	23
794	0.211	25.7	20.4	4.30	44.8	2,007	24
1,000	0.210	32.4	32.4	6.80	50.1	2,510	25
1,259	0.208	41.0	51.6	10.7	56.0	3,136	26
1,577	0.205	51.4	81.1	16.6	62.3	3,831	27
1,992	0.202	65.3	130	26.3	69.4	4,816	28
2,469	0.200	81.2	200	40.0	76.9	5,914	29
3,145	0.197	104	327	64.4	86.2	7,430	30
4,000	0.193	131	520	100	96	9,200	31
4,900	0.191	162	790	151	106	11,200	32
6,200	0.189	206	1,270	240	118	13,900	33
7,900	0.189	261	2,060	388	133	17,700	34
9,900	0.187	331	3,280	613	149	22,200	35
12,400	0.186	415	5,150	959	167	27,900	36
15,300	0.184	512	7,800	1,438	183	33,500	37
19,300	0.183	648	12,500	2,289	206	42,400	38
25,200	0.183	847	21,300	3,904	235	55,200	39
32,100	0.183	1,080	34,600	6,335	263	69,200	40
39,400	0.183	1,320	52,000	9,510	294	86,400	41
49,300	0.182	1,660	81,800	14,883	328	107,600	42

Physical and electrical properties of a range of wire sizes are given in the wire chart of table 9-1. These data are based on using a standard heavy film for the insulation. In the past, enamel insulations were used which required acid or abrasives for stripping the insulation from the wire ends to make electrical connections. Solderable insulations have been available for the last 10 years, so that the wire ends can be easily tinned. In the event litz wire is required, a cross reference between litz wire sizes and the solid equivalent is given in table 9-2. The convention for specifying litz wire is number of strands/wire size. For example, using the chart, a litz equivalent to No. 31 solid is 20 strands of No. 44, or 20/44. In general a large number of strands is desirable for more effective surface area.

The temperature coefficient of resistance for copper is 0.393% per degree Celsius. The DC resistance of a winding at a particular temperature is given by

$$R_{t_1} = R_t [1 + 0.00393 \, (t_1 - t)] \qquad (9\text{-}9)$$

where t is the initial temperature and t_1 is the final temperature, both in degrees Celsius. The maximum permitted temperature of most wire insulations is about 130°C.

To compute the number of turns for a required inductance, the inductance factor for the coil structure must be used. This factor is generally called A_L and is the nominal inductance per 1000 turns. Since inductance is proportional

TABLE 9-2   Stranded Wire Equivalent Chart

Solid Equivalent	Size per Strand										
	34	35	36	37	38	39	40	41	42	43	44
15	80	100									
16	64	80	100								
17	50	64	80	100							
18	40	50	64	80	100						
19	32	40	50	64	80	100					
20	25	32	40	50	64	80	100				
21	20	25	32	40	50	64	80	100			
22	16	20	25	32	40	50	64	80	100		
23	12	16	20	25	32	40	50	64	80	100	
24	10	12	16	20	25	32	40	50	64	80	100
25	8	10	12	16	20	25	32	40	50	64	80
26	6	8	10	12	16	20	25	32	40	50	64
27	5	6	8	10	12	16	20	25	32	40	50
28	4	5	6	8	10	12	16	20	25	32	40
29		4	5	6	8	10	12	16	20	25	32
30			4	5	6	8	10	12	16	20	25
31				4	5	6	8	10	12	16	20
32					4	5	6	8	10	12	16
33						4	5	6	8	10	12
34							4	5	6	8	10
35								4	5	6	8
36									4	5	6
37										4	5
38											4

to the number of turns squared, the required number of turns N for an inductance L is given by

$$N = 10^3 \sqrt{\frac{L}{A_L}} \qquad (9\text{-}10)$$

where L and A_L must be in identical units.

After the required number of turns is computed, a wire size must be chosen. For each winding structure an associated chart can be tabulated which indicates the maximum number of turns for each wire size. A wire size can then be chosen which results in the maximum utilization of the available winding cross-sectional area.

Coil winding methods are very diverse since the winding techniques depend upon the actual coil structure, the operating frequency range, etc. Coil winding techniques are discussed in the remainder of this chapter on an individual basis for each coil structure type.

9.2 LAMINATED INDUCTORS

Let us consider an inductor L at a filter cutoff ω_c having a loss R. The Q is given by $\omega_c L / R$. If we scale the filter to a cutoff of 0.1 ω_c, the inductor becomes $10L$, but R must not change to maintain the same Q. Therefore, low-frequency inductors generally require lower ohms per henry (Ω/H) than inductors for use at higher frequencies.

A very efficient coil structure for the range below approximately 500 Hz is a laminated inductor assembly. Laminated coils consist of a plastic or nylon bobbin containing the coil winding interleaved with F-shaped laminations of a high-permeability nickel alloy with an air gap in the center leg.

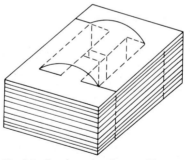

Fig. 9-2 Laminated coil assembly.

A typical lamination assembly is shown in figure 9-2. External clamps or a heavy varnish coating is used to keep the structure rigid. The air gap is maintained uniform and precise by a spacer made of paper or mylar cut to shape to fit the air gap. The laminations must not undergo any severe mechanical stresses during handling, since they have an oriented crystal structure which can be easily deformed, resulting in degraded performance.

Properties of Laminations

Laminations are made of nickel alloy materials that fall into three categories: silicon, 50% nickel, and 80% nickel. The last category is especially processed for high initial permeability and is particularly suited for filter inductors. Some commonly used trade names for 80% nickel alloy materials are Permalloy 80, Supermu 30, Supermu 40, 4–79 Mo-Permalloy, Hy Mu 80, and Mu Metal.

Nickel alloy laminations are available in a variety of forms such as EE, EI, UI, and the F shape. However, only the F shape is convenient for the introduction of a single gap in the center leg.

Physical size of the lamination is critical for optimum performance. Inductance stability is affected by material permeability variations with temperature and operating flux density. Larger gaps tend to isolate the resulting inductance from the permeability changes of the magnetic material. Permeability and core losses also vary with frequency. Decreasing the lamination thickness will extend the frequency range. Laminations are available in standard thicknesses of 0.004, 0.006, and 0.014 in, usually expressed in mils.

TABLE 9-3   Parameters of F Laminations

Shape	Stack Height for Square Cross Section, in	Window Area, in²	Core Cross-Sectional Area for Square Stack, in²	Length of Magnetic Path, in
F28–29	0.125	0.0394	0.0156	1.130
F156	0.156	0.0393	0.0243	1.190
F187	0.188	0.082	0.0353	1.625
F51	0.344	0.188	0.118	2.710
F21	0.500	0.254	0.250	3.260
F75	0.750	0.422	0.563	4.500

TABLE 9-4   Dimensions of F Laminations

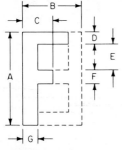

Shape	A	B	C	D	E	F	G	H
F28–29	0.500	0.438	0.219	0.0625	0.125	0.125	0.0625	0.375
F156	0.563	0.470	0.235	0.0781	0.125	0.156	0.0781	0.438
F187	0.750	0.625	0.322	0.094	0.188	0.188	0.094	0.563
F51	1.188	1.094	0.552	0.172	0.250	0.344	0.172	0.812
F21	1.625	1.318	0.659	0.250	0.312	0.500	0.250	1.125
F75	2.250	1.875	0.938	0.375	0.375	0.750	0.375	1.500

All dimensions in inches.

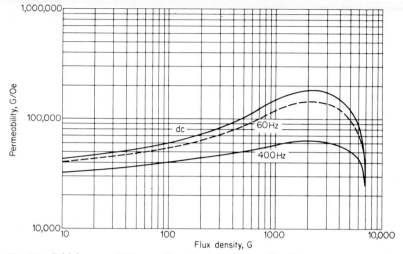

Fig. 9-3 Initial permeability vs. flux density of permalloy 80. (*Courtesy Magnetics Inc.*)

Table 9-3 lists some physical properties of actual laminations. The data are tabulated in order of increasing core cross-sectional area for a square stack, i.e., the center leg having equal width and thickness. The window area corresponds to the winding cross section. The window area divided by the total number of turns required determines the maximum wire cross section; so the wire size can be looked up in the wire chart of table 9-1. Since a bobbin will in all likelihood be used, the available winding cross section will be slightly less than the tabulated window area. The dimensions corresponding to the laminations listed in table 9-3 are given in table 9-4. Inductance can be computed from equations (9-5) and (9-8) using the tabulated parameters given in table 9-3, where the area and length of magnetic path must first be converted to square centimeters and centimeters, respectively. The initial permeability versus flux density is illustrated in figure 9-3. Clearly for flux densities above a few thousand gauss the material will saturate. Large variations of the initial permeability will occur during manufacture; so the gap will usually require adjustment for precise inductance values.

Designing from Q Curves

To design laminated inductors efficiently, it is helpful to use accumulated design data. Table 9-5 contains a wire chart indicating the maximum number of turns for each lamination covering a range of wire sizes. A set of Q curves is provided in figure 9-4. Using these curves, the particular lamination size, gap, and inductance factor A_L for the required Q at the operating frequency can be instantly determined.

As a general guideline, it is always desirable to operate on the rising portion of Q curves, since the falling portion corresponds to the region where the material's permeability changes with frequency, the core losses become significant, and self-resonance is approached. The effective inductance will then become

TABLE 9-5 Maximum Turns for Standard Laminations

AWG	F28–29	F156	F187	F51	F21	F75
22						315
23						397
24					248	500
25				250	313	630
26			135	317	394	794
27		84	171	400	496	1,000
28	93	106	215	503	625	1,260
29	117	134	271	635	785	1,588
30	148	169	341	800	991	2,000
31	186	212	430	1,005	1,250	2,520
32	224	268	541	1,270	1,573	3,176
33	295	337	682	1,600	1,982	4,002
34	370	425	860	2,180	2,500	5,042
35	465	535	1,083	2,540	3,150	6,353
36	590	674	1,365	3,200	3,980	8,005
37	740	850	1,720	4,030	5,000	10,086
38	930	1,071	2,167	5,100	6,300	12,710
39	1,170	1,349	2,730	6,400	7,938	16,010
40	1,480	1,700	3,440	8,064	10,000	
41	1,860	2,142	4,334	10,160		
42	2,240	2,698	5,461			
43	2,950	3,400				
44	3,700					

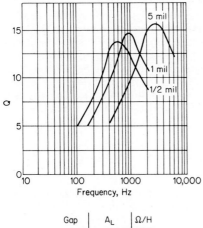

Gap	A_L	Ω/H
1/2 mil	470 mH	124
1 mil	330 mH	183
5 mil	130 mH	460

(a) F28-29 6 mil

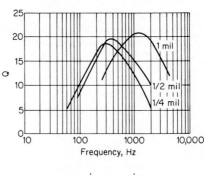

Gap	A_L	Ω/H
1/4 mil	1.08 H	52
1/2 mil	780 mH	71
1 mil	340 mH	160

(b) F156 6 mil

Fig. 9-4 Q curves for permalloy 80 laminations.

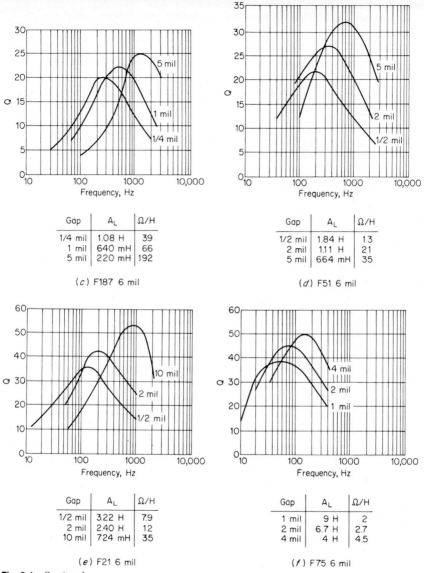

Gap	A_L	Ω/H
1/4 mil	1.08 H	39
1 mil	640 mH	66
5 mil	220 mH	192

(c) F187 6 mil

Gap	A_L	Ω/H
1/2 mil	1.84 H	13
2 mil	1.11 H	21
5 mil	664 mH	35

(d) F51 6 mil

Gap	A_L	Ω/H
1/2 mil	3.22 H	7.9
2 mil	2.40 H	12
10 mil	724 mH	35

(e) F21 6 mil

Gap	A_L	Ω/H
1 mil	9 H	2
2 mil	6.7 H	2.7
4 mil	4 H	4.5

(f) F75 6 mil

Fig. 9-4 *Continued*

frequency-dependent. When the Q rises linearly with frequency, the permeability remains constant and the losses are strictly due to the DC resistance of the winding.

Example 9-1

REQUIRED: Design a coil having an inductance of 30 H, a Q in excess of 20 at 50 Hz, and an overall size under 2 by 1½ by 1¼ in.

RESULT: We can determine from the Q curves of figure 9-4 that an F21 6-mil lamination will provide a Q of approximately 25 at 50 Hz using a ½-mil gap and according to table 9-4 will also meet the dimensional requirements.

Using an A_L (inductance per 1000 turns) of 3.22 H, the required number of turns is found from

$$N = 10^3 \sqrt{\frac{L}{A_L}} = 10^3 \sqrt{\frac{30}{3.22}} = 3050 \text{ turns} \qquad (9\text{-}10)$$

The corresponding wire size is determined from table 9-5 as AWG No. 35, thus completing the design.

9.3 TOROIDAL COILS

Toroidal cores are manufactured by pulverizing a magnetic alloy consisting of approximately 2% molybdenum, 81% nickel, and 17% iron into a fine powder, insulating the powder with a ceramic binder to form a uniformly distributed air gap, and then compressing it into a toroidal core at extremely high pressures. Finally the cores are coated with an insulating finish.

Molypermalloy powder cores (MPP cores) result in extremely stable inductive components for use below a few hundred kilohertz. Core losses are low over a wide range of available permeabilities. Inductance remains stable with large changes in flux density, frequency, temperature, and DC magnetization due to high resistivity, low hysteresis, and low eddy-current losses.

Characteristics of Cores

MPP cores are categorized according to size, permeability, and temperature stability. Generally the largest core size that physical and economical considerations permit should be chosen. Larger cores offer higher Q's, since flux density

TABLE 9-6   Toroidal Core Dimensions

OD, in	ID, in	HT, in	Cross Section, cm²	Path Length, cm	Window Area, circular mils	Wound OD, in	Coil HT, in
0.310	0.156	0.125	0.0615	1.787	18,200	$^{11}\!/_{32}$	$^{3}\!/_{16}$
0.500	0.300	0.187	0.114	3.12	75,600	$^{19}\!/_{32}$	$^{9}\!/_{32}$
0.650	0.400	0.250	0.192	4.11	140,600	$^{25}\!/_{32}$	$^{3}\!/_{8}$
0.800	0.500	0.250	0.226	5.09	225,600	1	$^{3}\!/_{8}$
0.900	0.550	0.300	0.331	5.67	277,700	$1^{3}\!/_{32}$	$^{1}\!/_{2}$
1.060	0.580	0.440	0.654	6.35	308,000	$1^{1}\!/_{4}$	$^{5}\!/_{8}$
1.350	0.920	0.350	0.454	8.95	788,500	$1^{5}\!/_{8}$	$^{5}\!/_{8}$
1.570	0.950	0.570	1.072	9.84	842,700	$1^{7}\!/_{8}$	$^{7}\!/_{8}$
2.000	1.250	0.530	1.250	12.73	1,484,000	$2^{3}\!/_{8}$	$1^{1}\!/_{8}$

NOTE: Core dimensions are before finish.

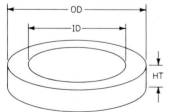

TABLE 9-7   **Electrical Properties for** $\mu = 125$

OD, in	A_L, mH	Ω/H
0.310	52	900
0.500	56	480
0.650	72	160
0.800	68	220
0.900	90	150
1.060	157	110
1.350	79	80
1.570	168	45
2.000	152	30

is lower due to the larger cross-sectional area, resulting in lower core losses, and the larger window area reduces the copper losses.

Cores range in size from an OD of 0.155 to 2.250 in. Table 9-6 contains physical data for some selected core sizes as well as the approximate overall dimensions for the wound coil.

Available core permeabilities range from 14 to 550. The lower permeabilities are more suitable for use at the higher frequencies, since the core losses are lower. Table 9-7 lists the A_L (inductance per 1000 turns) and ohms per henry for cores with a permeability of 125. For other permeabilities the A_L is directly proportional to μ and the ohms per henry is inversely proportional to μ. The ohms per henry corresponds to the DC resistance factor when the core window is approximately 50% utilized. Full window utilization is not possible, since a hole must be provided for a shuttle in the coil winding machine which applies the turns. The corresponding wire chart for each size core is given in table 9-8.

DC Bias and AC Flux Density   Under conditions of DC bias current, MPP cores may exhibit a reduction in permeability because of the effects of saturation. Figure 9-5 illustrates this effect, which is more pronounced for higher permeabilities. To use these curves, the magnetizing force for the design is computed

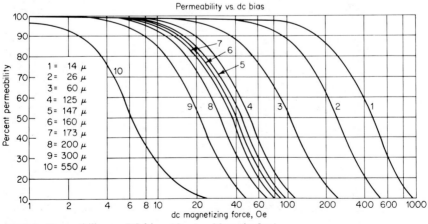

Fig. 9-5   Permeability vs. DC bias. *(Courtesy Magnetics Inc.)*

TABLE 9-8 Wire Capacities of Standard Toroidal Cores

Wire Size	Core OD, in								
	0.310	0.500	0.650	0.800	0.900	1.060	1.350	1.570	2.000
25		75	148	189	257	284	750	930	1,735
26		95	186	238	323	357	946	1,172	2,180
27		119	235	300	406	450	1,190	1,470	2,750
28		150	295	377	513	567	1,500	1,860	3,470
29		190	375	475	646	714	1,890	2,350	4,365
30	59	238	472	605	814	925	2,390	2,960	5,550
31	74	300	595	765	1,025	1,180	3,000	3,720	7,090
32	94	376	750	985	1,290	1,510	3,780	4,700	9,000
33	118	475	945	1,250	1,625	1,970	4,763	5,920	11,450
34	150	600	1,190	1,580	2,050	2,520	6,000	7,440	14,550
35	188	753	1,500	2,000	2,585	3,170	7,560	9,400	18,500
36	237	950	1,890	2,520	3,245	4,000	9,510	11,840	23,500
37	300	1,220	2,380	3,170	4,100	5,050	12,000	14,880	30,000
38	378	1,550	3,000	4,000	5,175	6,300	15,150	18,800	38,000
39	476	1,970	3,780	5,050	6,510	8,000	19,050	23,680	48,500
40	600	2,500	4,750	6,300	8,200	10,100	24,000	30,000	61,300
41	755								
42	950								
43	1,200								
44	1,510								
45	1,900								

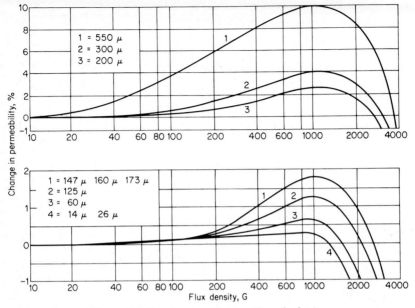

Fig. 9-6 Permeability vs. AC flux density. *(Courtesy Magnetics Inc.)*

using equation (9-2). If the reduction in permeability and resulting inductance decrease is no more than about 30%, the turns can be increased to compensate for the effect of the bias. If the decrease in permeability is more than 30%, the permeability will further decrease faster than N^2 if turns are added. A larger core is then required.

The core permeability will also change as a function of AC flux density. This effect is shown in figure 9-6. The flux density can be computed using equation (9-6). As the AC flux density is increased, the permeability will rise initially and then fall beyond approximately 2000 G. Flux density can be reduced by going to a larger core size.

The core losses can be assumed to be relatively constant for flux densities below 200 G. For higher excitations the Q may be adversely affected.

Temperature Stability   MPP cores are available in three categories of temperature stabilities: unstabilized, stabilized, and linear. Stabilizing techniques are based on the addition of a small amount of special compensating alloys having curie points within the temperature range of operation. (The curie point is the temperature where the material becomes nonmagnetic.) As each curie point is passed, the particles act as distributed air gaps which can be used to compensate for permeability changes of the basic alloy so as to maintain the inductance nearly constant.

Unstabilized cores, also called "A" stabilization, have no guaranteed limits for permeability variations with temperature. Typically the temperature coefficient of permeability is positive and ranges from 25 to 100 ppm/°C, but these limits are not guaranteed. Stabilized cores, on the other hand, have guaranteed limits over wide temperature ranges. These cores are available in four degrees of stabilization, M, W, D, and B. The limits are as given in table 9-9.

TABLE 9-9   Stabilized Cores

Stabilization	Temperature Range, °C	Inductance Stability, %
M	−65 to +125	±0.25
W	−55 to +85	±0.25
D	0 to +55	±0.1
B	+13 to +35	±0.1

Because of the nature of the compensation technique, the slope of inductance change with temperature can be either positive or negative within the temperature range of operation but will not exceed the guaranteed limits.

Polystyrene capacitors maintain a precise temperature coefficient of −120 ppm/°C from −55 to +85°C. Cores are available which provide a matching positive temperature coefficient so that a constant LC product can be maintained over the temperature range of operation. These cores are said to have linear temperature characteristics. Two types of linear temperature characteristics can be obtained. The $L6$ degree of stabilization has a positive temperature coefficient ranging between +20 and +75 ppm/°C from −55°C to +25°C and between +65 and +150 ppm/°C over the temperature range of +25 to +85°C. Ultralinear $K6$ cores have a guaranteed temperature coefficient of +120 ±40 ppm/°C from −55 to +85°C so that a very high degree of matching to a polystyrene capacitor can be obtained.

In general the temperature stability of MPP cores is affected by factors such as winding stresses and moisture. To minimize these factors, suitable precautions can be taken. Before adjustment, the coils should be temperature-cycled from −55 to +100°C at least once. After cooling to room temperature, the coils are adjusted and should be kept dry until encapsulated. Encapsulation compounds should be chosen carefully to minimize mechanical stresses. A common technique involves dipping the coils in a silicon rubber compound for a cushioning and sealing effect prior to encapsulation.

Winding Methods for Q Optimization

At low frequencies, the dominant losses are caused by the DC resistance of the winding. The major consideration then is to utilize the maximum possible winding area. Distributed capacity is of little consequence unless the inductance is extremely high, causing self-resonance to occur near the operating frequency range.

The most efficient method of packing the most turns on a toroidal core is to rotate the core continuously in the same direction in the winding machine until maximum capacity is obtained. This technique is called the "360° method."

At medium frequencies special winding techniques are required to minimize distributed capacity. By winding half the turns over a 180° sector of the core in a back-and-forth manner and then applying the remaining turns over the second half in the same fashion, the capacity will be reduced. This technique is called the "two-section method." If after winding half the turns the coil is removed from the machine, turned over, reinserted, and completed, the two starts can be tied together, and the resulting coil using the two finishes as terminals has even less capacity. This modified two-section winding structure is commonly referred to as "two-section reversed."

If the core is divided into four 90° quadrants and each sector is completed in a back-and-forth winding fashion using one-fourth the total turns, a four-section coil will be obtained. A four-section winding structure has lower distributed capacity than the two-section method. However, whereas the two-section and 360° methods correspond to the wire chart of table 9-8, the wire must be reduced one size for a four-section coil, resulting in more copper losses.

At frequencies near 50 kHz or higher, the distributed capacity becomes a serious limiting factor of the obtainable Q from both self-resonance and dielectric losses of the winding insulation. The optimum winding method for minimizing distributed capacity is "back-to-back progressive." Half the total number of turns are applied over a 180° sector of the core by gradually filling up a 30° sector at a time. The core is then removed from the machine, turned over, reinserted in the machine, and completed in the same manner. The two starts are joined and the two finish leads are used for the coil terminals. A barrier is frequently used to separate the two finish leads for further capacity reduction. Litz wire can be combined with the back-to-back progressive winding method to reduce skin-effect losses. As with a four-section winding the wire must be reduced one size from the wire chart of table 9-8. If the core is not removed from the winding machine after completion of a 180° sector but instead is continued for approximately 360°, a straight-progressive type of winding is obtained. Although slightly inferior to the back-to-back progressive, the reduction in winding time will frequently warrant its use.

When a coil includes a tap, a coefficient of magnetic coupling near unity is desirable to avoid leakage inductance. The 360° winding method will have a typical coefficient of coupling of about 0.99 for permeabilities of 125 and higher. A two-section winding has a coefficient of coupling near 0.8. The four-section and progressive winding methods result in coupling coefficients of approximately 0.3, which is usually unacceptable. A compromise between the 360° and the progressive method to improve coupling for tapped inductors involves applying the turns up to the tap in a straight-progressive fashion over the total core. The remaining turns are then distributed completely over the initial winding, also using the straight-progressive method.

In general, to obtain good coupling, the portion of the winding up to the tap should be in close proximity to the remainder of the winding. However, this results in higher distributed capacity; so a compromise may be desirable. Higher core permeability will increase the coupling but can sometimes result in excessive core loss. The higher the ratio of overall to tap inductance, the lower the corresponding coefficient of coupling. Inductance ratios of 10 or more should be avoided if possible.

Designing MPP Toroids from Q Curves

The Q curves given in figure 9-7 are based on empirical data using the 360° winding method. The range of distributed capacity is typically between 10 and 25 pF for cores under 0.500 in OD, 25 to 50 pF for cores between 0.500 and 1.500 in OD, and 50 to 80 pF for cores over 1.500 in OD.

Curves are presented for permeabilities of 60 and 125 and for a range of core sizes. For a given size and permeability the Q curves converge on the low-frequency portion of the curve, where the losses are determined almost exclusively by the DC resistance of the winding. The Q in this region can be approximated by

$$Q = \frac{2\pi f}{\Omega/H} \tag{9-11}$$

where f is the frequency of interest and Ω/H is the rated ohms per henry of the core as given by table 9-7 and modified for permeability if other than 125.

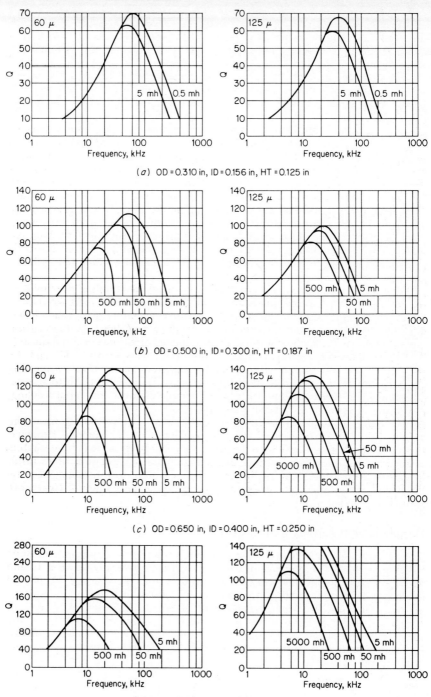

(a) OD = 0.310 in, ID = 0.156 in, HT = 0.125 in

(b) OD = 0.500 in, ID = 0.300 in, HT = 0.187 in

(c) OD = 0.650 in, ID = 0.400 in, HT = 0.250 in

(d) OD = 0.800 in, ID = 0.500 in, HT = 0.250 in

Fig. 9-7 *Q* curves of toroidal cores. *(Courtesy Magnetics Inc.)*

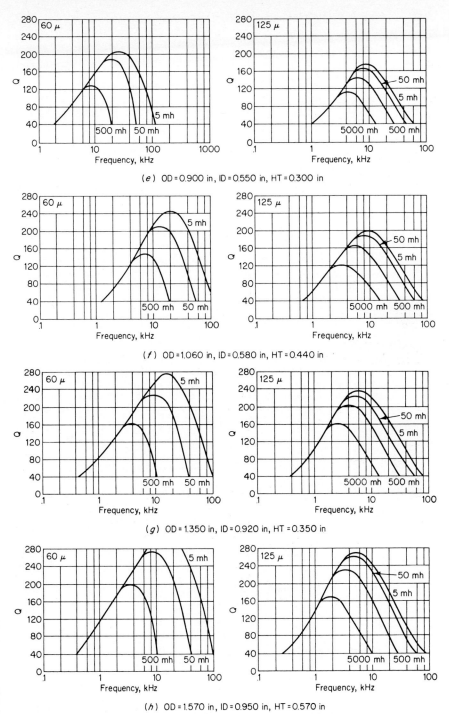

(e) OD = 0.900 in, ID = 0.550 in, HT = 0.300 in

(f) OD = 1.060 in, ID = 0.580 in, HT = 0.440 in

(g) OD = 1.350 in, ID = 0.920 in, HT = 0.350 in

(h) OD = 1.570 in, ID = 0.950 in, HT = 0.570 in

Fig. 9-7 *Continued*

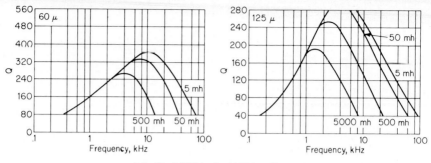

(*i*)   OD = 2.000 in, ID = 1.250 in, HT = 0.530 in

Fig. 9-7   *Continued*

As the frequency is increased, the curves start to diverge and reach a maximum at a frequency where the copper and core losses are equal. Beyond this region the core losses begin to dominate along with increased dielectric losses and self-resonance is approached; so the Q will roll off dramatically. It is always preferable to operate on the rising portion of Q curves, since the losses can be tightly controlled and the effective inductance remains relatively constant with frequency.

Example 9-2

REQUIRED: Design a toroidal inductor having an inductance of 1.5 H and a minimum Q of 55 at 1 kHz. The coil must pass a DC current of up to 10 mA and operate with AC signals as high as 10 V_{rms} with negligible effect.

RESULT: (*a*)   Using the Q curves of figure 9-7, a 1.060-in-diameter core having a μ of 125 will have a Q of approximately 60 at 1 kHz. The required number of turns is found from

$$N = 10^3 \sqrt{\frac{L}{A_L}} = 10^3 \sqrt{\frac{1.5}{0.157}} = 3090 \qquad (9\text{-}10)$$

where the value of A_L was given in table 9-7. The corresponding wire size is determined from table 9-8 as No. 35 AWG.

(*b*)   To estimate the effect of the DC current, compute the magnetizing force from

$$H = \frac{4\pi\, NI}{10\, mL} = \frac{4\pi \times 3090 \times 0.01}{10 \times 6.35} = 6.1 \text{ Oe} \qquad (9\text{-}2)$$

where mL, the magnetic path length, is obtained from table 9-6. According to figure 9-5 the permeability will remain essentially constant.

(*c*)   An excitation level of 10 V_{rms} at 1000 Hz results in a flux density of

$$B = \frac{E_{rms}}{4.44\, NfA \times 10^{-8}}$$

$$= \frac{10}{4.44 \times 3090 \times 1000 \times 0.654 \times 10^{-8}}$$

$$= 111 \text{ G} \qquad (9\text{-}6)$$

where A, the core's cross-sectional area, is also found in table 9-6. Figure 9-6 indicates that at the calculated flux density the permeability change will be of little consequence.

9.4 FERRITE POTCORES

Ferrites are ceramic structures created by combining iron oxide with oxides or carbonates of other metals such as manganese, nickel, or magnesium. The mixtures are pressed, fired in a kiln at very high temperatures, and machined into the required shapes.

The major advantage of ferrites over laminations and MPP cores is their high resistivity so that core losses are extremely low even at higher frequencies where eddy-current losses become critical. Additional properties such as high permeability and good stability with time and temperature have made ferrites the optimum core-material choice for frequencies from 10 kHz to well in the megahertz region.

The Potcore Structure

A typical potcore assembly is shown in figure 9-8. A winding supported on a bobbin is mounted in a set of symmetrical ferrite potcore halves. The assembly

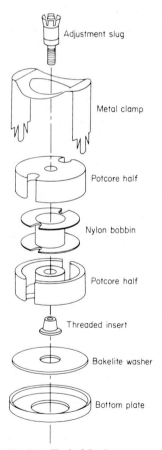

Adjustment slug

Metal clamp

Potcore half

Nylon bobbin

Potcore half

Threaded insert

Bakelite washer

Bottom plate

Fig. 9-8 Typical ferrite potcore assembly.

is held rigid by a metal clamp. An air gap is introduced in the center post of each half, since only the outside surfaces of the potcore halves mate with each other. By introducing an adjustment slug containing a ferrite sleeve, the effect of the gap can be partially neutralized as the slug is inserted into the gap region.

A ferrite potcore has a number of distinct advantages over other approaches. Since the wound coil is contained within the ferrite core, the structure is self-shielding, since stray magnetic fields are prevented from entering or leaving the structure.

Very high Q's and good temperature stability can be obtained by appropriate selection of materials and by controlling the effective permeability μ_e through the air gap. Fine adjustment of the effective permeability is accomplished using the adjustment slug.

Compared with other structures, ferrite potcores are more economical. Bobbins can be rapidly wound using multiple-bobbin winding machines in contrast to toroids, which must be individually wound. Assembly and mounting is easily accomplished using a variety of hardware available from the ferrite manufacturer. Printed circuit-type brackets and bobbins facilitate the use of potcores on printed circuit boards.

Potcores have been standardized into nine international sizes ranging from 9 by 5 to 42 by 29 mm, where these dimensions represent the diameter and height, respectively, of a potcore pair. These sizes are summarized in table 9-10.

TABLE 9-10   Standard Potcore Sizes

Manufacturer Core Size Designation	Diameter (max)		Height (max)		Cross Section,	Path Length,
	in	mm	in	mm	cm²	cm
905	0.366	9.3	0.212	5.4	0.101	1.25
1107	0.445	11.3	0.264	6.7	0.167	1.55
1408	0.559	14.2	0.334	8.5	0.251	1.98
1811	0.727	18.5	0.426	10.8	0.433	2.58
2213	0.858	21.8	0.536	13.6	0.635	3.15
2616	1.024	26.0	0.638	16.2	0.948	3.76
3019	1.201	30.5	0.748	19.0	1.38	4.52
3622	1.418	36.0	0.864	22.0	2.02	5.32
4229	1.697	43.1	1.162	29.5	2.66	6.81

Electrical Properties of Ferrite Potcores

Ferrite materials are available having a wide range of electrical properties. Let us first consider initial permeability as a function of frequency and temperature. Typical curves for a selection of materials from a particular manufacturer, Ferroxcube Corporation, are shown in figure 9-9.

A material must be chosen that provides uniform permeability over the frequency range of interest. It is evident from figure 9-9a that the lower-permeability materials are more suitable for higher-frequency ranges.

The temperature coefficient of permeability is also an important factor in LC tuned circuits and filters. The 3D3 and 4C4 permeabilities remain relatively flat over a wide temperature range, as indicated in figure 9-9b. However, the permeability magnitude is lower than that of the other types. The 3B7 material has a higher permeability but a more restricted temperature range. By proper

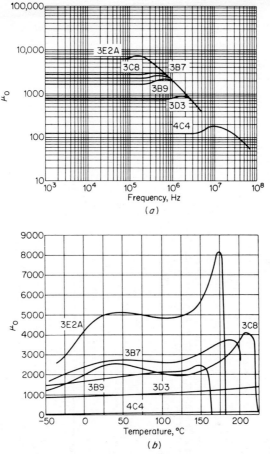

Fig. 9-9 Initial permeability vs. frequency and temperature. *(a)* initial permeability vs. frequency; *(b)* initial permeability vs. temperature. *(Courtesy Ferroxcube Corp.)*

selection of size and gap in conjunction with 3B9 type material, a positive linear temperature coefficient can be obtained to match the negative temperature coefficient of polystyrene capacitors to maintain a constant LC product.

The initial permeability, saturation flux density, and temperature factor for selected ferrite materials are given in table 9-11. The temperature factor is the temperature coefficient of permeability. This factor must be multiplied by the rated effective permeability of the core to obtain the core's temperature coefficient.

For a given core size, a wide range of A_L factors are available. This parameter is determined by the effective permeability of the core, which in turn is controlled by the initial permeability of the ferrite material, the core dimensions, and the size of the air gap introduced in the center post. Table 9-12 lists the effective permeability and the ohms per henry rating for all the standard core sizes where

TABLE 9-11 Basic Electrical Properties of Standard Ferrite Materials

	3B7	3B9	3D3	4C4
Initial permeability at 25°C	2300 ± 20%	1800 ± 20%	750 ± 20%	125 ± 20%
Saturation flux density at 25°C	3800 G	3200 G	3800 G	3000 G
Temperature factor	-0.6×10^{-6} min	$+0.9 \times 10^{-6}$ min	$+1.0 \times 10^{-6}$ min	-6.0×10^{-6} min
	$+0.6 \times 10^{-6}$ max	$+1.9 \times 10^{-6}$ max	$+3.0 \times 10^{-6}$ max	$+6.0 \times 10^{-6}$ max
	(+20 to +70°C)	(−30 to +70°C)	(−30 to +70°C)	(+5 to +55°C)

**TABLE 9-12   Electrical Properties of Standard
Core Sizes for A_L = 250 mH**

Size Designation	μ_e	Ω/H
905	230	700
1107	184	570
1408	156	380
1811	119	260
2213	98	190
2616	78	170
3019	64	140
3622	52	130
4229	51	80

the A_L is maintained constant at 250 mH. For other values the effective permeability is proportional to the rated A_L and the ohms per henry is inversely proportional to this same factor.

Potcores are also available having no gap. However, since the resulting electrical characteristics are essentially equivalent to the initial properties of the ferrite material, these cores should be restricted to applications where high permeability is the major requirement, such as for wide-band transformers.

Most ferrite materials saturate with AC flux densities in the region of 3000 to 4000 G. AC excitation levels should be kept below a few hundred gauss to avoid nonlinear saturation effects and to minimize core losses. Superimposing a DC bias will also cause saturation when the ampere-turns exceed a critical value dependent upon the core size, material, and gap. DC saturation curves are shown in figure 9-10 for the 1811, 2213, and 2616 sizes. For a given size and material, the cores corresponding to the lower A_L values have greater immunity to saturation, since the air gaps are larger.

For a specific potcore size and rated A_L, a variety of different adjusters are available featuring ranges of adjustment from a few percent to over 25%. For most applications, an adjustment range of about 10 or 15% will be satisfactory, since this should cover both the capacitor and core tolerances. The larger-range slugs should be avoided for precision requirements, since the adjustment may be too coarse.

To compensate for the effect of an adjuster, the nominal A_L rating of the core should be increased by a percentage equal to one-half the adjustment range for the actual turns computation. This is because the resulting A_L will be higher by one-half the range with the slug set for the midrange position.

Winding of Bobbins

Bobbins are normally wound by guiding the wire feed back and forth as the bobbin is rotated. In the event the inductor's self-resonant frequency becomes a problem, a multisection bobbin can be used to reduce distributed capacitance. One section at a time is filled to capacity. Most bobbin sizes are available with up to three sections.

A wire capacity chart for standard single-section bobbins is shown in table 9-13. Multisection bobbins have slightly less winding area than single-section bobbins; so the wire gauge may require reduction by one size in some cases.

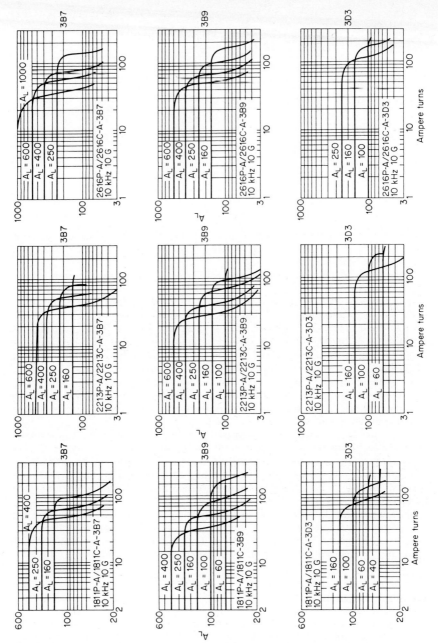

Fig. 9-10 A_L vs. DC bias for 1811, 2213, and 2616 size cores. (*Courtesy Ferroxcube Corp.*)

TABLE 9-13   Wire Capacities of Standard Bobbins

Wire Size	905	1107	1408	1811	2213	2616	3019	3622	4229
20							63	100	160
21						50	79	125	200
22						63	100	160	250
23					50	79	125	200	315
24					63	100	160	250	400
25				50	79	125	200	315	500
26				63	100	160	250	400	630
27			50	79	125	200	315	500	794
28			63	100	160	250	400	630	1,000
29			79	125	200	315	500	794	1,260
30			100	160	250	400	630	1,000	1,588
31		50	125	200	315	500	794	1,260	2,000
32		63	160	250	400	630	1,000	1,588	2,520
33		79	200	315	500	794	1,260	2,000	3,176
34	25	100	250	400	630	1,000	1,588	2,520	4,000
35	31	125	315	500	794	1,260	2,000	3,176	5,042
36	40	160	400	630	1,000	1,588	2,520	4,000	6,353
37	50	200	500	794	1,260	2,000	3,176	5,042	8,000
38	63	250	630	1,000	1,588	2,520	4,000	6,353	10,086
39	79	315	794	1,260	2,000	3,176	5,042	8,000	
40	100	400	1,000	1,588	2,520	4,000	6,353	10,086	
41	126	500	1,260	2,000	3,176	5,042	8,000		
42	159	630	1,588	2,520	4,000	6,353	10,086		
43	200	794	2,000	3,176	5,042	8,000			
44	252	1,000	2,520	4,000	6,353	10,086			
45	318	1,260	3,176	5,042	8,000				

Because of the air gap, the permeability is not uniform throughout the interior of the potcore. If the bobbin is only partially filled, the A_L will be reduced as a function of the relative winding height. The decrease in A_L for a 2616 size core can be approximated from the graph of figure 9-11. This effect becomes more pronounced as the gap is made larger.

To maximize Q over the frequency range of 50 kHz to 1 or 2 MHz, litz wire should be used. The litz wire equivalents to solid wire gauges can be determined from table 9-2.

Design of Potcores from Q Curves

Various methods are used by designers to determine the optimum core size, material, and gap for a particular application. Depending upon the requirements, some parameters are more significant than others. Generally, Q and temperature coefficient are of prime importance.

A set of standard Q curves facilitates the rapid selection of the optimum core for a given Q requirement. In general, larger cores provide higher Q, lower temperature coefficient, and better immunity from saturation. For a given core size, the lowest A_L which provides sufficient Q should be chosen to minimize temperature coefficient and saturation effects. Operation should be restricted to the rising portion of the curves to avoid the effects of core and dielectric losses as well as those of self-resonance.

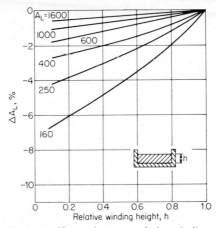

Fig. 9-11 Change in A_L vs. relative winding height for 2616 size potcore.

Selected Q curves are illustrated in figure 9-12. These curves represent measured Q's for specific turns using both solid and litz wire and single-section bobbins. In the event the actual design requirement involves excessive inductance values, multisection bobbins can be used to reduce distributed capacity.

Example 9-3

REQUIRED: Design a coil having an inductance of 1 mH with a minimum Q of 400 at 100 kHz. The inductor should be capable of supporting 50 mA of DC current.

RESULT: (a) Using the curves of figure 9-12, a 2616 size potcore having an A_L of 160 and 3B9 material (Ferroxcube 2616C-A160-3B9) will meet the Q requirements. The turns are found from

$$N = 10^3\sqrt{\frac{L}{A_L}} = 10^3\sqrt{\frac{1}{160}} = 79 \text{ turns} \qquad (9\text{-}10)$$

Using table 9-13, the required wire size is No. 23. For optimum Q, litz wire should be used. The litz equivalent is determined from table 9-2 as 100/43.

(b) The DC excitation results in an NI of 79 × 0.05, or 4. Using the curves of figure 9-10 the operating point is well below the saturation region.

9.5 HIGH-FREQUENCY COIL DESIGN

Powdered-Iron Toroids

Above 1 or 2 MHz, the core losses of most ferrite materials become prohibitive. Toroidal cores comprised of compressed iron powder then become desirable for use up to the VHF range.

These cores consist of finely divided iron particles which are insulated and then compressed at very high pressures into a toroidal form in a manner similar to MPP cores. A variety of iron powders are available suitable for use over

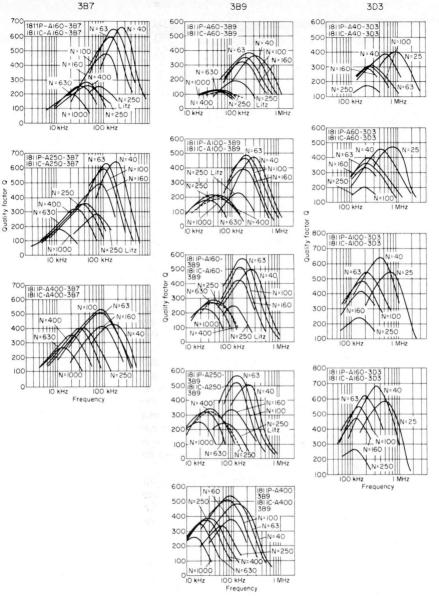

Fig. 9-12 Q curves for *(a)* 1811, *(b)* 2213, and *(c)* 2616 size potcores. *(Courtesy Ferroxcube Corp.)*

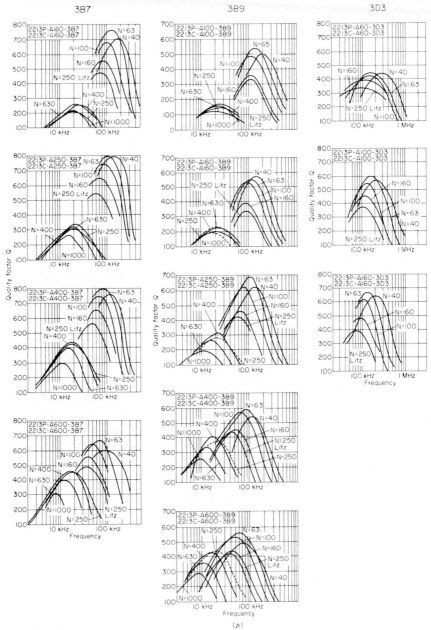

Fig. 9-12 *Continued*

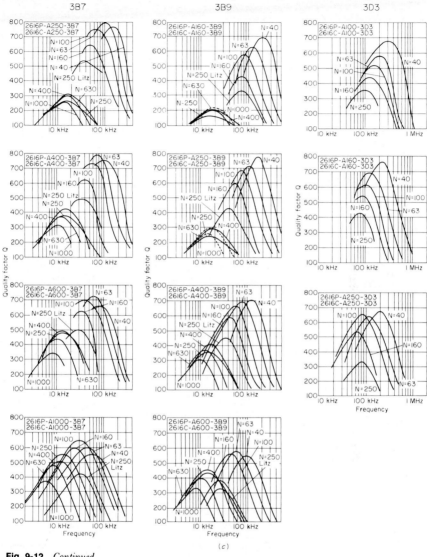

Fig. 9-12 *Continued*

different frequency bands. The more commonly used materials are listed in table 9-14 along with the nominal permeability, temperature coefficient, and maximum frequency of operation above which the core losses can become excessive. In general, a material should be selected that has the highest permeability at the frequency of operation.

For optimum Q, the distributed and stray capacities should be minimized. By maximizing permeability, the number of turns can be kept small. In addition, the turns should be applied as a single-layer winding using the largest wire

TABLE 9-14   Iron-Powder Core Materials

Material Designation	Nominal Permeability	Temperature Coefficient, ppm/°C	Maximum Frequency, MHz
GAF carbonyl HP	25	+140	2
GAF carbonyl C	25	+110	10
GAF carbonyl E	10	+40	40
GAF carbonyl TH	8.5	+20	40
GAF carbonyl SF	8	+30	150
GAF carbonyl W	7.5	+140	250

size that the core will accommodate. However, there is usually a point of diminishing return with increasing wire gauge; so unwieldy wire sizes should be avoided. The turns should be spread evenly around the core except that the start and finish should be separated by a small unwound sector. Litz wire can be used up to about 2 MHz. To approximate the required wire size, the inside circumference of the core can be computed and then divided by the number of turns to yield an approximate wire diameter. A gauge can then be chosen using table 9-1. Generally it should be about two sizes smaller than computed.

To minimize dielectric losses, teflon-coated wire is occasionally used, since this material has an extremely small power factor in comparison with the more commonly used polyurethane insulation. Also, wrapping the core with teflon tape prior to applying the winding helps to keep stray capacity low. If potting compounds are used, they should be very carefully chosen, since inductor Q's can be easily degraded upon impregnation by lossy materials.

Table 9-15 lists some physical properties of selected standard-size powdered-iron toroidal cores. For convenience the A_L is specified in terms of $\mu H/100$ turns. The required number of turns can then be computed from

$$N = 10^2 \sqrt{\frac{L}{A_L}} \tag{9-12}$$

where both L and A_L are in microhenrys. Some selected Q curves are illustrated in figure 9-13.

Air-Core Inductors

Coils containing no magnetic materials are said to be air-cored. Inductors of this form are useful above 10 MHz. Air has a μ of 1, will not saturate, and has no core losses; so the Q is strictly dependent upon the winding.

For a single-layer solenoid wound on a nonmagnetic material such as ceramic or phenolic, the inductance in henrys can be approximated within a reasonable accuracy by

$$L = N^2 \frac{r^2}{9r + 10l} \times 10^{-6} \tag{9-13}$$

where N is the number of turns, r is the radius (diameter/2), and l is the coil length, with r and l both in inches.

Air-core solenoid inductors have high leakage inductance; so they are easily affected by nearby metallic surfaces. A toroidal shape will have less leakage, since the magnetic field will be better contained. The expression for the inductance of an air-core toroid with a single-layer winding was given by equation (9-5) with $\mu = 1$.

TABLE 9-15 Powdered-Iron Toroidal Cores

OD, in	ID, in	HT, in	Cross Section, cm²	Path Length, cm	Material	A_L, μH/100 turns
0.120	0.062	0.035	0.0065	0.73	HP	28
					C	28
					E	11
					TH	10
					SF	9
					W	8
0.197	0.092	0.082	0.0278	1.57	HP	76
					C	76
					E	30
					TH	26
					SF	24
					W	23
0.309	0.156	0.125	0.0615	1.86	HP	105
					C	105
					E	42
					TH	36
					SF	33
					W	31
0.500	0.300	0.187	0.114	3.19	HP	118
					C	118
					E	47
					TH	40
					SF	38
					W	35
0.800	0.500	0.250	0.226	5.19	HP	146
					C	146
					E	58
					TH	50
					SF	47
					W	42
1.060	0.580	0.440	0.654	6.54	HP	327
					C	327
					E	130
					TH	111
					SF	104
					W	92
1.570	0.965	0.570	1.072	10.11	HP	345
					C	345
					E	138
					TH	117
					SF	110
					W	98

9.6 TRANSFORMER DESIGN TECHNIQUES

Transformers are used to match impedances, provide maximum power transfer, and couple signals while providing isolation from DC currents and voltages. They will also present balanced impedances for applications requiring this form of termination.

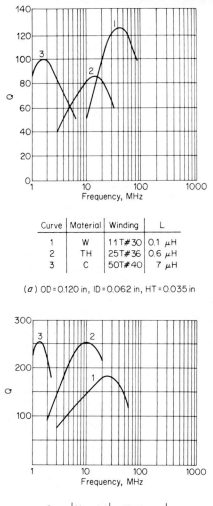

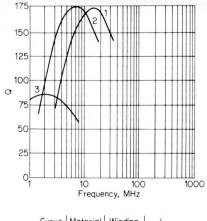

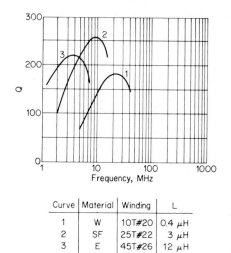

Curve	Material	Winding	L
1	W	11T#30	0.1 μH
2	TH	25T#36	0.6 μH
3	C	50T#40	7 μH

(a) OD = 0.120 in, ID = 0.062 in, HT = 0.035 in

Curve	Material	Winding	L
1	SF	17T#26	1 μH
2	E	29T#28	5 μH
3	C	50T#32	2 6 μH

(b) OD = 0.309 in, ID = 0.156 in, HT = 0.125 in

Curve	Material	Winding	L
1	W	10T#20	0.35 μH
2	SF	19T#20	1.4 μH
3	E	125T#15/44	73 μH

(c) OD = 0.500 in, ID = 0.300 in, HT = 0.187 in

Curve	Material	Winding	L
1	W	10T#20	0.4 μH
2	SF	25T#22	3 μH
3	E	45T#26	12 μH

(d) OD = 0.800 in, ID = 0.500 in, HT = 0.250 in

Fig. 9-13 Q curves of powdered-iron toroidal cores.

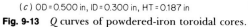

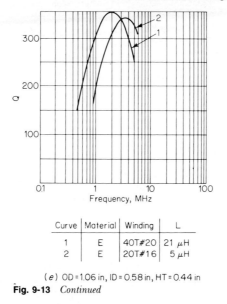

Curve	Material	Winding	L
1	E	40T#20	21 μH
2	E	20T#16	5 μH

(e) OD = 1.06 in, ID = 0.58 in, HT = 0.44 in

Fig. 9-13 *Continued*

Equivalent Circuits of Wide-Band Transformers

Figure 9-14a illustrates the linear equivalent circuit representation of a practical transformer. The secondary DC resistance is reflected into the primary, combined with the primary DC resistance, and represented by R_{DC}. Inductor L_K, the leakage inductance, is the primary inductance as seen with a short circuit on the transformer secondary. Capacitor C_T is the equivalent shunt distributed capacity. Inductance L_p is the primary inductance with the secondary open and is sometimes called the magnetizing inductance. This network is followed by

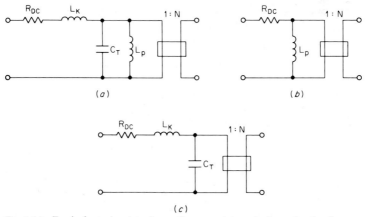

Fig. 9-14 Equivalent circuits of transformers: (a) equivalent circuit of transformer; (b) low-frequency equivalent circuit; (c) high-frequency equivalent circuit.

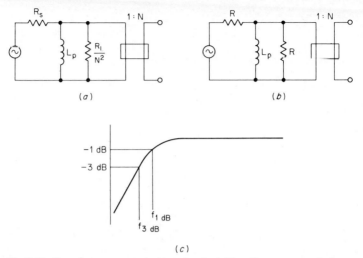

Fig. 9-15 Low-frequency equivalent circuit: (a) low-frequency equivalent circuit of terminated transformer; (b) circuit for $R_s = R_L = R$; (c) frequency response.

an ideal transformer with a turns ratio of 1 to N, which provides an impedance ratio of 1 to N^2.

This representation is based on low-level linear operation of the transformer core, i.e., constant permeability over the range of excitation. These conditions may not always be valid but are generally assumed as a first-order approximation. Careful design of the magnetics will ensure linearity.

To simplify analysis, the equivalent circuit of figure 9-14a can be reduced to the low- and high-frequency equivalent circuits of figure 9-14b and c, respectively. This approach is especially useful for analyzing wide-band transformers.

Let us first consider the low-frequency equivalent circuit of figure 9-14b. If the transformer is terminated with a load resistor R_L and is driven by a source having an impedance R_s, the load can be reflected into the primary, resulting in the circuit of figure 9-15a, where R_{DC} is neglected. (Inclusion of R_{DC} results in a flat insertion loss.) For a given source R_s, maximum power is delivered to the load if $N^2 = R_L/R_s$. Therefore, if we let $R_s = R_L/N^2 = R$, we obtain the circuit of figure 9-15b. This is equivalent to a first-order high-pass filter having equal source and load terminations followed by an ideal transformer which provides a voltage step-up of N.

The relative frequency response of a first-order high-pass filter is given by the general formula

$$A_{dB} = 10 \log \left[1 + \left(\frac{f_{3\,dB}}{f_x} \right)^2 \right] \tag{9-14}$$

where $f_{3\,dB}$ is the lower 3-dB point and f_x is the frequency of interest. Expressed in terms of the elements shown in figure 9-15b, the relative frequency response becomes

$$A_{dB} = 10 \log \left[1 + \left(\frac{R}{4\pi L_p f_x} \right)^2 \right] \tag{9-15}$$

The general formula for computing the value of L_p for a specified attenuation A_{dB} at a frequency f_{dB} can be derived from equation (9-15), which results in

$$L_p = \frac{R}{4\pi f_{dB}\sqrt{10^{0.1 A_{dB}} - 1}} \qquad (9\text{-}16)$$

For the specific cases of attenuation shown in figure 9-15c, equation (9-16) simplifies to

1 dB:
$$L_p = \frac{R}{2\pi f_{1\ dB}} \qquad (9\text{-}17)$$

3 dB:
$$L_p = \frac{R}{4\pi f_{3\ dB}} \qquad (9\text{-}18)$$

In the event N^2 is not equal to R_L/R_s, the condition for maximum power transfer, the value of R in equations (9-16) through (9-18) can be computed from

$$R = \frac{2 R_s R_L}{R_s N^2 + R_L} \qquad (9\text{-}19)$$

The high-frequency equivalent circuit was illustrated in figure 9-14c. The leakage inductance L_K and distributed capacity C_T form a second-order low-pass filter causing a roll-off in the high-frequency response.

The parameters L_K and C_T are parasitic elements rather than design parameters and should be minimized for optimum high-frequency response.

For low values of L_k, the coefficient of coupling K should be as high as possible. High K's are achieved by selecting materials which maintain high permeability at the higher operating frequencies. Winding methods should allow close proximity between primary and secondary, and high turns ratios should be avoided. Distributed capacity C_T can be reduced by minimizing the number of turns required through high-permeability materials and by using special winding techniques.

To obtain maximum flatness at the low-frequency end of a transformer's frequency range, the value of L_p should be as high as possible. However, if as a result the leakage inductance and distributed capacity become excessive, the high-frequency response will deteriorate; so a compromise may be required. Generally, the smaller the physical characteristics of the transformer which still meets the low-frequency requirements, the better will be the high-frequency response.

Pulse Transformers

Pulse transformers are required to pass square waves or pulse trains while maintaining the pulse shape within specified limits. A pulse can be represented by a series of sine waves as given by Fourier analysis theory. The transformer must have sufficient bandwidth to maintain fidelity of the output pulse.

The significant characteristics of a pulse are illustrated in figure 9-16. If we assume that a transformer behaves like a first-order network in the upper and lower cutoff regions, the pulse-shape distortion can be computed in terms of the transformer bandwidth. The rise time is defined as the time interval between the 10 and 90% points of full amplitude and is given in seconds by

$$T_r = \frac{0.35}{f_2} \qquad (9\text{-}20)$$

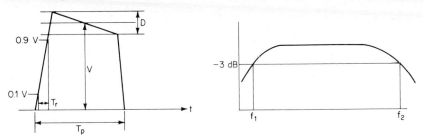

Fig. 9-16   Relationship between pulse rise time, droop, and bandwidth.

where f_2 is the upper 3-dB cutoff in hertz of the transformer. Clearly for fast rise times the transformer will require an extended high-frequency cutoff.

Pulse droop is defined using figure 9-16 as the ratio D/V expressed as a percentage. The capability to reproduce a pulse with minimum droop is directly related to the low-frequency response of the transformer. The percentage droop is given by

$$P_d = 200\pi T_p f_1 \qquad (9\text{-}21)$$

where T_p is the pulse width in seconds and f_1 is the lower 3-dB cutoff frequency. For minimum droop the low-frequency response should be extended as low as possible.

The pulse waveshape of figure 9-16 is somewhat of a simplification. In reality, overshoot and ringing may also occur. These transient effects may warrant only an empirical solution, but as a general guideline, the bandwidth should be as wide as possible for faithful pulse reproduction.

Magnetic-Core Selection for Transformers

Laminations   When a transformer's low-frequency bandwidth requirement extends below 100 Hz, laminations become the best choice, since materials such as Permalloy 80 have low-frequency initial permeabilities of over 30,000 (see figure 9-3). For maximum effective permeability a gapless structure is required. The F laminations discussed in section 9.2 are also available in an EI or EE shape. Referring to the lamination assembly illustrated in figure 9-2, the bobbin is stacked by alternating the E and I shapes with each layer. EE laminations consist of a large E and a small E. These shapes must also be alternated. This approach results in a nearly gapless configuration.

The inductance of a gapless lamination structure is given by

$$L_a = K_1 K_2 N^2 \mu_{ac} \qquad (9\text{-}22)$$

where K_1 is a stacking factor which is typically 0.9, K_2 is an inductance factor, N is the number of turns, and μ_{ac} is the effective permeability of the material at the frequency of operation. The inductance factors for EI and EE laminations corresponding to some of the sizes discussed in section 9.2 are given in table 9-16.

The corresponding winding data can be found in table 9-5.

Some common methods for winding transformer bobbins are illustrated in figure 9-17. Leakage inductance can be reduced by interleaving, i.e., splitting the primary or secondary into two windings and placing the other winding be-

**TABLE 9-16   Inductance Factors for
Permalloy 80 Laminations**

Size	K_2
EE 28–29	0.0444×10^{-8}
EI 187	0.0692×10^{-8}
EI 21	0.246×10^{-8}
EI 75	0.400×10^{-8}

tween these two sections. If the transformer has a $1:1$ turns ratio, bilfilar wire can be used. This wire consists of two conductors which are separately insulated but bonded together for close proximity in the magnetic circuit. This results in extremely low leakage inductance, but distributed capacity will be high.

Ferrite Cores   For wide-band transformer requirements a core material is desired that has high permeability at low frequencies and low losses throughout the entire frequency range of interest. These properties are inherent in ferrite materials. Ferrites offer the designer many advantages over alternative approaches for the frequency ranges where ferrites are suitable. They are available in a variety of shapes such as potcores, toroids, E cores, and I cores.

A potcore with no gap in the center post will form a nearly closed magnetic circuit. The effective permeability will then be almost as high as the initial permeability of the material. The potcore sizes and materials discussed in section 9.4 are also available in ungapped forms. The electrical properties are summarized in table 9-17. The wire chart of table 9-13 corresponds to these sizes.

Although a potcore may be of the gapless type, a small gap will always exist, since some slight flux leakage will occur because of construction and because the surfaces of the center posts of a potcore pair do not mate perfectly. As a result, the effective permeability shown in table 9-17 is less than the initial permeability of the material in most cases.

Ferrites are also available in a toroidal shape. Toroidal cores have the highest magnetic efficiency of any core structure, since the cross-sectional area is uniform throughout the core and the magnetic circuit is completely gapless. The effective permeability is then identical to the initial permeability of the material. Properties of some selected core sizes are given in table 9-18.

The choice of core material is based on the frequency range of operation. Using figure 9-9a, a material should be selected that has a high permeability which remains constant over the bandwidth of interest.

Choosing the type and size of core is a more difficult decision. Toroids feature a gapless construction which results in high efficiency. However, they are more difficult to wind than a potcore bobbin. Potcores are also easier to mount on printed circuit boards using available mounting hardware. They are generally

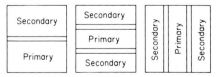

Fig. 9-17   Transformer winding methods.

TABLE 9-17   Summary of Ungapped Potcores

Size	Material	A_L, mH/1000 turns ±25%	μ_{eff}	Ω/H
905	3B7	1,229	1,217	132
	3B9	1,150	1,135	142
	3D3	665	650	247
	4C4	125	124	1,300
1107	3B7	1,800	1,340	78
	3B9	1,560	1,150	91
	3D3	865	635	165
	4C4	165	125	836
1408	3C8	2,800	1,780	33
	3B7	2,240	1,400	42
	3B9	1,910	1,190	49
	3D3	1,130	705	83
	4C4	200	125	470
1811	3C8	4,000	1,930	16
	3B7	3,680	1,740	18
	3B9	2,630	1,250	25
	3D3	1,550	735	42
	4C4	265	125	247
2213	3C8	5,200	2,040	9.1
	3B7	4,650	1,845	10
	3B9	3,250	1,275	15
	3D3	1,800	705	26
	4C4	300	120	155
2616	3C8	6,700	2,100	6.3
	3B7	6,000	1,880	7.0
	3B9	4,390	1,380	9.6
	3D3	2,340	735	18
	4C4	390	120	110
3019	3C8	8,300	2,180	4.1
	3B7	7,580	2,020	4.4
	3B9	5,750	1,480	6.0
3622	3C8	10,800	2,250	3.0
	3B7	9,660	2,000	3.4
	3B9	7,050	1,440	4.7
4229	3C8	11,500	2,350	1.7
	3B7	10,300	2,100	1.9

the best choice for transformers covering the frequency range from a few hundred hertz to a few megahertz. Gapped cores can be used in the event a DC current is superimposed along with the AC excitation.

The size of the potcore or toroid selected is usually determined by two factors: flux density and copper losses. For a given frequency and level of excitation and a required inductance value, a larger core size will result in reduced flux density. Also the required number of turns decreases with increasing core size and more bobbin area is available for the winding, which permits a larger wire gauge so that copper losses are reduced. However, cost and size limitations will usually dictate some degree of compromise.

TABLE 9-18 Ferrite Toroidal Cores

OD, in	ID, in	HT, in	Cross Section, cm²	Path Length, cm	Material	A_L, mH/1000 turns ±20%
0.375	0.187	0.125	0.076	2.16	4C4	55
					3D3	330
					3B7	1100
					3E2A	2135
0.500	0.281	0.188	0.133	3.03	4C4	70
					3D3	415
					3C8	1475
					3E2A	2750
0.870	0.540	0.250	0.270	5.52	4C4	75
					3C8	1650
					3E2A	3055
1.530	0.770	0.500	1.21	8.63	4C4	218
					3C8	4700
2.000	1.250	0.750	1.81	12.7	3C8	4800
2.900	1.530	0.500	2.21	17.1	3C8	4350

High-Frequency Transformers For transformers extending into the VHF region (30 to 300 MHz), powdered-iron core materials become the optimum choice. Core geometries can take the form of toroids, rods, or even potcores. Toroids are usually the best choice because of the gapless structure. A toroidal core can be chosen from the standard materials and sizes given in tables 9-14 and 9-15. The windings should usually take the form of a single layer for both primary and secondary. The larger winding is normally placed on the core first, and the secondary is then spread around the core, also in a single layer covering the entire primary so that the leakage inductance is minimized. Keeping the starts and finishes separated and using mylar or teflon tape over the core and between windings will serve to keep distributed capacity low.

Air-core transformers can be used beyond 50 or 100 MHz with good results. As with the powdered-iron cores, the geometry of the windings becomes extremely critical, so that leakage inductance and distributed capacity are at a minimum. They generally take the form of a solenoid, as discussed in section 9.5, under Air-Core Inductors.

Distribution of Copper Losses

For maximum transformer efficiency, the general practice has been to allocate equal cross-sectional area to the primary and secondary windings, although the actual number of turns may be unequal. This is accomplished by first doubling the number of turns for the larger winding and determining the corresponding wire size from the wire tables. Half the total turns and the wire size are then manipulated using the turns–wire gauge relationships outlined in section 9.1, under Design of Coil Windings, until the required number of turns for the smaller winding is obtained along with the resulting wire size. As a result the primary and secondary wire sizes can be determined so that both winding areas are equal. The DC resistance of each winding can also be estimated by manipulating the initial DC resistance value along with the turns and wire gauge. These techniques will be illustrated in example 9-4.

Equal cross-sectional areas for primary and secondary windings are not always practical or even necessary. When high turns ratios occur, equal winding areas may require impractically heavy wire gauges for the smaller winding; so a compromise may be required. Also, for high-frequency transformers, the copper losses are usually of little significance; so equal distribution of wire area is not important.

The Design Procedure

The initial steps in the design procedure are to select a core geometry and choose the core material. These decisions are interrelated, since some materials are available only in limited core configurations. The core material is usually determined by the low-frequency limit of the bandwidth requirement. The core size will depend upon the excitation level and allowable copper losses as well as both physical and economic considerations.

After the core is selected, a trial design is computed. Primary inductance is based on the relationships given by equations (9-16) through (9-18). The primary turns are then computed from the rated A_L of the core, and the secondary turns are based on the impedance ratio, i.e.,

$$N_s = N_p \sqrt{\frac{R_L}{R_s}} \tag{9-23}$$

The wire sizes are then determined using the guidelines of section 9.6, under Distribution of Copper Losses.

To verify that the core size is acceptable, the copper losses are estimated and the maximum flux density is calculated using the lowest frequency of operation. A prototype transformer can then be wound using one of the suggested winding methods and the frequency response can be measured in the same manner as a filter using the test circuit of figure 8-24.

If the low-frequency response has excessive roll-off, more inductance is required; so the turns should be increased. A roll-off at the high-frequency end can occur from distributed capacity, leakage inductance, or both. Distributed capacity as the dominant factor will usually cause a gentle roll-off, whereas leakage inductance may result in an initial rise in the response due to series resonance with distributed capacity prior to roll-off. Distributed capacity can be reduced by using a progressive winding method for toroids or segmented windings (split bobbins) for laminations or potcores. Leakage inductance is decreased by changing winding methods to obtain a higher coefficient of coupling. Both distributed capacity and leakage inductance can be reduced by using fewer turns provided that the low-frequency response remains acceptable.

The following example demonstrates the design of a wide-band transformer.

Example 9-4

REQUIRED: Design a transformer which satisfies the following requirements:
3 dB bandwidth   100 Hz to 30 kHz
$R_s = 600\ \Omega$ $R_L = 10\ k\Omega$
10 V_{rms} maximum across primary
Insertion loss of 1 dB maximum

RESULT: (a) Compute primary inductance for lower 3-dB cutoff at 100 Hz

$$L_P = \frac{R}{4\pi f_{3\,\text{dB}}} = \frac{600}{4\pi 100} = 477\text{mH} \qquad (9\text{-}18)$$

(b) Select a 2213 size ungapped potcore of 3C8 material from table 9-17. The tabulated A_L and Ω/H are 5200 and 9.1, respectively.

(c) Compute primary and secondary turns:

Primary:

$$N_p = 10^3 \sqrt{\frac{L_p}{A_L}} = 10^3 \sqrt{\frac{477}{5200}} = 303 \text{ turns} \qquad (9\text{-}10)$$

Secondary:

$$N_s = N_p \sqrt{\frac{R_L}{R_s}} = 303 \sqrt{\frac{10{,}000}{600}} = 1237 \text{ turns} \qquad (9\text{-}23)$$

(d) Determine primary and secondary wire sizes for equal winding cross-sectional areas and estimate copper losses:

1. *Double turns of larger winding and determine wire size and total resistance*
   $2 \times 1237 = 2474$ turns
   Wire gauge is No. 40 AWG (table 9-13)
   Total L is 31.8 H [equation (9-10)]
   Total resistance is 290 Ω (9.1 Ω/H)

2. *Determine secondary wire size and DC resistance from step 1*
   1237 turns No. 40, 145 Ω

3. *Manipulate turns to obtain wire size and DC resistance for primary winding.* (A factor of 1.26 relates the number of turns of adjacent wire sizes for equal areas. A factor of 1.26^2 or 1.58 relates the resistances for the turns corresponding to adjacent wire sizes.)

Turns	Wire Size	DC Resistance
1237 ⌐1.26	40	145 Ω ⌐1.58
982 ⌐	39	92 Ω ⌐
779	38	58 Ω
618	37	37 Ω
490	36	23 Ω
389	35	15 Ω
309	34	9.5 Ω

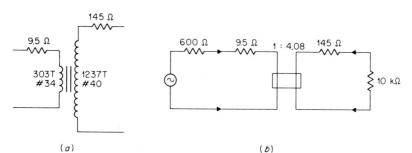

(a) (b)

Fig. 9-18 Transformer design of example 9-4: (a) transformer design; (b) midband equivalent circuit.

Primary consists of 303 turns of No. 34 AWG and has an estimated DC resistance of 9.5 Ω.

(e) The transformer design is shown in figure 9-18a. The midband equivalent circuit is represented in figure 9-18b using an ideal transformer and the source and load terminations. An insertion loss of approximately 0.26 dB can be computed due to the primary and secondary DC resistances.

REFERENCES

Arnold Engineering Co., "Arnold Iron Powder Cores," Marengo, Ill.

Ferroxcube Corp., "Linear Ferrite Magnetic Design Manual—Bulletin 550," Saugerties, N.Y.

Ferroxcube Corp., "Linear Ferrite Materials and Components," Saugerties, N.Y.

Magnetics Inc., "Ferrite Cores," Butler, Pa.

Magnetics Inc., "Magnetic Laminations," Catalog ML-303T, Butler, Pa.

Magnetics Inc., "Molypermalloy Power Cores," Catalog MPP-303S, Butler, Pa.

Micrometals, "Q Curves for Iron Powder Toroidal Cores," Anaheim, Calif.

Welsby, V. G., "The Theory and Design of Inductance Coils," Macdonald and Sons, London, 1950.

10

Component Selection for LC and Active Filters

10.1 CAPACITOR SELECTION

An extensive selection of capacitor types are available for the designer to choose from. They differ in terms of construction and electrical characteristics. The abundance of different capacitors often results in a dilemma in choosing the appropriate type for a specific application. Some of the factors to consider are stability, size, losses, voltage rating, tolerances, cost, and construction.

The initial step in the selection process is to determine the capacitor dielectric. These substances include air, glass, ceramic, mica, plastic films, aluminum, and tantalum. The type of mechanical construction must also be chosen. Since miniaturization is usually a prime consideration, the smallest possible capacitors will require thin dielectrics and efficient packaging without degrading performance.

Properties of Dielectrics

Capacitors in their most fundamental form consist of a pair of metallic plates or electrodes separated by an insulating substance called the dielectric. The capacitance in farads is given by

$$C = \frac{kA}{D} 8.85 \times 10^{-12} \qquad (10\text{-}1)$$

where k is the dielectric constant, A is the plate area in square meters, and D is the separation between plates in meters. The dielectric constant of air is 1. Since capacity is proportional to the dielectric constant, the choice of dielectric highly influences the physical size of the capacitor.

A practical capacitor can be represented by the equivalent circuit of figure 10-1a, where L_s is the series inductance, R_s is the series resistance, and R_p is the parallel resistance. L_s, R_s, and R_p are all parasitic elements. If we assume L_s negligible and R_p infinite, the equivalent impedance can be found by the vector addition of R_s and the capacitive reactance X_c, as shown in figure 10-1b.

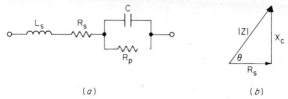

Fig. 10-1   Equivalent representation of a capacitor: (a) equivalent circuit; (b) simplified vector diagram.

An important characteristic of capacitors is dissipation factor. This parameter is the reciprocal of capacitor Q and is given by

$$d=\frac{1}{Q}=2\pi fCR_s=\cot\theta \qquad (10\text{-}2)$$

where f is the frequency of operation. Dissipation factor is an important figure of merit and should be as low as possible. This is especially true for selective LC filters where branch Q's should be high. Capacitor Q must be sufficiently higher than inductor Q so that the effective Q is not degraded.

Power factor is another figure of merit similar to dissipation factor. Referring to figure 10-1, the power factor is defined as

$$\text{PF}=\frac{R_s}{|Z|}=\cos\theta \qquad (10\text{-}3)$$

For capacitor Q in excess of 10, we can state $d = \text{PF} = 1/Q$. Dissipation factor or power factor is sometimes expressed as a percentage.

The shunt resistive element R_p in figure 10-1 is often referred to as insulation resistance and results from dielectric leakage currents. A commonly used figure of merit is the R_pC time constant, normally given in $\text{M}\Omega \times \mu\text{F}$. It is frequently convenient to combine all the losses in terms of a single resistor in parallel with C, which is given by

$$R=\frac{1}{2\pi fCd} \qquad (10\text{-}4)$$

Temperature coefficient (TC) is the rate of change of capacity with temperature and is usually given in parts per million per degree Celsius (ppm/°C). This parameter is extremely important when filter stability is critical. TC should either be minimized or of a specific nominal value so that cancellation will occur with an inductor or resistor's TC of the opposite sign. A dielectric material will also have an operating temperature range beyond which the material can undergo permanent molecular changes.

Another important parameter is retrace, which is defined as the capacity deviation from the initial value after temperature cycling. To maintain long-term stability in critical filters that are subjected to temperature variation, retrace should be small.

Table 10-1 summarizes some of the properties of the more commonly used dielectric materials when applied to capacitors.

TABLE 10-1 Properties of Capacitor Dielectrics

Capacitor Type	Dielectric Constant	TC, ppm/°C	Dissipation Factor, %	Insulation Resistance, MΩ-μF	Temperature Range, °C
Aluminum	8	+2500	10	100	−40 to +85
Ceramic (NPO)	30	±30	0.02	5×10^3	−55 to +125
Glass	5	+140	0.001	10^6	−55 to +125
Mica	6	±50	0.001	2.5×10^4	−55 to +150
Paper	3	±800	1.0	5×10^3	−55 to +125
Polycarbonate	3	±50	0.2	5×10^5	−55 to +125
Polyester (mylar)	3.2	+400	0.75	10^5	−55 to +125
Polypropylene	2.3	−200	0.2	10^5	−55 to +105
Polystyrene	2.5	−120	0.01	3.5×10^7	−55 to +85
Polysulfone	3.1	+80	0.3	10^5	−55 to +150
Porcelain	5	+120	0.1	5×10^5	−55 to +125
Tantalum	28	+800	4.0	20	−55 to +85
Teflon	2.1	−200	0.04	2.5×10^5	−70 to +250

Capacitor Construction

Capacitors are composed of interleaved layers of electrode and dielectric (except for the diffused types). With film-type capacitors such as polycarbonate, mylar, polypropylene, and polystyrene, the electrodes are either a conductive foil such as aluminum or a layer of metallization directly on the dielectric film. The electrode and dielectric combination is tightly rolled onto a core. The alternate layers of foil or metallization are slightly offset so that leads can be attached.

The assembly is heat-shrunk to form a tight package. It can then be sealed by coating with epoxy or by encapsulating in a molded package. A more economical form of sealing is to wrap the capacitor in a plastic film and fill the ends with epoxy. This method is called "wrap-and-fill construction."

The most economical form of construction involves inserting the wound capacitor in a polystyrene sleeve and heat shrinking the entire assembly to obtain a rigid package. However, the capacitor is not truly sealed and can be affected by humidity and cleaning solvents entering through the porous end seals. Increased protection against humidity or chemical effects is obtained by sealing the ends with epoxy in a manner similar to the wrap-and-fill construction. If a true hermetic seal is required for extremely harsh environments, the capacitor can be encased in a metal can having glass-to-metal end seals. A cutaway view of a film capacitor is illustrated in figure 10-2a.

Capacitors such as ceramic, mica, and porcelain contain a more rigid dielectric substance and cannot be rolled like the film capacitors. They are constructed in stacks or layers, as shown in figure 10-2b. Leads are attached to the end electrodes and the entire assembly is molded or dipped in epoxy. Ceramic-chip capacitors for use in hybrid circuits are constructed as in figure 10-2b and then directly bonded onto the metal substrate of the hybrid.

Capacitors having the highest possible capacitance per unit volume are the electrolytics made of either aluminum or tantalum. Basically, these capacitors consist of two electrodes immersed in a liquid electrolyte (figure 10-3). One or both electrodes are coated with an extremely thin oxide layer of aluminum or tantalum, forming a film having a high dielectric constant and good electrical characteristics. The electrolyte liquid makes contact between the film and the

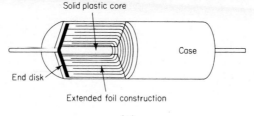

(a)

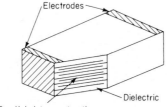

Parallel plate construction

(b)

Fig. 10-2 Capacitor construction: (a) cutaway view of film capacitor; (b) stacked construction.

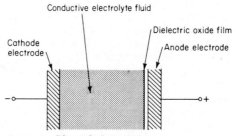

Fig. 10-3 Electrolytic capacitor.

electrodes. The entire unit is housed in a leakproof metal can. The dielectric film is "formed" by applying a DC voltage between cathode and anode, resulting in a permanent polarization of the electrodes. If both plates are "formed," a nonpolar unit will result having half the capacitance of the equivalent polarized type.

A solid-anode tantalum capacitor is composed of a sintered anode pellet on which is a tantalum oxide layer. The pellet is coated with a solid electrolyte of manganese dioxide which also becomes the cathode. This construction is superior electrically to the other forms of electrolytic construction.

Selecting Capacitors for Filter Applications

Film Capacitors   Polyester (mylar) capacitors are the smallest and most economical of the film types. They should be considered first for general-purpose filters operating below a few hundred kilohertz and at temperatures up to 125°C. They are available with either foil construction for normal applications or in metallized form, where size is restricted to values ranging from 1000 pF to a few microfarads.

Polycarbonate capacitors are slightly larger than mylar but have superior electrical properties, especially at higher temperatures. The dissipation factor remains low over a wide temperature range, and retrace characteristics are better than mylar. Polysufone is similar to polycarbonate but can be used up to 150°C.

Polystyrene capacitors have the best electrical properties of all film capacitors. Temperature coefficient is precisely controlled, almost perfectly linear, and has a nominal value of −120 ppm/°C. Because of the predictable temperature characteristics, these capacitors are highly suited for LC resonant circuits, where the inductors have a corresponding positive temperature coefficient. Capacity retrace is typically 0.1%. Losses are extremely small, resulting in a dissipation factor of approximately 0.01%. The maximum temperature, however, is limited to 85°C. Polypropylene capacitors are comparable in performance to polystyrene, although they have a slightly higher dissipation factor and temperature coefficient. Their maximum temperature rating, however, is 105°C. They are more economical than polystyrene and priced nearly competitive with mylar.

Figure 10-4 compares the capacitance and dissipation factor versus temperature for mylar, polycarbonate, and polystyrene capacitors. Clearly polystyrene is superior, although polycarbonate exhibits slightly less capacity variation with temperature over a limited temperature range.

Film capacitors are available with standard tolerances of 1, 2½, 5, 10, and 20%. For most applications the 2½ or 5% tolerances will be adequate. This is especially true when resonant frequency is adjustable by tuning the inductor or with a potentiometer in active filters. Precision capacitors tend to become more expensive with decreasing tolerance.

The first two significant figures for standard values are given in table 10-2. Frequently a design may require nonstandard values. It is nearly always far more economical to parallel a few off-the-shelf capacitors of standard values

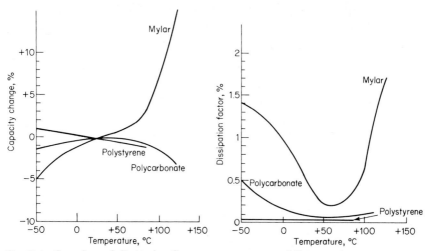

Fig. 10-4 Capacity and dissipation factor vs. temperature of film capacitors.

TABLE 10-2   Standard Capacitor Values

10	18	33	56
11	20	36	62
12	22	39	68
13	24	43	75
15	27	47	82
16	30	51	91

when possible than to order a custom-manufactured component unless the quantities involved are substantial.

If the differential voltage across a capacitor becomes excessive, the dielectric will break down and may become permanently damaged. Most film capacitors are available with DC voltage ratings ranging from 33 to 600 V or more. Since most filters process signals of a few volts, voltage rating is normally not a critical requirement. Since capacitor volume increases with voltage rating, unnecessarily high ratings should be avoided. However, polystyrene capacitors below 0.01 μF having a low voltage rating will have a tendency to change value from the printed circuit-board soldering process because of distortion of the dielectric film from heat conducted through the leads. Capacitors rated at 100 V or more will generally be immune, since the dielectric film will be sufficiently thick.

In addition to resistive losses the equivalent circuit of figure 10-1a contains a parasitic inductance L_s. As the frequency of operation is increased, series resonance of L_s and C is approached, resulting in a dramatic drop in impedance as shown in figure 10-5. Above self-resonance the reactance becomes inductive. The self-resonant frequency is highly dependent upon construction and value, but in general, film capacitors are limited to operation below a megahertz.

Mica Capacitors   Mica capacitors, although more costly than the film types, have unequaled electrical properties. Temperature coefficients as low as a few parts per million can be obtained. Retrace is better than 0.1% and the dissipation factor is typically 0.001%. Mica capacitors can operate to 150°C. Values above 10,000 pF become prohibitively expensive and should be avoided.

Mica plates are silvered on both sides to form electrodes, stacked, and thermally bonded together. This results in a mechanically stable assembly with characteristics indicative of the mica itself. A coating of epoxy resin is then applied

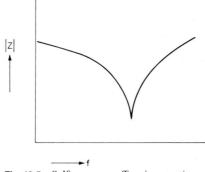

Fig. 10-5   Self-resonant effect in capacitors.

by dipping or other means. An alternate form of construction involves using eyelets to join together the mica plates. This form of construction has poor stability of capacity, since it may permit relative movement between plates. Silvered mica capacitors are also available in a molded package. These capacitors are comparable electrically with the epoxy-dipped variety. However, they are normally designed to operate at higher DC voltages ranging from 1000 to 2500 V and are not needed for most filter requirements.

The stacked construction of mica capacitors results in excellent performance well into the VHF region. The dissipation factor remains low at higher frequencies, and parasitic inductance is small.

A comprehensive MIL-type designation system is frequently used to specify mica capacitors. This nomenclature is illustrated below:

CMO5	E	D	510	J	O	3
Style		Working voltage		Tolerance		Vibration grade
	Characteristic		Capacitance		Temperature range	

Style:

The style is defined by the two-letter symbol CM followed by two digits to designate case size and lead configurations as follows:

Capacity Range, pF	Straight Leads	Crimped Leads
1–390	CM04	CM09
1–390	CM05	CM10
430–4,700	CM06	CM11
5,100–20,000	CM07	CM12
22,000–91,000	CM08	CM13

Characteristic:

The characteristic letter symbol defines the stability of capacity with temperature, and retrace limits.

Letter	Temperature Coefficient, ppm/°C	Retrace
B	Not specified	Not specified
C	±200	±(0.5% + 0.1 pF)
D	±100	±(0.3% + 0.1 pF)
E	−20 to +100	±(0.1% + 0.1 pF)
F	0 to +70	±(0.05% + 0.1 pF)

Working voltage:

The DC working voltage is specified in terms of a letter.

Letter	DC Voltage
Y	50
A	100
C	300
D	500

Capacitance:

The capacitance in picofarads is identified by a three-digit number, where the first two digits represent the two significant figures corresponding to table 10-2 and the third digit is the number of zeros to follow.

Tolerance:

Tolerances are indicated by a letter.

Letter	Tolerance
A	±1 pF
D	±0.5 pF
M	±20%
K	±10%
J	±5%
G	±2%
F	±1%
E	±0.5%

Temperature range:

A letter indicates the temperature range.

Letter	Temperature Range, °C
M	−55 to +70
N	−55 to +85
O	−55 to +125
P	−55 to +150

Vibration range:

The vibration range is identified as follows:

Number	Vibration Frequency, Hz
1	10–55
3	10–2000

Ceramic Capacitors   Ceramic capacitors are formed in stacks or layers using the construction illustrated in figure 10-2b. The ceramic plates are formed by first casting a film slurry of barium titanate and binders, which is then coated with a metallic ink to form electrodes. The plates are stacked into layers, fired in ovens, and separated into individual capacitors.

Ceramic capacitors are manufactured in a variety of shapes including monolithic chips which can be directly bonded to a metallized substrate.

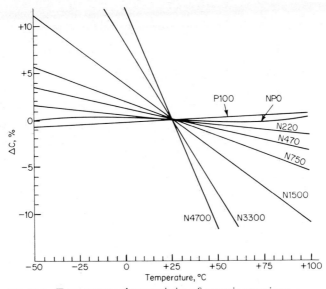

Fig. 10-6 Temperature characteristics of ceramic capacitors.

The basic dielectric material barium titanate has a dielectric constant k as high as 3000 at room temperature which can become as high as 10,000 at 125°C. The temperature at which the maximum k occurs is called the "curie point." By introducing additives, the curie point can be lowered, the temperature variation reduced, and negative temperature coefficients can be obtained.

Ceramic capacitors in disk form are available in specific temperature coefficients. Some typical values are shown in figure 10-6, where the P100 has a positive temperature coefficient of 100 ppm/°C and the negative types range from NPO (approximately zero temperature coefficient) through N4700 (4700 ppm/°C). These capacitors become very useful for temperature compensation of passive and active filter networks.

Capacities range from 0.5 pF to values as high as 2.2 μF. The DC working voltages may be as high as a few kilovolts or as little as 3 V for the larger capacitance values. Dissipation factor in most cases is less than 1%. Tolerances are normally 10 or 20%, which restricts this family of capacitors to very general applications.

Ceramic capacitors exhibit minimum parasitic inductance and little variation in dissipation factor from low audio to high radio frequencies. These properties make them particularly suited for RF decoupling and bypass applications in addition to temperature compensation. Using the feed-through type of construction to further minimize parasitic inductance, they can be used up to 10 GHz.

Electrolytic Capacitors   Aluminum electrolytic capacitors are intended for low-frequency bypassing or nonprecision timing and are generally unsuitable for active or passive filters. They have unsymmetrical and broad tolerances such as +80%/−20% and require a DC polarization. They also have poor stability and a shelf-life limitation. Large parasitic inductances and series resistances preclude usage at high frequencies.

Tantalum capacitors, on the other hand, can be used in low-frequency passive or active filters which require very large capacity values in a small volume. They have fairly high temperature coefficients (approximately +800 ppm/°C) and dissipation factors (approximately 4%). However, these limitations may not be serious when applied to low-selectivity filters. The high-frequency characteristics are superior to those of aluminum electrolytics. Tolerances of 10 or 20% are standard.

Tantalum capacitors are formed in a similar manner to aluminum electrolytics and are polarized. However, they have no shelf-life restrictions and can even operate indefinitely without DC polarization. A momentary reverse polarity usually will not damage the capacitor, which is not true in the case of the aluminum electrolytics.

A polarization voltage may not always be present. Nonpolar tantalums can be obtained but are somewhat more expensive and not always available off the shelf. If two identical tantalum capacitors are series-connected back to back as shown in figure 10-7, a nonpolar type will result. The total value will be $C/2$ corresponding to capacitors in series.

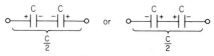

Fig. 10-7   Nonpolar tantalum capacitors.

Tantalum capacitors are available in three forms: dry foil, wet, and solid. The dry form consists of foil anodes and cathodes which are stacked or rolled using a paper spacer and impregnated with electrolyte. The wet forms are constructed using a porous slug of tantalum for the anode electrode and the silver-plated case as the cathode. The unit is filled with sulfuric acid for the electrolyte. Solid tantalums consist of a slab of compressed tantalum powder for the anode with a lead attached. A layer of tantalum pentoxide is formed on the surface for the dielectric, which is then connected to a lead to form the cathode, and the entire assembly is encapsulated in epoxy.

Trimmer Capacitors   In LC filters, particularly those for **RF** use, it is sometimes found more convenient to resonate a tuned circuit by adjusting capacity rather than inductance. A smaller trimmer is then placed in parallel with a fixed resonating capacitor. These trimmers usually consist of air, ceramic, mica, or glass as the dielectric.

Air capacitors consist of two sets of plates, one called the rotor, which is mounted on a shaft, and the other the stator, which is fixed. As the rotor is revolved, the plates intermesh without making contact, resulting in increasing capacity. For high capacity values the plate size and number must increase dramatically, since the dielectric constant of air is only 1. This becomes a serious limitation when the available room is restricted. Air trimmers range from maximum values of a few picofarads up to 500 pF.

Ceramic trimmers are smaller than air for comparable values due to increased dielectric constant. They are usually comprised of a single pair of ceramic disks joined at the center in a manner which permits rotation of one of the disks. A silvered region covers part of each disk forming the plates of the capacitor. As the disk is rotated, the silvered areas begin to overlap, resulting in increasing capacity. Ceramic trimmers are available with maximum capacity values up to 50 or 100 pF. Good performance is obtained well into the VHF frequency range.

Piston trimmers are comprised of a glass or quartz tube with an outside conductive coating corresponding to one electrode. The other plate or electrode is a piston which by rotation is inserted deeper into the outside tube, resulting in increased capacitance. Multiturn construction results in excellent resolution. Piston trimmers have the best electrical properties of all trimmer types but are the most costly. They are suitable for use even at microwave frequencies.

10.2 RESISTORS

A fundamental component of active filters are resistors. Sensitivity studies show that resistors are usually at least as important as capacitors; so their proper selection is crucial to the success of a particular design.

Resistors are formed by connecting leads across a resistive element. The resistance in ohms is determined by

$$R = \frac{\rho L}{A} \qquad (10\text{-}5)$$

where ρ is the resistivity of the element in ohm-centimeters, L is the length of the element in centimeters, and A is the cross-sectional area in square centimeters. By using materials of particular resistivities and special geometries, resistors can be manufactured having the desired properties. Resistors fall into one of two general categories: fixed or variable.

Fixed Resistors

Fixed resistors are normally classified as either carbon composition, carbon film, cermet film, metal film, or wirewound according to the resistive element. They are also grouped in terms of tolerance and wattage rating. Table 10-3 summarizes some typical properties of fixed resistors.

TABLE 10-3   Typical Properties of Fixed Resistors

Type	Range, Ω	Standard Tolerances, %	Wattage Rating	Temperature Coefficient, ppm/°C
Carbon composition	1–100 M	5, 10, 20	⅛, ¼, ½, 1, 2	±1000
Carbon film	1–10 M	2, 5	⅛, ¼, ½	±200
Cermet film	10–22 M	0.5, 1	¼, ½	±100
Metal film	0.1–1 M	0.1, 0.25, 0.5, 1	⅛, ¼, ½, 1, 2	±25
Wirewound	1–100 k	5, 10, 20	3, 5, 10, 20	±50

Carbon Composition   Probably the most widely used fixed resistor is the carbon composition type shown in figure 10-8. It consists of a solid cylinder composed mainly of carbon with wire leads which is molded under high pressure and temperature in an insulated jacket. By changing the proportion of carbon powder and filler, different resistance values can be obtained. Standard wattage ratings are 1/8, 1/4, 1/2, 1, and 2 W. Resistance values are maintained to about ±10% by the manufacturing process. Values within a ±5% tolerance band are obtained using automatic sorting equipment.

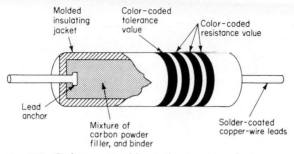

Fig. 10-8 Carbon composition resistor construction.

A series of standard values has been adopted by RETMA based on $10^{1/12}$ and $10^{1/24}$ and is widely used for 10 and 5% resistors. These preferred values are rounded to two significant figures which are given in table 10-4. All 10% values are also available in 5% tolerances but not conversely.

The value and tolerance of carbon composition resistors are indicated by a series of color-coded bands beginning on one end of the resistor body. The first and second bands determine the two significant figures, the third band the multiplier, and the fourth band the tolerance. Sometimes a fifth band is present to establish a reliability rating, as shown in table 10-5.

The power rating of resistors corresponds to the maximum power that can be continually dissipated with no permanent damage. This rating is normally applicable up to ambient temperatures of 70°C, above which derating is required. In general, a safety margin corresponding to a factor of 2 is desirable to obtain a high level of reliability.

Carbon composition resistors have a high temperature coefficient, typically 1000 ppm/°C. In addition, permanent resistance changes of a few percent will occur to poor retrace and aging. Generally, the lower resistance values exhibit better stability. Carbon composition resistors exhibit few parasitic effects and then only above 10 MHz.

Carbon composition resistors are also the most economical of the various resistor types. They are often used in general-purpose active filters, where stability is not a critical requirement.

TABLE 10-4   Preferred 5 and 10% Resistor Values

±10%	±5%	±10%	±5%
10	10	33	33
	11		36
12	12	39	39
	13		43
15	15	47	47
	16		51
18	18	56	56
	20		62
22	22	68	68
	24		75
27	27	82	82
	30		91

TABLE 10-5 Standard Color Code for Carbon Composition Resistors

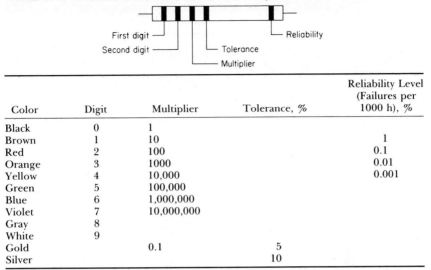

Color	Digit	Multiplier	Tolerance, %	Reliability Level (Failures per 1000 h), %
Black	0	1		
Brown	1	10		1
Red	2	100		0.1
Orange	3	1000		0.01
Yellow	4	10,000		0.001
Green	5	100,000		
Blue	6	1,000,000		
Violet	7	10,000,000		
Gray	8			
White	9			
Gold		0.1	5	
Silver			10	

Carbon Film Film resistors are manufactured by depositing a thin layer of a resistance element on a nonconductive substrate. This form of construction is illustrated in figure 10-9. The resistive element is usually spiral in form to increase the net resistance.

Carbon film resistors use this form of construction. They are manufactured by heating carbon-bearing gases so that a deposit of carbon film forms on the ceramic substrate. Temperature coefficients are in the region of 200 ppm/°C, which is a significant improvement over carbon composition. Resistance changes from aging and temperature cycling are typically specified at 1% or less. Normal tolerances are ±5%, although ±1% tolerance units can be obtained. The values are in accordance with table 10-4.

Carbon film resistors are especially suited for general-purpose applications requiring better performance than can be obtained using the carbon composition type at competitive pricing.

Metal Film Metal film resistors are without any doubt the most widely used type for precision requirements. They are manufactured by depositing nichrome

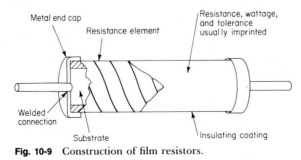

Fig. 10-9 Construction of film resistors.

alloys on a rod substrate. Exceptional characteristics can be obtained. Normal tolerances are ±1%, but tolerances of ±0.1%, ±0.25%, and ±0.5% are available. Temperature coefficients are typically 25 ppm/°C. Retrace and aging result in changes usually not exceeding 0.25%.

Metal film resistors are normally specified using a military numbering system as follows:

RN60	D	1023	F
Power	Characteristic	Value	Tolerance

Power:

Designation	Power Rating (70°C), W
RN50	$\frac{1}{10}$
RN55	$\frac{1}{8}$
RN60	$\frac{1}{4}$
RN65	$\frac{1}{2}$
RN70	$\frac{3}{4}$
RN75	1
RN80	2

Characteristic:

Letter	Temperature Coefficient, ppm/°C
B	±500
C	±50
D	±100
E	±25
F	±50

Value:

The value is given by four digits, where the first three digits are significant and are selected from the following table and the fourth digit is the number

10.0	**12.1**	**14.7**	**17.8**	**21.5**	**26.1**	**31.6**	**38.3**	**46.4**	**56.2**	**68.1**	**82.5**
10.1	12.3	14.9	18.0	21.8	26.4	32.0	38.8	47.0	56.9	69.0	83.5
10.2	**12.4**	**15.0**	**18.2**	**22.1**	**26.7**	**32.4**	**39.2**	**47.5**	**57.6**	**69.8**	**84.5**
10.4	12.6	15.2	18.4	22.3	27.1	32.8	39.7	48.1	58.3	70.6	85.6
10.5	**12.7**	**15.4**	**18.7**	**22.6**	**27.4**	**33.2**	**40.2**	**48.7**	**59.0**	**71.5**	**86.6**
10.6	12.9	15.6	18.9	22.9	27.7	33.6	40.7	49.3	59.7	72.3	87.6
10.7	**13.0**	**15.8**	**19.1**	**23.2**	**28.0**	**34.0**	**41.2**	**49.9**	**60.4**	**73.2**	**88.7**
10.9	13.2	16.0	19.3	23.4	28.4	34.4	41.7	50.5	61.2	74.1	89.8
11.0	**13.3**	**16.2**	**19.6**	**23.7**	**28.7**	**34.8**	**42.2**	**51.1**	**61.9**	**75.0**	**90.9**
11.1	13.5	16.4	19.8	24.0	29.1	35.2	42.7	51.7	62.6	75.9	92.0
11.3	**13.7**	**16.5**	**20.0**	**24.3**	**29.4**	**35.7**	**43.2**	**52.3**	**63.4**	**76.8**	**93.1**
11.4	13.8	16.7	20.3	24.6	29.8	36.1	43.7	53.0	64.2	77.7	94.2
11.5	**14.0**	**16.9**	**20.5**	**24.9**	**30.1**	**36.5**	**44.2**	**53.6**	**64.9**	**78.7**	**95.3**
11.7	14.2	17.2	20.8	25.2	30.5	37.0	44.8	54.2	65.7	79.6	96.5
11.8	**14.3**	**17.4**	**21.0**	**25.5**	**30.9**	**37.4**	**45.3**	**54.9**	**66.5**	**80.6**	**97.6**
12.0	14.5	17.6	21.3	25.8	31.2	37.9	45.9	55.6	67.3	81.6	98.8

of zeros. The bold values in this table specifically correspond to standard 1% values, and all listed numbers are standard for 0.1, 0.25, and 0.5% tolerances.

Tolerance:

Letter	Tolerance
B	±0.1%
C	±0.25%
D	±0.5%
F	±1%

Metal film resistors have many highly desirable features. In addition to low temperature coefficients and good long-term stability they exhibit the lowest noise attainable in resistors. Parasitic effects are minimal and have no significant effect below 10 MHz. Metal film resistors are rugged in design, have excellent immunity to environmental stress, and have high reliability. Although tolerances to 0.1% are available, 1% values are used almost exclusively because of their lower cost.

Cermet film resistors are manufactured by screening a layer of combined metal and ceramic or glass particles on a ceramic core and firing it at high temperatures. They can provide higher resistance values for a particular size than most other types, up to a few hundred megohms. However, they are somewhat inferior electrically to the metal film type. Their temperature coefficients are higher (typically 200 ppm/°C). Retrace and long-term stability are typically 0.5%. Tolerances of 1% are standard.

Precision resistors are also available in wirewound form. They consist essentially of resistance wire such as nichrome, wound on an insulated core. They are costlier than the metal film type and comparable in performance except for higher parasitics. Wirewound resistors are best suited for applications with higher power requirements.

In addition to the discrete resistors discussed, thin-film and thick-film resistive elements can be formed on chips for LSI and hybrid circuits. Thin-film resistors are obtained by evaporating, sputtering, or silk screening resistive elements on an insulating substrate. Thick-film resistors are produced by depositing resistive elements in ink form on the substrate using photographic techniques.

Variable Resistors

Variable resistors are commonly referred to as potentiometers or trimmers. They are classified by the type of resistance element such as carbon, cermet, or wirewound and also by whether they are single- or multiple-turn. The electrical properties of the three basic element types are given in table 10-6.

TABLE 10-6 Typical Properties of Potentiometers

Type	Range, Ω	Standard Tolerances,%	Wattage Rating	Temperature Coefficient, ppm/°C
Carbon composition	100–10 M	10, 20	0.5, 1, 2	±1000
Cermet	100–1 M	5, 10, 20	0.5, 1	±100
Wirewound	10–100 M	5, 10	0.5, 1, 5, 10	±100

Potentiometers are always three-terminal devices as depicted in figure 10-10a. (CW indicates direction of travel of the wiper for clockwise rotation.) Since most applications require a two-terminal variable resistor or rheostat rather than a voltage divider, the wiper is normally externally joined to one of the end terminals, as shown in figure 10-10b.

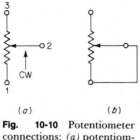

(a)   (b)

Fig.   10-10 Potentiometer connections: (a) potentiometer; (b) rheostat configuration.

Construction   The basic types of construction are the single-turn and multiturn forms. Single-turn potentiometers are rotary devices having a centrally located adjustment hub which contains the movable contact for the wiper. The movable contact rests upon the circular resistive element. The entire assembly may be exposed or enclosed in a plastic case containing PC pins or wire leads for the external connections. The adjustable hub is usually slotted for screwdriver access. Some types have a toothed thumbwheel for manual adjustment.

Single-turn trimpots require 270° of rotation to fully traverse the entire resistance element. Increasing the number of turns will improve the operator's ability to make very fine adjustments. Multiturn trimmers contain a threaded shaft. A threaded collar travels along this shaft, making contact with the resistive element. The collar functions as the wiper. This general construction is shown in figure 10-11.

Trimmer controls, especially the single-turn variety, are subject to movement of the wiper adjustment due to vibration. To assure stability, it is desirable to prevent movement by placing a small amount of a rigid sealer such as Glyptal on the adjuster after circuit alignment.

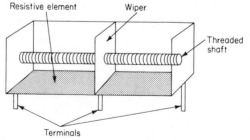

Fig.   10-11 Multiturn potentiometer construction.

Types of Resistance Elements Carbon composition potentiometers are formed by molding a carbon mixture on a nonconductive disk or base containing previously embedded leads. The wiper mechanism is then attached to complete the assembly.

Carbon composition potentiometers are available in both the single- and multiple-turn configurations. They are the most economical of all types and also have the poorest characteristics. TC is about 1000 ppm/°C. Retrace and long-term stability are poor. Therefore, carbon composition potentiometers find limited usage in filter circuits.

Cermet film potentiometers are the most commonly used type for filter networks. They are medium in cost, have a wide range of available values, and have good temperature characteristics and stability. Parasitic effects are minimal. They are manufactured by depositing cermet film on a nonconductive disk or base in thicknesses varying from 0.0005 to 0.005 in.

Wirewound potentiometers are formed by winding resistance wire (usually nichrome) on an insulated base. They have comparable temperature characteristics to cermet and are higher in cost. Their major attribute is high power capability. However, for most filter requirements this feature is of little importance.

Wirewound potentiometers have quite different adjustment characteristics than the other types. As the other type wipers are rotated, the resistance varies linearly with degrees of rotation, providing nearly infinite resolution. (Nonlinear tapers are also available, such as logarithmic.) The resistance of wirewound potentiometers, however, changes in discrete steps as the wiper moves from turn to turn. The resolution (or settability) therefore is not as good as the cermet or carbon composition types.

Ratings Since potentiometers are almost always used as variable elements, the overall tolerances are not critical and are generally 10 or 20%. For the same reason many different standard values are not required. Standard values are given by one significant figure, which is either 1, 2, or 5 and range from 10 Ω to 10 MΩ.

10.3 OPERATIONAL AMPLIFIERS

The versatility and low cost of integrated-circuit (IC) operational amplifiers have made them one of the most popular building blocks in the industry. The op amp is capable of performing many mathematical processes upon signals. For active filters, op amps are specifically used to provide gain and isolation.

IC op amps have evolved from the Fairchild μA 709 in the mid-sixties to the many different types available today having a variety of special features. This section reviews some of the essential characteristics, discusses some important considerations, and provides a survey of the most popular IC amplifier types. An extensive formal analysis is covered by many standard texts and will not be repeated here.

Review of Basic Operational-Amplifier Theory

A simplified equivalent circuit of an operational amplifier is shown in figure 10-12. The output voltage e_0 is the difference of the input voltages at the two input terminals amplified by amplifier gain A. A positive changing signal applied to the positive (+) input terminal results in a positive change at the output, whereas a positive changing signal applied to the negative input terminal (−) results in a negative change at the output. Hence the positive input terminal is called the "noninverting" input and the negative input terminal is referred to as the "inverting" input. If we consider the amplifier ideal, the input impedance R_i is infinite, the output impedance R_0 is zero, and the voltage gain A_0 is infinite.

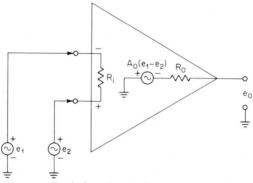

Fig. 10-12   Equivalent circuit of operational amplifier.

Negative feedback applied from the output to the inverting input results in zero differential input voltage if A_0 is infinite, and is called the "virtual ground effect." This property permits a wide variety of different amplifier configurations and simplifies circuit analysis.

Let us first consider the basic inverting amplifier circuit of figure 10-13. If we consider the amplifier ideal, the differential input voltage becomes zero be-

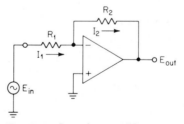

Fig. 10-13   Inverting amplifier.

cause of the negative feedback path through R_2. Therefore, the inverting input terminal is at ground potential. The currents through resistors R_1 and R_2 are

$$I_1 = \frac{E_{in}}{R_1} \tag{10-6}$$

and

$$I_2 = -\frac{E_{out}}{R_2} \tag{10-7}$$

If the amplifier input impedance is infinite, no current can flow into the inverting terminal; so $I_1 = I_2$. If we equate expressions (10-6) and (10-7) and solve for the overall transfer function, we obtain

$$\frac{E_{\text{out}}}{E_{\text{in}}} = -\frac{R_2}{R_1} \tag{10-8}$$

The circuit amplification is determined directly from the ratio of two resistors and is independent of the amplifier itself. Also the input impedance is R_1 and the output impedance is zero.

Multiple inputs can be summed at the inverting input terminal as a direct result of the virtual ground effect. The triple-input summing amplifier of figure 10-14 has the following output based on superposition:

$$E_{\text{out}} = -\frac{R_2}{R_{1a}} E_a - \frac{R_2}{R_{1b}} E_b - \frac{R_2}{R_{1c}} E_c \tag{10-9}$$

A noninverting amplifier can be configured using the circuit of figure 10-15. Since the differential voltage between the amplifier input terminal is zero, the voltage across R_1 is E_{in}. Since R_1 and R_2 form a voltage divider, we can state

$$\frac{E_{\text{out}}}{E_{\text{in}}} = \frac{R_1 + R_2}{R_1} \tag{10-10}$$

or the more popular form

$$\frac{E_{\text{out}}}{E_{\text{in}}} = 1 + \frac{R_2}{R_1} \tag{10-11}$$

where the input impedance is infinite and the output impedance is zero.

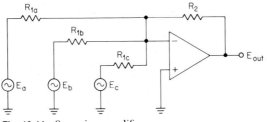

Fig. 10-14 Summing amplifier.

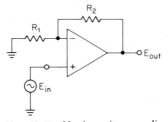

Fig. 10-15 Noninverting amplifier.

If we set R_1 to infinity and R_2 to zero, the gain becomes unity, which corresponds to the voltage follower configuration of figure 10-16.

The applications of operational amplifiers are by no means restricted to summing and amplification. If R_2 in figure 10-13, for example, were replaced by a capacitor, the circuit would serve as an integrator. Alternately a capacitor for R_1 would result in a differentiator. Nonlinear functions can be performed by introducing nonlinear elements in the feedback paths.

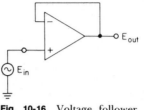

Fig. 10-16   Voltage follower.

Analysis of Nonideal Amplifiers

The fundamental equation for the closed-loop gain of the noninverting amplifier of figure 10-15 is given by

$$A_c = \frac{A_0}{1 + A_0\,\beta} \tag{10-12}$$

where A_0 is the amplifier's open-loop gain and β is the feedback factor. This expression should be familiar to those who have studied feedback systems or servo theory. The closed-loop gain of the inverting amplifier structure of figure 10-13 is expressed as

$$A_c = \frac{A_0\,(\beta - 1)}{1 + A_0\,\beta} \tag{10-13}$$

In both cases, the feedback factor, which corresponds to the portion of the output that is fed back to the input, is determined by

$$\beta = \frac{R_1}{R_1 + R_2} \tag{10-14}$$

Let us examine the term $1 + A_0\,\beta$ corresponding to the denominator of the closed-loop gain expressions. The open-loop gain of practical amplifiers is neither infinite nor real (zero phase shift). The magnitude and phase of A_0 will be a function of frequency. If at some frequency $A_0\,\beta$ were equal to -1, the denominator of equations (10-12) and (10-13) would vanish. The closed-loop gain then becomes infinite, which implies an oscillatory condition.

To prevent oscillations, the amplifier open-loop gain must be band-limited so that the product $A_0\,\beta$ is less than 1 below the frequency where the amplifier phase shift reaches 180°. This is achieved by introducing a gain roll-off beginning at low frequencies and continuing at a 6-dB per octave rate. This technique of ensuring stability is called "frequency compensation." It is evident from the closed-loop gain equations that for high closed-loop gains, β is diminished so that less frequency compensation will be required. Conversely the voltage follower will need the most compensation.

Effects of Finite Amplifier Gain The most critical factor in most op amp applications is the open-loop gain. In order to maintain stability, the open-loop gain must be band-limited. This is usually accomplished by introducing a real pole at a low frequency so that the gain rolls off at 6 dB per octave.

A typical open-loop gain plot is shown in figure 10-17a. The gain has two breakpoints. The low-frequency breakpoint is caused by a real pole resulting from the frequency compensation. The output phase lag increases to 45° at the breakpoint and asymptotically approaches 90° as the frequency is increased. The amplitude response rolls off at a rate of 6 dB per octave.

Another amplifier breakpoint occurs near 100 kHz. Above this second pole, the gain rolls off at 12 dB per octave, an additional 45° of phase shift occurs, and the asymptotic phase limit becomes 180°. The corresponding phase curve is shown in figure 10-17b.

Most operational amplifiers have a built-in frequency compensation network. These values correspond to the worst case for guaranteed stability, which is the voltage follower configuration. The penalty paid for this convenience is that in the case of high closed-loop gain ($\beta \ll 1$), the open-loop gain is less

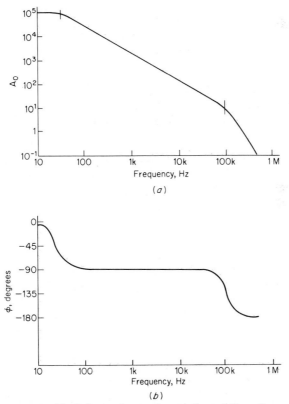

Fig. 10-17 Typical open-loop gain and phase shift vs. frequency: (a) open-loop gain vs. frequency; (b) open-loop phase shift vs. frequency.

than it really could be and yet retain stability; so some unnecessary closed-loop gain degradation will occur.

Amplifiers are frequently specified in terms of their unity gain-bandwidth product, i.e., the frequency at which the open-loop gain is unity.

To determine the effects of open-loop gain on closed-loop gain, let us substitute equation (10-14) into equations (10-12) and (10-13) for the noninverting and inverting amplifiers. The resulting gain expressions are

Noninverting amplifier:
$$A_c = \frac{1 + \dfrac{R_2}{R_1}}{\dfrac{1}{A_0}\left(1 + \dfrac{R_2}{R_1}\right) + 1} \qquad (10\text{-}15)$$

and

Inverting amplifier:
$$A_c = \frac{\dfrac{R_2}{R_1}}{\dfrac{1}{A_0}\left(1 + \dfrac{R_2}{R_1}\right) + 1} \qquad (10\text{-}16)$$

If A_0 were infinite, both denominators would reduce to unity. Equations (10-15) and (10-16) would then be equal to equations (10-11) and 10-8), the fundamental expressions for the gain of a noninverting and inverting amplifier.

To reduce the degrading effect of open-loop gain upon closed-loop gain, A_0 should be much higher than the desired A_c. Since the open-loop phase shift is usually 90° over most of the band of interest, the error term in the denominator of equations (10-15) and (10-16) is in quadrature with unity. This relationship minimizes the effect of open-loop gain upon closed-loop gain when the feedback network is purely resistive. An open-loop to closed-loop gain ratio of 10:1 will result in an error of only 0.5%, which is more than adequate for most requirements.

Finite open-loop gain also affects circuit input and output impedance. The input impedance of the noninverting amplifier can be derived as

$$R_{in} = (1 + A_0 \beta) R_i \qquad (10\text{-}17)$$

and the output impedance is given by

$$R_{out} = \frac{R_0}{A_0 \beta} \qquad (10\text{-}18)$$

where R_i and R_0 are the amplifiers' input and output impedance, respectively. Usually the closed-loop input and output impedance will have a negligible effect on circuit operation with moderate values of $A_0 \beta$.

Practical Amplifier Considerations

DC Offsets An inverting amplifier is shown in figure 10-18 with the addition of two bias currents I_a and I_b and an input offset voltage V_{dc}. These effects occur because of amplifier imperfections.

The input offset voltage results in an offset voltage at the output equal to V_{dc} times the closed-loop gain. The polarity of V_{dc} is random and is typically less than 10 mV.

The two bias currents are nearly equal except for a small difference or offset current I_0. In the circuit of figure 10-18, I_a produces an additional error voltage at the input given by $I_a\,R_{eq}$ where

$$R_{eq} = \frac{R_1\,R_2}{R_1 + R_2} \qquad (10\text{-}19)$$

The noninverting input bias current I_b has no effect. To minimize the effect of I_a, high values of R_{eq} should be avoided.

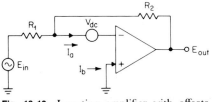

Fig. 10-18 Inverting amplifier with offsets.

A more commonly used approach involves introducing a resistor having the value R_{eq} between the noninverting input and ground as shown in figure 10-19. This has no effect on the overall gain. However, a DC offset voltage of $I_b\,R_{eq}$ is introduced at the noninverting input. Since the amplifier is a differential device, the net error voltage due to the offset currents is $(I_a - I_b)\,R_{eq}$ or $I_0\,R_{eq}$. Since I_a and I_b are each typically 80 nA and the offset current I_0 is in the range of 20 nA, a 4:1 reduction is obtained.

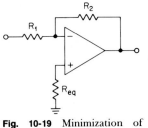

Fig. 10-19 Minimization of DC offsets due to bias currents.

In the case of the noninverting amplifier configuration of figure 10-15, the ratio R_2/R_1 is determined by gain equation (10-11). However, the actual values of R_1 and R_2 are nearly arbitrary and can be chosen so that their parallel combination (R_{eq}) is approximately equal to the DC loading on the noninverting input, i.e., the parallel combination of all resistances connected between the noninverting input and AC ground.

In general, for moderate closed-loop gains or AC-coupled circuits, the effects of DC offsets are of little consequence. For critical applications such as precision active low-pass filters for the recovery of low-level DC components, the methods discussed can be implemented. Some amplifiers will provide an input terminal for nulling of output DC offsets.

Slew-Rate Limiting   When an operational amplifier is used to provide a high-level sine wave at a high frequency, the output will tend to approach a triangular waveform. This effect is called "slew-rate limiting," and the slope of the triangular waveform is referred to as the "slew rate." Typical values range from 1 to 100 V/μs, depending upon the amplifier type. If the peak-to-peak output voltage is small, the effects of slew rate will be minimized. Bandwidth can also be extended by using the minimum frequency compensation required for the given closed-loop gain.

Power-Supply Considerations   Most IC op amps require dual supply voltages, typically ranging from ±5 to ±18 V. The actual voltage magnitude is not critical, provided that the output swing from the amplifier is a few volts less than the supply voltages to avoid clipping. Maximum supply-voltage ratings should not be exceeded, of course.

To ensure stability, both the positive and negative power-supply voltages should be adequately bypassed to ground for high frequencies. Bypass capacitors can be 0.1 μF ceramic or 10 μF tantalum in most cases. Regulated supplies are desirable but not required. Most amplifiers have a typical supply-voltage sensitivity of 30 μV/V; i.e., a 1-V power-supply variation will result in only a 30-μV change reflected at the amplifier input.

In many cases only a single positive supply voltage is available. Dual-voltage-type op amps can still be used by generating a reference voltage V_r which replaces the circuit's ground connections. The amplifier's negative power terminal is returned to ground and the positive power terminal is connected to the positive supply voltage.

The reference voltage should be midway between the positive supply voltage and ground and should be provided from a low-impedance source. A convenient means of generating V_r directly from the positive supply is shown in figure 10-20 using the voltage follower configuration.

Input signals referenced to ground must be AC-coupled and then superimposed upon V_r. The signals are decoupled at the output to restore the ground reference. A low-pass filter design using this method will not pass low-frequency components near DC.

This technique is illustrated in figure 10-21 as applied to some previous design examples. The circuit of figure 10-20 is used to generate V_r and all amplifiers are powered using single-ended power supplies.

Survey of Popular Amplifier Types   IC operational amplifiers range from economical general-purpose devices to the costlier high-performance units. Al-

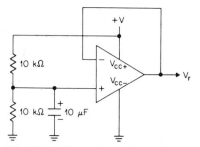

Fig. 10-20   Generation of a reference voltage.

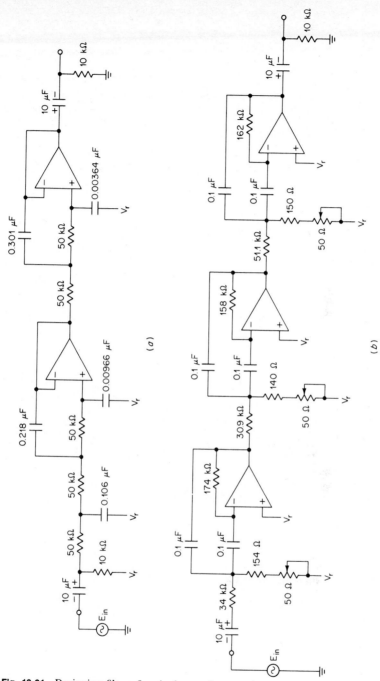

Fig. 10-21 Designing filters for single supply operation: (a) low-pass filter of example 3-7; (b) bandpass filter of example 5-12.

TABLE 10-7 Operational Amplifier Selection Guide

Type		Bias Current, nA	Offset Current, nA	Offset Voltage, mV	Unity Gain Bandwidth, MHz	Slew Rate at Unity Gain, V/μs	Comments (see Notes)
Single	μA 709	300	100	2	10	0.3	2
	μA 741	80	20	1	1	0.5	1, 3
	μA 748	80	20	1	1	0.5	2, 3
	LM 301	70	3	2	1	0.5	2, 3
	LM 307	70	3	2	1	0.5	1
	LF 355	30 pA	3 pA	3	2.5	5	1, 3, 4
	LF 356	30 pA	3 pA	3	4.5	12	1, 3, 4
	LF 357	30 pA	3 pA	3	20	50	1, 3, 4
Dual	MC 1458	80	20	1	1	0.5	1
	RC 4558	40	5	0.5	3	1	1
	μA 747	80	20	1	1	0.5	1, 3
	MC 1437	400	50	1	10	0.3	2
Quad	MC 3401	50			5	0.6	1
	MC 4741	80	20	2	1	0.5	1
	MC 3471	20 pA			10	20	1, 4

NOTES: 1. Internally compensated.
2. Externally compensated.
3. Offset null provision.
4. JFET input.

though many different types are available to choose from, certain units are more popular than others. These devices are listed in table 10-7 along with their typical characteristics.

The parameters in this table are intended as a general guideline and correspond to commercial-grade op amps at a 25°C ambient temperature. The manufacturer's data sheet will provide more detailed and specific information.

Single Op Amps   The first general-purpose IC op amp was the μA709. Although many major advances have been made, some by a few orders of magnitude, this amplifier still serves as a basic "workhorse." Three components are required for frequency compensation, as illustrated in figure 10-22a. The corresponding open- and closed-loop gains are shown in figure 10-22b and c, respectively.

The μA741 is a second-generation improvement over the μA709. The device features offset-voltage null capability, short-circuit protection, and a wider common-mode and differential voltage range than the μA709. Frequency compensation is built in, resulting in a unity gain bandwidth of 1 MHz. The μA748 is similar except that the frequency compensation is externally provided using a single capacitor. The frequency characteristics can then be tailored to different requirements. The LM301 is a high-performance version of the μA748. The LM307 is a high-performance version of the μA741 except that no offset adjustment capability is provided.

The LF355 through 357 series of op amps have extremely low bias and offset currents because of JFETs at the input. The LF357 has a unity gain bandwidth of 20 MHz and is especially suited for wide-band applications. Offset null adjustment capability is provided for all three devices.

Dual Op Amps Dual unit packages have evolved to reduce space and cost. The amplifiers are also electrically symmetrical. The MC1458 type consists of a pair of μA741 type amplifiers except that offset null capability is not provided. This dual amplifier in the eight-pin mini-DIP package is almost universally used for active filters in the audio range. Channel separation between amplifier halves is typically 120 dB. The RC4558 dual amplifier is similar except that the unity gain bandwidth is 3 MHz instead of 1 MHz.

The μA747 is also a dual 741 type except that offset adjustment capability is present. As a result the eight-pin mini-DIP package is not possible; so this device is normally provided in the 14-pin DIP configuration. The MC1437 is a dual 709 amplifier and also uses the 14-pin structure.

Quad Op Amps When several amplifiers in a single package is desired, a variety of quad-packaged op amps is available. Many of these devices are particularly aimed for the active filter market.

The MC3401 consists of four independent frequency-compensated amplifiers that are specifically designed to operate from a single supply. They differ from conventional op amps in that the noninverting input function is internally achieved by using a current mirror instead of a differential input configuration. The unity gain bandwidth is 5 MHz.

The MC4741 is a quad amplifier comparable in performance to the 741 type

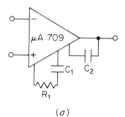

(a)

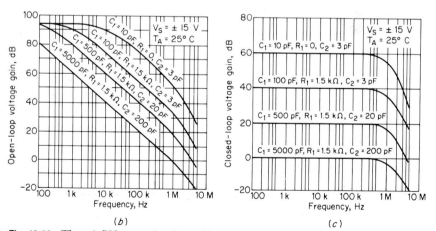

(b)   (c)

Fig. 10-22 The μA 709 operational amplifier: (a) basic configuration; (b) open-loop frequency response for various values of compensation; (c) frequency response for various closed-loop gains. (*Courtesy Fairchild Corp.*)

except that no offset null capability is provided. A 10-MHz unity gain bandwidth can be obtained using the MC3471 quad op amp featuring JFET inputs.

REFERENCES

Fairchild Semiconductor, "Linear Integrated Circuits Data Book," Fairchild Semiconductor, Mountain View, Calif.

Lindquist, C. S., "Active Network Design," Steward and Sons, Long Beach, Calif., 1977.

Stout, D. F., and Kaufman, M., "Handbook of Operational Amplifier Circuit Design," McGraw-Hill Book Company, New York, 1976.

Texas Instruments, "The Linear Control Circuits Data Book," Texas Instruments, Dallas, Tex., 1976.

11

Introduction to Digital Filters

11.1 THEORY OF DIGITAL FILTERS

The concept of digital filters differs dramatically from the conventional filters covered in this book. The term "digital filtering" involves a computational process performed upon a sequence of digital numbers representing an input signal. This digital sequence is obtained by sampling the input analog signal using an A/D converter. After processing, a D/A converter restores the analog signal format. A simplified block diagram is shown in figure 11-1.

Fig. 11-1 Simplified digital filter.

The intermediate computational process can be made to correspond to a desired form of filtering such as low-pass, high-pass, bandpass, and band-reject.

The digital approach to filtering offers several advantages over analog techniques. These include stable and repeatable performance, greater flexibility via programming capability, filtering at frequencies not practical for analog methods, and probably the most important attribute: the possibility of time sharing the major implementation elements.

Digital filters have not yet gained widespread usage, mainly because of the economics of implementation. Nevertheless the applications of digital filters are rapidly expanding as LSI technological improvements permit increased packing density and cost reduction.

This chapter provides a brief introduction to digital filtering. It is hoped that this discussion will motivate readers to further their knowledge in this area.

The Sampling Process and the z Transform

The analog signal $V_1(t)$ of figure 11-2a is a continuous function of time. If we sample this signal every T intervals, we obtain the train of pulses shown in figure 11-2b. The height of each pulse represents the amplitude of $V_1(t)$ at the sampling instant. These pulses can then be represented by binary numbers as shown in figure 11-2c. This entire process is performed by the A/D converter. The sampling rate must be at least twice the highest-frequency component contained in $V_1(t)$ so that the signal can be theoretically recovered from its discrete

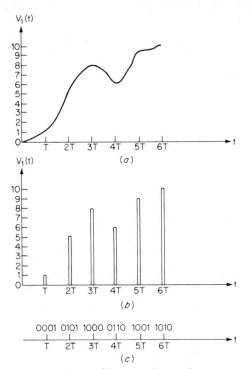

Fig. 11-2 The sampling operation: (a) input analog signal; (b) sample pulse train; (c) digitized samples.

samples. A sufficient number of bits should be used to obtain good accuracy. The input analog signal is normally first passed through a low-pass presampling filter to restrict the higher-frequency component.

Continuous time signals can be represented by the Laplace transform. In a similar manner the z transform can be used to mathematically describe a sampled-data system. The z transform for a sampled input signal is given by

$$V^*(z) = \sum_{n=0}^{\infty} V(nT)z^{-n} \qquad (11\text{-}1)$$

where $V(nT)$ are the discrete analog samples, z^{-n} is a delay operator, n is an integer representing the number of previous unit delays, and T is the sampling interval. The z transform is related to the Laplace transform by the substitution $z = e^{sT}$.

Difference Equations

In the linear analog domain, networks are described by transfer functions, which in turn can be expressed in the form of linear differential equations. In a similar manner digital filters can be described using linear difference equations. A difference equation defines an output in terms of the present input and previous values of input and output. The basic elements are delays, constant multipliers, and summing elements. The general form of a difference equation is

$$y(nT) = \sum_{i=0}^{N} A_i x(nT - iT) + \sum_{i=1}^{M} B_i y(nT - iT) \tag{11-2}$$

where $x(nT)$ represents the present input sample and $y(nT)$ is the present output sample. The previous input and output samples are represented by $x(iT)$ and $y(iT)$, respectively.

The z transform of this general difference equation can be derived as

$$y(z) = x(z) \sum_{i=0}^{N} A_i z^{-i} + y(z) \sum_{i=1}^{M} B_i z^{-i} \tag{11-3}$$

This equation mathematically states that the present output is the sum of the present and past inputs multiplied by each respective coefficient A_i plus the past outputs each multiplied by the respective coefficient B_i. Equation (11-3) can also be expressed in the form of a transfer function, which results in

$$T(z) = \frac{y(z)}{x(z)} = \frac{\displaystyle\sum_{i=0}^{N} A_i z^{-i}}{1 - \displaystyle\sum_{i=1}^{M} B_i z^{-i}} \tag{11-4}$$

First- and Second-Order Digital Filters

Let us now expand the general form of the difference equation with M and N set to 1. We can then obtain from equation (11-2) a first-order difference equation

$$y(nT) = A_0 x(nT) + A_1 x (nT - T) + B_1 y(nT - T) \tag{11-5}$$

If we perform the z transformation on this expression, the result becomes

$$E_o(z) = A_0 E_i(z) + A_1 z^{-1} E_i(z) + B_1 z^{-1} E_o(z) \tag{11-6}$$

where E_i and E_o are the input and output, respectively. We then solve for the first-order transfer function, which is

$$\frac{E_o(z)}{E_i(z)} = \frac{A_0 z + A_1}{z - B_1} \tag{11-7}$$

To realize a transfer function for a sampled-data system, certain elements are required to perform specific operations. These building blocks are represented in figure 11-3.

The unit delay operation shown in figure 11-3a performs the function of delaying the input samples by T seconds. Since all samples are uniformly spaced and occur at integer multiples of T, we frequently refer to T as a "unit delay" interval.

The summing element of figure 11-3b adds or subtracts two or more signals. Signs adjacent to each input indicate the operation performed.

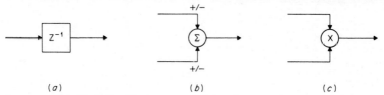

(a)   (b)   (c)

Fig. 11-3 Digital building blocks: (a) unit delay element; (b) summing element; (c) multiplier element.

The multiplier element of figure 11-3c provides an output which is the product of the two inputs. One of the inputs may be a constant.

Let us consider the configuration of figure 11-4. If we solve for E_a, the resulting nodal voltage is

$$E_a = E_i + B_1 E_a z^{-1} \tag{11-8}$$

The voltage at the output of the final summing element is found to be

$$E_o = A_o E_a + A_1 E_a z^{-1} \tag{11-9}$$

If we substitute equation (11-8) into equation (11-9) and solve for the overall transfer function E_o/E_i, we obtain equation (11-7). Figure 11-4 thus represents a first-order digital filter building block having one pole and one zero.

A second-order difference equation can also be obtained by expanding equation (11-2), which results in

$$y(nt) = A_o x(nT) + A_1 x(nT - T) + A_2 x(nT - 2T)$$
$$+ B_1 y(nT - T) + B_2 y(nT - 2T) \tag{11-10}$$

The corresponding z transform is

$$E_o(z) = A_o E_i(z) + A_1 z^{-1} E_i(z) + A_2 z^{-2} E_i(z) + B_1 z^{-1} E_o(z) + B_2 z^{-2} E_o(z) \tag{11-11}$$

This equation may also be rewritten in the form of a transfer function which is

$$\frac{E_o(z)}{E_i(z)} = \frac{A_o z^2 + A_1 z + A_2}{z^2 - B_1 z - B_2} \tag{11-12}$$

The second-order transfer function may be implemented by the circuit of figure 11-5. First- and second-order sections are cascaded to obtain a digital filter of the required complexity.

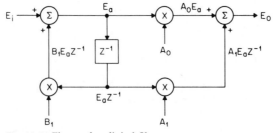

Fig. 11-4 First-order digital filter.

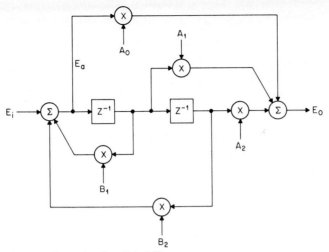

Fig. 11-5 Second-order digital filter.

The Second-Order Design Procedure

Let us consider a second-order analog transfer function of the general form

$$T(s) = \frac{\alpha_1 s + \alpha_2}{s^2 + \beta_1 s + \beta_2} \tag{11-13}$$

Leon and Bass (see references) derived a series of closed-form equations so that the A and B coefficients of the digital realization of figure 11-5 can be directly computed. These expressions are as follows:

$$A_0 = \alpha_1 + \frac{2\alpha_2}{\beta_1} \tag{11-14}$$

$$A_1 = 2e^{-T\beta_1/2} \left[\left(\frac{\alpha_1}{2} + \frac{\alpha_2}{\beta_1} \right) \cos \sqrt{\beta_2 - \frac{\beta_1^2}{4}} \right.$$
$$\left. + \left(\frac{\alpha_1}{\beta_1} \sqrt{\beta_2 - \frac{\beta_1^2}{4}} \right) \sin \sqrt{\beta_2 - \frac{\beta_1^2}{4}} \right] \tag{11-15}$$

$$A_2 = -2e^{-T\beta_1/2} \left(\cos \sqrt{\beta_2 - \frac{\beta_1^2}{4}} \right) \tag{11-16}$$

$$B_1 = -A_2 \tag{11-17}$$

$$B_2 = -e^{-T\beta_1} \tag{11-18}$$

The sampling interval T can be determined from

$$T = \frac{2\pi}{\omega_s} \tag{11-19}$$

where ω_s is the sampling rate in radians per second.

The Bandpass Transfer Function A second-order bandpass transfer function was defined in chapter 5 as

$$T(s) = \frac{Hs}{s^2 + \dfrac{\omega_r}{Q}s + \omega_r^2} \tag{5-48}$$

where ω_r is the radian resonant frequency, Q is the bandpass Q (i.e., $f_r/3$ dB BW), and H is a constant multiplier.

If we compare this transfer function with equation (11-13), the design parameters become

$$\alpha_1 = H \tag{11-20}$$

$$\alpha_2 = 0 \tag{11-21}$$

$$\beta_1 = \frac{\omega_r}{Q} \tag{11-22}$$

$$\beta_2 = \omega_r^2 \tag{11-23}$$

The resulting design equations are

$$A_0 = H \tag{11-24}$$

$$A_1 = 2e^{-T\omega_r/2Q}\left[\frac{H}{2}\cos\left(\omega_r\sqrt{1 - \frac{1}{4Q^2}}\right)\right.$$
$$\left. + \left(HQ\sqrt{1 - \frac{1}{4Q^2}}\right)\sin\left(\omega_r\sqrt{1 - \frac{1}{4Q^2}}\right)\right] \tag{11-25}$$

$$A_2 = -2e^{-T\omega_r/2Q}\left[\cos\left(\omega_r\sqrt{1 - \frac{1}{4Q^2}}\right)\right] \tag{11-26}$$

$$B_1 = -A_2 \tag{11-27}$$

$$B_2 = -e^{-T\omega_r/Q} \tag{11-28}$$

Equations (11-24) through (11-28) can now be directly used to calculate the digital coefficients for a second-order bandpass filter in terms of H, Q, and ω_r.

Example 11-1

REQUIRED: A second-order bandpass digital filter is required having the parameters:

$$\omega_r = 0.1$$
$$Q = 10$$
$$\omega_s = 1$$
$$H = 0.01$$

Determine the coefficients corresponding to the circuit of figure 11-5.

RESULT: The design parameters are computed as follows:

$$T = 2\pi \tag{11-19}$$

$$\alpha_1 = 0.01 \tag{11-20}$$

$$\alpha_2 = 0 \tag{11-21}$$

$$\beta_1 = 0.01 \tag{11-22}$$

$$\beta_2 = 0.01 \tag{11-23}$$

The coefficients can be found from

$$A_0 = 0.01 \tag{11-24}$$

$$A_1 = 0.028941 \tag{11-25}$$

$$A_2 = -1.92848 \tag{11-26}$$

$$B_1 = 1.92848 \tag{11-27}$$

$$B_2 = 0.939101 \tag{11-28}$$

11.2 NONRECURSIVE DIGITAL FILTERS

The digital filter configurations of the previous section are of the recursive type. A recursive realization is one in which the present output is a function of the present and past inputs and previous output values. A nonrecursive realization is one where the present output is a function of the present and past values of the input only.

A nonrecursive type of digital filter is shown in figure 11-6. This configuration in the analog form is sometimes called a transversal filter or tapped delay line. The output consists of the sum of the present output and N previous outputs, all weighted (multiplied) by individual coefficients.

A technique of determining the tap coefficients is the impulse response method. This approach assumes that the desired impulse response is known and is approximated in a piecewise manner by the finite sequence

$$h(n) \quad (n = 0, 1, 2, \ldots, N) \tag{11-29}$$

The circuit of figure 11-6 can then be used where the tap weights are the $h(n)$ coefficients.

A more desirable design procedure is not based on the impulse response but instead uses the desired frequency response as the starting point.

The basic premise of this method is the fact that all digital filters have a frequency response which is periodic with a period of ω_s, the sampling rate. This occurs because of the nature of the A/D converter and the sampling process.

A periodic function can be approximated by a Fourier series. If we specify a

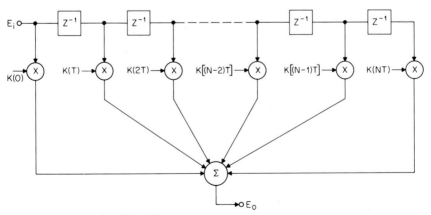

Fig. 11-6 Nonrecursive digital filter.

filter requirement in terms of magnitude only over the frequency range $0 \leq \omega \leq \omega_s/2$, a series of Fourier coefficients can be determined by the formula

$$\alpha_i = \frac{2}{\omega_s} \int_0^{\omega_s/2} A(\omega) \cos(iT\omega) \, d\omega \qquad (11\text{-}30)$$

where $A(\omega)$ is the magnitude function and T is the sampling interval ($T = 2\pi/\omega_s$). A total of $(2N+1)$ coefficients are calculated by letting $i = -N, -N+1, -N+2, \ldots, -1, 0, 1, 2, \ldots, N$.

The set of Fourier coefficients can be directly used as the design coefficients for the nonrecursive configuration of figure 11-6.

$$K(0) = \alpha_N$$
$$K(T) = \alpha_{N-1}$$
$$K(2T) = \alpha_{N-2}$$
$$\vdots$$
$$K[(N-2)T] = \alpha_2$$
$$K[(N-1)T] = \alpha_1$$
$$K(NT) = \alpha_0$$
$$K[(N+1)T] = \alpha_{-1}$$
$$K[(N+2)T] = \alpha_{-2}$$
$$\vdots$$
$$K[(2N-2)T] = \alpha_{-N+2}$$
$$K[(2N-1)T] = \alpha_{-N+1}$$
$$K[(2N)T] = \alpha_{-N}$$

These α coefficients have even symmetry, that is, $\alpha_i = \alpha_{-i}$. Almost half of the multiplier coefficients would be duplicated if the circuit of figure 11-6 were used. The circuit of figure 11-7 takes advantage of this symmetry and uses

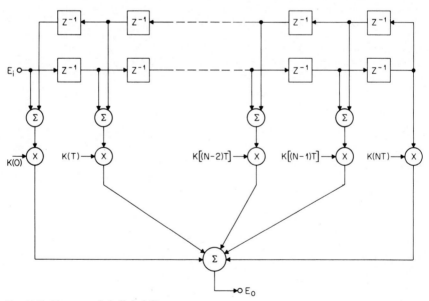

Fig. 11-7 Nonrecursive digital filter.

fewer multipliers at the expense of additional summing elements. However, multiplication is a relatively slow and costly implementation compared with the summing process.

Only α_0 through α_N need be calculated. These coefficients directly define the digital filter design coefficients, i.e., $K(0) = \alpha_N$ through $K(N) = \alpha_0$.

Digital filters designed in this manner exhibit a constant delay of NT s. This feature is usually highly desirable, since the filter's transient properties will be optimum.

Another feature is guaranteed stability. Since no feedback occurs, oscillatory conditions are impossible.

The accuracy of the match to the desired response is determined by the number of Fourier series terms, i.e., the magnitude of N. Since because of practical necessity the series must be truncated, some error will exist. In the case where an ideal rectangular response function is to be approximated, an effect known as Gibbs' phenomenon occurs where an overshoot appears in the frequency response. Even as N approaches infinity, the overshoot will be approximately 9%. This effect is shown in figure 11-8.

Techniques are available called "windowing." These methods involve modification of the Fourier coefficients in such a manner that the approximation is optimized.

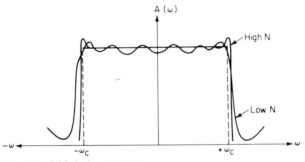

Fig. 11-8 Gibbs' phenomenon.

11.3 DIGITAL FILTER HARDWARE

Standard MSI chips can be interconnected to perform the various operations. LSI chips have recently been introduced that function as complete multipliers. Entire second-order recursive sections have also been integrated onto a single LSI chip by a few LSI manufacturers.

Since a large number of adders and multipliers would be required for a nonrecursive digital filter, time sharing of these devices becomes necessary. The configuration of figure 11-9 reduces the computational hardware to a single multiplier and summer. All the required tap coefficients are stored in a ROM and sequentially loaded into one input of a multiplier element. Simultaneously, the other input to the multiplier is stepped through the delay elements. The storage register in conjunction with the dual-input summing element performs the function of the multiple-input summer of figure 11-7.

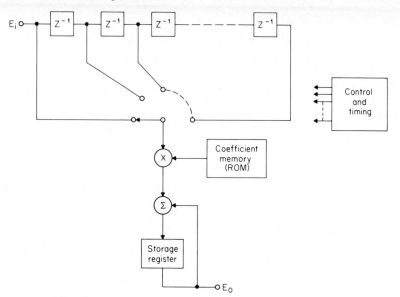

Fig. 11-9 Time-shared nonrecursive digital filter structure.

REFERENCES

Echard, J. D., and Boorstyn, R. R., "Digital Filtering for Radar Signal Processing Applications," IEEE Transactions on Audio and Electroacoustics, Vol. Au-20, No. 1, March 1972.

Leon, B. J., and Bass, S. C., "Practical Digital Filter Design," Lecture Notes, Center for Professional Advancement, East Brunswick, N.J., 1977.

Mick, J. R., "Digital Signal Processing Handbook," Advanced Micro Devices Inc., Sunnyvale, Calif., 1976.

Stanley, W. D., "Digital Signal Processing," Reston Publishing Co., Inc., Reston, Va., 1975.

12

Normalized Filter Design Tables

TABLE 12-1   Butterworth Pole Locations

Order n	Real Part $-\alpha$	Imaginary Part $\pm j\beta$
2	0.7071	0.7071
3	0.5000	0.8660
	1.0000	
4	0.9239	0.3827
	0.3827	0.9239
5	0.8090	0.5878
	0.3090	0.9511
	1.0000	
6	0.9659	0.2588
	0.7071	0.7071
	0.2588	0.9659
7	0.9010	0.4339
	0.6235	0.7818
	0.2225	0.9749
	1.0000	
8	0.9808	0.1951
	0.8315	0.5556
	0.5556	0.8315
	0.1951	0.9808
9	0.9397	0.3420
	0.7660	0.6428
	0.5000	0.8660
	0.1737	0.9848
	1.0000	
10	0.9877	0.1564
	0.8910	0.4540
	0.7071	0.7071
	0.4540	0.8910
	0.1564	0.9877

TABLE 12-2   Butterworth LC Element Values*

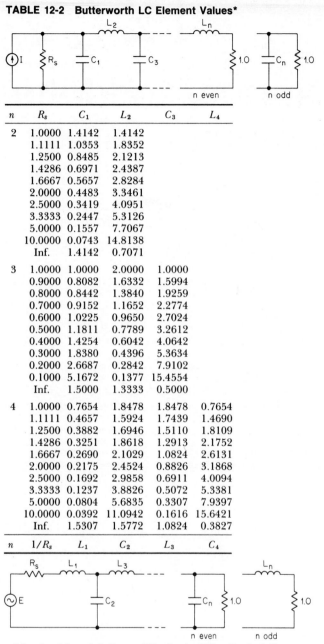

n	R_s	C_1	L_2	C_3	L_4
2	1.0000	1.4142	1.4142		
	1.1111	1.0353	1.8352		
	1.2500	0.8485	2.1213		
	1.4286	0.6971	2.4387		
	1.6667	0.5657	2.8284		
	2.0000	0.4483	3.3461		
	2.5000	0.3419	4.0951		
	3.3333	0.2447	5.3126		
	5.0000	0.1557	7.7067		
	10.0000	0.0743	14.8138		
	Inf.	1.4142	0.7071		
3	1.0000	1.0000	2.0000	1.0000	
	0.9000	0.8082	1.6332	1.5994	
	0.8000	0.8442	1.3840	1.9259	
	0.7000	0.9152	1.1652	2.2774	
	0.6000	1.0225	0.9650	2.7024	
	0.5000	1.1811	0.7789	3.2612	
	0.4000	1.4254	0.6042	4.0642	
	0.3000	1.8380	0.4396	5.3634	
	0.2000	2.6687	0.2842	7.9102	
	0.1000	5.1672	0.1377	15.4554	
	Inf.	1.5000	1.3333	0.5000	
4	1.0000	0.7654	1.8478	1.8478	0.7654
	1.1111	0.4657	1.5924	1.7439	1.4690
	1.2500	0.3882	1.6946	1.5110	1.8109
	1.4286	0.3251	1.8618	1.2913	2.1752
	1.6667	0.2690	2.1029	1.0824	2.6131
	2.0000	0.2175	2.4524	0.8826	3.1868
	2.5000	0.1692	2.9858	0.6911	4.0094
	3.3333	0.1237	3.8826	0.5072	5.3381
	5.0000	0.0804	5.6835	0.3307	7.9397
	10.0000	0.0392	11.0942	0.1616	15.6421
	Inf.	1.5307	1.5772	1.0824	0.3827

n	$1/R_s$	L_1	C_2	L_3	C_4

* Reprinted from A. I. Zverev, "Handbook of Filter Synthesis," John Wiley and Sons, New York, 1967.

TABLE 12-2 *(Continued)*

n	R_s	C_1	L_2	C_3	L_4	C_5	L_6	C_7
5	1.0000	0.6180	1.6180	2.0000	1.6180	0.6180		
	0.9000	0.4416	1.0265	1.9095	1.7562	1.3887		
	0.8000	0.4698	0.8660	2.0605	1.5443	1.7380		
	0.7000	0.5173	0.7313	2.2849	1.3326	2.1083		
	0.6000	0.5860	0.6094	2.5998	1.1255	2.5524		
	0.5000	0.6857	0.4955	3.0510	0.9237	3.1331		
	0.4000	0.8378	0.3877	3.7357	0.7274	3.9648		
	0.3000	1.0937	0.2848	4.8835	0.5367	5.3073		
	0.2000	1.6077	0.1861	7.1849	0.3518	7.9345		
	0.1000	3.1522	0.0912	14.0945	0.1727	15.7103		
	Inf.	1.5451	1.6944	1.3820	0.8944	0.3090		
6	1.0000	0.5176	1.4142	1.9319	1.9319	1.4142	0.5176	
	1.1111	0.2890	1.0403	1.3217	2.0539	1.7443	1.3347	
	1.2500	0.2445	1.1163	1.1257	2.2389	1.5498	1.6881	
	1.4286	0.2072	1.2363	0.9567	2.4991	1.3464	2.0618	
	1.6667	0.1732	1.4071	0.8011	2.8580	1.1431	2.5092	
	2.0000	0.1412	1.6531	0.6542	3.3687	0.9423	3.0938	
	2.5000	0.1108	2.0275	0.5139	4.1408	0.7450	3.9305	
	3.3333	0.0816	2.6559	0.3788	5.4325	0.5517	5.2804	
	5.0000	0.0535	3.9170	0.2484	8.0201	0.3628	7.9216	
	10.0000	0.0263	7.7053	0.1222	15.7855	0.1788	15.7375	
	Inf.	1.5529	1.7593	1.5529	1.2016	0.7579	0.2588	
7	1.0000	0.4450	1.2470	1.8019	2.0000	1.8019	1.2470	0.4450
	0.9000	0.2985	0.7111	1.4043	1.4891	2.1249	1.7268	1.2961
	0.8000	0.3215	0.6057	1.5174	1.2777	2.3338	1.5461	1.6520
	0.7000	0.3571	0.5154	1.6883	1.0910	2.6177	1.3498	2.0277
	0.6000	0.4075	0.4322	1.9284	0.9170	3.0050	1.1503	2.4771
	0.5000	0.4799	0.3536	2.2726	0.7512	3.5532	0.9513	3.0640
	0.4000	0.5899	0.2782	2.7950	0.5917	4.3799	0.7542	3.9037
	0.3000	0.7745	0.2055	3.6706	0.4373	5.7612	0.5600	5.2583
	0.2000	1.1448	0.1350	5.4267	0.2874	8.5263	0.3692	7.9079
	0.1000	2.2571	0.0665	10.7004	0.1417	16.8222	0.1823	15.7480
	Inf.	1.5576	1.7988	1.6588	1.3972	1.0550	0.6560	0.2225
n	$1/R_s$	L_1	C_2	L_3	C_4	L_5	C_6	L_7

TABLE 12-2   *(Continued)*

n	R_s	C_1	L_2	C_3	L_4	C_5	L_6	C_7	L_8	C_9	L_{10}
8	1.0000	0.3902	1.1111	1.6629	1.9616	1.9616	1.6629	1.1111	0.3902		
	1.1111	0.2075	0.7575	0.9925	1.6362	1.5900	2.1612	1.7092	1.2671		
	1.2500	0.1774	0.8199	0.8499	1.7779	1.3721	2.3874	1.5393	1.6246		
	1.4286	0.1513	0.9138	0.7257	1.9852	1.1760	2.6879	1.3490	2.0017		
	1.6667	0.1272	1.0455	0.6102	2.2740	0.9912	3.0945	1.1530	2.4524		
	2.0000	0.1042	1.2341	0.5003	2.6863	0.8139	3.6678	0.9558	3.0408		
	2.5000	0.0822	1.5201	0.3945	3.3106	0.6424	4.5308	0.7594	3.8825		
	3.3333	0.0608	1.9995	0.2919	4.3563	0.4757	5.9714	0.5650	5.2400		
	5.0000	0.0400	2.9608	0.1921	6.4523	0.3133	8.8538	0.3732	7.8952		
	10.0000	0.0198	5.8479	0.0949	12.7455	0.1547	17.4999	0.1846	15.7510		
	Inf.	1.5607	1.8246	1.7287	1.5283	1.2588	0.9371	0.5776	0.1951		
9	1.0000	0.3473	1.0000	1.5321	1.8794	2.0000	1.8794	1.5321	1.0000	0.3473	
	0.9000	0.2242	0.5388	1.0835	1.1859	1.7905	1.6538	2.1796	1.6930	1.2447	
	0.8000	0.2434	0.4623	1.1777	1.0200	1.9542	1.4336	2.4189	1.5318	1.6033	
	0.7000	0.2719	0.3954	1.3162	0.8734	2.1885	1.2323	2.7314	1.3464	1.9812	
	0.6000	0.3117	0.3330	1.5092	0.7361	2.5124	1.0410	3.1516	1.1533	2.4328	
	0.5000	0.3685	0.2735	1.7846	0.6046	2.9734	0.8565	3.7426	0.9579	3.0223	
	0.4000	0.4545	0.2159	2.2019	0.4775	3.6706	0.6771	4.6310	0.7624	3.8654	
	0.3000	0.5987	0.1600	2.9006	0.3539	4.8373	0.5022	6.1128	0.5680	5.2249	
	0.2000	0.8878	0.1054	4.3014	0.2333	7.1750	0.3312	9.0766	0.3757	7.8838	
	0.1000	1.7558	0.0521	8.5074	0.1153	14.1930	0.1638	17.9654	0.1862	15.7504	
	Inf.	1.5628	1.8424	1.7772	1.6202	1.4037	1.1408	0.8414	0.5155	0.1736	
10	1.0000	0.3129	0.9080	1.4142	1.7820	1.9754	1.9754	1.7820	1.4142	0.9080	0.3129
	1.1111	0.1614	0.5924	0.7853	1.3202	1.3230	1.8968	1.6956	2.1883	1.6785	1.2267
	1.2500	0.1388	0.6452	0.6762	1.4400	1.1420	2.0779	1.4754	2.4377	1.5245	1.5861
	1.4286	0.1190	0.7222	0.5797	1.6130	0.9802	2.3324	1.2712	2.7592	1.3431	1.9646
	1.6667	0.1004	0.8292	0.4891	1.8528	0.8275	2.6825	1.0758	3.1895	1.1526	2.4169
	2.0000	0.0825	0.9818	0.4021	2.1943	0.6808	3.1795	0.8864	3.7934	0.9588	3.0072
	2.5000	0.0652	1.2127	0.3179	2.7108	0.5384	3.9302	0.7018	4.7002	0.7641	3.8512
	3.3333	0.0484	1.5992	0.2358	3.5754	0.3995	5.1858	0.5211	6.2118	0.5700	5.2122
	5.0000	0.0319	2.3740	0.1556	5.3082	0.2636	7.7010	0.3440	9.2343	0.3775	7.8738
	10.0000	0.0158	4.7005	0.0770	10.5104	0.1305	15.2505	0.1704	18.2981	0.1872	15.7481
	Inf.	1.5643	1.8552	1.8121	1.6869	1.5100	1.2921	1.0406	0.7626	0.4654	0.1564
n	$1/R_s$	L_1	C_2	L_3	C_4	L_5	C_6	L_7	C_8	L_9	C_{10}

TABLE 12-3   Butterworth Uniform Dissipation Network*

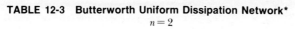

$n = 2$

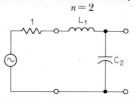

d	L_1	C_2	α_0, dB
0	0.7071	1.414	0
0.05	0.7609	1.410	0.614
0.10	0.8236	1.398	1.22
0.15	0.8974	1.374	1.83
0.20	0.9860	1.340	2.42
0.25	1.094	1.290	2.99
0.30	1.228	1.223	3.53
0.35	1.400	1.138	4.05
0.40	1.628	1.034	4.52
0.45	1.944	0.9083	4.94
0.50	2.414	0.7630	5.30
0.55	3.183	0.5989	5.59
0.60	4.669	0.4188	5.82
0.65	8.756	0.2267	5.96

d	C_2	L_1	α_0, dB

* By permission of P. R. Geffe.

TABLE 12-4   Butterworth Uniform Dissipation Network*

$$n = 3$$

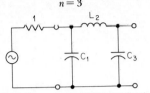

d	C_1	L_2	C_3	α_0, dB
0	0.5000	1.333	1.500	0
0.05	0.5405	1.403	1.457	0.868
0.10	0.5882	1.481	1.402	1.73
0.15	0.6452	1.567	1.334	2.60
0.20	0.7143	1.667	1.250	3.45
0.25	0.8000	1.786	1.149	4.30
0.30	0.9091	1.939	1.026	5.15
0.35	1.053	2.164	0.8743	5.98
0.40	1.250	2.581	0.6798	6.82
0.45	1.538	3.806	0.4126	7.66
d	L_3	C_2	L_1	α_0, dB

* By permission of P. R. Geffe.

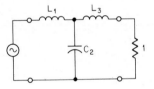

TABLE 12-5   Butterworth Uniform Dissipation Network*

$n = 4$

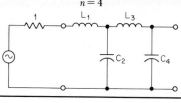

d	L_1	C_2	L_3	C_4	$\alpha_0,$ dB
0	0.3827	1.082	1.577	1.531	0
0.05	0.4144	1.156	1.636	1.454	1.13
0.10	0.4518	1.240	1.701	1.362	2.27
0.15	0.4967	1.339	1.777	1.250	3.39
0.20	0.5515	1.459	1.879	1.113	4.51
0.25	0.6199	1.609	2.039	0.9400	5.63
0.30	0.7077	1.812	2.384	0.7099	6.73
0.35	0.8243	2.124	3.848	0.3651	7.82
d	C_4	L_3	C_2	L_1	$\alpha_0,$ dB

* By permission of P. R. Geffe.

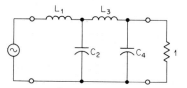

TABLE 12-6 Butterworth Uniform Dissipation Network*

$n = 5$

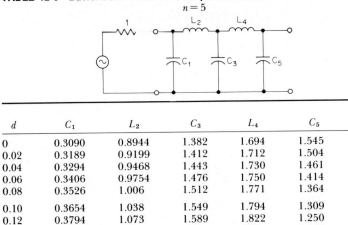

d	C_1	L_2	C_3	L_4	C_5	$\alpha_0,$ dB
0	0.3090	0.8944	1.382	1.694	1.545	0
0.02	0.3189	0.9199	1.412	1.712	1.504	0.562
0.04	0.3294	0.9468	1.443	1.730	1.461	1.12
0.06	0.3406	0.9754	1.476	1.750	1.414	1.69
0.08	0.3526	1.006	1.512	1.771	1.364	2.25
0.10	0.3654	1.038	1.549	1.794	1.309	2.81
0.12	0.3794	1.073	1.589	1.822	1.250	3.37
0.14	0.3943	1.111	1.633	1.854	1.184	3.93
0.16	0.4104	1.151	1.681	1.894	1.113	4.48
0.18	0.4281	1.195	1.734	1.946	1.034	5.04
0.20	0.4472	1.243	1.796	2.018	0.9452	5.59
0.22	0.4681	1.296	1.867	2.124	0.8434	6.15
0.24	0.4911	1.354	1.953	2.300	0.7242	6.70
0.26	0.5165	1.419	2.061	2.631	0.5798	7.25
0.28	0.5446	1.493	2.204	3.453	0.3965	7.79
0.30	0.5760	1.578	2.409	8.084	0.1476	8.34
d	L_5	C_4	L_3	C_2	L_1	$\alpha_0,$ dB

* By permission of P. R. Geffe.

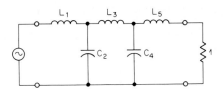

TABLE 12-7   Butterworth Uniform Dissipation Network*

$$n = 6$$

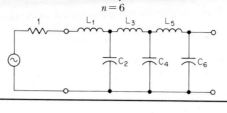

d	L_1	C_2	L_3	C_4	L_5	C_6	α_0, dB
0	0.2588	0.7579	1.202	1.553	1.759	1.533	0
0.02	0.2671	0.7804	1.232	1.581	1.727	1.502	0.671
0.04	0.2760	0.8043	1.264	1.611	1.786	1.446	1.34
0.06	0.2854	0.8297	1.297	1.643	1.802	1.386	2.01
0.08	0.2955	0.8569	1.333	1.679	1.821	1.321	2.68
0.10	0.3064	0.8860	1.372	1.714	1.844	1.250	3.35
0.12	0.3181	0.9172	1.413	1.755	1.874	1.171	4.02
0.14	0.3307	0.9508	1.458	1.802	1.917	1.083	4.69
0.16	0.3443	0.9871	1.508	1.860	1.979	0.9839	5.30
0.18	0.3594	1.027	1.558	1.923	2.080	0.8690	6.00
0.20	0.3754	1.070	1.621	2.008	2.258	0.7313	6.68
0.22	0.3931	1.117	1.690	2.122	2.646	0.5586	7.34
d	C_6	L_5	C_4	L_3	C_2	L_1	α_0, dB

* By permission of P. R. Geffe.

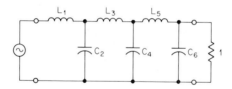

TABLE 12-8   Butterworth Uniform Dissipation Network*

$$n = 7$$

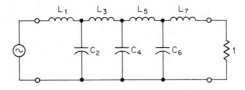

d	C_1	L_2	C_3	L_4	C_5	L_6	C_7	α_0, dB
0	0.2225	0.6560	1.054	1.397	1.659	1.799	1.588	0
0.02	0.2297	0.6759	1.084	1.428	1.684	1.808	1.496	0.781
0.04	0.2373	0.6972	1.114	1.461	1.712	1.818	1.428	1.56
0.06	0.2454	0.7198	1.146	1.496	1.742	1.832	1.354	2.34
0.08	0.2542	0.7440	1.180	1.533	1.775	1.851	1.274	3.12
0.10	0.2636	0.7699	1.217	1.573	1.813	1.878	1.184	3.90
0.12	0.2739	0.7980	1.254	1.614	1.860	1.923	1.085	4.68
0.14	0.2846	0.8281	1.294	1.659	1.910	1.992	0.9701	5.45
0.16	0.2966	0.8608	1.344	1.715	1.979	2.111	0.8350	6.23
0.18	0.3091	0.8960	1.394	1.778	2.073	2.356	0.6679	7.00
0.20	0.3232	0.9243	1.453	1.862	2.233	3.177	0.4220	7.77
d	L_7	C_6	L_5	C_4	L_3	C_2	L_1	α_0, dB

* By permission of P. R. Geffe.

TABLE 12-9   Butterworth Uniform Dissipation Network*

$n = 8$

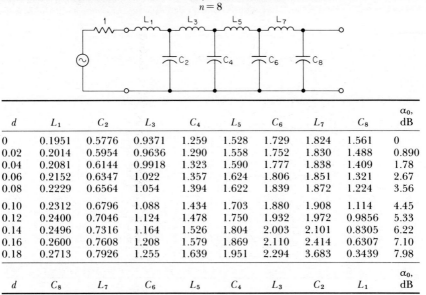

d	L_1	C_2	L_3	C_4	L_5	C_6	L_7	C_8	α_0, dB
0	0.1951	0.5776	0.9371	1.259	1.528	1.729	1.824	1.561	0
0.02	0.2014	0.5954	0.9636	1.290	1.558	1.752	1.830	1.488	0.890
0.04	0.2081	0.6144	0.9918	1.323	1.590	1.777	1.838	1.409	1.78
0.06	0.2152	0.6347	1.022	1.357	1.624	1.806	1.851	1.321	2.67
0.08	0.2229	0.6564	1.054	1.394	1.622	1.839	1.872	1.224	3.56
0.10	0.2312	0.6796	1.088	1.434	1.703	1.880	1.908	1.114	4.45
0.12	0.2400	0.7046	1.124	1.478	1.750	1.932	1.972	0.9856	5.33
0.14	0.2496	0.7316	1.164	1.526	1.804	2.003	2.101	0.8305	6.22
0.16	0.2600	0.7608	1.208	1.579	1.869	2.110	2.414	0.6307	7.10
0.18	0.2713	0.7926	1.255	1.639	1.951	2.294	3.683	0.3439	7.98

d	C_8	L_7	C_6	L_5	C_4	L_3	C_2	L_1	α_0, dB

* By permission of P. R. Geffe.

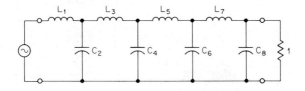

TABLE 12-10   Butterworth Uniform Dissipation Network*

$$n = 9$$

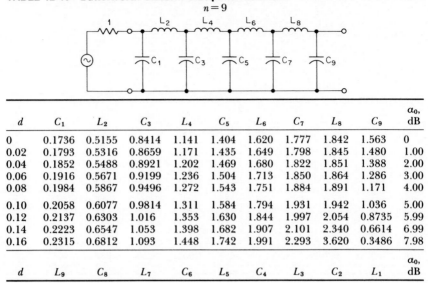

d	C_1	L_2	C_3	L_4	C_5	L_6	C_7	L_8	C_9	$\alpha_0,$ dB
0	0.1736	0.5155	0.8414	1.141	1.404	1.620	1.777	1.842	1.563	0
0.02	0.1793	0.5316	0.8659	1.171	1.435	1.649	1.798	1.845	1.480	1.00
0.04	0.1852	0.5488	0.8921	1.202	1.469	1.680	1.822	1.851	1.388	2.00
0.06	0.1916	0.5671	0.9199	1.236	1.504	1.713	1.850	1.864	1.286	3.00
0.08	0.1984	0.5867	0.9496	1.272	1.543	1.751	1.884	1.891	1.171	4.00
0.10	0.2058	0.6077	0.9814	1.311	1.584	1.794	1.931	1.942	1.036	5.00
0.12	0.2137	0.6303	1.016	1.353	1.630	1.844	1.997	2.054	0.8735	5.99
0.14	0.2223	0.6547	1.053	1.398	1.682	1.907	2.101	2.340	0.6614	6.99
0.16	0.2315	0.6812	1.093	1.448	1.742	1.991	2.293	3.620	0.3486	7.98
d	L_9	C_8	L_7	C_6	L_5	C_4	L_3	C_2	L_1	$\alpha_0,$ dB

* By permission of P. R. Geffe.

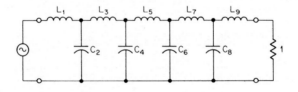

TABLE 12-11 **Butterworth Uniform Dissipation Network***

$n = 10$

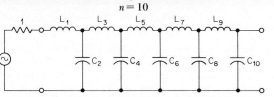

d	L_1	C_2	L_3	C_4	L_5	C_6	L_7	C_8	L_9	C_{10}	$\alpha_0,$ dB
0	0.1564	0.4654	0.7626	1.041	1.292	1.510	1.687	1.812	1.855	1.564	0
0.02	0.1614	0.4800	0.7854	1.069	1.324	1.541	1.714	1.831	1.855	1.471	1.11
0.04	0.1669	0.4956	0.8096	1.099	1.357	1.574	1.744	1.853	1.860	1.367	2.22
0.06	0.1726	0.5123	0.8353	1.132	1.392	1.610	1.777	1.882	1.875	1.249	3.33
0.08	0.1788	0.5301	0.8629	1.166	1.430	1.648	1.814	1.920	1.910	1.114	4.44
0.10	0.1854	0.5493	0.8924	1.203	1.471	1.692	1.860	1.976	1.991	0.9508	5.55
0.12	0.1926	0.5698	0.9242	1.243	1.516	1.741	1.918	2.067	2.201	0.7409	6.65
0.14	0.2003	0.5921	0.9584	1.286	1.566	1.798	1.997	2.239	3.051	0.4349	7.76
d	C_{10}	L_9	C_8	L_7	C_6	L_5	C_4	L_3	C_2	L_1	$\alpha_0,$ dB

* By permission of P. R. Geffe.

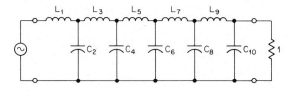

TABLE 12-12   Butterworth Lossy-L Network*
$$n = 2$$

d	L_1	C_2
0	0.7071	1.414
0.05	0.7330	1.364
0.10	0.7609	1.314
0.15	0.7910	1.264
0.20	0.8236	1.214
0.25	0.8589	1.164
0.30	0.8975	1.114
0.35	0.9397	1.064
0.40	0.9860	1.014
0.45	1.037	0.9642
0.50	1.094	0.9142
0.55	1.157	0.8642
0.60	1.228	0.8142
0.65	1.309	0.7642

* By permission of P. R. Geffe.

TABLE 12-13   Butterworth Lossy-L Network*
$$n = 3$$

d	C_1	L_2	C_3
0	0.5000	1.333	1.500
0.05	0.5128	1.403	1.390
0.10	0.5263	1.480	1.284
0.15	0.5405	1.565	1.182
0.20	0.5556	1.660	1.084
0.25	0.5714	1.766	0.9911
0.30	0.5882	1.885	0.9018
0.35	0.6061	2.021	0.8164
0.40	0.6250	2.177	0.7350
0.45	0.6452	2.358	0.6573

* By permission of P. R. Geffe.

TABLE 12-14   Butterworth Lossy-L Network*

$n = 4$

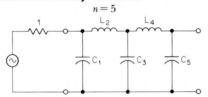

d	L_1	C_2	L_3	C_4
0	0.3827	1.082	1.577	1.531
0.05	0.3979	1.087	1.698	1.362
0.10	0.4144	1.091	1.834	1.205
0.15	0.4323	1.095	1.990	1.061
0.20	0.4518	1.098	2.170	0.9289
0.25	0.4732	1.100	2.380	0.8072
0.30	0.4967	1.102	2.628	0.6955
0.35	0.5227	1.102	2.926	0.5933

* By permission of P. R. Geffe.

TABLE 12-15   Butterworth Lossy-L Network*

$n = 5$

d	C_1	L_2	C_3	L_4	C_5
0	0.3090	0.8944	1.382	1.694	1.545
0.02	0.3129	0.9127	1.369	1.762	1.452
0.04	0.3168	0.9316	1.355	1.834	1.363
0.06	0.3209	0.9514	1.342	1.911	1.278
0.08	0.3251	0.9719	1.327	1.993	1.197
0.10	0.3294	0.9934	1.313	2.080	1.119
0.12	0.3338	1.016	1.298	2.173	1.046
0.14	0.3383	1.039	1.283	2.273	0.9754
0.16	0.3429	1.063	1.268	2.380	0.9086
0.18	0.3477	1.089	1.253	2.494	0.8450
0.20	0.3526	1.116	1.237	2.620	0.7844
0.22	0.3576	1.144	1.221	2.754	0.7269
0.24	0.3628	1.173	1.204	2.901	0.6721
0.26	0.3682	1.204	1.188	3.061	0.6201
0.28	0.3737	1.237	1.171	3.237	0.5076
0.30	0.3794	1.271	1.154	3.431	0.5236

* By permission of P. R. Geffe.

TABLE 12-16   Butterworth Lossy-L Network*

$$n = 6$$

d	L_1	C_2	L_3	C_4	L_5	C_6
0	0.2588	0.7579	1.202	1.553	1.759	1.553
0.02	0.2629	0.7631	1.235	1.519	1.850	1.436
0.04	0.2671	0.7683	1.271	1.485	1.947	1.326
0.06	0.2714	0.7736	1.308	1.451	2.052	1.223
0.08	0.2760	0.7789	1.347	1.417	2.165	1.125
0.10	0.2806	0.7843	1.388	1.383	2.228	1.034
0.12	0.2854	0.7897	1.432	1.349	2.421	0.9487
0.14	0.2904	0.7952	1.478	1.315	2.565	0.8684
0.16	0.2955	0.8007	1.527	1.281	2.723	0.7932
0.18	0.3009	0.8063	1.579	1.248	2.896	0.7227
0.20	0.3064	0.8118	1.634	1.214	3.807	0.6567
0.22	0.3121	0.8174	1.692	1.181	3.298	0.5949

* By permission of P. R. Geffe.

TABLE 12-17   Butterworth Lossy-L Network*

$$n = 7$$

d	C_1	L_2	C_3	L_4	C_5	L_6	C_7
0	0.2225	0.6560	1.054	1.397	1.659	1.799	1.588
0.02	0.2255	0.6688	1.053	1.449	1.602	1.913	1.417
0.04	0.2286	0.6822	1.051	1.504	1.546	2.038	1.288
0.06	0.2318	0.6960	1.048	1.564	1.490	2.173	1.167
0.08	0.2351	0.7104	1.045	1.627	1.436	2.322	1.056
0.10	0.2384	0.7255	1.043	1.694	1.382	2.484	0.9532
0.12	0.2419	0.7412	1.039	1.766	1.330	2.664	0.8581
0.14	0.2454	0.7575	1.036	1.842	1.278	2.862	0.7703
0.16	0.2491	0.7746	1.032	1.924	1.228	3.083	0.6892
0.18	0.2529	0.7924	1.028	2.013	1.178	3.330	0.6144
0.20	0.2568	0.8110	1.024	2.108	1.130	3.609	0.5454

* By permission of P. R. Geffe.

TABLE 12-18 Butterworth Lossy-L Network*

$n = 8$

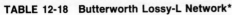

d	L_1	C_2	L_3	C_4	L_5	C_6	L_7	C_8
0	0.1951	0.5776	0.9371	1.259	1.528	1.729	1.824	1.561
0.02	0.1982	0.5829	0.9613	1.243	1.602	1.648	1.963	1.398
0.04	0.2014	0.5884	0.9868	1.227	1.680	1.569	2.116	1.249
0.06	0.2047	0.5939	1.014	1.211	1.764	1.493	2.285	1.113
0.08	0.2081	0.5996	1.042	1.194	1.856	1.419	2.472	0.9894
0.10	0.2116	0.6053	1.071	1.178	1.954	1.347	2.681	0.8768
0.12	0.2152	0.6111	1.102	1.160	2.061	1.278	2.914	0.7743
0.14	0.2190	0.6170	1.134	1.143	2.177	1.211	3.178	0.6810
0.16	0.2229	0.6231	1.169	1.124	2.302	1.147	3.477	0.5962
0.18	0.2270	0.6292	1.206	1.107	2.440	1.084	3.819	0.5191

* By permission of P. R. Geffe.

TABLE 12-19 Butterworth Lossy-L Network*

$n = 9$

d	C_1	L_2	C_3	L_4	C_5	L_6	C_7	L_8	C_9
0	0.1736	0.5155	0.8414	1.141	1.404	1.620	1.777	1.842	1.563
0.02	0.1761	0.5253	0.8432	1.180	1.371	1.716	1.672	2.006	1.377
0.04	0.1786	0.5354	0.8450	1.221	1.338	1.821	1.571	2.189	1.211
0.06	0.1812	0.5460	0.8467	1.264	1.304	1.934	1.474	2.393	1.061
0.08	0.1839	0.5570	0.8483	1.310	1.271	2.058	1.383	2.623	0.9261
0.10	0.1866	0.5684	0.8497	1.359	1.238	2.193	1.294	2.884	0.8054
0.12	0.1894	0.5802	0.8510	1.412	1.204	2.342	1.211	3.180	0.6971
0.14	0.1924	0.5926	0.8522	1.467	1.171	2.505	1.132	3.521	0.6001
0.16	0.1954	0.6054	0.8533	1.527	1.137	2.686	1.057	3.917	0.5132

* By permission of P. R. Geffe.

TABLE 12-20   Butterworth Lossy-L Network*

$n = 10$

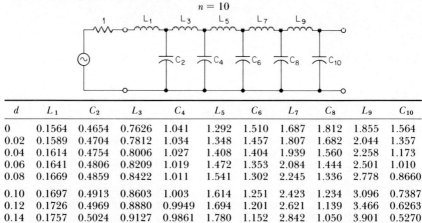

d	L_1	C_2	L_3	C_4	L_5	C_6	L_7	C_8	L_9	C_{10}
0	0.1564	0.4654	0.7626	1.041	1.292	1.510	1.687	1.812	1.855	1.564
0.02	0.1589	0.4704	0.7812	1.034	1.348	1.457	1.807	1.682	2.044	1.357
0.04	0.1614	0.4754	0.8006	1.027	1.408	1.404	1.939	1.560	2.258	1.173
0.06	0.1641	0.4806	0.8209	1.019	1.472	1.353	2.084	1.444	2.501	1.010
0.08	0.1669	0.4859	0.8422	1.011	1.541	1.302	2.245	1.336	2.778	0.8660
0.10	0.1697	0.4913	0.8603	1.003	1.614	1.251	2.423	1.234	3.096	0.7387
0.12	0.1726	0.4969	0.8880	0.9949	1.694	1.201	2.621	1.139	3.466	0.6263
0.14	0.1757	0.5024	0.9127	0.9861	1.780	1.152	2.842	1.050	3.901	0.5270

* By permission of P. R. Geffe.

TABLE 12-21   Butterworth Active Low-Pass Values*

Order n	C_1	C_2	C_3
2	1.414	0.7071	
3	3.546	1.392	0.2024
4	1.082	0.9241	
	2.613	0.3825	
5	1.753	1.354	0.4214
	3.235	0.3090	
6	1.035	0.9660	
	1.414	0.7071	
	3.863	0.2588	
7	1.531	1.336	0.4885
	1.604	0.6235	
	4.493	0.2225	
8	1.020	0.9809	
	1.202	0.8313	
	1.800	0.5557	
	5.125	0.1950	
9	1.455	1.327	0.5170
	1.305	0.7661	
	2.000	0.5000	
	5.758	0.1736	
10	1.012	0.9874	
	1.122	0.8908	
	1.414	0.7071	
	2.202	0.4540	
	6.390	0.1563	

* Reprinted from "Electronics," August 18, 1969, McGraw-Hill, Inc., 1969.

TABLE 12-22   0.01-dB Chebyshev Pole Locations

Order n	Real Part $-\alpha$	Imaginary Part $\pm j\beta$
2	0.6743	0.7075
3	0.4233	0.8663
	0.8467	
4	0.6762	0.3828
	0.2801	0.9241
5	0.5120	0.5879
	0.1956	0.9512
	0.6328	
6	0.5335	0.2588
	0.3906	0.7072
	0.1430	0.9660
7	0.4393	0.4339
	0.3040	0.7819
	0.1085	0.9750
	0.4876	
8	0.4268	0.1951
	0.3618	0.5556
	0.2418	0.8315
	0.08490	0.9808
9	0.3686	0.3420
	0.3005	0.6428
	0.1961	0.8661
	0.06812	0.9848
	0.3923	

TABLE 12-23   0.1-dB Chebyshev Pole Locations

Order n	Real Part $-\alpha$	Imaginary Part $\pm j\beta$
2	0.6104	0.7106
3	0.3490 0.6979	0.8684
4	0.2177 0.5257	0.9254 0.3833
5	0.3842 0.1468 0.4749	0.5884 0.9521
6	0.3916 0.2867 0.1049	0.2590 0.7077 0.9667
7	0.3178 0.2200 0.0785 0.3528	0.4341 0.7823 0.9755
8	0.3058 0.2592 0.1732 0.06082	0.1952 0.5558 0.8319 0.9812
9	0.2622 0.2137 0.1395 0.04845 0.2790	0.3421 0.6430 0.8663 0.9852

TABLE 12-24   0.25-dB Chebyshev Pole Locations

Order n	Real Part $-\alpha$	Imaginary Part $\pm j\beta$
2	0.5621	0.7154
3	0.3062	0.8712
	0.6124	
4	0.4501	0.3840
	0.1865	0.9272
5	0.3247	0.5892
	0.1240	0.9533
	0.4013	
6	0.3284	0.2593
	0.2404	0.7083
	0.08799	0.9675
7	0.2652	0.4344
	0.1835	0.7828
	0.06550	0.9761
	0.2944	
8	0.2543	0.1953
	0.2156	0.5561
	0.1441	0.8323
	0.05058	0.9817
9	0.2176	0.3423
	0.1774	0.6433
	0.1158	0.8667
	0.04021	0.9856
	0.2315	

TABLE 12-25   0.5-dB Chebyshev Pole Locations

Order n	Real Part $-\alpha$	Imaginary Part $\pm j\beta$
2	0.5129	0.7225
3	0.2683	0.8753
	0.5366	
4	0.3872	0.3850
	0.1605	0.9297
5	0.2767	0.5902
	0.1057	0.9550
	0.3420	
6	0.2784	0.2596
	0.2037	0.7091
	0.07459	0.9687
7	0.2241	0.4349
	0.1550	0.7836
	0.05534	0.9771
	0.2487	
8	0.2144	0.1955
	0.1817	0.5565
	0.1214	0.8328
	0.04264	0.9824
9	0.1831	0.3425
	0.1493	0.6436
	0.09743	0.8671
	0.03383	0.9861
	0.1949	

TABLE 12-26   1-dB Chebyshev Pole Locations

Order n	Real Part $-\alpha$	Imaginary Part $\pm j\beta$
2	0.4508	0.7351
3	0.2257	0.8822
	0.4513	
4	0.3199	0.3868
	0.1325	0.9339
5	0.2265	0.5918
	0.08652	0.9575
	0.2800	
6	0.2268	0.2601
	0.1660	0.7106
	0.06076	0.9707
7	0.1819	0.4354
	0.1259	0.7846
	0.04494	0.9785
	0.2019	
8	0.1737	0.1956
	0.1473	0.5571
	0.09840	0.8337
	0.03456	0.9836
9	0.1482	0.3427
	0.1208	0.6442
	0.07884	0.8679
	0.02739	0.9869
	0.1577	

TABLE 12-27   0.01-dB Chebyshev LC Element Values*

n	R_s	C_1	L_2	C_3	L_4
2	1.1007	1.3472	1.4829		
	1.1111	1.2472	1.5947		
	1.2500	0.9434	1.9974		
	1.4286	0.7591	2.3442		
	1.6667	0.6091	2.7496		
	2.0000	0.4791	3.2772		
	2.5000	0.3634	4.0328		
	3.3333	0.2590	5.2546		
	5.0000	0.1642	7.6498		
	10.0000	0.0781	14.7492		
	Inf.	1.4118	0.7415		
3	1.0000	1.1811	1.8214	1.1811	
	0.9000	1.0917	1.6597	1.4802	
	0.8000	1.0969	1.4431	1.8057	
	0.7000	1.1600	1.2283	2.1653	
	0.6000	1.2737	1.0236	2.5984	
	0.5000	1.4521	0.8294	3.1644	
	0.4000	1.7340	0.6452	3.9742	
	0.3000	2.2164	0.4704	5.2800	
	0.2000	3.1934	0.3047	7.8338	
	0.1000	6.1411	0.1479	15.3899	
	Inf.	1.5012	1.4330	0.5905	
4	1.1000	0.9500	1.9382	1.7608	1.0457
	1.1111	0.8539	1.9460	1.7439	1.1647
	1.2500	0.6182	2.0749	1.5417	1.6170
	1.4286	0.4948	2.2787	1.3336	2.0083
	1.6667	0.3983	2.5709	1.1277	2.4611
	2.0000	0.3156	2.9943	0.9260	3.0448
	2.5000	0.2418	3.6406	0.7293	3.8746
	3.3333	0.1744	4.7274	0.5379	5.2085
	5.0000	0.1121	6.9102	0.3523	7.8126
	10.0000	0.0541	13.4690	0.1729	15.5100
	Inf.	1.5287	1.6939	1.3122	0.5229
n	$1/R_s$	L_1	C_2	L_3	C_4

* Reprinted from A. I. Zverev, "Handbook of Filter Synthesis," John Wiley and Sons, New York, 1967.

TABLE 12-27 *(Continued)*

n	R_s	C_1	L_2	C_3	L_4	C_5	L_6	C_7
5	1.0000	0.9766	1.6849	2.0366	1.6849	0.9766		
	0.9000	0.8798	1.4558	2.1738	1.6412	1.2739		
	0.8000	0.8769	1.2350	2.3785	1.4991	1.6066		
	0.7000	0.9263	1.0398	2.6582	1.3228	1.9772		
	0.6000	1.0191	0.8626	3.0408	1.1345	2.4244		
	0.5000	1.1658	0.6985	3.5835	0.9421	3.0092		
	0.4000	1.3983	0.5442	4.4027	0.7491	3.8453		
	0.3000	1.7966	0.3982	5.7721	0.5573	5.1925		
	0.2000	2.6039	0.2592	8.5140	0.3679	7.8257		
	0.1000	5.0406	0.1266	16.7406	0.1819	15.6126		
	Inf.	1.5466	1.7950	1.6449	1.2365	0.4883		
6	1.1007	0.8514	1.7956	1.8411	2.0266	1.6312	0.9372	
	1.1111	0.7597	1.7817	1.7752	2.0941	1.6380	1.0533	
	1.2500	0.5445	1.8637	1.4886	2.4025	1.5067	1.5041	
	1.4286	0.4355	2.0383	1.2655	2.7346	1.3318	1.8987	
	1.6667	0.3509	2.2978	1.0607	3.1671	1.1451	2.3568	
	2.0000	0.2786	2.6781	0.8671	3.7683	0.9536	2.9483	
	2.5000	0.2139	3.2614	0.6816	4.6673	0.7606	3.7899	
	3.3333	0.1547	4.2448	0.5028	6.1631	0.5676	5.1430	
	5.0000	0.0997	6.2227	0.3299	9.1507	0.3760	7.7852	
	10.0000	0.0483	12.1707	0.1623	18.1048	0.1865	15.5950	
	Inf.	1.5510	1.8471	1.7897	1.5976	1.1904	0.4686	
7	1.0000	0.9127	1.5947	2.0021	1.8704	2.0021	1.5947	0.9127
	0.9000	0.8157	1.3619	2.0886	1.7217	2.2017	1.5805	1.2060
	0.8000	0.8111	1.1504	2.2618	1.5252	2.4647	1.4644	1.5380
	0.7000	0.8567	0.9673	2.5158	1.3234	2.8018	1.3066	1.9096
	0.6000	0.9430	0.8025	2.8720	1.1237	3.2496	1.1310	2.3592
	0.5000	1.0799	0.6502	3.3822	0.9276	3.8750	0.9468	2.9478
	0.4000	1.2971	0.5072	4.1563	0.7350	4.8115	0.7584	3.7900
	0.3000	1.6692	0.3716	5.4540	0.5459	6.3703	0.5682	5.1476
	0.2000	2.4235	0.2423	8.0565	0.3604	9.4844	0.3776	7.8019
	0.1000	4.7006	0.1186	15.8718	0.1784	18.8179	0.1879	15.6523
	Inf.	1.5593	1.8671	1.8657	1.7651	1.5633	1.1610	0.4564
n	$1/R_s$	L_1	C_2	L_3	C_4	L_5	C_6	L_7

TABLE 12-27 *(Continued)*

n	R_s	C_1	L_2	C_3	L_4	C_5	L_6	C_7	L_8	C_9	L_{10}
8	1.1007	0.8145	1.7275	1.7984	2.0579	1.8695	1.9796	1.5694	0.8966		
	1.1111	0.7248	1.7081	1.7239	2.1019	1.8259	2.0595	1.5827	1.0111		
	1.2500	0.5176	1.7772	1.4315	2.3601	1.5855	2.4101	1.4754	1.4597		
	1.4286	0.4138	1.9422	1.2141	2.6686	1.3723	2.7734	1.3142	1.8544		
	1.6667	0.3336	2.1896	1.0169	3.0808	1.1660	3.2393	1.1369	2.3136		
	2.0000	0.2650	2.5533	0.8313	3.6598	0.9639	3.8820	0.9518	2.9073		
	2.5000	0.2036	3.1118	0.6537	4.5303	0.7653	4.8393	0.7627	3.7524		
	3.3333	0.1474	4.0539	0.4826	5.9828	0.5697	6.4287	0.5718	5.1118		
	5.0000	0.0951	5.9495	0.3170	8.8889	0.3770	9.6002	0.3804	7.7668		
	10.0000	0.0462	11.6509	0.1562	17.6067	0.1870	19.1009	0.1895	15.6158		
	Inf.	1.5588	1.8848	1.8988	1.8556	1.7433	1.5391	1.1412	0.4483		
9	1.0000	0.8854	1.5513	1.9614	1.8616	2.0717	1.8616	1.9614	1.5513	0.8854	
	0.9000	0.7886	1.3192	2.0330	1.6941	2.2249	1.7402	2.1774	1.5478	1.1764	
	0.8000	0.7834	1.1127	2.1959	1.4930	2.4614	1.5603	2.4565	1.4423	1.5076	
	0.7000	0.8273	0.9353	2.4404	1.2924	2.7808	1.3662	2.8093	1.2927	1.8793	
	0.6000	0.9109	0.7761	2.7852	1.0962	3.2140	1.1688	3.2747	1.1233	2.3295	
	0.5000	1.0436	0.6290	3.2805	0.9045	3.8249	0.9710	3.9223	0.9436	2.9193	
	0.4000	1.2542	0.4910	4.0329	0.7167	4.7444	0.7739	4.8900	0.7582	3.7637	
	0.3000	1.6151	0.3599	5.2951	0.5325	6.2792	0.5780	6.4989	0.5697	5.1254	
	0.2000	2.3468	0.2349	7.8274	0.3518	9.3504	0.3835	9.7114	0.3797	7.7882	
	0.1000	4.5556	0.1150	15.4334	0.1743	18.5641	0.1908	19.3382	0.1895	15.6645	
	Inf.	1.5646	1.8884	1.9242	1.8977	1.8425	1.7261	1.5217	1.1273	0.4427	
10	1.1007	0.7970	1.6930	1.7690	2.0395	1.8827	2.0724	1.8529	1.9472	1.5380	0.8773
	1.1111	0.7083	1.6714	1.6921	2.0763	1.8281	2.1308	1.8167	2.0310	1.5541	0.9910
	1.2500	0.5049	1.7353	1.4005	2.3184	1.5706	2.4371	1.5953	2.3952	1.4574	1.4381
	1.4286	0.4037	1.8958	1.1871	2.6178	1.3552	2.7830	1.3895	2.7685	1.3027	1.8327
	1.6667	0.3255	2.1375	0.9942	3.0205	1.1497	3.2370	1.1863	3.2448	1.1300	2.2923
	2.0000	0.2586	2.4932	0.8128	3.5878	0.9497	3.8698	0.9849	3.9004	0.9484	2.8867
	2.5000	0.1988	3.0398	0.6394	4.4418	0.7538	4.8173	0.7849	4.8757	0.7617	3.7333
	3.3333	0.1440	3.9619	0.4723	5.8678	0.5612	6.3951	0.5863	6.4939	0.5722	5.0955
	5.0000	0.0930	5.8175	0.3103	8.7220	0.3715	9.5486	0.3893	9.7217	0.3814	7.7563
	10.0000	0.0451	11.3993	0.1530	17.2866	0.1844	19.0046	0.1938	19.3905	0.1904	15.6234
	Inf.	1.5625	1.8978	1.9323	1.9288	1.8907	1.8309	1.7128	1.5088	1.1173	0.4386
n	$1/R_s$	L_1	C_2	L_3	C_4	L_5	C_6	L_7	C_8	L_9	C_{10}

TABLE 12-28   0.1-dB Chebyshev LC Element Values*

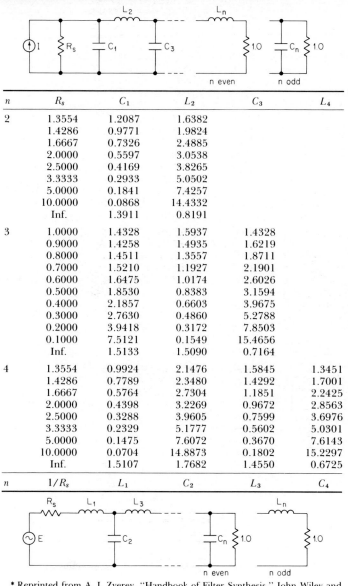

n	R_s	C_1	L_2	C_3	L_4
2	1.3554	1.2087	1.6382		
	1.4286	0.9771	1.9824		
	1.6667	0.7326	2.4885		
	2.0000	0.5597	3.0538		
	2.5000	0.4169	3.8265		
	3.3333	0.2933	5.0502		
	5.0000	0.1841	7.4257		
	10.0000	0.0868	14.4332		
	Inf.	1.3911	0.8191		
3	1.0000	1.4328	1.5937	1.4328	
	0.9000	1.4258	1.4935	1.6219	
	0.8000	1.4511	1.3557	1.8711	
	0.7000	1.5210	1.1927	2.1901	
	0.6000	1.6475	1.0174	2.6026	
	0.5000	1.8530	0.8383	3.1594	
	0.4000	2.1857	0.6603	3.9675	
	0.3000	2.7630	0.4860	5.2788	
	0.2000	3.9418	0.3172	7.8503	
	0.1000	7.5121	0.1549	15.4656	
	Inf.	1.5133	1.5090	0.7164	
4	1.3554	0.9924	2.1476	1.5845	1.3451
	1.4286	0.7789	2.3480	1.4292	1.7001
	1.6667	0.5764	2.7304	1.1851	2.2425
	2.0000	0.4398	3.2269	0.9672	2.8563
	2.5000	0.3288	3.9605	0.7599	3.6976
	3.3333	0.2329	5.1777	0.5602	5.0301
	5.0000	0.1475	7.6072	0.3670	7.6143
	10.0000	0.0704	14.8873	0.1802	15.2297
	Inf.	1.5107	1.7682	1.4550	0.6725
n	$1/R_s$	L_1	C_2	L_3	C_4

* Reprinted from A. I. Zverev, "Handbook of Filter Synthesis," John Wiley and Sons, New York, 1967.

TABLE 12-28   *(Continued)*

n	R_s	C_1	L_2	C_3	L_4	C_5	L_6	C_7
5	1.0000	1.3013	1.5559	2.2411	1.5559	1.3013		
	0.9000	1.2845	1.4329	2.3794	1.4878	1.4883		
	0.8000	1.2998	1.2824	2.5819	1.3815	1.7384		
	0.7000	1.3580	1.1170	2.8679	1.2437	2.0621		
	0.6000	1.4694	0.9469	3.2688	1.0846	2.4835		
	0.5000	1.6535	0.7777	3.8446	0.9126	3.0548		
	0.4000	1.9538	0.6119	4.7193	0.7333	3.8861		
	0.3000	2.4765	0.4509	6.1861	0.5503	5.2373		
	0.2000	3.5457	0.2950	9.1272	0.3659	7.8890		
	0.1000	6.7870	0.1447	17.9569	0.1820	15.7447		
	Inf.	1.5613	1.8069	1.7659	1.4173	0.6507		
6	1.3554	0.9419	2.0797	1.6581	2.2473	1.5344	1.2767	
	1.4286	0.7347	2.2492	1.4537	2.5437	1.4051	1.6293	
	1.6667	0.5422	2.6003	1.1830	3.0641	1.1850	2.1739	
	2.0000	0.4137	3.0679	0.9575	3.7119	0.9794	2.7936	
	2.5000	0.3095	3.7652	0.7492	4.6512	0.7781	3.6453	
	3.3333	0.2195	4.9266	0.5514	6.1947	0.5795	4.9962	
	5.0000	0.1393	7.2500	0.3613	9.2605	0.3835	7.6184	
	10.0000	0.0666	14.2200	0.1777	18.4267	0.1901	15.3495	
	Inf.	1.5339	1.8838	1.8306	1.7485	1.3937	0.6383	
7	1.0000	1.2615	1.5196	2.2392	1.6804	2.2392	1.5196	1.2615
	0.9000	1.2422	1.3946	2.3613	1.5784	2.3966	1.4593	1.4472
	0.8000	1.2550	1.2449	2.5481	1.4430	2.6242	1.3619	1.6967
	0.7000	1.3100	1.0826	2.8192	1.2833	2.9422	1.2326	2.0207
	0.6000	1.4170	0.9169	3.2052	1.1092	3.3841	1.0807	2.4437
	0.5000	1.5948	0.7529	3.7642	0.9276	4.0150	0.9142	3.0182
	0.4000	1.8853	0.5926	4.6179	0.7423	4.9702	0.7384	3.8552
	0.3000	2.3917	0.4369	6.0535	0.5557	6.5685	0.5569	5.2167
	0.2000	3.4278	0.2862	8.9371	0.3692	9.7697	0.3723	7.8901
	0.1000	6.5695	0.1405	17.6031	0.1838	19.3760	0.1862	15.8127
	Inf.	1.5748	1.8577	1.9210	1.8270	1.7340	1.3786	0.6307
n	$1/R_s$	L_1	C_2	L_3	C_4	L_5	C_6	L_7

TABLE 12-28 *(Continued)*

n	R_s	C_1	L_2	C_3	L_4	C_5	L_6	C_7	L_8	C_9	L_{10}
8	1.3554	0.9234	2.0454	1.6453	2.2826	1.6841	2.2300	1.5091	1.2515		
	1.4286	0.7186	2.2054	1.4350	2.5554	1.4974	2.5422	1.3882	1.6029		
	1.6667	0.5298	2.5459	1.1644	3.0567	1.2367	3.0869	1.1769	2.1477		
	2.0000	0.4042	3.0029	0.9415	3.6917	1.0118	3.7619	0.9767	2.7690		
	2.5000	0.3025	3.6859	0.7365	4.6191	0.7990	4.7388	0.7787	3.6240		
	3.3333	0.2147	4.8250	0.5421	6.1483	0.5930	6.3423	0.5820	4.9811		
	5.0000	0.1364	7.1050	0.3554	9.1917	0.3917	9.5260	0.3863	7.6164		
	10.0000	0.0652	13.9469	0.1749	18.3007	0.1942	19.0437	0.1922	15.3880		
	Inf.	1.5422	1.9106	1.9008	1.9252	1.8200	1.7231	1.3683	0.6258		
9	1.0000	1.2446	1.5017	2.2220	1.6829	2.2957	1.6829	2.2220	1.5017	1.2446	
	0.9000	1.2244	1.3765	2.3388	1.5756	2.4400	1.5870	2.3835	1.4444	1.4297	
	0.8000	1.2361	1.2276	2.5201	1.4365	2.6561	1.4572	2.6168	1.3505	1.6788	
	0.7000	1.2898	1.0670	2.7856	1.2751	2.9647	1.3019	2.9422	1.2248	2.0029	
	0.6000	1.3950	0.9035	3.1653	1.1008	3.3992	1.1304	3.3937	1.0761	2.4264	
	0.5000	1.5701	0.7419	3.7166	0.9198	4.0244	0.9494	4.0377	0.9121	3.0020	
	0.4000	1.8566	0.5840	4.5594	0.7359	4.9750	0.7630	5.0118	0.7382	3.8412	
	0.3000	2.3560	0.4307	5.9781	0.5509	6.5700	0.5736	6.6413	0.5579	5.2068	
	0.2000	3.3781	0.2822	8.8291	0.3661	9.7699	0.3827	9.9047	0.3737	7.8891	
	0.1000	6.4777	0.1386	17.3994	0.1823	19.3816	0.1912	19.6976	0.1873	15.8393	
	Inf.	1.5804	1.8727	1.9584	1.9094	1.9229	1.8136	1.7150	1.3611	0.6223	
10	1.3554	0.9146	2.0279	1.6346	2.2777	1.6963	2.2991	1.6805	2.2155	1.4962	1.2397
	1.4286	0.7110	2.1837	1.4231	2.5425	1.5002	2.5915	1.5000	2.5322	1.3789	1.5903
	1.6667	0.5240	2.5194	1.1536	3.0362	1.2349	3.1229	1.2444	3.0839	1.1717	2.1351
	2.0000	0.3998	2.9713	0.9326	3.6647	1.0089	3.7923	1.0214	3.7669	0.9741	2.7572
	2.5000	0.2993	3.6476	0.7295	4.5843	0.7962	4.7673	0.8090	4.7547	0.7779	3.6136
	3.3333	0.2124	4.7758	0.5370	6.1022	0.5907	6.3734	0.6020	6.3758	0.5822	4.9735
	5.0000	0.1350	7.0347	0.3522	9.1248	0.3902	9.5681	0.3987	9.5942	0.3871	7.6148
	10.0000	0.0646	13.8141	0.1734	18.1739	0.1935	19.1282	0.1981	19.2158	0.1929	15.4052
	Inf.	1.5460	1.9201	1.9216	1.9700	1.9102	1.9194	1.8083	1.7090	1.3559	0.6198
n	$1/R_s$	L_1	C_2	L_3	C_4	L_5	C_6	L_7	C_8	L_9	C_{10}

TABLE 12-29   0.25-dB Chebyshev LC Element Values

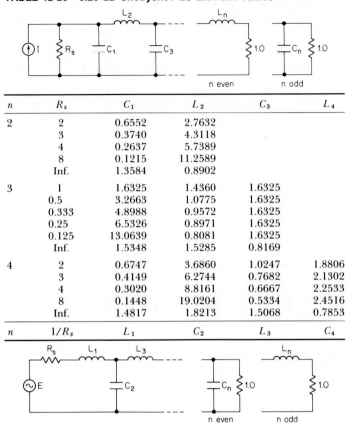

n	R_s	C_1	L_2	C_3	L_4
2	2	0.6552	2.7632		
	3	0.3740	4.3118		
	4	0.2637	5.7389		
	8	0.1215	11.2589		
	Inf.	1.3584	0.8902		
3	1	1.6325	1.4360	1.6325	
	0.5	3.2663	1.0775	1.6325	
	0.333	4.8988	0.9572	1.6325	
	0.25	6.5326	0.8971	1.6325	
	0.125	13.0639	0.8081	1.6325	
	Inf.	1.5348	1.5285	0.8169	
4	2	0.6747	3.6860	1.0247	1.8806
	3	0.4149	6.2744	0.7682	2.1302
	4	0.3020	8.8161	0.6667	2.2533
	8	0.1448	19.0204	0.5334	2.4516
	Inf.	1.4817	1.8213	1.5068	0.7853
n	$1/R_s$	L_1	C_2	L_3	C_4

TABLE 12-29   *(Continued)*

n	R_s	C_1	L_2	C_3	L_4	C_5	L_6	C_7
5	1	1.5046	1.4436	2.4050	1.4436	1.5046		
	0.5	3.0103	0.7218	3.6080	1.4436	1.5046		
	0.333	4.5149	0.4812	4.8100	1.4436	1.5046		
	0.25	6.0196	0.3615	6.0130	1.4436	1.5046		
	0.125	12.0402	0.1807	10.8230	1.4436	1.5046		
	Inf.	1.5765	1.7822	1.8225	1.4741	0.7523		
6	2	0.6867	3.2074	0.9308	3.8102	1.2163	1.7088	
	3	0.4330	5.0976	0.5392	6.0963	1.0804	1.8393	
	4	0.3173	6.9486	0.3821	8.2530	1.0221	1.8987	
	8	0.1539	14.3100	0.1762	16.7193	0.9393	1.9868	
	Inf.	1.5060	1.9221	1.8191	1.8329	1.4721	0.7610	
7	1	1.5120	1.4169	2.4535	1.5350	2.4535	1.4169	1.5120
	0.5	3.024	0.7085	4.9069	1.1515	2.4535	1.4169	1.5120
	0.333	4.5361	0.4723	7.3596	1.0230	2.4535	1.4169	1.5120
	0.25	6.0471	0.3542	9.8120	0.9593	2.4535	1.4169	1.5120
	0.125	12.0952	0.1776	19.6251	0.8631	2.4535	1.4169	1.5120
	Inf.	1.6009	1.8287	1.9666	1.8234	1.8266	1.4629	0.7555
n	$1/R_s$	L_1	C_2	L_3	C_4	L_5	C_6	L_7

TABLE 12-30   0.5-dB Chebyshev LC Element Values*

n	R_s	C_1	L_2	C_3	L_4	C_5	L_6
2	1.9841	0.9827	1.9497				
	2.0000	0.9086	2.1030				
	2.5000	0.5635	3.1647				
	3.3333	0.3754	4.4111				
	5.0000	0.2282	6.6995				
	10.0000	0.1052	13.3221				
	Inf.	1.3067	0.9748				
3	1.0000	1.8636	1.2804	1.8636			
	0.9000	1.9175	1.2086	2.0255			
	0.8000	1.9965	1.1203	2.2368			
	0.7000	2.1135	1.0149	2.5172			
	0.6000	2.2889	0.8937	2.8984			
	0.5000	2.5571	0.7592	3.4360			
	0.4000	2.9854	0.6146	4.2416			
	0.3000	3.7292	0.4633	5.5762			
	0.2000	5.2543	0.3087	8.2251			
	0.1000	9.8899	0.1534	16.1177			
	Inf.	1.5720	1.5179	0.9318			
4	1.9841	0.9202	2.5864	1.3036	1.8258		
	2.0000	0.8452	2.7198	1.2383	1.9849		
	2.5000	0.5162	3.7659	0.8693	3.1205		
	3.3333	0.3440	5.1196	0.6208	4.4790		
	5.0000	0.2100	7.7076	0.3996	6.9874		
	10.0000	0.0975	15.3520	0.1940	14.2616		
	Inf.	1.4361	1.8888	1.5211	0.9129		

TABLE 12-30   *(Continued)*

n	R_s	C_1	L_2	C_3	L_4	C_5	L_6
5	1.0000	1.8068	1.3025	2.6914	1.3025	1.8068	
	0.9000	1.8540	1.2220	2.8478	1.2379	1.9701	
	0.8000	1.9257	1.1261	3.0599	1.1569	2.1845	
	0.7000	2.0347	1.0150	3.3525	1.0582	2.4704	
	0.6000	2.2006	0.8901	3.7651	0.9420	2.8609	
	0.5000	2.4571	0.7537	4.3672	0.8098	3.4137	
	0.4000	2.8692	0.6091	5.2960	0.6640	4.2447	
	0.3000	3.5877	0.4590	6.8714	0.5075	5.6245	
	0.2000	5.0639	0.3060	10.0537	0.3430	8.3674	
	0.1000	9.5560	0.1525	19.6465	0.1731	16.5474	
	Inf.	1.6299	1.7400	1.9217	1.5138	0.9034	
6	1.9841	0.9053	2.5774	1.3675	2.7133	1.2991	1.7961
	2.0000	0.8303	2.7042	1.2912	2.8721	1.2372	1.9557
	2.5000	0.5056	3.7219	0.8900	4.1092	0.8808	3.1025
	3.3333	0.3370	5.0554	0.6323	5.6994	0.6348	4.4810
	5.0000	0.2059	7.6145	0.4063	8.7319	0.4121	7.0310
	10.0000	0.0958	15.1862	0.1974	17.6806	0.2017	14.4328
	Inf.	1.4618	1.9799	1.7803	1.9253	1.5077	0.8981
n	$1/R_s$	L_1	C_2	L_3	C_4	L_5	C_6

* Reprinted from A. I. Zverev, "Handbook of Filter Synthesis," John Wiley and Sons, New York, 1967.

TABLE 12-30 *(Continued)*

n	R_s	C_1	L_2	C_3	L_4	C_5	L_6	C_7	L_8	C_9	L_{10}
7	1.0000	1.7896	1.2961	2.7177	1.3848	2.7177	1.2961	1.7896			
	0.9000	1.8348	1.2146	2.8691	1.3080	2.8829	1.2335	1.9531			
	0.8000	1.9045	1.1182	3.0761	1.2149	3.1071	1.1546	2.1681			
	0.7000	2.0112	1.0070	3.3638	1.1050	3.4163	1.0582	2.4554			
	0.6000	2.1744	0.8824	3.7717	0.9786	3.8524	0.9441	2.8481			
	0.5000	2.4275	0.7470	4.3695	0.8377	4.4886	0.8137	3.4050			
	0.4000	2.8348	0.6035	5.2947	0.6846	5.4698	0.6690	4.2428			
	0.3000	3.5456	0.4548	6.8674	0.5221	7.1341	0.5129	5.6350			
	0.2000	5.0070	0.3034	10.0491	0.3524	10.4959	0.3478	8.4041			
	0.1000	9.4555	0.1513	19.6486	0.1778	20.6314	0.1761	16.6654			
	Inf.	1.6464	1.7772	2.0306	1.7892	1.9239	1.5034	0.8948			
8	1.9841	0.8998	2.5670	1.3697	2.7585	1.3903	2.7175	1.2938	1.7852		
	2.0000	0.8249	2.6916	1.2919	2.9134	1.3160	2.8800	1.2331	1.9449		
	2.5000	0.5017	3.6988	0.8878	4.1404	0.9184	4.1470	0.8815	3.0953		
	3.3333	0.3344	5.0234	0.6304	5.7323	0.6577	5.7761	0.6370	4.4807		
	5.0000	0.2044	7.5682	0.4052	8.7771	0.4257	8.8833	0.4146	7.0453		
	10.0000	0.0951	15.1014	0.1969	17.7747	0.2081	18.0544	0.2035	14.4924		
	Inf.	1.4710	2.0022	1.8248	2.0440	1.7911	1.9218	1.5003	0.8926		

n	$1/R_s$	L_1	C_2	L_3	C_4	L_5	C_6	L_7	C_8	L_9	C_{10}
9	1.0000	1.7822	1.2921	2.7162	1.3922	2.7734	1.3922	2.7162	1.2921	1.7822	
	0.9000	1.8267	1.2103	2.8658	1.3135	2.9353	1.3165	2.8834	1.2302	1.9458	
	0.8000	1.8955	1.1139	3.0709	1.2189	3.1565	1.2246	3.1102	1.1523	2.1611	
	0.7000	2.0013	1.0028	3.3565	1.1075	3.4635	1.1157	3.4232	1.0568	2.4489	
	0.6000	2.1634	0.8786	3.7621	0.9801	3.8985	0.9900	3.8647	0.9436	2.8426	
	0.5000	2.4150	0.7436	4.3573	0.8385	4.5355	0.8493	4.5087	0.8140	3.4010	
	0.4000	2.8203	0.6008	5.2792	0.6850	5.5207	0.6957	5.5023	0.6700	4.2416	
	0.3000	3.5279	0.4528	6.8474	0.5223	7.1951	0.5318	7.1876	0.5142	5.6390	
	0.2000	4.9830	0.3021	10.0212	0.3526	10.5818	0.3600	10.5925	0.3491	8.4189	
	0.1000	9.4131	0.1507	19.5995	0.1779	20.8006	0.1822	20.8588	0.1770	16.7140	
	Inf.	1.6533	1.7890	2.0570	1.8383	2.0481	1.7910	1.9199	1.4981	0.8911	
10	1.9841	0.8972	2.5610	1.3683	2.7631	1.4009	2.7795	1.3927	2.7148	1.2908	1.7801
	2.0000	0.8223	2.6845	1.2901	2.9166	1.3246	2.9390	1.3191	2.8783	1.2306	1.9398
	2.5000	0.4999	3.6868	0.8858	4.1383	0.9216	4.2020	0.9238	4.1540	0.8812	3.0919
	3.3333	0.3332	5.0071	0.6289	5.7274	0.6594	5.8399	0.6631	5.7948	0.6376	4.4804
	5.0000	0.2037	7.5446	0.4042	8.7695	0.4266	8.9727	0.4300	8.9249	0.4154	7.0518
	10.0000	0.0948	15.0578	0.1965	17.7624	0.2086	18.2313	0.2107	18.1644	0.2041	14.5199
	Inf.	1.4753	2.0107	1.8386	2.0733	1.8432	2.0494	1.7904	1.9183	1.4965	0.8900

TABLE 12-31   1-dB Chebyshev LC Element Values

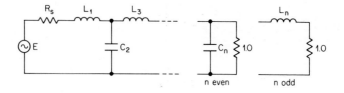

n	R_s	C_1	L_2	C_3	L_4
2	3	0.5723	3.1317		
	4	0.3653	4.6002		
	8	0.1571	9.6582		
	Inf.	1.2128	1.1093		
3	1	2.2160	1.0883	2.2160	
	0.5	4.4309	0.8168	2.2160	
	0.333	6.6469	0.7259	2.2160	
	0.25	8.8619	0.6799	2.2160	
	0.125	17.7248	0.6120	2.2160	
	Inf.	1.6522	1.4595	1.1080	
4	3	0.6529	4.4110	0.8140	2.5346
	4	0.4517	7.0825	0.6118	2.8484
	8	0.2085	17.1639	0.4275	3.2811
	Inf.	1.3499	2.0102	1.4879	1.1057
n	$1/R_s$	L_1	C_2	L_3	C_4

n	R_s	C_1	L_2	C_3	L_4	C_5	L_6	C_7
5	1	2.2072	1.1279	3.1025	1.1279	2.2072		
	0.5	4.4144	0.5645	4.6532	1.1279	2.2072		
	0.333	6.6216	0.3763	6.2050	1.1279	2.2072		
	0.25	8.8288	0.2822	7.7557	1.1279	2.2072		
	0.125	17.6565	0.1406	13.9606	1.1279	2.2072		
	Inf.	1.7213	1.6448	2.0614	1.4928	1.1031		
6	3	0.6785	3.8725	0.7706	4.7107	0.9692	2.4060	
	4	0.4810	5.6441	0.4759	7.3511	0.8494	2.5820	
	8	0.2272	12.3095	0.1975	16.740	0.7256	2.7990	
	Inf.	1.3775	2.0969	1.6896	2.0744	1.4942	1.1022	
7	1	2.2043	1.1311	3.1472	1.1942	3.1472	1.1311	2.2043
	0.5	4.4075	0.5656	6.2934	0.8951	3.1472	1.1311	2.2043
	0.333	6.6118	0.3774	9.4406	0.7955	3.1472	1.1311	2.2043
	0.25	8.8151	0.2828	12.5879	0.7466	3.1472	1.1311	2.2043
	0.125	17.6311	0.1414	25.175	0.6714	3.1472	1.1311	2.2043
	Inf.	1.7414	1.6774	2.1554	1.7028	2.0792	1.4943	1.1016
n	$1/R_s$	L_1	C_2	L_3	C_4	L_5	C_6	L_7

TABLE 12-32   0.1-dB Chebyshev Uniform Dissipation Network

n	d	C_1	L_2	C_3	L_4	C_5	L_6	C_7
2	0.0172	1.3855	0.8433					
	0.0257	1.3816	0.8550					
	0.0515	1.3680	0.8939					
3	0.024	1.4848	1.5390	0.7556				
	0.036	1.4696	1.5543	0.7765				
	0.072	1.4168	1.6015	0.8473				
4	0.0275	1.4375	1.7978	1.5103	0.7266			
	0.0412	1.3975	1.8148	1.5394	0.7570			
	0.0824	1.2556	1.8767	1.6353	0.8637			
5	0.0294	1.4558	1.8064	1.8280	1.4933	0.7194		
	0.0441	1.3945	1.8076	1.8643	1.5352	0.7591		
	0.0881	1.1449	1.8416	2.0209	1.6839	0.9123		
6	0.0305	1.3672	1.8874	1.8612	1.8361	1.4907	0.7224	
	0.0457	1.2645	1.8973	1.8842	1.8907	1.5454	0.7738	
	0.0915	0.6579	2.3639	2.1574	2.1803	1.7574	0.9825	
7	0.0312	1.3628	1.8252	1.9694	1.8797	1.8455	1.4963	0.7316
	0.0468	1.2079	1.8220	2.0207	1.9213	1.9192	1.5646	0.7957
n	d	L_1	C_2	L_3	C_4	L_5	C_6	L_7

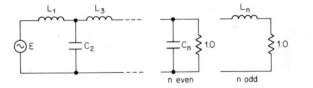

TABLE 12-33   0.25-dB Chebyshev Uniform Dissipation Network

n	d	C_1	L_2	C_3	L_4	C_5	L_6	C_7
2	0.0209	1.3504	0.9157					
	0.0313	1.3376	0.9413					
	0.0626	1.3120	1.0004					
3	0.0266	1.5022	1.5548	0.8733				
	0.0399	1.4834	1.5674	0.9046				
	0.0798	1.4220	1.6062	1.0149				
4	0.0292	1.3894	1.8590	1.5593	0.8651			
	0.0439	1.3370	1.8818	1.5866	0.9107			
	0.0877	1.1444	1.9764	1.6823	1.0839			
5	0.0306	1.4599	1.7670	1.8976	1.5503	0.8503		
	0.0459	1.3881	1.7604	1.9455	1.5917	0.9102		
	0.0919	1.0397	1.8181	2.2035	1.7528	1.1497		
6	0.0314	1.3054	1.9347	1.8339	1.9443	1.5697	0.8883	
	0.0471	1.1696	1.9560	1.8541	2.0218	1.6259	0.9700	
7	0.0319	1.3584	1.7680	2.0376	1.8610	1.9707	1.5820	0.9091
	0.0479	1.1264	1.7722	2.1452	1.9132	2.0814	1.6541	1.0125

n	d	L_1	C_2	L_3	C_4	L_5	C_6	L_7

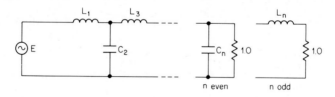

TABLE 12-34   0.5-dB Chebyshev Uniform Dissipation Network

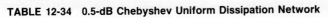

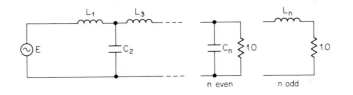

n	d	C_1	L_2	C_3	L_4	C_5	L_6	C_7
2	0.0240	1.2855	1.0228					
	0.0360	1.2730	1.0478					
	0.0720	1.2313	1.1340					
3	0.0286	1.5376	1.5341	1.0122				
	0.0428	1.5189	1.5423	1.0589				
	0.0856	1.4489	1.5621	1.2247				
4	0.0305	1.3205	1.9413	1.5631	1.0275			
	0.0457	1.2549	1.9741	1.5850	1.0964			
	0.0915	0.9991	2.1359	1.6692	1.3707			
5	0.0315	1.5031	1.6980	2.0264	1.5773	1.0529		
	0.0472	1.4162	1.6768	2.0995	1.6133	1.1482		
	0.0944	0.7139	2.0994	2.7297	1.8007	1.5751		
6	0.0320	1.2200	2.0123	1.7707	2.0758	1.5927	1.0858	
	0.0480	1.0389	2.0612	1.7895	2.1976	1.6448	1.2117	
7	0.0324	1.3659	1.6801	2.1488	1.8047	2.1230	1.6090	1.1228
	0.0485	0.9024	1.8171	2.4475	1.8985	2.3126	1.6811	1.2856
n	d	L_1	C_2	L_3	C_4	L_5	C_6	L_7

TABLE 12-35   1-dB Chebyshev Uniform Dissipation Network

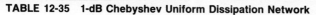

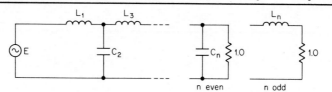

n	d	C_1	L_2	C_3	L_4	C_5	L_6	C_7
2	0.0274	1.1762	1.1811					
	0.0411	1.1020	1.2201					
	0.0821	1.0898	1.3564					
3	0.0304	1.6215	1.4562	1.2328				
	0.0457	1.6029	1.4518	1.3062				
	0.0913	1.5241	1.4387	1.5908				
4	0.0317	1.2015	2.0965	1.5037	1.2847			
	0.0475	1.1141	2.1544	1.5132	1.3984			
	0.0950	0.7434	2.5398	1.5890	1.9049			
5	0.0322	1.5869	1.5610	2.2258	1.5269	1.3419		
	0.0484	1.4680	1.5218	2.3591	1.5507	1.5052		
6	0.0326	1.0736	2.1789	1.6385	2.3150	1.5454	1.4052	
	0.0489	0.7747	2.3611	1.6958	2.5658	1.5894	1.6293	
7	0.0328	1.3610	1.5258	2.4016	1.6865	2.4067	1.5634	1.4749
n	d	L_1	C_2	L_3	C_4	L_5	C_6	L_7

**TABLE 12-36   0.01-dB Chebyshev Active
Low-Pass Values**

Order n	C_1	C_2
2	1.4826	0.7042
4	1.4874	1.1228
	3.5920	0.2985
6	1.8900	1.5249
	2.5820	0.5953
	7.0522	0.1486
8	2.3652	1.9493
	2.7894	0.8196
	4.1754	0.3197
	11.8920	0.08672

TABLE 12-37   0.1-dB Chebyshev Active Low-Pass Values*

Order n	C_1	C_2	C_3
2	1.638	0.6955	
3	6.653	1.825	0.1345
4	1.900	1.241	
	4.592	0.2410	
5	4.446	2.520	0.3804
	6.810	0.1580	
6	2.553	1.776	
	3.487	0.4917	
	9.531	0.1110	
7	5.175	3.322	0.5693
	4.546	0.3331	
	12.73	0.08194	
8	3.270	2.323	
	3.857	0.6890	
	5.773	0.2398	
	16.44	0.06292	
9	6.194	4.161	0.7483
	4.678	0.4655	
	7.170	0.1812	
	20.64	0.04980	
10	4.011	2.877	
	4.447	0.8756	
	5.603	0.3353	
	8.727	0.1419	
	25.32	0.04037	

* Reprinted from "Electronics," August 18, 1969, McGraw-Hill, Inc., 1969.

TABLE 12-38   0.25-dB Chebyshev Active Low-Pass Values*

Order n	C_1	C_2	C_3
2	1.778	0.6789	
3	8.551	2.018	0.1109
4	2.221	1.285	
	5.363	0.2084	
5	5.543	2.898	0.3425
	8.061	0.1341	
6	3.044	1.875	
	4.159	0.4296	
	11.36	0.09323	
7	6.471	3.876	0.5223
	5.448	0.2839	
	15.26	0.06844	
8	3.932	2.474	
	4.638	0.6062	
	6.942	0.2019	
	19.76	0.05234	
9	7.766	4.891	0.6919
	5.637	0.3983	
	8.639	0.1514	
	24.87	0.04131	
10	4.843	3.075	
	5.368	0.7725	
	6.766	0.2830	
	10.53	0.1181	
	30.57	0.03344	

* Reprinted from "Electronics," August 18, 1969, McGraw-Hill, Inc., 1969.

TABLE 12-39   0.5-dB Chebyshev Active Low-Pass Values*

Order n	C_1	C_2	C_3
2	1.950	0.6533	
3	11.23	2.250	0.0895
4	2.582	1.300	
	6.233	0.1802	
5	6.842	3.317	0.3033
	9.462	0.1144	
6	3.592	1.921	
	4.907	0.3743	
	13.40	0.07902	
7	7.973	4.483	0.4700
	6.446	0.2429	
	18.07	0.05778	
8	4.665	2.547	
	5.502	0.5303	
	8.237	0.1714	
	23.45	0.04409	
9	9.563	5.680	0.6260
	6.697	0.3419	
	10.26	0.1279	
	29.54	0.03475	
10	5.760	3.175	
	6.383	0.6773	
	8.048	0.2406	
	12.53	0.09952	
	36.36	0.02810	

* Reprinted from "Electronics," August 18, 1969, McGraw-Hill, Inc., 1969.

TABLE 12-40   1-dB Chebyshev Active Low-Pass Values*

Order n	C_1	C_2	C_3
2	2.218	0.6061	
3	16.18	2.567	0.06428
4	3.125	1.269	
	7.546	0.1489	
5	8.884	3.935	0.2540
	11.55	0.09355	
6	4.410	1.904	
	6.024	0.3117	
	16.46	0.06425	
7	10.29	5.382	0.4012
	7.941	0.1993	
	22.25	0.04684	
8	5.756	2.538	
	6.792	0.4435	
	10.15	0.1395	
	28.94	0.03568	
9	12.33	6.853	0.5382
	8.281	0.2813	
	12.68	0.1038	
	36.51	0.02808	
10	7.125	3.170	
	7.897	0.5630	
	9.952	0.1962	
	15.50	0.08054	
	44.98	0.02269	

* Reprinted from "Electronics," August 18, 1969, McGraw-Hill, Inc., 1969.

TABLE 12-41   Bessel Pole Locations

Order n	Real Part $-\alpha$	Imaginary Part $\pm j\beta$
2	1.1030	0.6368
3	1.0509	1.0025
	1.3270	
4	1.3596	0.4071
	0.9877	1.2476
5	1.3851	0.7201
	0.9606	1.4756
	1.5069	
6	1.5735	0.3213
	1.3836	0.9727
	0.9318	1.6640
7	1.6130	0.5896
	1.3797	1.1923
	0.9104	1.8375
	1.6853	
8	1.7627	0.2737
	0.8955	2.0044
	1.3780	1.3926
	1.6419	0.8253
9	1.8081	0.5126
	1.6532	1.0319
	1.3683	1.5685
	0.8788	2.1509
	1.8575	

TABLE 12-42 Bessel LC Element Values*

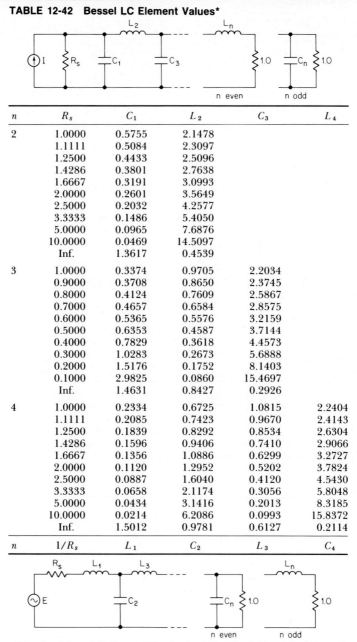

n	R_s	C_1	L_2	C_3	L_4
2	1.0000	0.5755	2.1478		
	1.1111	0.5084	2.3097		
	1.2500	0.4433	2.5096		
	1.4286	0.3801	2.7638		
	1.6667	0.3191	3.0993		
	2.0000	0.2601	3.5649		
	2.5000	0.2032	4.2577		
	3.3333	0.1486	5.4050		
	5.0000	0.0965	7.6876		
	10.0000	0.0469	14.5097		
	Inf.	1.3617	0.4539		
3	1.0000	0.3374	0.9705	2.2034	
	0.9000	0.3708	0.8650	2.3745	
	0.8000	0.4124	0.7609	2.5867	
	0.7000	0.4657	0.6584	2.8575	
	0.6000	0.5365	0.5576	3.2159	
	0.5000	0.6353	0.4587	3.7144	
	0.4000	0.7829	0.3618	4.4573	
	0.3000	1.0283	0.2673	5.6888	
	0.2000	1.5176	0.1752	8.1403	
	0.1000	2.9825	0.0860	15.4697	
	Inf.	1.4631	0.8427	0.2926	
4	1.0000	0.2334	0.6725	1.0815	2.2404
	1.1111	0.2085	0.7423	0.9670	2.4143
	1.2500	0.1839	0.8292	0.8534	2.6304
	1.4286	0.1596	0.9406	0.7410	2.9066
	1.6667	0.1356	1.0886	0.6299	3.2727
	2.0000	0.1120	1.2952	0.5202	3.7824
	2.5000	0.0887	1.6040	0.4120	4.5430
	3.3333	0.0658	2.1174	0.3056	5.8048
	5.0000	0.0434	3.1416	0.2013	8.3185
	10.0000	0.0214	6.2086	0.0993	15.8372
	Inf.	1.5012	0.9781	0.6127	0.2114
n	$1/R_s$	L_1	C_2	L_3	C_4

* Reprinted from A. I. Zverev, "Handbook of Filter Synthesis," John Wiley and Sons, New York, 1967.

TABLE 12-42   *(Continued)*

n	R_s	C_1	L_2	C_3	L_4	C_5	L_6	C_7
5	1.0000	0.1743	0.5072	0.8040	1.1110	2.2582		
	0.9000	0.1926	0.4542	0.8894	0.9945	2.4328		
	0.8000	0.2154	0.4016	0.9959	0.8789	2.6497		
	0.7000	0.2447	0.3494	1.1323	0.7642	2.9272		
	0.6000	0.2836	0.2977	1.3138	0.6506	3.2952		
	0.5000	0.3380	0.2465	1.5672	0.5382	3.8077		
	0.4000	0.4194	0.1958	1.9464	0.4270	4.5731		
	0.3000	0.5548	0.1457	2.5768	0.3174	5.8433		
	0.2000	0.8251	0.0964	3.8352	0.2095	8.3747		
	0.1000	1.6349	0.0478	7.6043	0.1036	15.9487		
	Inf.	1.5125	1.0232	0.7531	0.4729	0.1618		
6	1.0000	0.1365	0.4002	0.6392	0.8538	1.1126	2.2645	
	1.1111	0.1223	0.4429	0.5732	0.9456	0.9964	2.4388	
	1.2500	0.1082	0.4961	0.5076	1.0600	0.8810	2.6554	
	1.4286	0.0943	0.5644	0.4424	1.2069	0.7665	2.9325	
	1.6667	0.0804	0.6553	0.3775	1.4022	0.6530	3.3001	
	2.0000	0.0666	0.7824	0.3131	1.6752	0.5405	3.8122	
	2.5000	0.0530	0.9725	0.2492	2.0837	0.4292	4.5770	
	3.3333	0.0395	1.2890	0.1859	2.7633	0.3193	5.8467	
	5.0000	0.0261	1.9209	0.1232	4.1204	0.2110	8.3775	
	10.0000	0.0130	3.8146	0.0612	8.1860	0.1045	15.9506	
	Inf.	1.5124	1.0329	0.8125	0.6072	0.3785	0.1287	
7	1.0000	0.1106	0.3259	0.5249	0.7020	0.8690	1.1052	2.2659
	0.9000	0.1224	0.2923	0.5815	0.6302	0.9630	0.9899	2.4396
	0.8000	0.1372	0.2589	0.6521	0.5586	1.0803	0.8754	2.6556
	0.7000	0.1562	0.2257	0.7428	0.4873	1.2308	0.7618	2.9319
	0.6000	0.1815	0.1927	0.8634	0.4163	1.4312	0.6491	3.2984
	0.5000	0.2168	0.1599	1.0321	0.3457	1.7111	0.5374	3.8090
	0.4000	0.2698	0.1274	1.2847	0.2755	2.1304	0.4269	4.5718
	0.3000	0.3579	0.0951	1.7051	0.2058	2.8280	0.3177	5.8380
	0.2000	0.5338	0.0630	2.5448	0.1365	4.2214	0.2100	8.3623
	0.1000	1.0612	0.0313	5.0616	0.0679	8.3967	0.1040	15.9166
	Inf.	1.5087	1.0293	0.8345	0.6752	0.5031	0.3113	0.1054
n	$1/R_s$	L_1	C_2	L_3	C_4	L_5	C_6	L_7

TABLE 12-42 *(Continued)*

n	R_s	C_1	L_2	C_3	L_4	C_5	L_6	C_7	L_8	C_9	L_{10}
8	1.0000	0.0919	0.2719	0.4409	0.5936	0.7303	0.8695	1.0956	2.2656		
	1.1111	0.0825	0.3013	0.3958	0.6580	0.6559	0.9639	0.9813	2.4388		
	1.2500	0.0731	0.3380	0.3509	0.7385	0.5817	1.0816	0.8678	2.6541		
	1.4286	0.0637	0.3850	0.3061	0.8418	0.5078	1.2328	0.7552	2.9295		
	1.6667	0.0545	0.4477	0.2616	0.9794	0.4342	1.4340	0.6435	3.2949		
	2.0000	0.0452	0.5354	0.2173	1.1718	0.3608	1.7153	0.5329	3.8041		
	2.5000	0.0360	0.6667	0.1732	1.4599	0.2878	2.1367	0.4233	4.5645		
	3.3333	0.0269	0.8852	0.1294	1.9396	0.2151	2.8380	0.3151	5.8271		
	5.0000	0.0179	1.3218	0.0859	2.8981	0.1429	4.2389	0.2083	8.3441		
	10.0000	0.0089	2.6307	0.0427	5.7710	0.0711	8.4376	0.1032	15.8768		
	Inf.	1.5044	1.0214	0.8392	0.7081	0.5743	0.4253	0.2616	0.0883		
9	1.0000	0.0780	0.2313	0.3770	0.5108	0.6306	0.7407	0.8639	1.0863	2.2649	
	0.9000	0.0864	0.2077	0.4180	0.4588	0.6994	0.6655	0.9578	0.9730	2.4376	
	0.8000	0.0970	0.1841	0.4691	0.4069	0.7854	0.5905	1.0750	0.8604	2.6524	
	0.7000	0.1105	0.1607	0.5348	0.3553	0.8957	0.5157	1.2255	0.7488	2.9271	
	0.6000	0.1286	0.1373	0.6222	0.3038	1.0427	0.4411	1.4258	0.6380	3.2915	
	0.5000	0.1538	0.1141	0.7445	0.2525	1.2483	0.3667	1.7059	0.5283	3.7993	
	0.4000	0.1916	0.0910	0.9278	0.2014	1.5563	0.2926	2.1256	0.4197	4.5578	
	0.3000	0.2545	0.0680	1.2329	0.1506	2.0692	0.2189	2.8241	0.3124	5.8171	
	0.2000	0.3803	0.0452	1.8426	0.1000	3.0941	0.1455	4.2196	0.2065	8.3276	
	0.1000	0.7573	0.0225	3.6704	0.0498	6.1666	0.0725	8.4023	0.1023	15.8408	
	Inf.	1.5006	1.0127	0.8361	0.7220	0.6142	0.4963	0.3654	0.2238	0.0754	
10	1.0000	0.0672	0.1998	0.3270	0.4454	0.5528	0.6493	0.7420	0.8561	1.0781	2.2641
	1.1111	0.0604	0.2216	0.2937	0.4941	0.4967	0.7205	0.6668	0.9492	0.9656	2.4365
	1.2500	0.0536	0.2488	0.2606	0.5548	0.4408	0.8093	0.5918	1.0654	0.8539	2.6508
	1.4286	0.0467	0.2836	0.2275	0.6327	0.3850	0.9233	0.5170	1.2147	0.7430	2.9249
	1.6667	0.0400	0.3301	0.1945	0.7366	0.3294	1.0753	0.4423	1.4134	0.6331	3.2885
	2.0000	0.0332	0.3951	0.1617	0.8818	0.2739	1.2879	0.3678	1.6913	0.5242	3.7953
	2.5000	0.0265	0.4924	0.1290	1.0995	0.2186	1.6064	0.2936	2.1076	0.4164	4.5521
	3.3333	0.0198	0.6546	0.0965	1.4620	0.1635	2.1369	0.2197	2.8007	0.3099	5.8087
	5.0000	0.0132	0.9786	0.0641	2.1864	0.1087	3.1971	0.1461	4.1854	0.2049	8.3137
	10.0000	0.0066	1.9499	0.0319	4.3583	0.0542	6.3759	0.0728	8.3359	0.1015	15.8108
	Inf.	1.4973	1.0045	0.8297	0.7258	0.6355	0.5401	0.4342	0.3182	0.1942	0.0653
n	$1/R_s$	L_1	C_2	L_3	C_4	L_5	C_6	L_7	C_8	L_9	C_{10}

TABLE 12-43 Bessel Active Low-Pass Values

Order n	C_1	C_2	C_3
2	0.9066	0.6800	
3	1.423	0.9880	0.2538
4	0.7351 1.012	0.6746 0.3900	
5	1.010 1.041	0.8712 0.3100	0.3095
6	0.6352 0.7225 1.073	0.6100 0.4835 0.2561	
7	0.8532 0.7250 1.100	0.7792 0.4151 0.2164	0.3027
8	0.5673 0.6090 0.7257 1.116	0.5540 0.4861 0.3590 0.1857	
9	0.7564 0.6048 0.7307 1.137	0.7070 0.4352 0.3157 0.1628	0.2851
10	0.5172 0.5412 0.6000 0.7326 1.151	0.5092 0.4682 0.3896 0.2792 0.1437	

TABLE 12-44   Linear Phase with Equiripple Error of 0.05° Pole Locations

Order n	Real Part $-\alpha$	Imaginary Part $\pm j\beta$
2	1.0087	0.6680
3	0.8541	1.0725
	1.0459	
4	0.9648	0.4748
	0.7448	1.4008
5	0.8915	0.8733
	0.6731	1.7085
	0.9430	
6	0.8904	0.4111
	0.8233	1.2179
	0.6152	1.9810
7	0.8425	0.7791
	0.7708	1.5351
	0.5727	2.2456
	0.8615	
8	0.8195	0.3711
	0.7930	1.1054
	0.7213	1.8134
	0.5341	2.4761
9	0.7853	0.7125
	0.7555	1.4127
	0.6849	2.0854
	0.5060	2.7133
	0.7938	
10	0.7592	0.3413
	0.7467	1.0195
	0.7159	1.6836
	0.6475	2.3198
	0.4777	2.9128

TABLE 12-45   Linear Phase with Equiripple Error of 0.5° Pole Locations

Order n	Real Part $-\alpha$	Imaginary Part $\pm j\beta$
2	0.8590	0.6981
3	0.6969	1.1318
	0.8257	
4	0.7448	0.5133
	0.6037	1.4983
5	0.6775	0.9401
	0.5412	1.8256
	0.7056	
6	0.6519	0.4374
	0.6167	1.2963
	0.4893	2.0982
7	0.6190	0.8338
	0.5816	1.6453
	0.4598	2.3994
	0.6283	
8	0.5791	0.3857
	0.5665	1.1505
	0.5303	1.8914
	0.4184	2.5780
9	0.5688	0.7595
	0.5545	1.5089
	0.5179	2.2329
	0.4080	2.9028
	0.5728	
10	0.5249	0.3487
	0.5193	1.0429
	0.5051	1.7261
	0.4711	2.3850
	0.3708	2.9940

TABLE 12-46   Linear Phase with Equiripple Error of 0.05° LC Element Values*

n	R_s	C_1	L_2	C_3	L_4
2	1.0000	0.6480	2.1085		
	1.1111	0.5703	2.2760		
	1.2500	0.4955	2.4817		
	1.4286	0.4235	2.7422		
	1.6667	0.3544	3.0848		
	2.0000	0.2880	3.5589		
	2.5000	0.2244	4.2630		
	3.3333	0.1637	5.4270		
	5.0000	0.1059	7.7400		
	10.0000	0.0513	14.6480		
	Inf.	1.3783	0.4957		
3	1.0000	0.4328	1.0427	2.2542	
	0.9000	0.4745	0.9330	2.4258	
	0.8000	0.5262	0.8238	2.6400	
	0.7000	0.5925	0.7153	2.9146	
	0.6000	0.6805	0.6078	3.2795	
	0.5000	0.8032	0.5015	3.7884	
	0.4000	0.9865	0.3967	4.5487	
	0.3000	1.2910	0.2938	5.8106	
	0.2000	1.8983	0.1931	8.3253	
	0.1000	3.7161	0.0950	15.8472	
	Inf.	1.5018	0.9328	0.3631	
4	1.0000	0.3363	0.7963	1.1428	2.2459
	1.1111	0.2993	0.8810	1.0212	2.4241
	1.2500	0.2631	0.9865	0.9012	2.6445
	1.4286	0.2275	1.1216	0.7826	2.9254
	1.6667	0.1926	1.3009	0.6657	3.2970
	2.0000	0.1584	1.5509	0.5502	3.8138
	2.5000	0.1250	1.9244	0.4364	4.5844
	3.3333	0.0923	2.5448	0.3242	5.8626
	5.0000	0.0606	3.7818	0.2139	8.4091
	10.0000	0.0298	7.4845	0.1058	16.0266
	Inf.	1.5211	1.0444	0.7395	0.2925
n	$1/R_s$	L_1	C_2	L_3	C_4

* Reprinted from A. I. Zverev, "Handbook of Filter Synthesis," John Wiley and Sons, New York, 1967.

TABLE 12-46 *(Continued)*

n	R_s	C_1	L_2	C_3	L_4	C_5	L_6	C_7
5	1.0000	0.2751	0.6541	0.8892	1.1034	2.2873		
	0.9000	0.3031	0.5868	0.9841	0.9904	2.4589		
	0.8000	0.3380	0.5197	1.1026	0.8774	2.6733		
	0.7000	0.3827	0.4529	1.2548	0.7648	2.9484		
	0.6000	0.4420	0.3865	1.4575	0.6526	3.3144		
	0.5000	0.5248	0.3204	1.7408	0.5410	3.8254		
	0.4000	0.6486	0.2549	2.1651	0.4302	4.5896		
	0.3000	0.8544	0.1899	2.8713	0.3205	5.8595		
	0.2000	1.2649	0.1257	4.2817	0.2120	8.3922		
	0.1000	2.4940	0.0624	8.5082	0.1051	15.9739		
	Inf.	1.5144	1.0407	0.8447	0.6177	0.2456		
6	1.0000	0.2374	0.5662	0.7578	0.8760	1.1163	2.2448	
	1.1111	0.2120	0.6272	0.6799	0.9726	0.9977	2.4214	
	1.2500	0.1870	0.7032	0.6023	1.0931	0.8807	2.6396	
	1.4286	0.1622	0.8008	0.5253	1.2475	0.7652	2.9174	
	1.6667	0.1378	0.9306	0.4487	1.4530	0.6512	3.2849	
	2.0000	0.1138	1.1118	0.3725	1.7401	0.5387	3.7958	
	2.5000	0.0901	1.3830	0.2969	2.1698	0.4277	4.5579	
	3.3333	0.0669	1.8340	0.2217	2.8849	0.3182	5.8220	
	5.0000	0.0441	2.7343	0.1472	4.3129	0.2103	8.3408	
	10.0000	0.0218	5.4312	0.0732	8.5924	0.1041	15.8769	
	Inf.	1.5050	1.0306	0.8554	0.7283	0.5389	0.2147	
7	1.0000	0.2085	0.4999	0.6653	0.7521	0.8749	1.0671	2.2845
	0.9000	0.2302	0.4488	0.7374	0.6768	0.9687	0.9580	2.4538
	0.8000	0.2573	0.3978	0.8274	0.6013	1.0861	0.8489	2.6655
	0.7000	0.2919	0.3470	0.9431	0.5258	1.2369	0.7400	2.9375
	0.6000	0.3380	0.2964	1.0972	0.4503	1.4381	0.6314	3.2996
	0.5000	0.4023	0.2461	1.3127	0.3749	1.7196	0.5235	3.8051
	0.4000	0.4986	0.1960	1.6356	0.2995	2.1416	0.4163	4.5613
	0.3000	0.6585	0.1463	2.1734	0.2242	2.8445	0.3101	5.8180
	0.2000	0.9778	0.0970	3.2480	0.1492	4.2496	0.2052	8.3246
	0.1000	1.9340	0.0482	6.4698	0.0744	8.4623	0.1017	15.8281
	Inf.	1.4988	1.0071	0.8422	0.7421	0.6441	0.4791	0.1911
n	$1/R_s$	L_1	C_2	L_3	C_4	L_5	C_6	L_7

TABLE 12-46 *(Continued)*

n	R_s	C_1	L_2	C_3	L_4	C_5	L_6	C_7	L_8	C_9	L_{10}
8	1.0000	0.1891	0.4543	0.6031	0.6750	0.7590	0.8427	1.0901	2.2415		
	1.1111	0.1691	0.5035	0.5415	0.7500	0.6813	0.9362	0.9735	2.4176		
	1.2500	0.1494	0.5650	0.4802	0.8435	0.6041	1.0527	0.8588	2.6349		
	1.4286	0.1298	0.6438	0.4191	0.9637	0.5272	1.2019	0.7459	2.9113		
	1.6667	0.1105	0.7487	0.3583	1.1237	0.4508	1.4004	0.6345	3.2767		
	2.0000	0.0914	0.8953	0.2978	1.3475	0.3748	1.6776	0.5247	3.7846		
	2.5000	0.0725	1.1148	0.2376	1.6827	0.2991	2.0927	0.4164	4.5418		
	3.3333	0.0539	1.4801	0.1776	2.2411	0.2237	2.7833	0.3096	5.7978		
	5.0000	0.0356	2.2095	0.1180	3.3568	0.1488	4.1627	0.2046	8.3004		
	10.0000	0.0176	4.3954	0.0588	6.7021	0.0742	8.2969	0.1013	15.7878		
	Inf.	1.4953	1.0018	0.8264	0.7396	0.6688	0.5858	0.4369	0.1743		
9	1.0000	0.1718	0.4146	0.5498	0.6132	0.6774	0.7252	0.8450	1.0447	2.2834	
	0.9000	0.1900	0.3724	0.6097	0.5519	0.7513	0.6529	0.9352	0.9382	2.4512	
	0.8000	0.2125	0.3302	0.6846	0.4905	0.8436	0.5805	1.0481	0.8314	2.6613	
	0.7000	0.2415	0.2882	0.7807	0.4291	0.9624	0.5079	1.1933	0.7247	2.9315	
	0.6000	0.2800	0.2463	0.9088	0.3676	1.1207	0.4352	1.3870	0.6184	3.2914	
	0.5000	0.3337	0.2046	1.0880	0.3062	1.3424	0.3624	1.6581	0.5125	3.7941	
	0.4000	0.4141	0.1631	1.3565	0.2448	1.6749	0.2897	2.0647	0.4075	4.5462	
	0.3000	0.5478	0.1219	1.8038	0.1834	2.2289	0.2170	2.7420	0.3035	5.7960	
	0.2000	0.8148	0.0809	2.6977	0.1222	3.3369	0.1445	4.0960	0.2007	8.2890	
	0.1000	1.6146	0.0403	5.3782	0.0610	6.6602	0.0721	8.1556	0.0995	15.7520	
	Inf.	1.4907	0.9845	0.8116	0.7197	0.6646	0.6089	0.5359	0.4003	0.1598	
10	1.0000	0.1601	0.3867	0.5125	0.5702	0.6243	0.6557	0.7319	0.8178	1.0767	2.2387
	1.1111	0.1433	0.4288	0.4604	0.6336	0.5609	0.7290	0.6567	0.9089	0.9608	2.4151
	1.2500	0.1267	0.4812	0.4084	0.7127	0.4977	0.8205	0.5820	1.0221	0.8471	2.6323
	1.4286	0.1102	0.5486	0.3567	0.8143	0.4348	0.9380	0.5079	1.1672	0.7354	2.9082
	1.6667	0.0939	0.6383	0.3051	0.9498	0.3721	1.0944	0.4342	1.3600	0.6254	3.2727
	2.0000	0.0778	0.7637	0.2537	1.1392	0.3096	1.3131	0.3609	1.6291	0.5170	3.7791
	2.5000	0.0618	0.9515	0.2024	1.4232	0.2473	1.6408	0.2880	2.0320	0.4102	4.5340
	3.3333	0.0460	1.2641	0.1515	1.8961	0.1852	2.1866	0.2154	2.7022	0.3049	5.7860
	5.0000	0.0304	1.8885	0.1007	2.8416	0.1232	3.2775	0.1433	4.0406	0.2014	8.2806
	10.0000	0.0151	3.7600	0.0502	5.6766	0.0615	6.5485	0.0714	8.0520	0.0997	15.7441
	Inf.	1.4905	0.9858	0.8018	0.7123	0.6540	0.6141	0.5669	0.5003	0.3741	0.1494
n	$1/R_s$	L_1	C_2	L_3	C_4	L_5	C_6	L_7	C_8	L_9	C_{10}

TABLE 12-47   Linear Phase with Equiripple Error of 0.5° LC Element Values*

n	R_s	C_1	L_2	C_3	L_4
2	1.0000	0.8245	1.9800		
	1.1111	0.7166	2.1640		
	1.2500	0.6160	2.3850		
	1.4286	0.5216	2.6603		
	1.6667	0.4327	3.0181		
	2.0000	0.3489	3.5088		
	2.5000	0.2700	4.2329		
	3.3333	0.1956	5.4242		
	5.0000	0.1258	7.7842		
	10.0000	0.0606	14.8185		
	Inf.	1.4022	0.5821		
3	1.0000	0.5534	1.0218	2.4250	
	0.9000	0.6059	0.9213	2.5929	
	0.8000	0.6710	0.8197	2.8046	
	0.7000	0.7540	0.7173	3.0787	
	0.6000	0.8639	0.6141	3.4462	
	0.5000	1.0168	0.5105	3.9625	
	0.4000	1.2448	0.4068	4.7385	
	0.3000	1.6231	0.3034	6.0326	
	0.2000	2.3771	0.2008	8.6197	
	0.1000	4.6332	0.0994	16.3738	
	Inf.	1.5495	0.9820	0.4506	
4	1.0000	0.4526	0.7967	1.2669	2.0504
	1.1111	0.3996	0.8889	1.1137	2.2502
	1.2500	0.3486	1.0028	0.9699	2.4866
	1.4286	0.2995	1.1481	0.8333	2.7788
	1.6667	0.2521	1.3405	0.7024	3.1576
	2.0000	0.2062	1.6083	0.5762	3.6769
	2.5000	0.1618	2.0079	0.4542	4.4438
	3.3333	0.1190	2.6708	0.3358	5.7080
	5.0000	0.0777	3.9916	0.2207	8.2168
	10.0000	0.0380	7.9426	0.1088	15.7068
	Inf.	1.4944	1.0715	0.7889	0.3708
n	$1/R_s$	L_1	C_2	L_3	C_4

* Reprinted from A. I. Zverev, "Handbook of Filter Synthesis," John Wiley and Sons, New York, 1967.

TABLE 12-47 *(Continued)*

n	R_s	C_1	L_2	C_3	L_4	C_5	L_6	C_7
5	1.0000	0.3658	0.6768	0.9513	1.0113	2.4446		
	0.9000	0.4027	0.6099	1.0486	0.9157	2.6062		
	0.8000	0.4485	0.5427	1.1700	0.8182	2.8114		
	0.7000	0.5069	0.4752	1.3260	0.7189	3.0787		
	0.6000	0.5843	0.4074	1.5341	0.6181	3.4387		
	0.5000	0.6921	0.3395	1.8253	0.5160	3.9462		
	0.4000	0.8530	0.2714	2.2623	0.4130	4.7108		
	0.3000	1.1201	0.2033	2.9908	0.3094	5.9881		
	0.2000	1.6524	0.1352	4.4478	0.2057	8.5444		
	0.1000	3.2454	0.0674	8.8185	0.1024	16.2117		
	Inf.	1.5327	1.0180	0.8740	0.6709	0.3182		
6	1.0000	0.3313	0.5984	0.8390	0.7964	1.2734	2.0111	
	1.1111	0.2934	0.6667	0.7446	0.8985	1.1050	2.2282	
	1.2500	0.2571	0.7515	0.6542	1.0223	0.9549	2.4742	
	1.4286	0.2219	0.8600	0.5666	1.1787	0.8164	2.7718	
	1.6667	0.1876	1.0040	0.4812	1.3848	0.6859	3.1529	
	2.0000	0.1541	1.2051	0.3976	1.6709	0.5615	3.6720	
	2.5000	0.1216	1.5058	0.3155	2.0972	0.4420	4.4362	
	3.3333	0.0898	2.0058	0.2347	2.8044	0.3266	5.6935	
	5.0000	0.0589	3.0038	0.1553	4.2137	0.2146	8.1871	
	10.0000	0.0290	5.9928	0.0771	8.4320	0.1058	15.6296	
	Inf.	1.4849	1.0430	0.8427	0.7651	0.5972	0.2844	
7	1.0000	0.2826	0.5332	0.7142	0.6988	0.9219	0.9600	2.4404
	0.9000	0.3118	0.4802	0.7896	0.6322	1.0137	0.8718	2.5953
	0.8000	0.3481	0.4271	0.8836	0.5649	1.1287	0.7809	2.7936
	0.7000	0.3945	0.3739	1.0043	0.4967	1.2768	0.6875	3.0535
	0.6000	0.4560	0.3206	1.1650	0.4277	1.4750	0.5919	3.4051
	0.5000	0.5416	0.2671	1.3899	0.3580	1.7531	0.4947	3.9025
	0.4000	0.6695	0.2136	1.7271	0.2874	2.1714	0.3961	4.6534
	0.3000	0.8819	0.1601	2.2890	0.2163	2.8700	0.2969	5.9091
	0.2000	1.3054	0.1066	3.4127	0.1445	4.2690	0.1974	8.4236
	0.1000	2.5731	0.0532	6.7835	0.0724	8.4691	0.0983	15.9666
	Inf.	1.5079	0.9763	0.8402	0.7248	0.6741	0.5305	0.2532
n	$1/R_s$	L_1	C_2	L_3	C_4	L_5	C_6	L_7

TABLE 12-47 (Continued)

n	R_s	C_1	L_2	C_3	L_4	C_5	L_6	C_7	L_8	C_9	L_{10}
8	1.0000	0.2718	0.4999	0.6800	0.6312	0.8498	0.7447	1.3174	1.9626		
	1.1111	0.2408	0.5567	0.6045	0.7116	0.7452	0.8529	1.1169	2.2146		
	1.2500	0.2114	0.6271	0.5324	0.8086	0.6506	0.9780	0.9551	2.4766		
	1.4286	0.1828	0.7173	0.4622	0.9315	0.5612	1.1331	0.8117	2.7837		
	1.6667	0.1549	0.8373	0.3934	1.0939	0.4753	1.3355	0.6795	3.1715		
	2.0000	0.1276	1.0049	0.3256	1.3201	0.3920	1.6148	0.5550	3.6960		
	2.5000	0.1009	1.2559	0.2589	1.6580	0.3107	2.0297	0.4362	4.4654		
	3.3333	0.0747	1.6734	0.1930	2.2194	0.2311	2.7164	0.3220	5.7294		
	5.0000	0.0492	2.5074	0.1279	3.3400	0.1530	4.0835	0.2114	8.2345		
	10.0000	0.0242	5.0066	0.0636	6.6971	0.0760	8.1733	0.1042	15.7101		
	Inf.	1.4915	1.0265	0.8169	0.7548	0.6709	0.6318	0.4995	0.2387		
9	1.0000	0.2347	0.4493	0.5914	0.5747	0.7027	0.6552	0.8944	0.9255	2.4332	
	0.9000	0.2594	0.4045	0.6547	0.5193	0.7754	0.5943	0.9809	0.8427	2.5822	
	0.8000	0.2900	0.3597	0.7336	0.4635	0.8662	0.5322	1.0895	0.7566	2.7745	
	0.7000	0.3291	0.3148	0.8348	0.4073	0.9829	0.4690	1.2299	0.6673	3.0283	
	0.6000	0.3810	0.2699	0.9695	0.3505	1.1388	0.4046	1.4183	0.5753	3.3734	
	0.5000	0.4533	0.2249	1.1580	0.2932	1.3572	0.3392	1.6834	0.4812	3.8629	
	0.4000	0.5613	0.1799	1.4405	0.2355	1.6854	0.2727	2.0828	0.3855	4.6032	
	0.3000	0.7407	0.1348	1.9111	0.1772	2.2331	0.2054	2.7508	0.2889	5.8424	
	0.2000	1.0986	0.0898	2.8522	0.1185	3.3299	0.1373	4.0895	0.1921	8.3246	
	0.1000	2.1702	0.0448	5.6749	0.0594	6.6230	0.0688	8.1099	0.0956	15.7718	
	Inf.	1.4888	0.9495	0.8044	0.6892	0.6589	0.5952	0.5645	0.4475	0.2141	
10	1.0000	0.2359	0.4369	0.5887	0.5428	0.7034	0.5827	0.8720	0.6869	1.4317	1.8431
	1.1111	0.2081	0.4866	0.5218	0.6141	0.6141	0.6729	0.7394	0.8187	1.1397	2.1907
	1.2500	0.1827	0.5480	0.4601	0.6972	0.5376	0.7708	0.6394	0.9483	0.9616	2.4734
	1.4286	0.1582	0.6267	0.3999	0.8024	0.4651	0.8922	0.5487	1.1042	0.8122	2.7907
	1.6667	0.1343	0.7314	0.3407	0.9416	0.3948	1.0514	0.4631	1.3052	0.6777	3.1847
	2.0000	0.1108	0.8777	0.2823	1.1356	0.3263	1.2719	0.3811	1.5809	0.5525	3.7138
	2.5000	0.0877	1.0969	0.2247	1.4258	0.2591	1.6003	0.3017	1.9888	0.4338	4.4876
	3.3333	0.0651	1.4619	0.1676	1.9085	0.1931	2.1451	0.2242	2.6628	0.3200	5.7573
	5.0000	0.0429	2.1910	0.1112	2.8724	0.1280	3.2311	0.1483	4.0033	0.2100	8.2726
	10.0000	0.0212	4.3764	0.0553	5.7612	0.0636	6.4830	0.0736	8.0118	0.1035	15.7776
	Inf.	1.4973	1.0192	0.8005	0.7312	0.6498	0.6331	0.5775	0.5501	0.4369	0.2091
n	$1/R_s$	L_1	C_2	L_3	C_4	L_5	C_6	L_7	C_8	L_9	C_{10}

TABLE 12-48   Linear Phase with Equiripple Error of 0.05° Active Low-Pass Values

Order n	C_1	C_2
2	0.9914	0.6891
4	1.0365	0.8344
	1.3426	0.2959
6	1.1231	0.9257
	1.2146	0.3810
	1.6255	0.1430
8	1.2203	1.0126
	1.2610	0.4285
	1.3864	0.1894
	1.8723	0.08324
10	1.3172	1.0957
	1.3392	0.4676
	1.3968	0.2139
	1.5444	0.1116
	2.0934	0.05483

TABLE 12-49   Linear Phase with Equiripple Error of 0.5° Active Low-Pass Values

Order n	C_1	C_2
2	1.1641	0.7011
4	1.3426	0.9103
	1.6565	0.2314
6	1.5340	1.0578
	1.6215	0.2993
	2.0437	0.1054
8	1.7268	1.1962
	1.7652	0.3445
	1.8857	0.1374
	2.3901	0.06134
10	1.9051	1.3218
	1.9257	0.3826
	1.9798	0.1562
	2.1227	0.07971
	2.6969	0.04074

TABLE 12-50   Transitional Gaussian to 6-dB Pole Locations

Order n	Real Part $-\alpha$	Imaginary Part $\pm j\beta$
3	0.9622 0.9776	1.2214
4	0.7940 0.6304	0.5029 1.5407
5	0.6190 0.3559 0.6650	0.8254 1.5688
6	0.5433 0.4672 0.2204	0.3431 0.9991 1.5067
7	0.4580 0.3649 0.1522 0.4828	0.5932 1.1286 1.4938
8	0.4222 0.3833 0.2878 0.1122	0.2640 0.7716 1.2066 1.4798
9	0.3700 0.3230 0.2309 0.08604 0.3842	0.4704 0.9068 1.2634 1.4740
10	0.3384 0.3164 0.2677 0.1849 0.06706	0.2101 0.6180 0.9852 1.2745 1.4389

TABLE 12-51   Transitional Gaussian to 12-dB Pole Locations

Order n	Real Part $-\alpha$	Imaginary Part $\pm j\beta$
3	0.9360	1.2168
	0.9630	
4	0.9278	1.6995
	0.9192	0.5560
5	0.8075	0.9973
	0.7153	2.0532
	0.8131	
6	0.7019	0.4322
	0.6667	1.2931
	0.4479	2.1363
7	0.6155	0.7703
	0.5486	1.5154
	0.2905	2.1486
	0.6291	
8	0.5441	0.3358
	0.5175	0.9962
	0.4328	1.6100
	0.1978	2.0703
9	0.4961	0.6192
	0.4568	1.2145
	0.3592	1.7429
	0.1489	2.1003
	0.5065	
10	0.4535	0.2794
	0.4352	0.8289
	0.3886	1.3448
	0.2908	1.7837
	0.1136	2.0599

TABLE 12-52   Transitional Gaussian to 6-dB LC Element Values*

n	R_s	C_1	L_2	C_3	L_4	C_5	L_6	C_7
3	1.0000	0.4042	0.8955	2.3380				
	0.9000	0.4440	0.8038	2.5027				
	0.8000	0.4935	0.7121	2.7088				
	0.7000	0.5568	0.6205	2.9739				
	0.6000	0.6407	0.5292	3.3275				
	0.5000	0.7575	0.4384	3.8223				
	0.4000	0.9319	0.3482	4.5635				
	0.3000	1.2213	0.2590	5.7972				
	0.2000	1.7980	0.1709	8.2605				
	0.1000	3.5236	0.0845	15.6391				
	Inf.	1.4742	0.8328	0.3446				
4	1.0000	0.4198	0.7832	1.1598	2.1427			
	1.1111	0.3720	0.8717	1.0279	2.3286			
	1.2500	0.3256	0.9816	0.9010	2.5539			
	1.4286	0.2804	1.1220	0.7781	2.8367			
	1.6667	0.2365	1.3083	0.6587	3.2069			
	2.0000	0.1938	1.5678	0.5424	3.7179			
	2.5000	0.1524	1.9552	0.4289	4.4761			
	3.3333	0.1122	2.5982	0.3180	5.7296			
	5.0000	0.0733	3.8797	0.2095	8.2221			
	10.0000	0.0359	7.7138	0.1035	15.6717			
	Inf.	1.4871	1.0222	0.7656	0.3510			
5	1.0000	0.4544	0.8457	1.0924	1.0774	2.4138		
	0.9000	0.4991	0.7622	1.2046	0.9769	2.5746		
	0.8000	0.5543	0.6781	1.3452	0.8739	2.7797		
	0.7000	0.6247	0.5936	1.5263	0.7687	3.0475		
	0.6000	0.7179	0.5086	1.7683	0.6615	3.4087		
	0.5000	0.8476	0.4233	2.1077	0.5527	3.9183		
	0.4000	1.0411	0.3379	2.6176	0.4428	4.6863		
	0.3000	1.3621	0.2526	3.4680	0.3321	5.9690		
	0.2000	2.0019	0.1677	5.1693	0.2212	8.5360		
	0.1000	3.9166	0.0833	10.2723	0.1103	16.2354		
	Inf.	1.5392	1.0993	1.0203	0.8269	0.3824		
6	1.0000	0.5041	0.9032	1.2159	1.0433	1.4212	2.0917	
	1.1111	0.4427	1.0079	1.0739	1.1892	1.2274	2.3324	
	1.2500	0.3853	1.1364	0.9415	1.3611	1.0620	2.5935	
	1.4286	0.3306	1.2999	0.8145	1.5753	0.9111	2.9053	
	1.6667	0.2779	1.5162	0.6914	1.8557	0.7692	3.3032	
	2.0000	0.2271	1.8169	0.5713	2.2433	0.6333	3.8456	
	2.5000	0.1780	2.2654	0.4534	2.8200	0.5016	4.6459	
	3.3333	0.1308	3.0091	0.3376	3.7758	0.3730	5.9662	
	5.0000	0.0853	4.4902	0.2235	5.6803	0.2468	8.5904	
	10.0000	0.0416	8.9199	0.1109	11.3810	0.1225	16.4352	
	Inf.	1.5664	1.2166	1.1389	1.1010	0.8844	0.4062	

TABLE 12-52   *(Continued)*

n	R_s	C_1	L_2	C_3	L_4	C_5	L_6	C_7
7	1.0000	0.4918	0.9232	1.2146	1.1224	1.3154	1.1407	2.5039
	0.9000	0.5403	0.8318	1.3393	1.0196	1.4426	1.0434	2.6575
	0.8000	0.6001	0.7399	1.4950	0.9141	1.6040	0.9401	2.8593
	0.7000	0.6760	0.6474	1.6952	0.8061	1.8144	0.8317	3.1285
	0.6000	0.7763	0.5545	1.9626	0.6956	2.0986	0.7190	3.4967
	0.5000	0.9157	0.4613	2.3373	0.5829	2.5004	0.6029	4.0203
	0.4000	1.1236	0.3681	2.9002	0.4684	3.1072	0.4844	4.8129
	0.3000	1.4685	0.2750	3.8389	0.3524	4.1232	0.3644	6.1397
	0.2000	2.1560	0.1823	5.7166	0.2354	6.1604	0.2433	8.7977
	0.1000	4.2137	0.0905	11.3483	0.1178	12.2787	0.1217	16.7743
	Inf.	1.5950	1.2166	1.2240	1.1784	1.1260	0.8975	0.4110
n	$1/R_s$	L_1	C_2	L_3	C_4	L_5	C_6	L_7

* Reprinted from A. I. Zverev, "Handbook of Filter Synthesis," John Wiley and Sons, New York, 1967.

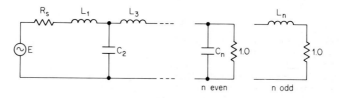

TABLE 12-52 *(Continued)*

n	R_s	C_1	L_2	C_3	L_4	C_5	L_6	C_7	L_8	C_9	L_{10}
8	1.0502	0.5031	0.9699	1.2319	1.1324	1.4262	1.0449	1.6000	1.9285		
	1.1111	0.4586	1.0338	1.1286	1.2497	1.2635	1.2099	1.3372	2.2286		
	1.2500	0.3964	1.1670	0.9831	1.4404	1.0842	1.4259	1.1197	2.5453		
	1.4286	0.3392	1.3351	0.8487	1.6698	0.9299	1.6706	0.9502	2.8771		
	1.6667	0.2848	1.5571	0.7195	1.9674	0.7863	1.9808	0.7989	3.2846		
	2.0000	0.2325	1.8656	0.5939	2.3776	0.6487	2.4039	0.6569	3.8326		
	2.5000	0.1822	2.3255	0.4710	2.9870	0.5151	3.0295	0.5204	4.6374		
	3.3333	0.1337	3.0879	0.3504	3.9962	0.3840	4.0640	0.3874	5.9636		
	5.0000	0.0872	4.6062	0.2318	6.0063	0.2547	6.1243	0.2566	8.5995		
	10.0000	0.0425	9.1467	0.1150	12.0217	0.1268	12.2919	0.1276	16.4808		
	Inf.	1.5739	1.2698	1.2325	1.2633	1.2017	1.1404	0.9066	0.4148		
9	1.0000	0.4979	0.9367	1.2371	1.1589	1.3845	1.1670	1.3983	1.1422	2.5277	
	0.9000	0.5475	0.8439	1.3648	1.0517	1.5194	1.0673	1.5233	1.0527	2.6698	
	0.8000	0.6083	0.7505	1.5238	0.9424	1.6894	0.9625	1.6850	0.9540	2.8635	
	0.7000	0.6854	0.6567	1.7278	0.8306	1.9103	0.8527	1.8996	0.8472	3.1279	
	0.6000	0.7870	0.5624	1.9998	0.7165	2.2081	0.7383	2.1929	0.7342	3.4938	
	0.5000	0.9280	0.4679	2.3811	0.6002	2.6288	0.6202	2.6105	0.6166	4.0174	
	0.4000	1.1383	0.3732	2.9537	0.4820	3.2641	0.4993	3.2438	0.4960	4.8118	
	0.3000	1.4873	0.2788	3.9087	0.3625	4.3274	0.3763	4.3062	0.3734	6.1429	
	0.2000	2.1829	0.1848	5.8191	0.2421	6.4590	0.2518	6.4381	0.2496	8.8109	
	0.1000	4.2652	0.0918	11.5490	0.1211	12.8601	0.1262	12.8438	0.1250	16.8186	
	Inf.	1.6014	1.2508	1.2817	1.2644	1.2805	1.2103	1.1456	0.9096	0.4160	
10	1.1372	0.4682	1.0839	1.1516	1.2991	1.3293	1.2748	1.4216	1.1730	1.5040	2.1225
	1.1372	0.4682	1.0839	1.1516	1.2991	1.3293	1.2748	1.4216	1.1730	1.5040	2.1225
	1.2500	0.4087	1.1987	1.0148	1.4855	1.1389	1.5155	1.1705	1.4593	1.1798	2.5537
	1.4286	0.3489	1.3718	0.8744	1.7253	0.9733	1.7813	0.9908	1.7344	0.9878	2.9155
	1.6667	0.2928	1.6000	0.7409	2.0334	0.8219	2.1124	0.8338	2.0664	0.8275	3.3380
	2.0000	0.2389	1.9169	0.6114	2.4574	0.6776	2.5622	0.6868	2.5129	0.6799	3.8995
	2.5000	0.1872	2.3893	0.4848	3.0868	0.5377	3.2264	0.5451	3.1699	0.5387	4.7218
	3.3333	0.1373	3.1723	0.3606	4.1290	0.4007	4.3241	0.4065	4.2549	0.4011	6.0762
	5.0000	0.0895	4.7317	0.2385	6.2048	0.2657	6.5094	0.2698	6.4154	0.2659	8.7681
	10.0000	0.0437	9.3953	0.1183	12.4165	0.1322	13.0503	0.1345	12.8837	0.1323	16.8178
	Inf.	1.6077	1.3178	1.2927	1.3406	1.3070	1.3160	1.2409	1.1733	0.9311	0.4257
n	$1/R_s$	L_1	C_2	L_3	C_4	L_5	C_6	L_7	C_8	L_9	C_{10}

TABLE 12-53   Transitional Gaussian to 12-dB LC Element Values*

n even   n odd

n	R_s	C_1	L_2	C_3	L_4	C_5	L_6	C_7
3	1.0000	0.4152	0.9050	2.3452				
	0.9000	0.4560	0.8126	2.5101				
	0.8000	0.5067	0.7202	2.7166				
	0.7000	0.5715	0.6278	2.9825				
	0.6000	0.6573	0.5356	3.3372				
	0.5000	0.7769	0.4438	3.8336				
	0.4000	0.9554	0.3526	4.5775				
	0.3000	1.2517	0.2623	5.8157				
	0.2000	1.8420	0.1732	8.2884				
	0.1000	3.6083	0.0856	15.6955				
	Inf.	1.4800	0.8440	0.3527				
4	1.0000	0.3097	0.6545	1.0598	2.1518			
	1.1111	0.2757	0.7262	0.9418	2.3289			
	1.2500	0.2423	0.8156	0.8268	2.5459			
	1.4286	0.2096	0.9300	0.7146	2.8203			
	1.6667	0.1775	1.0821	0.6050	3.1814			
	2.0000	0.1461	1.2944	0.4980	3.6812			
	2.5000	0.1153	1.6118	0.3934	4.4241			
	3.3333	0.0853	2.1393	0.2913	5.6532			
	5.0000	0.0560	3.1917	0.1916	8.0979			
	10.0000	0.0276	6.3425	0.0944	15.4048			
	Inf.	1.4585	0.9300	0.6294	0.2707			
5	1.0000	0.2909	0.5837	0.8112	0.9660	2.3745		
	0.9000	0.3207	0.5253	0.8961	0.8707	2.5377		
	0.8000	0.3577	0.4667	1.0019	0.7746	2.7433		
	0.7000	0.4051	0.4081	1.1379	0.6777	3.0092		
	0.6000	0.4680	0.3495	1.3192	0.5804	3.3650		
	0.5000	0.5556	0.2908	1.5727	0.4827	3.8642		
	0.4000	0.6865	0.2322	1.9528	0.3850	4.6138		
	0.3000	0.9038	0.1738	2.5859	0.2875	5.8631		
	0.2000	1.3372	0.1155	3.8515	0.1907	8.3597		
	0.1000	2.6347	0.0575	7.6464	0.0947	15.8420		
	Inf.	1.4953	0.9388	0.7587	0.5724	0.2592		

TABLE 12-53   *(Continued)*

n	R_s	C_1	L_2	C_3	L_4	C_5	L_6	C_7
6	1.0000	0.3164	0.6070	0.7962	0.7880	1.1448	2.1154	
	1.1111	0.2813	0.6750	0.7108	0.8826	1.0087	2.3076	
	1.2500	0.2470	0.7597	0.6273	0.9994	0.8804	2.5365	
	1.4286	0.2135	0.8681	0.5452	1.1481	0.7580	2.8209	
	1.6667	0.1807	1.0123	0.4644	1.3451	0.6402	3.1908	
	2.0000	0.1487	1.2136	0.3847	1.6194	0.5263	3.6993	
	2.5000	0.1174	1.5148	0.3060	2.0292	0.4157	4.4522	
	3.3333	0.0868	2.0154	0.2282	2.7098	0.3080	5.6952	
	5.0000	0.0570	3.0146	0.1513	4.0679	0.2028	8.1654	
	10.0000	0.0280	6.0071	0.0753	8.1355	0.1002	15.5460	
	Inf.	1.4732	0.9894	0.8129	0.7484	0.5979	0.2752	
7	1.0000	0.3207	0.6267	0.8091	0.7753	0.9241	0.9649	2.3829
	0.9000	0.3534	0.5641	0.8946	0.7016	1.0176	0.8750	2.5374
	0.8000	0.3940	0.5015	1.0015	0.6270	1.1350	0.7824	2.7351
	0.7000	0.4458	0.4387	1.1388	0.5513	1.2867	0.6876	2.9937
	0.6000	0.5146	0.3758	1.3218	0.4747	1.4899	0.5910	3.3428
	0.5000	0.6102	0.3128	1.5781	0.3972	1.7755	0.4931	3.8355
	0.4000	0.7531	0.2498	1.9626	0.3189	2.2054	0.3943	4.5779
	0.3000	0.9902	0.1869	2.6034	0.2399	2.9236	0.2951	5.8175
	0.2000	1.4630	0.1243	3.8851	0.1603	4.3624	0.1961	8.2974
	0.1000	2.8780	0.0619	7.7299	0.0803	8.6825	0.0976	15.7326
	Inf.	1.4861	0.9693	0.8643	0.8040	0.7689	0.6157	0.2826
n	$1/R_s$	L_1	C_2	L_3	C_4	L_5	C_6	L_7

* Reprinted from A. I. Zverev, "Handbook of Filter Synthesis," John Wiley and Sons, New York, 1967.

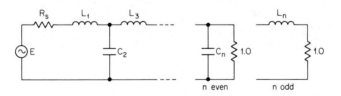

TABLE 12-53 (Continued)

n	R_s	C_1	L_2	C_3	L_4	C_5	L_6	C_7	L_8	C_9	L_{10}
8	1.0000	0.3449	0.6565	0.8686	0.8028	0.9701	0.8182	1.2503	2.0612		
	1.1111	0.3053	0.7304	0.7729	0.9044	0.8550	0.9339	1.0753	2.2930		
	1.2500	0.2674	0.8221	0.6810	1.0276	0.7489	1.0694	0.9272	2.5445		
	1.4286	0.2308	0.9394	0.5913	1.1837	0.6480	1.2378	0.7929	2.8444		
	1.6667	0.1952	1.0953	0.5034	1.3898	0.5504	1.4579	0.6673	3.2267		
	2.0000	0.1604	1.3128	0.4168	1.6764	0.4553	1.7620	0.5476	3.7473		
	2.5000	0.1265	1.6381	0.3314	2.1045	0.3621	2.2143	0.4323	4.5145		
	3.3333	0.0934	2.1788	0.2471	2.8155	0.2702	2.9640	0.3203	5.7789		
	5.0000	0.0613	3.2575	0.1638	4.2343	0.1794	4.4583	0.2111	8.2899		
	10.0000	0.0301	6.4874	0.0814	8.4843	0.0894	8.9330	0.1044	15.7917		
	Inf.	1.4974	1.0324	0.8943	0.8908	0.8494	0.8098	0.6452	0.2955		
9	1.0000	0.3318	0.6500	0.8467	0.8167	0.9426	0.8239	0.9857	0.9630	2.4140	
	0.9000	0.3657	0.5852	0.9363	0.7389	1.0390	0.7492	1.0803	0.8785	2.5608	
	0.8000	0.4078	0.5201	1.0480	0.6602	1.1599	0.6725	1.2003	0.7894	2.7524	
	0.7000	0.4614	0.4550	1.1914	0.5806	1.3160	0.5936	1.3568	0.6963	3.0070	
	0.6000	0.5324	0.3897	1.3825	0.4999	1.5251	0.5127	1.5681	0.6001	3.3538	
	0.5000	0.6312	0.3243	1.6500	0.4183	1.8192	0.4301	1.8667	0.5015	3.8460	
	0.4000	0.7787	0.2590	2.0512	0.3358	2.2620	0.3460	2.3177	0.4016	4.5897	
	0.3000	1.0234	0.1938	2.7200	0.2526	3.0021	0.2607	3.0729	0.3009	5.8332	
	0.2000	1.5113	0.1288	4.0574	0.1687	4.4850	0.1744	4.5875	0.2000	8.3219	
	0.1000	2.9716	0.0641	8.0692	0.0845	8.9384	0.0875	9.1379	0.0996	15.7849	
	Inf.	1.4917	0.9908	0.9105	0.8770	0.8910	0.8457	0.8022	0.6376	0.2917	
10	1.0139	0.3500	0.6698	0.8817	0.8148	1.0183	0.7949	1.0929	0.7508	1.4303	1.8322
	1.1111	0.3092	0.7364	0.7856	0.9200	0.8864	0.9293	0.9147	0.9187	1.1138	2.2110
	1.2500	0.2701	0.8290	0.6907	1.0477	0.7734	1.0708	0.7890	1.0712	0.9404	2.4914
	1.4286	0.2328	0.9474	0.5992	1.2075	0.6681	1.2428	0.6777	1.2503	0.7978	2.8009
	1.6667	0.1968	1.1046	0.5099	1.4179	0.5671	1.4663	0.5734	1.4791	0.6690	3.1853
	2.0000	0.1616	1.3239	0.4221	1.7103	0.4689	1.7746	0.4734	1.7920	0.5483	3.7034
	2.5000	0.1274	1.6517	0.3355	2.1467	0.3728	2.2329	0.3761	2.2552	0.4326	4.4640
	3.3333	0.0941	2.1966	0.2501	2.8714	0.2781	2.9923	0.2806	3.0215	0.3206	5.7162
	5.0000	0.0617	3.2837	0.1658	4.3171	0.1845	4.5061	0.1863	4.5478	0.2114	8.2022
	10.0000	0.0303	6.5385	0.0824	8.6473	0.0919	9.0397	0.0929	9.1181	0.1046	15.6293
	Inf.	1.4826	1.0350	0.9134	0.9263	0.9061	0.9159	0.8654	0.8190	0.6502	0.2973
n	$1/R_s$	L_1	C_2	L_3	C_4	L_5	C_6	L_7	C_8	L_9	C_{10}

TABLE 12-54   Transitional Gaussian to 6-dB Active Low-Pass Values

Order n	C_1	C_2
4	1.2594	0.8989
	1.5863	0.2275
6	1.8406	1.3158
	2.1404	0.3841
	4.5372	0.09505
8	2.3685	1.7028
	2.6089	0.5164
	3.4746	0.1870
	8.9127	0.05094
10	2.9551	2.1329
	3.1606	0.6564
	3.7355	0.2568
	5.4083	0.1115
	14.9120	0.03232

TABLE 12-55   Transitional Gaussian to 12-dB Active Low-Pass Values

Order n	C_1	C_2
4	1.0778	0.2475
	1.0879	0.7965
6	1.4247	1.0330
	1.5000	0.3150
	2.2326	0.09401
8	1.8379	1.3309
	1.9324	0.4106
	2.3105	0.1557
	5.0556	0.04573
10	2.2051	1.5984
	2.2978	0.4965
	2.5733	0.1983
	3.4388	0.08903
	8.8028	0.02669

TABLE 12-56 Elliptic-Function LC Element Values

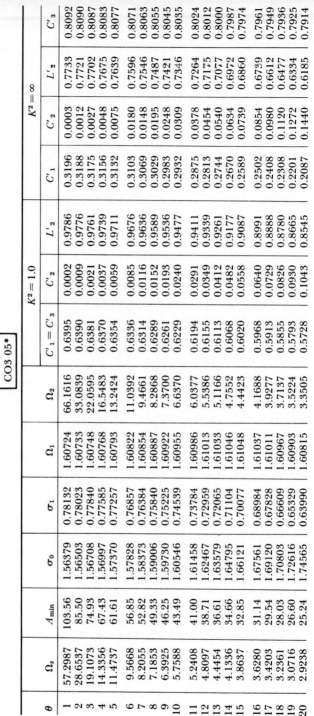

CO3 05*

θ	Ω_s	A_{min}	σ_0	σ_1	Ω_1	Ω_2	$K^2=1.0$			$K^2=\infty$			
							$C'_1=C'_3$	C'_2	L'_2	C'_1	C'_2	L'_2	C'_3
1	57.2987	103.56	1.56379	0.78132	1.60724	66.1616	0.6395	0.0002	0.9786	0.3196	0.0003	0.7733	0.8092
2	28.6537	85.50	1.56503	0.78023	1.60733	33.0839	0.6390	0.0009	0.9776	0.3188	0.0012	0.7721	0.8090
3	19.1073	74.93	1.56708	0.77840	1.60748	22.0595	0.6381	0.0021	0.9761	0.3175	0.0027	0.7702	0.8087
4	14.3356	67.43	1.56997	0.77585	1.60768	16.5483	0.6376	0.0037	0.9739	0.3156	0.0048	0.7675	0.8083
5	11.4737	61.61	1.57370	0.77257	1.60793	13.2424	0.6354	0.0059	0.9711	0.3132	0.0075	0.7639	0.8077
6	9.5668	56.85	1.57828	0.76857	1.60822	11.0392	0.6336	0.0085	0.9676	0.3103	0.0108	0.7596	0.8071
7	8.2055	52.82	1.58373	0.76384	1.60854	9.4661	0.6314	0.0116	0.9636	0.3069	0.0148	0.7546	0.8063
8	7.1853	49.33	1.59006	0.75840	1.60887	8.2868	0.6289	0.0152	0.9589	0.3029	0.0195	0.7487	0.8055
9	6.3925	46.25	1.59730	0.75225	1.60922	7.3700	0.6261	0.0193	0.9536	0.2983	0.0248	0.7421	0.8045
10	5.7588	43.49	1.60546	0.74539	1.60955	6.6370	0.6229	0.0240	0.9477	0.2932	0.0309	0.7346	0.8035
11	5.2408	41.00	1.61458	0.73784	1.60986	6.0377	0.6194	0.0291	0.9411	0.2875	0.0378	0.7264	0.8024
12	4.8097	38.71	1.62467	0.72959	1.61013	5.5386	0.6155	0.0349	0.9339	0.2813	0.0454	0.7175	0.8012
13	4.4454	36.61	1.63579	0.72065	1.61033	5.1166	0.6113	0.0412	0.9261	0.2744	0.0540	0.7077	0.8000
14	4.1336	34.66	1.64795	0.71104	1.61046	4.7552	0.6068	0.0482	0.9177	0.2670	0.0634	0.6972	0.7987
15	3.8637	32.85	1.66121	0.70077	1.61048	4.4423	0.6020	0.0558	0.9087	0.2589	0.0739	0.6860	0.7974
16	3.6280	31.14	1.67561	0.68984	1.61037	4.1688	0.5968	0.0640	0.8991	0.2502	0.0854	0.6739	0.7961
17	3.4203	29.54	1.69120	0.67828	1.61011	3.9277	0.5913	0.0729	0.8888	0.2408	0.0980	0.6612	0.7949
18	3.2361	28.03	1.70803	0.66609	1.60967	3.7137	0.5855	0.0826	0.8780	0.2308	0.1120	0.6477	0.7936
19	3.0716	26.60	1.72616	0.65329	1.60903	3.5224	0.5793	0.0930	0.8665	0.2201	0.1272	0.6334	0.7925
20	2.9238	25.24	1.74565	0.63990	1.60815	3.3505	0.5728	0.1043	0.8545	0.2087	0.1440	0.6185	0.7914

θ	Ω_s	A_{min}	σ_0	σ_1	Ω_1	Ω_2	$L'_1 = L'_3$	L'_2	C'_2	L'_1	L'_2	C'_2	L'_3
21	2.7904	23.95	1.76659	0.62595	1.60701	3.1951	0.5661	0.1164	0.8418	0.1965	0.1625	0.6028	0.7905
22	2.6695	22.71	1.78903	0.61144	1.60558	3.0541	0.5590	0.1294	0.8286	0.1836	0.1828	0.5865	0.7897
23	2.5593	21.53	1.81308	0.59641	1.60383	2.9256	0.5515	0.1434	0.8148	0.1699	0.2052	0.5695	0.7891
24	2.4586	20.40	1.83881	0.58089	1.60172	2.8079	0.5438	0.1585	0.8004	0.1553	0.2298	0.5519	0.7887
25	2.3662	19.31	1.86632	0.56489	1.59923	2.6999	0.5358	0.1747	0.7855	0.1399	0.2571	0.5337	0.7887
26	2.2812	18.27	1.89572	0.54847	1.59633	2.6003	0.5275	0.1921	0.7700	0.1236	0.2873	0.5149	0.7889
27	2.2027	17.26	1.92713	0.53164	1.59298	2.5083	0.5189	0.2108	0.7540	0.1063	0.3208	0.4955	0.7896
28	2.1301	16.30	1.96066	0.51445	1.58916	2.4231	0.5100	0.2309	0.7375	0.0880	0.3580	0.4757	0.7908
29	2.0627	15.37	1.99644	0.49694	1.58485	2.3438	0.5009	0.2526	0.7205	0.0687	0.3997	0.4555	0.7924
30	2.0000	14.47	2.03461	0.47915	1.58000	2.2701	0.4915	0.2760	0.7031	0.0483	0.4463	0.4348	0.7947
31	1.9416	13.61	2.07532	0.46114	1.57461	2.2012	0.4819	0.3012	0.6852	0.0268	0.4986	0.4139	0.7977
32	1.8871	12.77	2.11873	0.44294	1.56865	2.1368	0.4720	0.3284	0.6669	0.0040	0.5577	0.3927	0.8014
33	1.8361	11.97	2.16500	0.42462	1.56210	2.0765	0.4619	0.3578	0.6482	-0.0201	0.6244	0.3714	0.8061
34	1.7883	11.20	2.21430	0.40622	1.55494	2.0199	0.4516	0.3896	0.6291	-0.0456	0.7003	0.3500	0.8117
35	1.7434	10.46	2.26682	0.38781	1.54717	1.9666	0.4411	0.4241	0.6097	-0.0725	0.7868	0.3286	0.8185

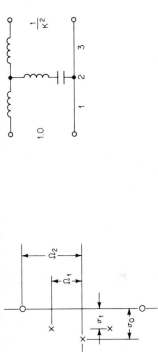

* Adapted from A. I. Zverev, "Handbook of Filter Synthesis," John Wiley and Sons, New York, 1967.

TABLE 12-56   *(Continued)*

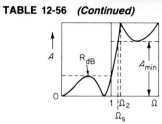

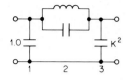

CO3 20*

							$K^2 = 1.0$			$K^2 = \infty$			
θ	Ω_s	A_{min}	σ_0	σ_1	Ω_1	Ω_2	$C'_1 = C'_3$	C'_2	L'_2	C'_1	C'_2	L'_2	C'_3
1	57.2987	115.77	0.84082	0.42027	1.13143	66.1616	1.1893	0.0002	1.1540	0.5946	0.0002	1.1713	1.1717
2	28.6537	97.70	0.84114	0.42001	1.13150	33.0839	1.1889	0.0008	1.1533	0.5940	0.0008	1.1704	1.1715
3	19.1073	87.13	0.84169	0.41958	1.13160	22.0595	1.1881	0.0018	1.1522	0.5932	0.0018	1.1690	1.1710
4	14.3356	79.63	0.84246	0.41899	1.13175	16.5483	1.1870	0.0032	1.1507	0.5920	0.0031	1.1670	1.1704
5	11.4737	73.81	0.84345	0.41822	1.13194	13.2424	1.1856	0.0050	1.1488	0.5904	0.0049	1.1645	1.1696
6	9.5668	69.05	0.84466	0.41728	1.13218	11.0392	1.1839	0.0072	1.1464	0.5885	0.0071	1.1614	1.1686
7	8.2055	65.03	0.84609	0.41617	1.13245	9.4661	1.1819	0.0098	1.1436	0.5862	0.0096	1.1577	1.1675
8	7.1853	61.54	0.84776	0.41489	1.13276	8.2868	1.1796	0.0128	1.1404	0.5836	0.0126	1.1535	1.1662
9	6.3925	58.46	0.84965	0.41344	1.13311	7.3700	1.1770	0.0162	1.1367	0.5807	0.0160	1.1487	1.1647
10	5.7588	55.70	0.85177	0.41182	1.13350	6.6370	1.1740	0.0200	1.1326	0.5773	0.0199	1.1434	1.1630
11	5.2408	53.20	0.85413	0.41003	1.13392	6.0377	1.1708	0.0243	1.1281	0.5737	0.0241	1.1374	1.1611
12	4.8097	50.92	0.85673	0.40807	1.13438	5.5386	1.1672	0.0290	1.1231	0.5696	0.0288	1.1310	1.1591
13	4.4454	48.82	0.85957	0.40593	1.13486	5.1166	1.1634	0.0342	1.1177	0.5653	0.0340	1.1239	1.1570
14	4.1336	46.87	0.86266	0.40363	1.13538	4.7552	1.1592	0.0398	1.1119	0.5605	0.0396	1.1163	1.1546
15	3.8637	45.05	0.86600	0.40115	1.13592	4.4423	1.1547	0.0458	1.1057	0.5554	0.0457	1.1082	1.1521
16	3.6280	43.35	0.86959	0.39851	1.13649	4.1688	1.1500	0.0524	1.0990	0.5500	0.0523	1.0994	1.1495
17	3.4203	41.75	0.87345	0.39569	1.13709	3.9277	1.1449	0.0594	1.0919	0.5441	0.0595	1.0902	1.1467
18	3.2361	40.23	0.87759	0.39270	1.13770	3.7137	1.1395	0.0669	1.0844	0.5379	0.0671	1.0803	1.1437
19	3.0716	38.80	0.88199	0.38954	1.13833	3.5224	1.1338	0.0749	1.0764	0.5314	0.0753	1.0700	1.1407
20	2.9238	37.44	0.88668	0.38621	1.13897	3.3505	1.1278	0.0834	1.0681	0.5244	0.0841	1.0590	1.1374
21	2.7904	36.14	0.89167	0.38272	1.13963	3.1951	1.1215	0.0925	1.0593	0.5171	0.0935	1.0475	1.1340
22	2.6695	34.90	0.89695	0.37905	1.14029	3.0541	1.1149	0.1021	1.0500	0.5094	0.1035	1.0355	1.1305
23	2.5593	33.71	0.90254	0.37521	1.14096	2.9256	1.1080	0.1123	1.0404	0.5013	0.1142	1.0229	1.1269
24	2.4586	32.57	0.90845	0.37120	1.14162	2.8079	1.1008	0.1231	1.0303	0.4928	0.1256	1.0098	1.1232
25	2.3662	31.47	0.91469	0.36702	1.14228	2.6999	1.0933	0.1345	1.0199	0.4839	0.1377	0.9961	1.1193
26	2.2812	30.41	0.92127	0.36268	1.14294	2.6003	1.0855	0.1466	1.0090	0.4746	0.1506	0.9819	1.1153
27	2.2027	29.39	0.92820	0.35817	1.14358	2.5083	1.0773	0.1593	0.9976	0.4649	0.1643	0.9672	1.1113
28	2.1301	28.41	0.93550	0.35349	1.14420	2.4231	1.0689	0.1728	0.9859	0.4548	0.1789	0.9519	1.1071
29	2.0627	27.45	0.94318	0.34864	1.14480	2.3438	1.0602	0.1869	0.9738	0.4443	0.1944	0.9362	1.1029
30	2.0000	26.53	0.95125	0.34364	1.14538	2.2701	1.0512	0.2019	0.9612	0.4333	0.2110	0.9199	1.0985

TABLE 12-56 (Continued)

							$K^2 = 1.0$			$K^2 = \infty$			
θ	Ω_s	A_{min}	σ_0	σ_1	Ω_1	Ω_2	$C'_1 = C'_3$	C'_2	L'_2	C'_1	C'_2	L'_2	C'_3
31	1.9416	25.63	0.95973	0.33847	1.14592	2.2012	1.0420	0.2176	0.9483	0.4219	0.2285	0.9030	1.0942
32	1.8871	24.76	0.96863	0.33313	1.14643	2.1368	1.0324	0.2343	0.9349	0.4101	0.2473	0.8857	1.0897
33	1.8361	23.92	0.97799	0.32764	1.14689	2.0765	1.0225	0.2518	0.9212	0.3978	0.2672	0.8679	1.0853
34	1.7883	23.09	0.98780	0.32199	1.14730	2.0199	1.0123	0.2702	0.9070	0.3851	0.2885	0.8496	1.0808
35	1.7434	22.29	0.99810	0.31619	1.14766	1.9666	1.0019	0.2897	0.8925	0.3719	0.3112	0.8308	1.0762
36	1.7013	21.51	1.00890	0.31023	1.14796	1.9165	0.9912	0.3103	0.8776	0.3582	0.3355	0.8116	1.0717
37	1.6616	20.74	1.02024	0.30412	1.14819	1.8692	0.9802	0.3320	0.8623	0.3441	0.3614	0.7919	1.0672
38	1.6243	20.00	1.03213	0.29786	1.14835	1.8245	0.9689	0.3549	0.8466	0.3294	0.3892	0.7718	1.0627
39	1.5890	19.27	1.04460	0.29147	1.14842	1.7823	0.9573	0.3791	0.8305	0.3142	0.4191	0.7512	1.0583
40	1.5557	18.56	1.05768	0.28493	1.14841	1.7423	0.9455	0.4047	0.8141	0.2985	0.4511	0.7303	1.0540
41	1.5243	17.86	1.07140	0.27825	1.14830	1.7044	0.9334	0.4318	0.7973	0.2823	0.4856	0.7089	1.0497
42	1.4945	17.18	1.08579	0.27145	1.14810	1.6684	0.9210	0.4605	0.7801	0.2655	0.5228	0.6872	1.0456
43	1.4663	16.52	1.10089	0.26452	1.14778	1.6343	0.9084	0.4909	0.7627	0.2481	0.5629	0.6651	1.0416
44	1.4396	15.86	1.11673	0.25747	1.14735	1.6018	0.8955	0.5232	0.7448	0.2301	0.6064	0.6427	1.0378
45	1.4142	15.22	1.13336	0.25031	1.14679	1.5710	0.8823	0.5576	0.7267	0.2115	0.6535	0.6200	1.0341
46	1.3902	14.60	1.15082	0.24304	1.14611	1.5415	0.8689	0.5942	0.7082	0.1923	0.7048	0.5971	1.0307
47	1.3673	13.98	1.16915	0.23567	1.14528	1.5135	0.8553	0.6331	0.6895	0.1725	0.7607	0.5739	1.0276
48	1.3456	13.38	1.18840	0.22821	1.14432	1.4868	0.8415	0.6747	0.6705	0.1519	0.8217	0.5505	1.0248
49	1.3250	12.79	1.20862	0.22067	1.14320	1.4613	0.8274	0.7192	0.6511	0.1307	0.8886	0.5270	1.0223
50	1.3054	12.22	1.22988	0.21306	1.14192	1.4369	0.8131	0.7668	0.6316	0.1087	0.9621	0.5034	1.0202
51	1.2868	11.65	1.25221	0.20539	1.14048	1.4137	0.7986	0.8179	0.6118	0.0860	1.0431	0.4797	1.0185
52	1.2690	11.10	1.27570	0.19766	1.13887	1.3914	0.7839	0.8728	0.5918	0.0625	1.1327	0.4560	1.0173
53	1.2521	10.56	1.30040	0.18990	1.13709	1.3702	0.7690	0.9319	0.5716	0.0382	1.2320	0.4323	1.0166
54	1.2361	10.03	1.32639	0.18211	1.13512	1.3498	0.7539	0.9958	0.5512	0.0130	1.3426	0.4088	1.0165
55	1.2208	9.51	1.35374	0.17431	1.13297	1.3303	0.7387	1.0648	0.5306	−0.0131	1.4662	0.3854	1.0171
56	1.2062	9.01	1.38253	0.16652	1.13064	1.3117	0.7233	1.1397	0.5100	−0.0401	1.6046	0.3622	1.0184
57	1.1924	8.51	1.41284	0.15873	1.12811	1.2938	0.7078	1.2210	0.4892	−0.0681	1.7605	0.3393	1.0205
58	1.1792	8.03	1.44478	0.15098	1.12540	1.2767	0.6921	1.3097	0.4684	−0.0971	1.9366	0.3168	1.0235
59	1.1666	7.57	1.47842	0.14328	1.12249	1.2603	0.6764	1.4065	0.4476	−0.1272	2.1364	0.2947	1.0274
60	1.1547	7.11	1.51387	0.13565	1.11939	1.2446	0.6606	1.5127	0.4268	−0.1584	2.3640	0.2731	1.0323
θ	Ω_s	A_{min}	σ_0	σ_1	Ω_1	Ω_2	$L'_1 = L'_3$	L'_2	C'_2	L'_1	L'_2	C'_2	L'_3

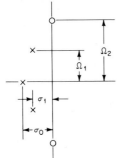

* Adapted from A. I. Zverev, "Handbook of Filter Synthesis," John Wiley and Sons, New York, 1967.

TABLE 12-56 *(Continued)*

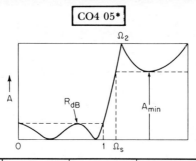

CO4 05*

θ	Ω_s	A_{min}	σ_1	σ_3	Ω_1	Ω_2	Ω_3
C	∞	∞	−0.4050275	−0.9778230	1.3452476	∞	0.5572198
6.0	10.350843	88.5	−0.4016789	−0.9797660	1.3444158	11.367741	0.5610746
7.0	8.876727	83.1	−0.4004697	−0.9804681	1.3441135	9.747389	0.5624723
8.0	7.771760	78.5	−0.3990745	−0.9812783	1.3437635	8.532615	0.5640888
9.0	6.912894	74.4	−0.3974934	−0.9821970	1.3433651	7.588226	0.5659256
10.0	6.226301	70.7	−0.3957264	−0.9832241	1.3429179	6.833109	0.5679847
11.0	5.664999	67.3	−0.3937735	−0.9843599	1.3424210	6.215646	0.5702682
12.0	5.197666	64.3	−0.3916348	−0.9856046	1.3418737	5.701423	0.5727782
13.0	4.802620	61.5	−0.3893103	−0.9869583	1.3412750	5.266618	0.5755174
14.0	4.464371	58.9	−0.3868003	−0.9884214	1.3406240	4.894214	0.5784885
15.0	4.171563	56.5	−0.3841047	−0.9899941	1.3399195	4.571732	0.5816946
16.0	3.915678	54.2	−0.3812237	−0.9916767	1.3391603	4.289813	0.5851389
17.0	3.690200	52.1	−0.3781575	−0.9934696	1.3383450	4.041300	0.5888249
18.0	3.490065	50.1	−0.3749063	−0.9953733	1.3374723	3.820626	0.5927565
19.0	3.311272	48.1	−0.3714704	−0.9973882	1.3365405	3.623399	0.5969377
20.0	3.150622	46.3	−0.3678500	−0.9995149	1.3355479	3.446101	0.6013728
21.0	3.005526	44.6	−0.3640455	−1.0017540	1.3344925	3.285888	0.6060665
22.0	2.873864	42.9	−0.3600572	−1.0041063	1.3333724	3.140431	0.6110236
23.0	2.753885	41.3	−0.3558857	−1.0065726	1.3321853	3.007807	0.6162494
24.0	2.644133	39.8	−0.3515315	−1.0091538	1.3309287	2.886413	0.6217494
25.0	2.543380	38.4	−0.3469952	−1.0118510	1.3296002	2.774903	0.6275294
26.0	2.450592	37.0	−0.3422776	−1.0146654	1.3281970	2.672139	0.6335958
27.0	2.364885	35.6	−0.3373795	−1.0175982	1.3267160	2.577149	0.6399549
28.0	2.285502	34.3	−0.3323020	−1.0206511	1.3251539	2.489103	0.6466138
29.0	2.211792	33.0	−0.3270460	−1.0238258	1.3235074	2.407283	0.6535797
30.0	2.143189	31.8	−0.3216130	−1.0271242	1.3217728	2.331070	0.6608602
31.0	2.079202	30.6	−0.3160044	−1.0305485	1.3199459	2.259921	0.6684634
32.0	2.019399	29.4	−0.3102219	−1.0341012	1.3180227	2.193363	0.6763979
33.0	1.963403	28.3	−0.3042674	−1.0377851	1.3159984	2.130982	0.6846724
34.0	1.910879	27.2	−0.2981431	−1.0416034	1.3138683	2.072410	0.6932966
35.0	1.861534	26.1	−0.2918515	−1.0455595	1.3116273	2.017322	0.7022800
36.0	1.815103	25.1	−0.2853955	−1.0496574	1.3092698	1.965429	0.7116332
37.0	1.771354	24.0	−0.2787783	−1.0539018	1.3067901	1.916475	0.7213667
38.0	1.730076	23.0	−0.2720036	−1.0582975	1.3041820	1.870229	0.7314921
39.0	1.691083	22.1	−0.2650754	−1.0628501	1.3014390	1.826485	0.7420210
40.0	1.654204	21.1	−0.2579985	−1.0675662	1.2985544	1.785057	0.7529658
θ	Ω_s	A_{min}	σ_1	σ_3	Ω_1	Ω_2	Ω_3

* Adapted from A. I. Zverev, "Handbook of Filter Synthesis," John Wiley and Sons, New York, 1967.

TABLE 12-56 *(Continued)*

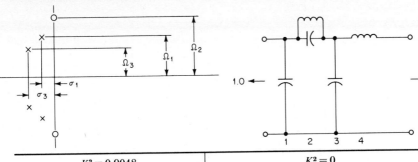

$K^2 = 0.9048$					$K^2 = 0$				
C_1	C_2	L_2	C_3	L_4	C_1	C_2	L_2	C_3	L_4
0.7231	0.000000	1.207	1.334	0.6543	0.36157	0.000000	0.90444	1.16498	1.04995
0.7174	0.006461	1.198	1.330	0.6549	0.35333	0.00867	0.89233	1.16304	1.05291
0.7154	0.008813	1.194	1.329	0.6552	0.35034	0.01185	0.88795	1.16236	1.05398
0.7130	0.01154	1.190	1.327	0.6555	0.34687	0.01556	0.88290	1.16158	1.05522
0.7103	0.01464	1.186	1.325	0.6558	0.34293	0.01980	0.87717	1.16072	1.05662
0.7073	0.01814	1.181	1.323	0.6561	0.33851	0.02460	0.87076	1.15976	1.05820
0.7040	0.02202	1.176	1.321	0.6565	0.33360	0.02997	0.86368	1.15873	1.05993
0.7003	0.02630	1.170	1.318	0.6569	0.32819	0.03594	0.85592	1.15763	1.06184
0.6963	0.03100	1.163	1.316	0.6574	0.32227	0.04254	0.84748	1.15646	1.06391
0.6920	0.03612	1.156	1.313	0.6579	0.31584	0.04980	0.83836	1.15523	1.06616
0.6874	0.04166	1.148	1.310	0.6584	0.30888	0.05774	0.82856	1.15396	1.06857
0.6824	0.04766	1.140	1.306	0.6590	0.30139	0.06642	0.81808	1.15265	0.07115
0.6771	0.05411	1.132	1.303	0.6596	0.29334	0.07588	0.80692	1.15132	1.07390
0.6715	0.06103	1.122	1.299	0.6603	0.28473	0.08616	0.79508	1.14998	1.07682
0.6655	0.06845	1.113	1.295	0.6610	0.27554	0.09733	0.78256	1.14865	1.07991
0.6592	0.07637	1.103	1.291	0.6617	0.26575	0.10945	0.76935	1.14734	1.08316
0.6526	0.08482	1.092	1.286	0.6624	0.25534	0.12260	0.75547	1.14607	1.08658
0.6456	0.09383	1.081	1.282	0.6632	0.24428	0.13685	0.74091	1.14486	1.09016
0.6383	0.1034	1.069	1.277	0.6641	0.23257	0.15232	0.72568	1.14374	1.09391
0.6306	0.1136	1.057	1.272	0.6649	0.22015	0.16911	0.70977	1.14274	1.09780
0.6226	0.1244	1.044	1.267	0.6658	0.20702	0.18735	0.69320	1.14187	1.10185
0.6143	0.1359	1.030	1.262	0.6668	0.19313	0.20719	0.67596	1.14118	1.10605
0.6055	0.1481	1.017	1.256	0.6677	0.17844	0.22880	0.65807	1.14071	1.11038
0.5964	0.1611	1.002	1.250	0.6687	0.16291	0.25238	0.63954	1.14049	1.11485
0.5870	0.1748	0.9872	1.244	0.6698	0.14651	0.27816	0.62037	1.14057	1.11943
0.5772	0.1894	0.9717	1.238	0.6708	0.12916	0.30642	0.60057	1.14101	1.12412
0.5670	0.2049	0.9558	1.232	0.6719	0.11082	0.33749	0.58017	1.14186	1.12890
0.5564	0.2213	0.9393	1.226	0.6730	0.09142	0.37172	0.55919	1.14320	1.13375
0.5455	0.2388	0.9223	1.219	0.6742	0.07088	0.40959	0.53763	1.14511	1.13865
0.5341	0.2573	0.9048	1.212	0.6753	0.04911	0.45163	0.51554	1.14767	1.14358
0.5224	0.2771	0.8868	1.205	0.6765	0.02602	0.49848	0.49294	1.15100	1.14850
0.5103	0.2982	0.8683	1.198	0.6777	0.00149	0.55094	0.46988	1.15520	1.15338
0.4978	0.3206	0.8492	1.191	0.6789	−0.02462	0.60995	0.44638	1.16042	1.15818
0.4848	0.3446	0.8297	1.184	0.6801	−0.05244	0.67668	0.42250	1.16681	1.16286
0.4715	0.3702	0.8098	1.177	0.6813	−0.08216	0.75258	0.39830	1.17457	1.16735
0.4577	0.3976	0.7893	1.169	0.6825	−0.11399	0.83946	0.37385	1.18392	1.17160
L_1	L_2	C_2	L_3	C_4	L_1	L_2	C_2	L_3	C_4

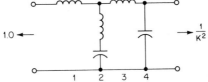

TABLE 12-56 (Continued)

CO4 20*

θ	T	Ω_s	A_{min}	Ω_2	$K^2 = 0.6667$					$K^2 = 0$				
					C_1	C_2	L_2	C_3	L_4	C_1	C_2	L_2	C_3	L_4
T	∞	∞	∞	∞	1.265	0.000000	1.291	1.936	0.8434	0.63253	0.00000	1.27782	1.54262	1.28323
06		10.350843	100.7	11.367741	1.260	0.006028	1.284	1.932	0.8437					
07		8.876727	95.3	9.747390	1.258	0.008216	1.281	1.930	0.8439					
08		7.771760	90.7	8.532615	1.255	0.01074	1.278	1.928	0.8440					
09		6.912894	86.6	7.588226	1.253	0.01362	1.275	1.926	0.8442					
10		6.226301	82.9	6.833109	1.250	0.01685	1.271	1.924	0.8443					
11		5.664999	79.6	6.215646	1.247	0.02043	1.267	1.921	0.8445	0.61292	0.02076	1.24651	1.53268	1.28756
12		5.197666	76.5	5.701423	1.243	0.02436	1.263	1.918	0.8448	0.60918	0.02480	1.24056	1.53081	1.28838
13		4.802620	73.7	5.266618	1.239	0.02866	1.258	1.915	0.8450	0.60509	0.02921	1.23409	1.52878	1.28928
14		4.464371	71.1	4.894214	1.235	0.03333	1.253	1.912	0.8453	0.60068	0.03402	1.22710	1.52660	1.29025
15		4.171563	68.7	4.571731	1.231	0.03837	1.247	1.908	0.8456	0.59592	0.03923	1.21959	1.52427	1.29129
16		3.915678	66.4	4.289813	1.226	0.04380	1.241	1.904	0.8459	0.59082	0.04485	1.21156	1.52179	1.29240
17		3.690200	64.3	4.041300	1.221	0.04961	1.234	1.900	0.8462	0.58539	0.05090	1.20302	1.51916	1.29359
18		3.490065	62.3	3.820626	1.216	0.05581	1.227	1.895	0.8465	0.57960	0.05738	1.19395	1.51639	1.29485
19		3.311272	60.3	3.623399	1.210	0.06242	1.220	1.891	0.8469	0.57347	0.06431	1.18437	1.51348	1.29619
20		3.150622	58.5	3.446101	1.204	0.06944	1.213	1.886	0.8473	0.56698	0.07171	1.17427	1.51043	1.29760

θ	Ω_s	A_min	Ω_2	L_1	L_2	C_2	L_3	C_4	L_1	L_2	C_2	L_3	C_4
21	3.005526	56.8	3.285888	1.198	0.07689	1.205	1.881	0.8477	0.56014	0.07959	1.16365	1.50725	1.29908
22	2.873864	55.1	3.140431	1.191	0.08476	1.196	1.875	0.8481	0.55294	0.08798	1.15251	1.50394	1.30064
23	2.753885	53.6	3.007807	1.184	0.09309	1.187	1.870	0.8485	0.54537	0.09689	1.14085	1.50050	1.30227
24	2.644133	52.0	2.886413	1.177	0.1019	1.178	1.864	0.8490	0.53744	0.10634	1.12867	1.49694	1.30397
25	2.543380	50.6	2.774903	1.169	0.1111	1.169	1.858	0.8494	0.52913	0.11637	1.11597	1.49326	1.30575
26	2.450592	49.2	2.672139	1.161	0.1209	1.159	1.851	0.8499	0.52044	0.12700	1.10276	1.48947	1.30760
27	2.364885	47.8	2.577149	1.153	0.1311	1.148	1.845	0.8505	0.51136	0.13826	1.08902	1.48558	1.30952
28	2.285502	46.5	2.489103	1.145	0.1419	1.138	1.838	0.8510	0.50190	0.15018	1.07477	1.48159	1.31151
29	2.211792	45.2	2.407283	1.136	0.1532	1.126	1.831	0.8516	0.49203	0.16279	1.06001	1.47750	1.31358
30	2.143189	44.0	2.331070	1.127	0.1651	1.115	1.824	0.8521	0.48176	0.17615	1.04472	1.47333	1.31571
31	2.079202	42.8	2.259921	1.117	0.1775	1.103	1.816	0.8527	0.47107	0.19030	1.02893	1.46908	1.31792
32	2.019399	41.6	2.193363	1.108	0.1906	1.091	1.808	0.8533	0.45996	0.20527	1.01262	1.46476	1.32019
33	1.963403	40.5	2.130982	1.097	0.2043	1.078	1.800	0.8540	0.44842	0.22114	0.99580	1.46039	1.32253
34	1.910879	39.4	2.072410	1.087	0.2186	1.065	1.792	0.8546	0.43644	0.23796	0.97847	1.45596	1.32494
35	1.861534	38.3	2.017322	1.076	0.2337	1.051	1.784	0.8553	0.42400	0.25580	0.96063	1.45150	1.32740
36	1.815103	37.2	1.965429	1.065	0.2495	1.038	1.775	0.8560	0.41109	0.27473	0.94228	1.44701	1.32993
37	1.771354	36.2	1.916475	1.054	0.2661	1.023	1.766	0.8567	0.39771	0.29484	0.92344	1.44251	1.33251
38	1.730076	35.2	1.870229	1.042	0.2835	1.009	1.757	0.8574	0.38383	0.31623	0.90409	1.43801	1.33515
39	1.691083	34.2	1.826485	1.030	0.3017	0.9936	1.748	0.8581	0.36944	0.33899	0.88426	1.43352	1.33783
40	1.654204	33.3	1.785057	1.017	0.3208	0.9782	1.738	0.8589	0.35453	0.36326	0.86393	1.42908	1.34056

* Adapted from A. I. Zverev, "Handbook of Filter Synthesis," John Wiley and Sons, New York, 1967.

TABLE 12-56 *(Continued)*

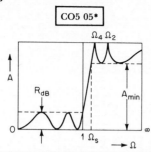

CO5 05*

θ	Ω_s	A_{min}	σ_0	σ_1	σ_3	Ω_1	Ω_2	Ω_3	Ω_4
C	∞	∞	0.80639	−0.24919	−0.65238	1.2218	∞	0.7551	∞
2.0	28.6537	167.86	0.80679	−0.24892	−0.65232	1.2216	48.7389	0.7554	30.1274
3.0	19.1073	150.25	0.80730	−0.24860	−0.65225	1.2214	32.4927	0.7559	20.0893
4.0	14.3356	137.74	0.80801	−0.24815	−0.65215	1.2212	24.3697	0.7566	15.0716
5.0	11.4737	128.04	0.80893	−0.24757	−0.65202	1.2210	19.4959	0.7575	12.0620
6.0	9.5668	120.11	0.81005	−0.24686	−0.65186	1.2207	16.2468	0.7588	10.0565
7.0	8.2055	113.40	0.81138	−0.24602	−0.65167	1.2203	13.9260	0.7599	8.6247
8.0	7.1853	107.59	0.81292	−0.24505	−0.65144	1.2199	12.1854	0.7613	7.5516
9.0	6.3925	102.45	0.81468	−0.24396	−0.65118	1.2194	10.8316	0.7630	6.7175
10.0	5.7588	97.86	0.81664	−0.24274	−0.65088	1.2188	9.7486	0.7649	6.0507
11.0	5.2408	93.69	0.81883	−0.24140	−0.65055	1.2182	8.8625	0.7670	5.5057
12.0	4.8097	89.89	0.82124	−0.23993	−0.65017	1.2176	8.1241	0.7693	5.0520
13.0	4.4454	86.39	0.82387	−0.23834	−0.64975	1.2168	7.4993	0.7718	4.6684
14.0	4.1336	83.14	0.82673	−0.23663	−0.64928	1.2160	6.9638	0.7744	4.3401
15.0	3.8637	80.11	0.82982	−0.23479	−0.64877	1.2152	6.4997	0.7774	4.0559
16.0	3.6280	77.27	0.83315	−0.23284	−0.64820	1.2143	6.0936	0.7805	3.8076
17.0	3.4203	74.60	0.83672	−0.23077	−0.64758	1.2133	5.7353	0.7838	3.5888
18.0	3.2361	72.08	0.84055	−0.22859	−0.64690	1.2123	5.4168	0.7873	3.3946
19.0	3.0716	69.69	0.84463	−0.22629	−0.64616	1.2112	5.1318	0.7911	3.2212
20.0	2.9238	67.41	0.84897	−0.22387	−0.64534	1.2100	4.8753	0.7950	3.0654
21.0	2.7904	65.25	0.85358	−0.22135	−0.64446	1.2088	4.6433	0.7992	2.9246
22.0	2.6695	63.18	0.85847	−0.21872	−0.64350	1.2075	4.4323	0.8036	2.7970
23.0	2.5593	61.20	0.86364	−0.21598	−0.64245	1.2061	4.2397	0.8082	2.6807
24.0	2.4586	59.29	0.86911	−0.21313	−0.64131	1.2047	4.0631	0.8131	2.5743
25.0	2.3662	57.46	0.87488	−0.21019	−0.64008	1.2032	3.9007	0.8181	2.4767
26.0	2.2812	55.70	0.88097	−0.20714	−0.63875	1.2016	3.7507	0.8234	2.3868
27.0	2.2027	54.00	0.88739	−0.20400	−0.63731	1.2000	3.6119	0.8289	2.3038
28.0	2.1301	52.35	0.89414	−0.20076	−0.63575	1.1983	3.4829	0.8347	2.2270
29.0	2.0627	50.76	0.90126	−0.19742	−0.63407	1.1965	3.3629	0.8407	2.1556
30.0	2.0000	49.22	0.90873	−0.19400	−0.63225	1.1947	3.2508	0.8469	2.0892

TABLE 12-56   (Continued)

θ	Ω_s	$A_{\min}$	σ_0	σ_1	σ_3	Ω_1	Ω_2	Ω_3	Ω_4
31.0	1.9416	47.72	0.91660	−0.19049	−0.63029	1.1928	3.1460	0.8533	2.0274
32.0	1.8871	46.27	0.92486	−0.18690	−0.62818	1.1908	3.0476	0.8600	1.9695
33.0	1.8361	44.85	0.93355	−0.18322	−0.62590	1.1887	2.9553	0.8669	1.9154
34.0	1.7883	43.47	0.94267	−0.17947	−0.62345	1.1866	2.8683	0.8741	1.8646
35.0	1.7434	42.13	0.95225	−0.17565	−0.62081	1.1844	2.7864	0.8815	1.8170
36.0	1.7013	40.81	0.96232	−0.17175	−0.61798	1.1821	2.7089	0.8891	1.7722
37.0	1.6616	39.53	0.97290	−0.16778	−0.61493	1.1797	2.6356	0.8970	1.7299
38.0	1.6243	38.28	0.98402	−0.16375	−0.61166	1.1773	2.5662	0.9051	1.6901
39.0	1.5890	37.05	0.99571	−0.15966	−0.60815	1.1748	2.5003	0.9135	1.6525
40.0	1.5557	35.85	1.00800	−0.15552	−0.60438	1.1722	2.4377	0.9220	1.6170
41.0	1.5243	34.67	1.02092	−0.15132	−0.60034	1.1695	2.3781	0.9308	1.5833
42.0	1.4945	33.52	1.03453	−0.14707	−0.59601	1.1668	2.3213	0.9399	1.5515
43.0	1.4663	32.38	1.04885	−0.14277	−0.59137	1.1640	2.2672	0.9491	1.5213
44.0	1.4396	31.27	1.06395	−0.13844	−0.58641	1.1611	2.2154	0.9586	1.4926
45.0	1.4142	30.17	1.07986	−0.13407	−0.58110	1.1581	2.1660	0.9683	1.4654
46.0	1.3902	29.09	1.09665	−0.12967	−0.57542	1.1551	2.1187	0.9782	1.4396
47.0	1.3673	28.03	1.11439	−0.12524	−0.56935	1.1519	2.0733	0.9883	1.4150
48.0	1.3456	26.99	1.13313	−0.12079	−0.56287	1.1487	2.0299	0.9985	1.3916
49.0	1.3250	25.95	1.15297	−0.11632	−0.55595	1.1455	1.9881	1.0090	1.3693
50.0	1.3054	24.94	1.17399	−0.11184	−0.54858	1.1421	1.9480	1.0195	1.3481
51.0	1.2868	23.93	1.19628	−0.10735	−0.54072	1.1387	1.9095	1.0302	1.3279
52.0	1.2690	22.94	1.21995	−0.10286	−0.53235	1.1352	1.8724	1.0410	1.3087
53.0	1.2521	21.96	1.24513	−0.09838	−0.52344	1.1316	1.8366	1.0519	1.2903
54.0	1.2361	20.99	1.27194	−0.09390	−0.51398	1.1280	1.8021	1.0628	1.2728
55.0	1.2208	20.04	1.30055	−0.08944	−0.50393	1.1243	1.7689	1.0738	1.2561
56.0	1.2062	19.09	1.33113	−0.08500	−0.49326	1.1205	1.7368	1.0847	1.2402
57.0	1.1924	18.15	1.36386	−0.08058	−0.48196	1.1167	1.7057	1.0956	1.2250
58.0	1.1792	17.23	1.39898	−0.07620	−0.47000	1.1128	1.6757	1.1064	1.2104
59.0	1.1666	16.31	1.43673	−0.07186	−0.45737	1.1088	1.6467	1.1170	1.1966
60.0	1.1547	15.40	1.47740	−0.06757	−0.44404	1.1048	1.6185	1.1274	1.1834
θ	Ω_s	$A_{\min}$	σ_0	σ_1	σ_3	Ω_1	Ω_2	Ω_3	Ω_4

* Adapted from A. I. Zverev, "Handbook of Filter Synthesis," John Wiley and Sons, New York, 1967.

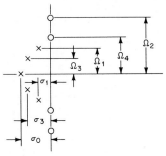

TABLE 12-56 *(Continued)*

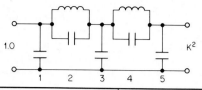

	$K^2 = 1.0$							$K^2 = \infty$					
θ	C_1	C_2	L_2	C_3	C_4	L_4	C_5	C_1	C_2	L_2	C_3	C_4	L_4	C_5
C	0.7664	0.0000	1.3100	1.5880	0.0000	1.3100	0.7664	0.3832	0.0000	0.9671	1.2835	0.0000	1.3983	1.2042
2.0	0.7661	0.0003	1.3099	1.5877	0.0008	1.3091	0.7656	0.3828	0.0004	0.9666	1.2827	0.0008	1.3971	1.2038
3.0	0.7658	0.0007	1.3095	1.5868	0.0018	1.3075	0.7646	0.3823	0.0010	0.9659	1.2818	0.0018	1.3955	1.2031
4.0	0.7654	0.0012	1.3088	1.5855	0.0033	1.3054	0.7633	0.3816	0.0017	0.9649	1.2804	0.0032	1.3934	1.2023
5.0	0.7648	0.0020	1.3080	1.5839	0.0052	1.3026	0.7615	0.3807	0.0027	0.9637	1.2787	0.0049	1.3907	1.2011
6.0	0.7641	0.0029	1.3070	1.5820	0.0076	1.2993	0.7594	0.3796	0.0039	0.9621	1.2766	0.0071	1.3874	1.1997
7.0	0.7632	0.0039	1.3058	1.5796	0.0103	1.2953	0.7569	0.3783	0.0054	0.9603	1.2741	0.0097	1.3835	1.1981
8.0	0.7623	0.0051	1.3044	1.5770	0.0135	1.2907	0.7540	0.3768	0.0070	0.9582	1.2713	0.0127	1.3789	1.1962
9.0	0.7612	0.0065	1.3028	1.5739	0.0172	1.2855	0.7507	0.3750	0.0089	0.9558	1.2680	0.0161	1.3738	1.1941
10.0	0.7600	0.0080	1.3011	1.5706	0.0213	1.2797	0.7470	0.3731	0.0110	0.9531	1.2644	0.0200	1.3681	1.1917
11.0	0.7586	0.0098	1.2991	1.5669	0.0259	1.2733	0.7429	0.3710	0.0134	0.9502	1.2604	0.0242	1.3618	1.1891
12.0	0.7572	0.0116	1.2970	1.5628	0.0309	1.2663	0.7384	0.3686	0.0160	0.9470	1.2561	0.0289	1.3548	1.1863
13.0	0.7556	0.0137	1.2947	1.5584	0.0364	1.2586	0.7335	0.3660	0.0188	0.9435	1.2513	0.0341	1.3473	1.1831
14.0	0.7538	0.0159	1.2922	1.5536	0.0424	1.2504	0.7283	0.3633	0.0219	0.9397	1.2462	0.0396	1.3392	1.1798
15.0	0.7519	0.0183	1.2895	1.5485	0.0489	1.2416	0.7226	0.3603	0.0253	0.9356	1.2408	0.0457	1.3305	1.1762
16.0	0.7499	0.0209	1.2866	1.5431	0.0559	1.2321	0.7165	0.3570	0.0289	0.9312	1.2349	0.0522	1.3212	1.1723
17.0	0.7478	0.0236	1.2836	1.5374	0.0635	1.2221	0.7101	0.3536	0.0328	0.9265	1.2287	0.0592	1.3113	1.1682
18.0	0.7455	0.0266	1.2803	1.5313	0.0716	1.2115	0.7032	0.3499	0.0370	0.9216	1.2222	0.0667	1.3009	1.1639
19.0	0.7431	0.0297	1.2768	1.5249	0.0802	1.2002	0.6959	0.3460	0.0414	0.9163	1.2153	0.0747	1.2898	1.1593
20.0	0.7406	0.0330	1.2732	1.5182	0.0895	1.1884	0.6883	0.3419	0.0462	0.9108	1.2080	0.0833	1.2782	1.1545
21.0	0.7379	0.0365	1.2694	1.5112	0.0994	1.1760	0.6802	0.3375	0.0513	0.9050	1.2004	0.0923	1.2660	1.1495
22.0	0.7350	0.0402	1.2653	1.5038	0.1099	1.1630	0.6717	0.3329	0.0566	0.8988	1.1924	0.1020	1.2532	1.1442
23.0	0.7321	0.0441	1.2611	1.4962	0.1210	1.1494	0.6628	0.3281	0.0623	0.8924	1.1841	0.1122	1.2399	1.1387
24.0	0.7290	0.0482	1.2567	1.4882	0.1329	1.1353	0.6534	0.3230	0.0684	0.8857	1.1755	0.1231	1.2260	1.1329
25.0	0.7257	0.0524	1.2520	1.4800	0.1454	1.1205	0.6437	0.3176	0.0748	0.8786	1.1666	0.1346	1.2115	1.1269
26.0	0.7223	0.0569	1.2472	1.4715	0.1588	1.1052	0.6335	0.3120	0.0816	0.8713	1.1573	0.1467	1.1964	1.1207
27.0	0.7187	0.0617	1.2421	1.4627	0.1729	1.0893	0.6229	0.3061	0.0888	0.8637	1.1477	0.1595	1.1809	1.1143
28.0	0.7150	0.0666	1.2369	1.4537	0.1879	1.0729	0.6118	0.2999	0.0963	0.8557	1.1377	0.1731	1.1647	1.1076
29.0	0.7112	0.0718	1.2314	1.4444	0.2038	1.0559	0.6003	0.2935	0.1043	0.8475	1.1275	0.1875	1.1480	1.1007
30.0	0.7072	0.0772	1.2257	1.4348	0.2206	1.0383	0.5884	0.2867	0.1128	0.8389	1.1170	0.2026	1.1308	1.0936

TABLE 12-56   *(Continued)*

	$K^2 = 1.0$							$K^2 = \infty$						
θ	C_1	C_2	L_2	C_3	C_4	L_4	C_5	C_1	C_2	L_2	C_3	C_4	L_4	C_5
31.0	0.7030	0.0828	1.2198	1.4250	0.2384	1.0201	0.5760	0.2797	0.1217	0.8300	1.1061	0.2186	1.1130	1.0863
32.0	0.6987	0.0887	1.2136	1.4150	0.2574	1.0015	0.5631	0.2723	0.1312	0.8208	1.0950	0.2355	1.0947	1.0788
33.0	0.6942	0.0948	1.2073	1.4048	0.2774	0.9822	0.5498	0.2647	0.1411	0.8112	1.0836	0.2534	1.0758	1.0710
34.0	0.6896	0.1012	1.2007	1.3943	0.2988	0.9625	0.5360	0.2567	0.1517	0.8013	1.0719	0.2722	1.0565	1.0630
35.0	0.6847	0.1078	1.1938	1.3837	0.3214	0.9422	0.5217	0.2484	0.1628	0.7911	1.0600	0.2922	1.0366	1.0549
36.0	0.6798	0.1148	1.1867	1.3729	0.3455	0.9214	0.5070	0.2397	0.1746	0.7806	1.0478	0.3134	1.0162	1.0465
37.0	0.6746	0.1220	1.1794	1.3619	0.3712	0.9001	0.4917	0.2306	0.1870	0.7697	1.0354	0.3357	0.9953	1.0379
38.0	0.6693	0.1295	1.1717	1.3508	0.3985	0.8782	0.4759	0.2212	0.2002	0.7585	1.0227	0.3595	0.9738	1.0291
39.0	0.6637	0.1374	1.1638	1.3396	0.4278	0.8559	0.4595	0.2114	0.2142	0.7469	1.0098	0.3847	0.9519	1.0202
40.0	0.6580	0.1456	1.1556	1.3282	0.4590	0.8331	0.4426	0.2012	0.2290	0.7350	0.9967	0.4115	0.9295	1.0111
41.0	0.6521	0.1541	1.1472	1.3168	0.4924	0.8099	0.4252	0.1905	0.2447	0.7227	0.9834	0.4399	0.9067	1.0017
42.0	0.6460	0.1630	1.1384	1.3053	0.5283	0.7862	0.4072	0.1794	0.2614	0.7100	0.9699	0.4703	0.8833	0.9923
43.0	0.6397	0.1722	1.1292	1.2937	0.5669	0.7620	0.3885	0.1678	0.2791	0.6970	0.9563	0.5027	0.8595	0.9826
44.0	0.6332	0.1819	1.1198	1.2822	0.6085	0.7375	0.3693	0.1558	0.2981	0.6836	0.9425	0.5374	0.8353	0.9728
45.0	0.6265	0.1920	1.1099	1.2706	0.6535	0.7125	0.3494	0.1432	0.3183	0.6697	0.9285	0.5745	0.8106	0.9629
46.0	0.6195	0.2025	1.0997	1.2591	0.7022	0.6871	0.3288	0.1300	0.3398	0.6555	0.9145	0.6143	0.7855	0.9528
47.0	0.6124	0.2135	1.0891	1.2478	0.7550	0.6614	0.3075	0.1163	0.3630	0.6409	0.9003	0.6572	0.7599	0.9426
48.0	0.6050	0.2251	1.0780	1.2365	0.8126	0.6354	0.2855	0.1019	0.3878	0.6259	0.8861	0.7035	0.7430	0.9323
49.0	0.5973	0.2372	1.0665	1.2254	0.8756	0.6090	0.2628	0.0869	0.4144	0.6104	0.8718	0.7536	0.7077	0.9219
50.0	0.5894	0.2498	1.0545	1.2145	0.9446	0.5824	0.2392	0.0712	0.4432	0.5946	0.8575	0.8079	0.6810	0.9114
51.0	0.5813	0.2632	1.0420	1.2039	1.0206	0.5556	0.2147	0.0548	0.4743	0.5783	0.8432	0.8671	0.6540	0.9008
52.0	0.5729	0.2772	1.0289	1.1937	1.1047	0.5285	0.1894	0.0375	0.5080	0.5615	0.8290	0.9317	0.6267	0.8903
53.0	0.5642	0.2920	1.0152	1.1838	1.1980	0.5013	0.1631	0.0194	0.5447	0.5443	0.8148	1.0027	0.5990	0.8797
54.0	0.5552	0.3076	1.0008	1.1744	1.3020	0.4740	0.1357	0.0004	0.5847	0.5266	0.8007	1.0808	0.5711	0.8691
55.0	0.5460	0.3242	0.9858	1.1656	1.4187	0.4467	0.1073	-0.0197	0.6285	0.5085	0.7868	1.1672	0.5430	0.8586
56.0	0.5364	0.3417	0.9700	1.1575	1.5502	0.4194	0.0777	-0.0408	0.6768	0.4899	0.7731	1.2633	0.5147	0.8482
57.0	0.5265	0.3605	0.9533	1.1501	1.6993	0.3921	0.0468	-0.0631	0.7301	0.4708	0.7597	1.3707	0.4862	0.8380
58.0	0.5163	0.3805	0.9358	1.1437	1.8693	0.3651	0.0145	-0.0867	0.7893	0.4512	0.7465	1.4914	0.4576	0.8280
59.0	0.5057	0.4020	0.9174	1.1382	2.0645	0.3382	-0.0191	-0.1117	0.8554	0.4312	0.7338	1.6279	0.4290	0.8182
60.0	0.4948	0.4251	0.8979	1.1340	2.2902	0.3117	-0.0545	-0.1382	0.9295	0.4107	0.7215	1.7833	0.4004	0.8087
θ	L_1	L_2	C_2	L_3	L_4	C_4	L_5	L_1	L_2	C_2	L_3	L_4	C_4	L_5

TABLE 12-56   *(Continued)*

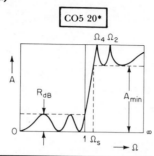

CO5 20*

θ	Ω_s	A_{min}	σ_0	σ_1	σ_3	Ω_1	Ω_2	Ω_3	Ω_4
C	∞	∞	0.47472	−0.14670	−0.38406	1.0528	∞	0.6506	∞
1.0	57.2987	210.17	0.47476	−0.14667	−0.38405	1.0527	97.4775	0.6507	60.2470
2.0	28.6537	180.07	0.47489	−0.14659	−0.38404	1.0527	48.7389	0.6508	30.1274
3.0	19.1073	162.45	0.47511	−0.14646	−0.38402	1.0527	32.4927	0.6511	20.0893
4.0	14.3356	149.95	0.47542	−0.14628	−0.38399	1.0527	24.3697	0.6514	15.0716
5.0	11.4737	140.25	0.47582	−0.14605	−0.38396	1.0526	19.4959	0.6519	12.0620
6.0	9.5668	132.32	0.47631	−0.14577	−0.38392	1.0526	16.2468	0.6524	10.0565
7.0	8.2055	125.61	0.47689	−0.14543	−0.38387	1.0526	13.9260	0.6531	8.6247
8.0	7.1853	119.80	0.47757	−0.14505	−0.38381	1.0525	12.1854	0.6538	7.5516
9.0	6.3925	114.66	0.47833	−0.14461	−0.38375	1.0524	10.8316	0.6547	6.7175
10.0	5.7588	110.06	0.47918	−0.14412	−0.38367	1.0524	9.7486	0.6557	6.0507
11.0	5.2408	105.90	0.48013	−0.14358	−0.38359	1.0523	8.8625	0.6567	5.5057
12.0	4.8097	102.10	0.48117	−0.14299	−0.38349	1.0522	8.1241	0.6579	5.0520
13.0	4.4454	98.59	0.48231	−0.14235	−0.38338	1.0521	7.4993	0.6592	4.6684
14.0	4.1336	95.34	0.48355	−0.14166	−0.38327	1.0520	6.9638	0.6606	4.3401
15.0	3.8637	92.32	0.48488	−0.14092	−0.38314	1.0519	6.4997	0.6621	4.0559
16.0	3.6280	89.48	0.48631	−0.14013	−0.38299	1.0518	6.0936	0.6637	3.8076
17.0	3.4203	86.81	0.48785	−0.13929	−0.38283	1.0517	5.7353	0.6654	3.5888
18.0	3.2361	84.29	0.48949	−0.13840	−0.38266	1.0515	5.4168	0.6672	3.3946
19.0	3.0716	81.89	0.49123	−0.13746	−0.38247	1.0514	5.1318	0.6691	3.2212
20.0	2.9238	79.62	0.49308	−0.13647	−0.38227	1.0512	4.8753	0.6712	3.0654
21.0	2.7904	77.46	0.49503	−0.13543	−0.38204	1.0511	4.6433	0.6733	2.9246
22.0	2.6695	75.39	0.49710	−0.13434	−0.38180	1.0509	4.4323	0.6756	2.7970
23.0	2.5593	73.40	0.49929	−0.13320	−0.38153	1.0507	4.2397	0.6780	2.6807
24.0	2.4586	71.50	0.50159	−0.13201	−0.38125	1.0505	4.0631	0.6805	2.5743
25.0	2.3662	69.67	0.50401	−0.13078	−0.38093	1.0503	3.9007	0.6831	2.4767
26.0	2.2812	67.91	0.50655	−0.12950	−0.38060	1.0501	3.7507	0.6858	2.3868
27.0	2.2027	66.21	0.50922	−0.12817	−0.38023	1.0499	3.6119	0.6887	2.3038
28.0	2.1301	64.56	0.51201	−0.12679	−0.37983	1.0496	3.4829	0.6916	2.2270
29.0	2.0627	62.97	0.51494	−0.12536	−0.37941	1.0494	3.3629	0.6947	2.1556
30.0	2.0000	61.43	0.51801	−0.12389	−0.37895	1.0491	3.2508	0.6980	2.0892

TABLE 12-56   *(Continued)*

θ	Ω_s	$A_{\min}$	σ_0	σ_1	σ_3	Ω_1	Ω_2	Ω_3	Ω_4
31.0	1.9416	59.93	0.52122	−0.12238	−0.37845	1.0488	3.1460	0.7013	2.0274
32.0	1.8871	58.47	0.52458	−0.12081	−0.37791	1.0486	3.0476	0.7048	1.9695
33.0	1.8361	57.06	0.52808	−0.11920	−0.37734	1.0483	2.9553	0.7084	1.9154
34.0	1.7883	55.68	0.53175	−0.11755	−0.37672	1.0479	2.8683	0.7122	1.8846
35.0	1.7434	54.33	0.53557	−0.11585	−0.37605	1.0476	2.7864	0.7160	1.8170
36.0	1.7013	53.02	0.53956	−0.11411	−0.37533	1.0473	2.7089	0.7200	1.7722
37.0	1.6616	51.74	0.54373	−0.11233	−0.37456	1.0469	2.6356	0.7242	1.7299
38.0	1.6243	50.49	0.54808	−0.11050	−0.37372	1.0465	2.5662	0.7285	1.6901
39.0	1.5890	49.26	0.55261	−0.10863	−0.37283	1.0461	2.5003	0.7329	1.6525
40.0	1.5557	48.06	0.55734	−0.10672	−0.37187	1.0457	2.4377	0.7375	1.6170
41.0	1.5243	46.88	0.56228	−0.10477	−0.37085	1.0453	2.3781	0.7422	1.5833
42.0	1.4945	45.72	0.56743	−0.10278	−0.36974	1.0448	2.3213	0.7471	1.5515
43.0	1.4663	44.59	0.57281	−0.10075	−0.36856	1.0444	2.2672	0.7521	1.5213
44.0	1.4396	43.47	0.57842	−0.09868	−0.36829	1.0439	2.2154	0.7573	1.4926
45.0	1.4142	42.38	0.58428	−0.09658	−0.36593	1.0434	2.1660	0.7626	1.4654
46.0	1.3902	41.30	0.59039	−0.09444	−0.36448	1.0429	2.1187	0.7681	1.4396
47.0	1.3673	40.23	0.59678	−0.09226	−0.36291	1.0424	2.0733	0.7738	1.4150
48.0	1.3456	39.19	0.60345	−0.09004	−0.36124	1.0418	2.0299	0.7796	1.3916
49.0	1.3250	38.15	0.61043	−0.08780	−0.35945	1.0412	1.9881	0.7856	1.3693
50.0	1.3054	37.13	0.61773	−0.08552	−0.35753	1.0406	1.9480	0.7917	1.3481
51.0	1.2868	36.12	0.62537	−0.08321	−0.35548	1.0400	1.9095	0.7980	1.3279
52.0	1.2690	35.13	0.63336	−0.08087	−0.35329	1.0393	1.8724	0.8045	1.3087
53.0	1.2521	34.14	0.64174	−0.07850	−0.35094	1.0387	1.8366	0.8112	1.2903
54.0	1.2361	33.17	0.65053	−0.07610	−0.34842	1.0380	1.8021	0.8180	1.2728
55.0	1.2208	32.20	0.65975	−0.07368	−0.34573	1.0373	1.7689	0.8250	1.2561
56.0	1.2062	31.25	0.66943	−0.07123	−0.34285	1.0365	1.7368	0.8322	1.2402
57.0	1.1924	30.30	0.67962	−0.06876	−0.33977	1.0358	1.7057	0.8395	1.2250
58.0	1.1792	29.36	0.69033	−0.06627	−0.33647	1.0350	1.6757	0.8470	1.2104
59.0	1.1666	28.42	0.70163	−0.06376	−0.33295	1.0342	1.6467	0.8547	1.1966
60.0	1.1547	27.49	0.71354	−0.06124	−0.32917	1.0333	1.6185	0.8626	1.1834
θ	Ω_s	$A_{\min}$	σ_0	σ_1	σ_3	Ω_1	Ω_2	Ω_3	Ω_4

* Adapted from A. I. Zverev, "Handbook of Filter Synthesis," John Wiley and Sons, New York, 1967.

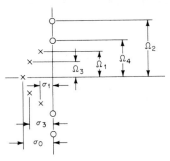

TABLE 12-56 *(Continued)*

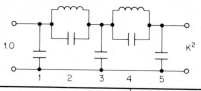

| | $K^2 = 1.0$ | | | | | | | $K^2 = \infty$ | | | | | | |
θ	C_1	C_2	L_2	C_3	C_4	L_4	C_5	C_1	C_2	L_2	C_3	C_4	L_4	C_5
C	1.302	0.0000	1.346	2.129	0.0000	1.346	1.302	0.6510	0.0000	1.3234	1.6362	0.0000	1.6265	1.4246
1.0	1.30183	0.00008	1.34548	2.12835	0.00020	1.34532	1.30170	0.6509	0.0001	1.3233	1.6360	0.0002	1.6263	1.4244
2.0	1.30163	0.00031	1.34523	2.12770	0.00082	1.34459	1.30112	0.6507	0.0003	1.3229	1.6355	0.0007	1.6254	1.4240
3.0	1.30130	0.00070	1.34483	2.12660	0.00184	1.34339	1.30016	0.6503	0.0007	1.3223	1.6345	0.0015	1.6240	1.4233
4.0	1.30084	0.00125	1.34426	2.12507	0.00328	1.34170	1.29881	0.6498	0.0013	1.3215	1.6332	0.0027	1.6220	1.4224
5.0	1.30024	0.00196	1.34353	2.12311	0.00513	1.33954	1.29708	0.6491	0.0020	1.3205	1.6315	0.0042	1.6195	1.4211
6.0	1.29951	0.00282	1.34264	2.12070	0.00740	1.33689	1.29496	0.6483	0.0029	1.3193	1.6294	0.0061	1.6164	1.4196
7.0	1.29865	0.00384	1.34159	2.11786	0.01008	1.33376	1.29246	0.6473	0.0039	1.3178	1.6270	0.0083	1.6128	1.4178
8.0	1.29766	0.00502	1.34037	2.11459	0.01318	1.33015	1.28957	0.6462	0.0051	1.3161	1.6241	0.0109	1.6085	1.4158
9.0	1.29653	0.00637	1.33899	2.11088	0.01671	1.32607	1.28630	0.6450	0.0065	1.3141	1.6209	0.0138	1.6038	1.4135
10.0	1.29527	0.00787	1.33744	2.10675	0.02067	1.32150	1.28264	0.6436	0.0080	1.3120	1.6174	0.0171	1.5984	1.4109
11.0	1.29387	0.00953	1.33573	2.10217	0.02506	1.31646	1.27859	0.6420	0.0097	1.3096	1.6134	0.0207	1.5926	1.4080
12.0	1.29234	0.01136	1.33386	2.09717	0.02989	1.31094	1.27417	0.6403	0.0116	1.3069	1.6091	0.0247	1.5861	1.4048
13.0	1.29067	0.01335	1.33182	2.09172	0.03516	1.30495	1.26936	0.6384	0.0136	1.3041	1.6044	0.0291	1.5791	1.4014
14.0	1.28887	0.01551	1.32961	2.08588	0.04089	1.29848	1.26416	0.6364	0.0159	1.3010	1.5993	0.0338	1.5716	1.3977
15.0	1.28693	0.01783	1.32724	2.07959	0.04707	1.29154	1.25858	0.6343	0.0182	1.2976	1.5939	0.0389	1.5635	1.3938
16.0	1.28485	0.02033	1.32470	2.07288	0.05371	1.28413	1.25261	0.6319	0.0208	1.2941	1.5881	0.0444	1.5548	1.3895
17.0	1.28263	0.02300	1.32199	2.06574	0.06084	1.27625	1.24627	0.6295	0.0236	1.2903	1.5819	0.0502	1.5456	1.3850
18.0	1.28027	0.02584	1.31911	2.05819	0.06844	1.26790	1.23953	0.6268	0.0265	1.2862	1.5754	0.0565	1.5359	1.3803
19.0	1.27778	0.02885	1.31607	2.05021	0.07655	1.25909	1.23241	0.6240	0.0296	1.2820	1.5685	0.0632	1.5256	1.3753
20.0	1.27514	0.03205	1.31285	2.04182	0.08515	1.24981	1.22491	0.6211	0.0329	1.2774	1.5612	0.0703	1.5148	1.3700
21.0	1.27236	0.03542	1.30945	2.03301	0.09428	1.24007	1.21703	0.6180	0.0364	1.2727	1.5536	0.0778	1.5035	1.3644
22.0	1.26943	0.03898	1.30589	2.02379	0.10393	1.22987	1.20876	0.6147	0.0402	1.2677	1.5456	0.0857	1.4916	1.3586
23.0	1.26636	0.04272	1.30215	2.01416	0.11414	1.21921	1.20010	0.6112	0.0441	1.2624	1.5373	0.0941	1.4791	1.3525
24.0	1.26314	0.04666	1.29823	2.00412	0.12490	1.20809	1.19107	0.6076	0.0482	1.2569	1.5286	0.1029	1.4662	1.3461
25.0	1.25978	0.05079	1.29413	1.99368	0.13625	1.19652	1.18164	0.6038	0.0525	1.2512	1.5195	0.1122	1.4527	1.3395
26.0	1.25262	0.05511	1.28985	1.98283	0.14819	1.18450	1.17183	0.5998	0.0571	1.2452	1.5101	0.1220	1.4387	1.3327
27.0	1.25259	0.05963	1.28540	1.97159	0.16075	1.17203	1.16164	0.5957	0.0619	1.2389	1.5004	0.1323	1.4242	1.3256
28.0	1.24877	0.06436	1.28075	1.95995	0.17396	1.15911	1.15106	0.5914	0.0669	1.2324	1.4903	0.1431	1.4092	1.3182
29.0	1.22480	0.06930	1.27592	1.94792	0.18783	1.14576	1.14010	0.5869	0.0721	1.2256	1.4798	0.1544	1.3936	1.3105
30.0	1.24067	0.07446	1.27091	1.93550	0.20239	1.13196	1.12874	0.5822	0.0777	1.2186	1.4690	0.1663	1.3776	1.3027

TABLE 12-56 *(Continued)*

	$K^2=1.0$							$K^2=\infty$						
θ	C_1	C_2	L_2	C_3	C_4	L_4	C_5	C_1	C_2	L_2	C_3	C_4	L_4	C_5
31.0	1.23638	0.07983	1.26570	1.92270	0.21768	1.11772	1.11700	0.5773	0.0834	1.2113	1.4579	0.1788	1.3610	1.2945
32.0	1.23192	0.08543	1.26030	1.90952	0.23371	1.10305	1.10487	0.5723	0.0894	1.2037	1.4464	0.1918	1.3439	1.2861
33.0	1.22731	0.09126	1.25470	1.89596	0.25054	1.08795	1.09235	0.5670	0.0957	1.1959	1.4346	0.2055	1.3264	1.2775
34.0	1.22252	0.09732	1.24890	1.88203	0.26819	1.07242	1.07944	0.5616	0.1023	1.1878	1.4225	0.2198	1.3083	1.2686
35.0	1.21757	0.10363	1.24290	1.86773	0.28671	1.05648	1.06614	0.5559	0.1092	1.1794	1.4100	0.2348	1.2898	1.2595
36.0	1.21244	0.11019	1.23669	1.85307	0.30614	1.04011	1.05244	0.5500	0.1164	1.1707	1.3972	0.2506	1.2707	1.2501
37.0	1.20714	0.11701	1.23028	1.83806	0.32654	1.02332	1.03835	0.5439	0.1239	1.1618	1.3841	0.2671	1.2512	1.2405
38.0	1.20166	0.12410	1.22364	1.82269	0.34795	1.00613	1.02386	0.5376	0.1318	1.1525	1.3707	0.2843	1.2313	1.2306
39.0	1.19600	0.13146	1.21679	1.80698	0.37044	0.98853	1.00897	0.5311	0.1400	1.1429	1.3570	0.3024	1.2108	1.2205
40.0	1.19015	0.13911	1.20971	1.79093	0.39408	0.97053	0.99368	0.5244	0.1485	1.1331	1.3429	0.3214	1.1899	1.2102
41.0	1.18411	0.14706	1.20241	1.77455	0.41894	0.95213	0.97798	0.5174	0.1575	1.1229	1.3286	0.3413	1.1686	1.1996
42.0	1.17787	0.15532	1.19486	1.75784	0.44510	0.93335	0.96187	0.5102	0.1668	1.1124	1.3139	0.3623	1.1467	1.1888
43.0	1.17144	0.16389	1.18708	1.74081	0.47265	0.91417	0.94535	0.5027	0.1766	1.1016	1.2989	0.3843	1.1245	1.1778
44.0	1.16480	0.17280	1.17904	1.72347	0.50170	0.89462	0.92841	0.4949	0.1868	1.0905	1.2837	0.4074	1.1018	1.1665
45.0	1.15794	0.18206	1.17075	1.70583	0.53236	0.87470	0.91105	0.4869	0.1975	1.0790	1.2681	0.4317	1.0787	1.1551
46.0	1.15088	0.19169	1.16219	1.68789	0.56476	0.85441	0.89326	0.4787	0.2087	1.0672	1.2523	0.4573	1.0551	1.1433
47.0	1.14359	0.20169	1.15336	1.66967	0.59903	0.83376	0.87504	0.4701	0.2205	1.0550	1.2362	0.4844	1.0311	1.1314
48.0	1.13607	0.21210	1.14425	1.65117	0.63534	0.81276	0.85638	0.4612	0.2328	1.0425	1.2198	0.5129	1.0068	1.1193
49.0	1.12831	0.22293	1.13484	1.63241	0.67386	0.79141	0.83727	0.4520	0.2457	1.0296	1.2032	0.5431	0.9820	1.1069
50.0	1.12031	0.23421	1.12513	1.61339	0.71481	0.76973	0.81771	0.4425	0.2593	1.0163	1.1863	0.5750	0.9568	1.0943
51.0	1.11206	0.24596	1.11509	1.59413	0.75841	0.74773	0.79768	0.4327	0.2736	1.0026	1.1691	0.6089	0.9313	1.0815
52.0	1.10354	0.25821	1.10473	1.57465	0.80492	0.72541	0.77717	0.4225	0.2886	0.9884	1.1517	0.6449	0.9053	1.0685
53.0	1.09476	0.27099	1.09401	1.55494	0.85465	0.70278	0.75619	0.4120	0.3044	0.9739	1.1341	0.6833	0.8790	1.0553
54.0	1.08569	0.28433	1.08293	1.53504	0.90794	0.67986	0.73470	0.4010	0.3211	0.9589	1.1162	0.7242	0.8524	1.0419
55.0	1.07633	0.29828	1.07147	1.51496	0.96518	0.65667	0.71270	0.3897	0.3388	0.9435	1.0981	0.7679	0.8254	1.0283
56.0	1.06666	0.31288	1.05960	1.49471	1.02684	0.63320	0.69016	0.3779	0.3574	0.9275	1.0798	0.8147	0.7981	1.0144
57.0	1.05668	0.32817	1.04731	1.47431	1.09344	0.60949	0.66709	0.3657	0.3772	0.9111	1.0613	0.8550	0.7705	1.0004
58.0	1.04636	0.34422	1.03456	1.45379	1.16561	0.58554	0.64344	0.3530	0.3983	0.8942	1.0426	0.9192	0.7425	0.9863
59.0	1.03570	0.36109	1.02134	1.43317	1.24407	0.56138	0.61920	0.3398	0.4207	0.8767	1.0237	0.9777	0.7143	0.9719
60.0	1.02467	0.37885	1.00760	1.41247	1.32969	0.53702	0.59435	0.3261	0.4446	0.8587	1.0046	1.0412	0.6858	0.9574
θ	L_1	L_2	C_2	L_3	L_4	C_4	L_5	L_1	L_2	C_2	L_3	L_4	C_4	L_5

$$1.0 \qquad \frac{1}{K^2}$$

(ladder network, nodes labeled 1 2 3 4 5)

TABLE 12-56 *(Continued)*

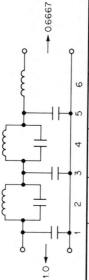

CO6 20 b*

θ / T	Ω_s	A_min	C_1	C_2	L_2	Ω_2	C_3	C_4	L_4	Ω_4	C_5	L_6	θ / T
	∞	∞	1.322	0.0000	1.373	∞	2.203	0.0000	1.469	∞	2.059	0.8816	
16	3.751039	112.5	1.299	0.0250	1.344	5.452491	2.142	0.0468	1.412	3.888329	2.017	0.8828	16
17	3.535748	109.3	1.296	0.0283	1.341	5.133037	2.135	0.0530	1.405	3.664543	2.012	0.8830	17
18	3.344698	106.3	1.293	0.0318	1.337	4.849152	2.126	0.0596	1.397	3.465915	2.006	0.8831	18
19	3.174064	103.4	1.290	0.0355	1.333	4.595218	2.118	0.0666	1.389	3.288476	2.000	0.8833	19
20	3.020785	100.7	1.286	0.0395	1.328	4.366743	2.108	0.0740	1.380	3.129050	1.993	0.8835	20
21	2.882384	98.1	1.283	0.0436	1.324	4.160091	2.099	0.0818	1.371	2.985065	1.987	0.8837	21
22	2.756834	95.6	1.279	0.0480	1.319	3.972284	2.089	0.0901	1.362	2.854418	1.979	0.8839	22
23	2.642462	93.3	1.275	0.0527	1.314	3.800865	2.078	0.0989	1.352	2.735370	1.972	0.8841	23
24	2.537873	91.0	1.270	0.0576	1.309	3.643786	2.067	0.1081	1.341	2.626475	1.964	0.8843	24
25	2.441895	88.8	1.266	0.0627	1.303	3.499325	2.055	0.1177	1.331	2.526516	1.956	0.8845	25
26	2.353536	86.7	1.261	0.0680	1.297	3.366027	2.043	0.1279	1.320	2.434463	1.948	0.8848	26
27	2.271953	84.6	1.256	0.0736	1.291	3.242651	2.031	0.1385	1.308	2.349441	1.939	0.8850	27
28	2.196422	82.6	1.251	0.0795	1.285	3.128134	2.018	0.1497	1.296	2.270699	1.930	0.8853	28
29	2.126320	80.7	1.246	0.0857	1.279	3.021559	2.005	0.1613	1.284	2.197588	1.921	0.8855	29
30	2.061105	78.9	1.240	0.0921	1.272	2.922132	1.991	0.1735	1.271	2.129549	1.911	0.8858	30
31	2.000308	77.1	1.235	0.0988	1.265	2.829162	1.977	0.1863	1.257	2.066092	1.901	0.8861	31
32	1.943517	75.3	1.229	0.1057	1.258	2.742042	1.962	0.1996	1.244	2.006790	1.891	0.8864	32
33	1.890370	73.6	1.223	0.1130	1.250	2.660241	1.947	0.2136	1.230	1.951268	1.881	0.8867	33
34	1.840548	72.0	1.216	0.1206	1.243	2.583290	1.931	0.2281	1.215	1.899195	1.870	0.8870	34
35	1.793769	70.4	1.210	0.1285	1.235	2.510772	1.915	0.2433	1.200	1.850277	1.859	0.8873	35

36	0.8877	1.847	1.804254	1.185	0.2592	1.899	2.442318	1.226	0.1367	1.203	68.8	1.749781	36
37	0.8880	1.835	1.760893	1.169	0.2758	1.882	2.377598	1.218	0.1452	1.196	67.3	1.708362	37
38	0.8884	1.823	1.719987	1.153	0.2931	1.864	2.316318	1.209	0.1541	1.189	65.8	1.669312	38
39	0.8887	1.811	1.681350	1.137	0.3112	1.847	2.258212	1.200	0.1634	1.181	64.3	1.632449	39
40	0.8891	1.798	1.644814	1.120	0.3301	1.828	2.203043	1.191	0.1730	1.174	62.8	1.597615	40
41	0.8895	1.785	1.610227	1.103	0.3498	1.810	2.150595	1.181	0.1830	1.166	61.4	1.564662	41
42	0.8898	1.771	1.577454	1.085	0.3704	1.791	2.100673	1.172	0.1934	1.158	60.0	1.533460	42
43	0.8902	1.758	1.546370	1.067	0.3920	1.771	2.053102	1.161	0.2043	1.149	58.7	1.503888	43
44	0.8906	1.744	1.516862	1.049	0.4145	1.751	2.007720	1.151	0.2155	1.141	57.3	1.475840	44
45	0.8910	1.729	1.488829	1.030	0.4381	1.731	1.964382	1.140	0.2272	1.132	56.0	1.449216	45
46	0.8915	1.715	1.462178	1.011	0.4628	1.710	1.922953	1.130	0.2394	1.123	54.7	1.423927	46
47	0.8919	1.700	1.436822	0.9910	0.4888	1.689	1.883312	1.118	0.2521	1.113	53.4	1.399891	47
48	0.8923	1.684	1.412684	0.9711	0.5160	1.668	1.845347	1.107	0.2653	1.103	52.2	1.377032	48
49	0.8928	1.669	1.389693	0.9508	0.5446	1.646	1.808954	1.095	0.2791	1.093	50.9	1.355282	49
50	0.8932	1.653	1.367782	0.9302	0.5747	1.623	1.774040	1.083	0.2935	1.083	49.7	1.334577	50
51	0.8937	1.637	1.346891	0.9092	0.6063	1.600	1.740516	1.070	0.3084	1.073	48.5	1.314859	51
52	0.8942	1.620	1.326965	0.8878	0.6397	1.577	1.708301	1.057	0.3241	1.062	47.3	1.296076	52
53	0.8946	1.603	1.307952	0.8661	0.6749	1.554	1.677322	1.044	0.3404	1.050	46.1	1.278176	53
54	0.8951	1.586	1.289805	0.8440	0.7122	1.530	1.647510	1.031	0.3574	1.039	45.0	1.261116	54
55	0.8956	1.568	1.272479	0.8216	0.7517	1.506	1.618799	1.017	0.3752	1.027	43.8	1.244853	55
56	0.8961	1.551	1.255935	0.7989	0.7936	1.481	1.591131	1.003	0.3939	1.015	42.7	1.229348	56
57	0.8966	1.532	1.240135	0.7758	0.8382	1.456	1.564449	0.9881	0.4135	1.002	41.5	1.214564	57
58	0.8971	1.514	1.225044	0.7523	0.8857	1.431	1.538703	0.9732	0.4340	0.9894	40.4	1.200469	58
59	0.8976	1.495	1.210630	0.7286	0.9365	1.405	1.513843	0.9578	0.4556	0.9760	39.3	1.187032	59
60	0.8981	1.476	1.196863	0.7045	0.9909	1.379	1.489825	0.9420	0.4783	0.9623	38.1	1.174224	60
61	0.8987	1.456	1.183715	0.6801	1.049	1.353	1.466607	0.9258	0.5022	0.9481	37.0	1.162017	61
62	0.8992	1.436	1.171161	0.6554	1.112	1.326	1.444148	0.9091	0.5274	0.9335	35.9	1.150388	62
63	0.8997	1.416	1.159176	0.6304	1.181	1.299	1.422411	0.8920	0.5541	0.9184	34.8	1.139313	63
64	0.9002	1.395	1.147737	0.6051	1.255	1.272	1.401362	0.8743	0.5824	0.9028	33.7	1.128771	64
65	0.9008	1.374	1.136826	0.5795	1.335	1.244	1.380967	0.8562	0.6125	0.8867	32.6	1.118742	65
66	0.9013	1.352	1.126421	0.5536	1.424	1.216	1.361196	0.8374	0.6445	0.8700	31.5	1.109208	66
67	0.9018	1.330	1.116505	0.5274	1.521	1.188	1.342017	0.8182	0.6787	0.8528	30.4	1.100151	67
68	0.9023	1.308	1.107063	0.5010	1.629	1.160	1.323405	0.7982	0.7153	0.8349	29.3	1.091555	68
69	0.9028	1.285	1.098078	0.4744	1.748	1.131	1.305331	0.7777	0.7547	0.8163	28.2	1.083407	69
70	0.9032	1.261	1.089536	0.4475	1.883	1.102	1.287771	0.7564	0.7972	0.7970	27.1	1.075391	70

TABLE 12-56 (Continued)

θ (T)	Ω_s	A_{min}	C_1	C_2	L_2	Ω_2	C_3	C_4	L_4	Ω_4	C_5	L_6	θ (T)
T	∞	∞	1.322	0.0000	1.373	∞	2.203	0.0000	1.469	∞	2.059	0.8816	T
71	1.068397	26.0	0.7769	0.8433	0.7344	1.270700	1.073	2.034	0.4204	1.081425	1.237	0.9037	71
72	1.061511	24.9	0.7560	0.8936	0.7116	1.254095	1.044	2.206	0.3931	1.073732	1.213	0.9040	72
73	1.055024	23.7	0.7341	0.9487	0.6878	1.237933	1.015	2.405	0.3657	1.066446	1.188	0.9044	73
74	1.048925	22.6	0.7112	1.010	0.6631	1.222193	0.9860	2.634	0.3381	1.059558	1.162	0.9047	74
75	1.043207	21.5	0.6872	1.077	0.6374	1.206854	0.9568	2.905	0.3105	1.053059	1.135	0.9049	75
76	1.037860	20.3	0.6620	1.153	0.6104	1.191893	0.9278	3.226	0.2828	1.046940	1.107	0.9050	76
77	1.032878	19.1	0.6353	1.239	0.5822	1.177291	0.8991	3.615	0.2552	1.041196	1.079	0.9049	77
78	1.028255	17.9	0.6071	1.338	0.5525	1.163026	0.8706	4.093	0.2277	1.035818	1.050	0.9047	78
79	1.023985	16.6	0.5770	1.453	0.5211	1.149076	0.8427	4.695	0.2005	1.030804	1.019	0.9042	79
80	1.020064	15.4	0.5450	1.590	0.4879	1.135418	0.8156	5.471	0.1736	1.026148	0.9868	0.9033	80
81	1.016487	14.1	0.5105	1.755	0.4526	1.122029	0.7895	6.502	0.1473	1.021849	0.9529	0.9020	81
82	1.013253	12.7	0.4732	1.960	0.4149	1.108880	0.7650	7.925	0.1218	1.017905	0.9170	0.9001	82
83	1.010360	11.4	0.4325	2.223	0.3745	1.095939	0.7426	9.982	0.0974	1.014316	0.8784	0.8972	83
84	1.007808	9.9	0.3876	2.576	0.3309	1.083168	0.7234	13.14	0.0744	1.011085	0.8365	0.8930	84
85	1.005599	8.5	0.3377	3.075	0.2838	1.070517	0.7089	18.40	0.0535	1.008216	0.7898	0.8870	85
θ	Ω_s	A_{min}	L_1	C_2	L_2	Ω_2	L_3	L_4	C_4	Ω_4	L_5	C_6	θ

1.0 → 1500

* Reprinted from R. Saal and E. Ulbrich, "On the Design of Filters by Synthesis," IRE Transactions on Circuit Theory, December 1958.

COO6 20 c*

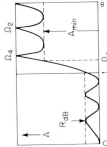

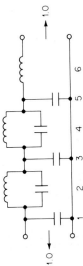

θ/T	Ω_s	A_min	C_1	C_2	L_2	Ω_2	C_3	C_4	L_4	Ω_4	C_5	L_6	θ/T
	∞	∞	1.159	0.0000	1.529	∞	1.838	0.0000	1.838	∞	1.529	1.159	
16	3.878298	112.5	1.138	0.0209	1.500	5.644802	1.790	0.0350	1.769	4.020935	1.500	1.158	16
17	3.655090	109.3	1.135	0.0237	1.496	5.314073	1.784	0.0396	1.761	3.788961	1.496	1.158	17
18	3.456975	106.3	1.132	0.0266	1.492	5.020165	1.777	0.0445	1.751	3.583033	1.492	1.158	18
19	3.279996	103.4	1.129	0.0297	1.488	4.757266	1.770	0.0497	1.742	3.399040	1.488	1.158	19
20	3.120982	100.7	1.125	0.0330	1.483	4.520722	1.763	0.0552	1.731	3.233693	1.483	1.158	20
21	2.977369	98.1	1.122	0.0365	1.478	4.306769	1.756	0.0611	1.720	3.084330	1.479	1.158	21
22	2.847060	95.6	1.118	0.0401	1.473	4.112326	1.748	0.0673	1.709	2.948774	1.474	1.157	22
23	2.728322	93.3	1.114	0.0440	1.468	3.934847	1.739	0.0738	1.697	2.825225	1.469	1.157	23
24	2.619709	91.0	1.110	0.0480	1.463	3.772213	1.731	0.0807	1.685	2.712184	1.464	1.157	24
25	2.520009	88.8	1.106	0.0523	1.457	3.622641	1.722	0.0879	1.672	2.608393	1.458	1.157	25
26	2.428196	86.7	1.102	0.0568	1.451	3.484624	1.712	0.0955	1.658	2.512785	1.452	1.157	26
27	2.343395	84.6	1.097	0.0614	1.445	3.356877	1.702	0.1035	1.644	2.424454	1.446	1.156	27
28	2.264858	82.6	1.092	0.0663	1.439	3.238301	1.692	0.1118	1.630	2.342621	1.440	1.156	28
29	2.191939	80.7	1.087	0.0714	1.432	3.127945	1.682	0.1205	1.615	2.266617	1.433	1.156	29
30	2.124078	78.9	1.082	0.0767	1.425	3.024987	1.671	0.1297	1.599	2.195860	1.427	1.156	30
31	2.060787	77.1	1.077	0.0822	1.418	2.928712	1.660	0.1392	1.583	2.129845	1.420	1.155	31
32	2.001642	75.3	1.071	0.0880	1.410	2.838492	1.648	0.1492	1.567	2.068129	1.413	1.155	32
33	1.946266	73.6	1.065	0.0940	1.403	2.753776	1.636	0.1597	1.550	2.010323	1.405	1.155	33
34	1.894331	72.0	1.059	0.1003	1.395	2.674079	1.624	0.1706	1.532	1.956085	1.398	1.154	34
35	1.845543	70.4	1.053	0.1068	1.386	2.598969	1.611	0.1820	1.514	1.905110	1.390	1.154	35

TABLE 12-56 *(Continued)*

θ / T	Ω_s	A_{min}	C_1	C_2	L_2	Ω_2	C_3	C_4	L_4	Ω_4	C_5	L_6	θ / T
∞	∞	∞	1.159	0.0000	1.529	∞	1.838	0.0000	1.838	∞	1.529	1.159	∞
36	1.799643	68.8	1.047	0.1135	1.378	2.528063	1.598	0.1939	1.496	1.857129	1.382	1.154	36
37	1.756398	67.3	1.040	0.1206	1.369	2.461022	1.585	0.2063	1.477	1.811902	1.374	1.153	37
38	1.715603	65.8	1.033	0.1279	1.360	2.397538	1.571	0.2192	1.457	1.769212	1.365	1.153	38
39	1.677070	64.3	1.026	0.1355	1.351	2.337337	1.557	0.2328	1.437	1.728868	1.356	1.152	39
40	1.640634	62.8	1.019	0.1434	1.341	2.280174	1.543	0.2469	1.417	1.690696	1.348	1.152	40
41	1.608142	61.4	1.012	0.1516	1.332	2.225824	1.528	0.2617	1.396	1.654538	1.338	1.151	41
42	1.573460	60.0	1.004	0.1601	1.321	2.174087	1.513	0.2772	1.374	1.620254	1.329	1.151	42
43	1.542462	58.7	0.9963	0.1689	1.311	2.124779	1.498	0.2993	1.352	1.587714	1.319	1.150	43
44	1.513038	57.3	0.9882	0.1781	1.300	2.077734	1.482	0.3103	1.330	1.556804	1.309	1.150	44
45	1.485086	56.0	0.9798	0.1877	1.289	2.032800	1.466	0.3280	1.307	1.527416	1.299	1.149	45
46	1.458511	54.7	0.9712	0.1976	1.278	1.989839	1.450	0.3465	1.284	1.499453	1.289	1.148	46
47	1.433230	53.4	0.9624	0.2079	1.266	1.948725	1.433	0.3659	1.260	1.472828	1.278	1.148	47
48	1.409164	52.2	0.9533	0.2187	1.255	1.909340	1.416	0.3863	1.235	1.447459	1.267	1.147	48
49	1.386241	50.9	0.9439	0.2298	1.242	1.871578	1.399	0.4078	1.211	1.423273	1.256	1.146	49
50	1.364398	49.7	0.9343	0.2414	1.230	1.835340	1.381	0.4303	1.185	1.400200	1.245	1.146	50
51	1.343572	48.5	0.9244	0.2535	1.217	1.800536	1.363	0.4540	1.160	1.378179	1.234	1.145	51
52	1.323710	47.3	0.9142	0.2661	1.204	1.767082	1.345	0.4790	1.133	1.357152	1.222	1.144	52
53	1.304759	46.1	0.9037	0.2792	1.190	1.734901	1.327	0.5054	1.107	1.337064	1.210	1.143	53
54	1.286672	45.0	0.8929	0.2929	1.176	1.703919	1.308	0.5333	1.080	1.317868	1.197	1.142	54
55	1.269406	43.8	0.8819	0.3072	1.162	1.674071	1.289	0.5628	1.052	1.299518	1.185	1.141	55
56	1.252921	42.7	0.8705	0.3221	1.147	1.645294	1.269	0.5941	1.024	1.281971	1.172	1.140	56
57	1.237179	41.5	0.8587	0.3377	1.132	1.617530	1.249	0.6274	0.9957	1.265189	1.159	1.139	57
58	1.222145	40.4	0.8466	0.3541	1.116	1.590725	1.229	0.6629	0.9668	1.249136	1.145	1.138	58
59	1.207787	39.3	0.8342	0.3712	1.100	1.564828	1.209	0.7008	0.9375	1.233777	1.131	1.137	59
60	1.194077	38.1	0.8214	0.3892	1.084	1.539791	1.188	0.7413	0.9077	1.219083	1.117	1.136	60
61	1.180985	37.0	0.8081	0.4081	1.067	1.515571	1.167	0.7848	0.8775	1.205023	1.103	1.134	61
62	1.168486	35.9	0.7945	0.4280	1.049	1.492126	1.146	0.8317	0.8468	1.191572	1.088	1.133	62
63	1.156557	34.8	0.7804	0.4490	1.032	1.469414	1.125	0.8823	0.8157	1.178704	1.074	1.131	63
64	1.145175	33.7	0.7659	0.4712	1.013	1.447401	1.103	0.9372	0.7843	1.166396	1.058	1.130	64
65	1.134320	32.6	0.7509	0.4947	0.9940	1.426049	1.081	0.9970	0.7524	1.154626	1.043	1.128	65
66	1.123973	31.5	0.7354	0.5196	0.9744	1.405326	1.059	1.062	0.7201	1.143375	1.026	1.126	66
67	1.114116	30.4	0.7193	0.5462	0.9542	1.385199	1.037	1.134	0.6874	1.132624	1.010	1.125	67
68	1.104733	29.3	0.7027	0.5746	0.9332	1.365637	1.014	1.213	0.6543	1.122356	0.9932	1.123	68
69	1.095809	28.2	0.6854	0.6050	0.9115	1.346613	0.9915	1.301	0.6208	1.112555	0.9759	1.120	69
70	1.087329	27.1	0.6674	0.6377	0.8891	1.328096	0.9686	1.400	0.5870	1.103207	0.9582	1.118	70

θ	Ω_s	A_min	L_1	L_2	C_2	Ω_2	L_3	L_4	C_4	Ω_4	L_5	C_6	θ
71	1.079282	26.0	0.6488	0.6730	0.8657	1.310060	0.9456	1.511	0.5528	1.094297	0.9399	1.116	71
72	1.071656	24.9	0.6293	0.7114	0.8415	1.292478	0.9225	1.636	0.5184	1.085815	0.9211	1.113	72
73	1.064439	23.7	0.6089	0.7533	0.8162	1.275324	0.8994	1.780	0.4836	1.077747	0.9017	1.110	73
74	1.057623	22.6	0.5876	0.7994	0.7898	1.258571	0.8762	1.947	0.4486	1.070085	0.8816	1.107	74
75	1.051198	21.5	0.5652	0.8503	0.7621	1.242193	0.8530	2.141	0.4134	1.062820	0.8608	1.104	75
76	1.045158	20.3	0.5417	0.9073	0.7331	1.226164	0.8299	2.372	0.3781	1.055943	0.8393	1.100	76
77	1.039495	19.1	0.5168	0.9716	0.7025	1.210456	0.8071	2.650	0.3426	1.049447	0.8168	1.096	77
78	1.034204	17.9	0.4905	1.045	0.6701	1.195041	0.7845	2.990	0.3072	1.043327	0.7932	1.091	78
79	1.029281	16.6	0.4624	1.130	0.6358	1.179887	0.7625	3.415	0.2720	1.037578	0.7685	1.086	79
80	1.024722	15.4	0.4323	1.230	0.5991	1.164960	0.7411	3.961	0.2370	1.032198	0.7423	1.080	80
81	1.020525	14.1	0.3999	1.350	0.5598	1.150224	0.7206	4.677	0.2026	1.027183	0.7144	1.073	81
82	1.016691	12.7	0.3648	1.499	0.5174	1.135632	0.7016	5.659	0.1690	1.022536	0.6845	1.064	82
83	1.013219	11.4	0.3263	1.687	0.4715	1.121129	0.6845	7.062	0.1366	1.018256	0.6518	1.055	83
84	1.010114	9.9	0.2837	1.938	0.4214	1.106645	0.6702	9.190	0.1058	1.014351	0.6158	1.043	84
85	1.007381	8.5	0.2958	2.288	0.3664	1.092084	0.6603	12.67	0.0772	1.010827	0.5750	1.027	85
θ	Ω_s	A_min	L_1	L_2	C_2	Ω_2	L_3	L_4	C_4	Ω_4	L_5	C_6	θ

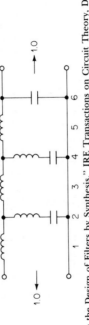

* Reprinted from R. Saal and E. Ulbrich, "On the Design of Filters by Synthesis," IRE Transactions on Circuit Theory, December 1958.

TABLE 12-56 (Continued)

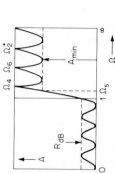

CO7 20*

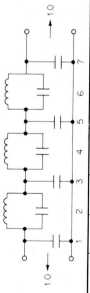

θ / T	Ω_s	A_{min}	C_1	C_2	L_2	Ω_2	C_3	C_4	L_4	Ω_4	C_5	C_6	L_6	Ω_6	C_7	θ / T
∞	∞	∞	1.335	0.0000	1.389	∞	2.240	0.0000	1.515	∞	2.240	0.0000	1.389	∞	1.335	∞
26	2.281172	105.4	1.310	0.0290	1.358	5.038750	2.100	0.1353	1.357	2.333900	2.049	0.0955	1.281	2.859592	1.247	26
27	2.202689	103.0	1.308	0.0314	1.355	4.848897	2.089	0.1465	1.345	2.253156	2.034	0.1034	1.272	2.756829	1.240	27
28	2.130054	100.7	1.306	0.0339	1.353	4.672457	2.078	0.1582	1.332	2.178409	2.019	0.1117	1.263	2.661529	1.233	28
29	2.062665	98.5	1.304	0.0364	1.350	4.508037	2.066	0.1704	1.319	2.109040	2.003	0.1204	1.254	2.572921	1.226	29
30	2.000000	96.3	1.302	0.0391	1.347	4.354434	2.054	0.1833	1.305	2.044515	1.987	0.1295	1.245	2.490337	1.218	30
31	1.941604	94.2	1.299	0.0420	1.344	4.210595	2.042	0.1966	1.292	1.984368	1.970	0.1390	1.235	2.413194	1.210	31
32	1.887080	92.2	1.297	0.0449	1.341	4.075602	2.029	0.2106	1.277	1.928190	1.952	0.1490	1.225	2.340984	1.202	32
33	1.836078	90.2	1.294	0.0479	1.338	3.948647	2.016	0.2252	1.262	1.875623	1.934	0.1593	1.214	2.273259	1.193	33
34	1.788292	88.3	1.292	0.0511	1.335	3.829016	2.002	0.2404	1.247	1.826351	1.916	0.1702	1.204	2.209625	1.184	34
35	1.743347	86.4	1.289	0.0544	1.332	3.716076	1.988	0.2562	1.232	1.780095	1.897	0.1815	1.193	2.149731	1.175	35
36	1.701302	84.6	1.286	0.0578	1.328	3.609267	1.973	0.2727	1.216	1.736606	1.878	0.1932	1.181	2.093268	1.165	36
37	1.661640	82.8	1.283	0.0614	1.324	3.508087	1.959	0.2900	1.199	1.695662	1.858	0.2055	1.169	2.039957	1.155	37
38	1.624269	81.0	1.280	0.0650	1.321	3.412086	1.943	0.3079	1.183	1.657065	1.837	0.2183	1.157	1.989552	1.145	38
39	1.589016	79.3	1.277	0.0689	1.317	3.320862	1.928	0.3267	1.165	1.620638	1.817	0.2317	1.145	1.941830	1.135	39
40	1.555724	77.6	1.274	0.0728	1.313	3.234050	1.912	0.3462	1.148	1.586220	1.795	0.2456	1.132	1.896591	1.124	40
41	1.524253	76.0	1.270	0.0770	1.308	3.151325	1.895	0.3666	1.130	1.553668	1.773	0.2601	1.119	1.853653	1.113	41
42	1.494477	74.3	1.267	0.0812	1.304	3.072388	1.879	0.3879	1.112	1.522851	1.751	0.2753	1.105	1.812855	1.102	42
43	1.466279	72.8	1.263	0.0857	1.300	2.996969	1.862	0.4101	1.093	1.493651	1.728	0.2911	1.092	1.774048	1.090	43
44	1.439557	71.2	1.259	0.0903	1.295	2.924824	1.844	0.4332	1.074	1.465961	1.705	0.3076	1.077	1.737098	1.078	44
45	1.414214	69.7	1.255	0.0950	1.290	2.855727	1.826	0.4575	1.055	1.439683	1.682	0.3248	1.063	1.701881	1.066	45

46	1.053	1.668286	1.048	0.3428	1.657	1.414728	1.035	0.4828	1.808	2.789476	1.285	0.1000	1.251	68.2	1.390164	46
47	1.040	1.636211	1.033	0.3617	1.633	1.391016	1.015	0.5093	1.789	2.725881	1.280	0.1051	1.247	66.7	1.367327	47
48	1.027	1.605563	1.017	0.3814	1.608	1.368471	0.9944	0.5370	1.770	2.664770	1.275	0.1105	1.243	65.2	1.345633	48
49	1.013	1.576255	1.001	0.4020	1.583	1.347026	0.9736	0.5661	1.751	2.605984	1.269	0.1160	1.238	63.7	1.325013	49
50	0.9992	1.548208	0.9850	0.4235	1.557	1.326618	0.9525	0.5965	1.731	2.549377	1.264	0.1217	1.234	62.3	1.305407	50
51	0.9848	1.521349	0.9684	0.4462	1.531	1.307190	0.9310	0.6286	1.711	2.494813	1.258	0.1277	1.229	60.9	1.286760	51
52	0.9699	1.495612	0.9514	0.4699	1.504	1.288687	0.9093	0.6622	1.690	2.442167	1.252	0.1339	1.224	59.5	1.269018	52
53	0.9547	1.470934	0.9340	0.4948	1.477	1.271063	0.8872	0.6977	1.669	2.391323	1.246	0.1404	1.219	58.1	1.252136	53
54	0.9391	1.447259	0.9163	0.5211	1.450	1.254270	0.8648	0.7351	1.648	2.342170	1.239	0.1471	1.213	56.8	1.236068	54
55	0.9230	1.424533	0.8981	0.5487	1.422	1.238269	0.8420	0.7745	1.626	2.294610	1.232	0.1541	1.208	55.4	1.220775	55
56	0.9065	1.402707	0.8796	0.5778	1.394	1.223020	0.8190	0.8163	1.604	2.248546	1.225	0.1614	1.202	54.1	1.206218	56
57	0.8896	1.381735	0.8607	0.6085	1.365	1.208487	0.7957	0.8605	1.581	2.203891	1.218	0.1690	1.196	52.7	1.192363	57
58	0.8722	1.361575	0.8414	0.6411	1.336	1.194638	0.7721	0.9075	1.558	2.160560	1.211	0.1770	1.190	51.4	1.179178	58
59	0.8543	1.342188	0.8217	0.6755	1.307	1.181442	0.7482	0.9576	1.535	2.118476	1.203	0.1853	1.183	50.1	1.166633	59
60	0.8360	1.323537	0.8016	0.7121	1.278	1.168869	0.7240	1.011	1.511	2.077565	1.195	0.1939	1.177	48.8	1.154701	60
61	0.8171	1.305587	0.7811	0.7510	1.248	1.156895	0.6995	1.068	1.487	2.037756	1.186	0.2030	1.170	47.5	1.143354	61
62	0.7976	1.288307	0.7602	0.7925	1.218	1.145494	0.6748	1.129	1.463	1.998983	1.177	0.2125	1.163	46.2	1.132570	62
63	0.7776	1.271668	0.7389	0.8369	1.188	1.134644	0.6498	1.195	1.438	1.961181	1.168	0.2225	1.155	44.9	1.122326	63
64	0.7570	1.255641	0.7171	0.8845	1.157	1.124323	0.6245	1.267	1.412	1.924292	1.159	0.2331	1.147	43.7	1.112602	64
65	0.7357	1.240200	0.6949	0.9357	1.126	1.114512	0.5990	1.344	1.386	1.888255	1.149	0.2441	1.139	42.4	1.103378	65
66	0.7138	1.225322	0.6722	0.9909	1.095	1.105192	0.5732	1.428	1.360	1.853014	1.138	0.2559	1.130	41.1	1.094636	66
67	0.6911	1.210984	0.6490	1.051	1.064	1.096346	0.5472	1.520	1.333	1.818515	1.127	0.2682	1.121	39.8	1.086360	67
68	0.6676	1.197165	0.6254	1.116	1.032	1.087959	0.5209	1.622	1.306	1.784703	1.116	0.2814	1.112	38.5	1.078535	68
69	0.6433	1.183845	0.6013	1.187	1.001	1.080016	0.4945	1.734	1.278	1.751526	1.104	0.2953	1.101	37.2	1.071145	69
70	0.6181	1.171007	0.5767	1.265	0.9689	1.072504	0.4678	1.859	1.250	1.718931	1.091	0.3102	1.091	35.9	1.064178	70
71	0.5920	1.158633	0.5516	1.351	0.9371	1.065409	0.4409	1.998	1.221	1.686865	1.077	0.3262	1.080	34.6	1.057621	71
72	0.5647	1.146708	0.5259	1.446	0.9051	1.058721	0.4138	2.156	1.192	1.655277	1.063	0.3433	1.068	33.3	1.051462	72
73	0.5363	1.135217	0.4997	1.553	0.8731	1.052428	0.3865	2.336	1.162	1.624111	1.048	0.3618	1.055	32.0	1.045692	73
74	0.5066	1.124147	0.4729	1.673	0.8412	1.046522	0.3591	2.543	1.131	1.593311	1.032	0.3818	1.042	30.7	1.040299	74
75	0.4754	1.113485	0.4455	1.810	0.8093	1.040993	0.3315	2.784	1.100	1.562818	1.014	0.4037	1.028	29.3	1.035276	75

TABLE 12-56 (Continued)

θ	Ωs	Amin	C1	C2	L2	Ω2	C3	C4	L4	Ω4	C5	C6	L6	Ω6	C7	θ
T	∞	∞	1.335	0.0000	1.389	∞	2.240	0.0000	1.515	∞	2.240	0.0000	1.389	∞	1.335	T
76	1.030614	27.9	1.013	0.4278	0.9953	1.532571	1.069	3.068	0.3038	1.035833	0.7776	1.968	0.4175	1.103221	0.4426	76
77	1.026304	26.5	0.9960	0.4544	0.9749	1.502499	1.036	3.408	0.2760	1.031035	0.7460	2.151	0.3888	1.093345	0.4079	77
78	1.022341	25.1	0.9782	0.4841	0.9527	1.472529	1.004	3.822	0.2483	1.026592	0.7148	2.368	0.3595	1.083849	0.3710	78
79	1.018717	23.6	0.9588	0.5177	0.9282	1.442574	0.9699	4.337	0.2205	1.022499	0.6841	2.628	0.3295	1.074724	0.3316	79
80	1.015427	22.1	0.9376	0.5562	0.9011	1.412537	0.9356	4.994	0.1929	1.018751	0.6540	2.946	0.2987	1.065966	0.2892	80
81	1.012465	20.6	0.9142	0.6011	0.8707	1.382299	0.9006	5.858	0.1656	1.015345	0.6248	3.346	0.2672	1.057569	0.2431	81
82	1.009828	18.9	0.8881	0.6545	0.8363	1.351718	0.8648	7.036	0.1387	1.012276	0.5968	3.863	0.2350	1.049533	0.1926	82
83	1.007510	17.3	0.8587	0.7197	0.7967	1.320610	0.8283	8.723	0.1125	1.009543	0.5706	4.559	0.2021	1.041856	0.1363	83
84	1.005508	15.5	0.8252	0.8023	0.7504	1.288733	0.7911	11.29	0.0873	1.007145	0.5470	5.545	0.1685	1.034542	0.0725	84
85	1.003820	13.6	0.7863	0.9121	0.6953	1.255747	0.7533	15.55	0.0636	1.005081	0.5275	7.042	0.1345	1.027600	-0.0016	85
θ	Ωs	Amin	L1	L2	C2	Ω2	L3	L4	C4	Ω4	L5	L6	C6	Ω6	L7	θ

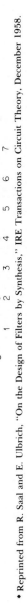

* Reprinted from R. Saal and E. Ulbrich, "On the Design of Filters by Synthesis," IRE Transactions on Circuit Theory, December 1958.

CO8 20 b*

→ 0.6667

1.0 →

θ, T	Ω_s	A_{min}	C_1	C_2	L_2	Ω_2	C_3	C_4	L_4	Ω_4	C_5	C_6	L_6	Ω_6	C_7	L_8	T, θ
∞	∞	∞	1.343	0.0000	1.398	∞	2.261	0.0000	1.538	∞	2.307	0.0000	1.507	∞	2.097	0.8954	∞
31	1.974165	111.4	1.289	0.0601	1.331	3.534655	2.051	0.1878	1.319	2.008837	2.028	0.1353	1.348	2.341711	1.977	0.8980	31
32	1.918381	109.1	1.285	0.0644	1.326	3.422880	2.037	0.2011	1.305	1.951720	2.010	0.1448	1.337	2.272128	1.969	0.8982	32
33	1.866186	106.8	1.281	0.0687	1.321	3.317819	2.023	0.2150	1.291	1.898264	1.992	0.1548	1.327	2.206887	1.961	0.8984	33
34	1.817268	104.6	1.277	0.0733	1.317	3.218876	2.008	0.2294	1.276	1.848149	1.973	0.1651	1.315	2.145605	1.952	0.8986	34
35	1.771347	102.4	1.273	0.0781	1.311	3.125526	1.993	0.2445	1.261	1.801092	1.953	0.1759	1.304	2.087945	1.943	0.8987	35
36	1.728178	100.3	1.268	0.0830	1.306	3.037300	1.978	0.2602	1.245	1.756840	1.933	0.1871	1.292	2.033606	1.934	0.8989	36
37	1.687539	98.3	1.264	0.0881	1.301	2.953780	1.962	0.2765	1.229	1.715168	1.912	0.1988	1.280	1.982320	1.925	0.8991	37
38	1.649233	96.3	1.259	0.0934	1.295	2.874592	1.946	0.2936	1.213	1.675876	1.891	0.2110	1.267	1.933848	1.916	0.8993	38
39	1.613085	94.3	1.255	0.0990	1.289	2.799400	1.929	0.3113	1.196	1.638784	1.869	0.2236	1.254	1.887974	1.906	0.8995	39
40	1.578935	92.4	1.250	0.1047	1.283	2.727903	1.912	0.3298	1.179	1.603728	1.847	0.2368	1.241	1.844505	1.896	0.8997	40
41	1.546640	90.5	1.245	0.1107	1.277	2.659827	1.895	0.3491	1.161	1.570563	1.824	0.2505	1.228	1.803267	1.885	0.9000	41
42	1.516070	88.7	1.239	0.1169	1.271	2.594925	1.877	0.3692	1.143	1.539156	1.801	0.2647	1.214	1.764103	1.875	0.9002	42
43	1.487108	86.8	1.234	0.1233	1.264	2.532974	1.859	0.3902	1.125	1.509389	1.777	0.2796	1.200	1.726867	1.864	0.9004	43
44	1.459648	85.1	1.228	0.1300	1.257	2.473768	1.840	0.4120	1.106	1.481151	1.753	0.2950	1.185	1.691432	1.852	0.9006	44
45	1.433592	83.3	1.222	0.1369	1.250	2.417121	1.821	0.4348	1.087	1.454344	1.728	0.3110	1.170	1.657678	1.841	0.9009	45
46	1.408853	81.6	1.217	0.1441	1.243	2.362865	1.801	0.4586	1.068	1.428878	1.703	0.3278	1.155	1.625498	1.829	0.9011	46
47	1.385348	79.9	1.210	0.1516	1.235	2.310842	1.781	0.4835	1.048	1.404670	1.677	0.3452	1.139	1.594793	1.817	0.9014	47
48	1.363006	78.2	1.204	0.1594	1.228	2.260911	1.761	0.5095	1.028	1.381645	1.651	0.3634	1.123	1.565471	1.805	0.9016	48
49	1.341757	76.5	1.197	0.1674	1.220	2.212939	1.740	0.5368	1.008	1.359734	1.624	0.3823	1.107	1.537451	1.792	0.9019	49
50	1.321539	74.9	1.191	0.1758	1.211	2.166805	1.719	0.5653	0.9869	1.338873	1.597	0.4021	1.090	1.510657	1.779	0.9021	50
51	1.302296	73.3	1.184	0.1845	1.203	2.122397	1.698	0.5951	0.9658	1.319003	1.569	0.4227	1.073	1.485017	1.766	0.9024	51
52	1.283974	71.7	1.177	0.1936	1.194	2.079568	1.676	0.6265	0.9444	1.300068	1.541	0.4443	1.055	1.460459	1.752	0.9028	52
53	1.266526	70.1	1.169	0.2030	1.185	2.038353	1.653	0.6594	0.9226	1.282028	1.512	0.4668	1.037	1.436948	1.738	0.9029	53
54	1.249906	68.6	1.161	0.2129	1.176	1.998531	1.630	0.6941	0.9006	1.264827	1.483	0.4904	1.019	1.414404	1.724	0.9032	54
55	1.234073	67.0	1.153	0.2231	1.167	1.960064	1.607	0.7306	0.8782	1.248426	1.453	0.5152	1.001	1.392784	1.710	0.9035	55

TABLE 12-56 (Continued)

θ / T	Ωs	Amin	C1 / L1	C2 / L2	L2 / C2	Ω2	C3 / L3	C4 / L4	L4 / C4	Ω4	C5 / L5	C6 / L6	L6 / C6	Ω6	C7 / L7	Ls / C8	θ / T
T	∞	∞	1.343	0.0000	1.398	∞	2.261	0.0000	1.538	∞	2.307	0.0000	1.507	∞	2.097	0.8954	T
56	1.218988	65.5	1.145	0.2338	1.157	1.922874	1.583	0.7691	0.8555	1.232786	1.423	0.5412	0.9816	1.372040	1.695	0.9038	56
57	1.204616	64.0	1.136	0.2450	1.146	1.886889	1.559	0.8098	0.8326	1.217872	1.393	0.5684	0.9622	1.352130	1.680	0.9041	57
58	1.190925	62.5	1.128	0.2566	1.136	1.852022	1.535	0.8529	0.8093	1.203647	1.362	0.5971	0.9425	1.333008	1.664	0.9044	58
59	1.177883	61.0	1.119	0.2688	1.125	1.818264	1.509	0.8986	0.7857	1.190084	1.331	0.6273	0.9223	1.314647	1.648	0.9046	59
60	1.165463	59.5	1.109	0.2816	1.114	1.785501	1.484	0.9472	0.7619	1.177152	1.299	0.6592	0.9017	1.297002	1.632	0.9049	60
61	1.153638	58.0	1.099	0.2950	1.102	1.753696	1.458	0.9990	0.7378	1.164824	1.267	0.6929	0.8808	1.280043	1.615	0.9053	61
62	1.142384	56.5	1.089	0.3090	1.090	1.722794	1.432	1.054	0.7134	1.153075	1.234	0.7286	0.8593	1.263739	1.598	0.9056	62
63	1.131677	55.0	1.079	0.3238	1.078	1.692746	1.405	1.114	0.6887	1.141883	1.201	0.7665	0.8375	1.248062	1.581	0.9059	63
64	1.121498	53.6	1.068	0.3393	1.065	1.663502	1.377	1.177	0.6638	1.131225	1.168	0.8069	0.8152	1.232986	1.563	0.9062	64
65	1.111827	52.1	1.056	0.3557	1.052	1.635018	1.350	1.246	0.6387	1.121082	1.134	0.8499	0.7925	1.218485	1.544	0.9065	65
66	1.102644	50.6	1.044	0.3730	1.038	1.607250	1.321	1.320	0.6132	1.111435	1.100	0.8959	0.7693	1.204537	1.526	0.9068	66
67	1.093934	49.2	1.032	0.3913	1.023	1.580155	1.292	1.401	0.5876	1.102268	1.065	0.9452	0.7457	1.191121	1.506	0.9071	67
68	1.085681	47.7	1.019	0.4108	1.008	1.553693	1.263	1.489	0.5617	1.093563	1.030	0.9984	0.7215	1.178215	1.487	0.9075	68
69	1.077870	46.2	1.006	0.4315	0.9929	1.527823	1.233	1.585	0.5356	1.085307	0.9947	1.056	0.6969	1.165802	1.466	0.9078	69
70	1.070487	44.8	0.9917	0.4536	0.9766	1.502508	1.202	1.692	0.5092	1.077485	0.9590	1.118	0.6717	1.153865	1.445	0.9081	70
71	1.063520	43.3	0.9769	0.4772	0.9597	1.477710	1.171	1.809	0.4826	1.070085	0.9230	1.186	0.6460	1.142387	1.424	0.9084	71
72	1.056959	41.8	0.9614	0.5026	0.9419	1.453390	1.140	1.941	0.4559	1.063096	0.8867	1.261	0.6197	1.131353	1.402	0.9087	72
73	1.050791	40.3	0.9450	0.5300	0.9232	1.429511	1.107	2.089	0.4289	1.056506	0.8501	1.343	0.5928	1.120750	1.379	0.9091	73
74	1.045007	38.7	0.9277	0.5598	0.9036	1.406035	1.074	2.257	0.4017	1.050305	0.8131	1.434	0.5653	1.110565	1.355	0.9094	74
75	1.039599	37.2	0.9094	0.5922	0.8829	1.382921	1.040	2.449	0.3743	1.044487	0.7759	1.537	0.5371	1.100787	1.331	0.9097	75
76	1.034558	35.6	0.8899	0.6278	0.8611	1.360130	1.006	2.671	0.3468	1.039040	0.7384	1.652	0.5082	1.091406	1.305	0.9100	76
77	1.029877	34.0	0.8691	0.6670	0.8379	1.337618	0.9701	2.931	0.3191	1.033960	0.7007	1.784	0.4785	1.082412	1.279	0.9103	77
78	1.025550	32.4	0.8469	0.7108	0.8131	1.315338	0.9337	3.241	0.2913	1.029240	0.6628	1.936	0.4480	1.073796	1.251	0.9105	78
79	1.021570	30.7	0.8228	0.7601	0.7866	1.293239	0.8964	3.616	0.2633	1.024874	0.6247	2.114	0.4166	1.065554	1.222	0.9108	79
80	1.017932	29.0	0.7968	0.8162	0.7581	1.271262	0.8579	4.079	0.2352	1.020857	0.5864	2.326	0.3843	1.057678	1.192	0.9110	80
81	1.014633	27.2	0.7683	0.8812	0.7271	1.249341	0.8183	4.667	0.2071	1.017188	0.5482	2.584	0.3509	1.050154	1.159	0.9111	81
82	1.011669	25.3	0.7368	0.9577	0.6931	1.227394	0.7774	5.436	0.1790	1.013862	0.5100	2.907	0.3162	1.043012	1.125	0.9111	82
83	1.009036	23.3	0.7017	1.050	0.6555	1.205319	0.7351	6.484	0.1509	1.010878	0.4719	3.323	0.2803	1.036219	1.088	0.9110	83
84	1.006735	21.3	0.6619	1.165	0.6133	1.182984	0.6911	7.989	0.1231	1.008238	0.4342	3.884	0.2428	1.029789	1.047	0.9106	84
85	1.004764	19.0	0.6159	1.315	0.5649	1.160208	0.6453	10.32	0.0958	1.005943	0.3973	4.688	0.2036	1.023728	1.003	0.9098	85
θ / T	Ωs	Amin	L1	L2	C2	Ω2	L3	L4	C4	Ω4	L5	L6	C6	Ω6	L7	C8	θ / T

1.0 →   ← 1.500

Node positions: 1 2 3 4 5 6 7 8

* Reprinted from R. Saal and E. Ulbrich, "On the Design of Filters by Synthesis," IRE Transactions on Circuit Theory, December 1958.

CO8 20 c*

θ/T	Ω_s	A_{min}	C_1	C_2	L_2	Ω_2	C_3	C_4	L_4	Ω_4	C_5	C_6	L_6	Ω_6	C_7	L_8
	∞	∞	1.215	0.0000	1.523	∞	1.963	0.0000	1.840	∞	1.840	0.0000	1.963	∞	1.523	1.215
31	2.007273	111.4	1.159	0.0527	1.457	3.607866	1.782	0.1508	1.589	2.042921	1.628	0.1002	1.755	2.384868	1.442	1.210
32	1.950201	109.1	1.155	0.0564	1.452	3.494036	1.770	0.1614	1.573	1.984498	1.614	0.1073	1.741	2.313817	1.437	1.210
33	1.896788	106.8	1.151	0.0602	1.448	3.387051	1.758	0.1725	1.556	1.929808	1.600	0.1147	1.727	2.247191	1.431	1.210
34	1.846713	104.6	1.148	0.0642	1.443	3.286306	1.746	0.1841	1.539	1.878522	1.585	0.1223	1.713	2.184601	1.425	1.210
35	1.799694	102.4	1.143	0.0683	1.438	3.191263	1.733	0.1962	1.521	1.830353	1.570	0.1304	1.697	2.125700	1.420	1.209
36	1.755478	100.3	1.138	0.0726	1.432	3.101445	1.720	0.2088	1.503	1.785042	1.555	0.1387	1.682	2.070184	1.413	1.208
37	1.713841	98.3	1.134	0.0770	1.427	3.016426	1.706	0.2219	1.485	1.742360	1.539	0.1474	1.666	2.017779	1.407	1.208
38	1.674581	96.3	1.129	0.0816	1.421	2.935824	1.692	0.2355	1.466	1.702102	1.522	0.1565	1.650	1.968241	1.401	1.207
39	1.637519	94.3	1.124	0.0864	1.416	2.859298	1.678	0.2497	1.446	1.664085	1.506	0.1659	1.633	1.921349	1.394	1.207
40	1.602492	92.4	1.119	0.0914	1.410	2.786539	1.663	0.2645	1.426	1.628143	1.489	0.1757	1.616	1.876908	1.387	1.206
41	1.569355	90.5	1.114	0.0965	1.404	2.717269	1.648	0.2799	1.406	1.594127	1.471	0.1859	1.598	1.834738	1.380	1.206
42	1.537975	88.7	1.109	0.1018	1.397	2.651235	1.633	0.2959	1.385	1.561901	1.453	0.1965	1.580	1.794680	1.373	1.205
43	1.508233	86.8	1.103	0.1073	1.391	2.588210	1.617	0.3127	1.364	1.531345	1.435	0.2075	1.562	1.756586	1.365	1.205
44	1.480020	85.1	1.097	0.1131	1.384	2.527985	1.601	0.3301	1.342	1.502346	1.416	0.2190	1.543	1.720325	1.358	1.204
45	1.453236	83.3	1.092	0.1190	1.377	2.470370	1.585	0.3483	1.320	1.474804	1.397	0.2310	1.523	1.685775	1.350	1.203
46	1.427793	81.6	1.085	0.1251	1.370	2.415192	1.568	0.3673	1.297	1.448627	1.378	0.2435	1.503	1.652826	1.342	1.203
47	1.403607	79.9	1.079	0.1315	1.362	2.362292	1.551	0.3871	1.274	1.423730	1.358	0.2565	1.483	1.621377	1.334	1.202
48	1.380603	78.2	1.073	0.1381	1.355	2.311524	1.534	0.4078	1.251	1.400037	1.338	0.2701	1.462	1.591337	1.325	1.201
49	1.358712	76.5	1.066	0.1450	1.347	2.262754	1.516	0.4295	1.227	1.377477	1.317	0.2842	1.441	1.562620	1.317	1.200
50	1.337870	74.9	1.059	0.1521	1.339	2.215859	1.498	0.4521	1.203	1.355985	1.296	0.2990	1.419	1.535148	1.308	1.200
51	1.318020	73.3	1.052	0.1595	1.330	2.170723	1.479	0.4759	1.178	1.335501	1.275	0.3144	1.397	1.508850	1.299	1.199
52	1.299107	71.7	1.045	0.1672	1.322	2.127183	1.460	0.5008	1.153	1.315965	1.253	0.3305	1.374	1.483645	1.290	1.198
53	1.281092	70.1	1.037	0.1751	1.313	2.085317	1.441	0.5269	1.128	1.297345	1.231	0.3474	1.351	1.459515	1.280	1.197
54	1.263899	68.6	1.029	0.1834	1.304	2.044857	1.421	0.5543	1.102	1.279574	1.208	0.3651	1.328	1.436360	1.270	1.196
55	1.247516	67.0	1.021	0.1920	1.294	2.005778	1.401	0.5832	1.076	1.262617	1.186	0.3836	1.304	1.414144	1.260	1.195

TABLE 12-56 (Continued)

θ	T	L_8	C_7	Ω_6	L_6	C_6	C_5	Ω_4	L_4	C_4	C_3	Ω_2	L_2	C_2	C_1	A_{min}	Ω_s	T	θ
		1.215	1.523	∞	1.963	0.0000	1.840	∞	1.840	0.0000	1.963	∞	1.523	0.0000	1.215	∞	∞		
56	56	1.194	1.250	1.392816	1.279	0.4030	1.162	1.246432	1.049	0.6137	1.381	1.967999	1.285	0.2010	1.013	65.5	1.231893	56	56
57	57	1.193	1.240	1.372332	1.254	0.4234	1.139	1.230984	1.022	0.6458	1.360	1.931448	1.274	0.2103	1.004	64.0	1.216995	57	57
58	58	1.192	1.229	1.352645	1.228	0.4449	1.115	1.216235	0.9944	0.6798	1.339	1.896030	1.264	0.2201	0.9956	62.5	1.202788	58	58
59	59	1.190	1.218	1.333732	1.202	0.4676	1.090	1.202161	0.9666	0.7159	1.318	1.861754	1.253	0.2302	0.9864	61.0	1.189241	59	59
60	60	1.189	1.207	1.315541	1.176	0.4914	1.066	1.188726	0.9384	0.7541	1.296	1.828485	1.242	0.2408	0.9770	59.5	1.176326	60	60
61	61	1.188	1.195	1.298043	1.149	0.5167	1.041	1.175903	0.9099	0.7948	1.274	1.796189	1.231	0.2519	0.9672	58.0	1.164014	61	61
62	62	1.186	1.183	1.281207	1.121	0.5434	1.015	1.163668	0.8810	0.8383	1.251	1.764813	1.219	0.2635	0.9570	56.5	1.152282	62	62
63	63	1.185	1.171	1.265004	1.093	0.5718	0.9897	1.151997	0.8517	0.8848	1.228	1.734303	1.206	0.2756	0.9465	55.0	1.141107	63	63
64	64	1.183	1.159	1.249405	1.064	0.6020	0.9636	1.140869	0.8220	0.9346	1.205	1.704611	1.193	0.2884	0.9356	53.6	1.130466	64	64
65	65	1.182	1.146	1.234386	1.035	0.6342	0.9373	1.130262	0.7921	0.9883	1.181	1.675689	1.180	0.3018	0.9243	52.1	1.120340	65	65
66	66	1.180	1.133	1.219923	1.005	0.6687	0.9106	1.120158	0.7617	1.046	1.157	1.647491	1.166	0.3159	0.9125	50.6	1.110711	66	66
67	67	1.178	1.120	1.205992	0.9744	0.7056	0.8835	1.110539	0.7311	1.109	1.132	1.619975	1.152	0.3308	0.9003	49.2	1.101561	67	67
68	68	1.176	1.106	1.192574	0.9433	0.7454	0.8561	1.101389	0.7000	1.178	1.107	1.593097	1.137	0.3465	0.8876	47.7	1.092874	68	68
69	69	1.174	1.091	1.179649	0.9115	0.7884	0.8285	1.092694	0.6687	1.253	1.082	1.566816	1.121	0.3632	0.8743	46.2	1.084636	69	69
70	70	1.172	1.077	1.167198	0.8791	0.8350	0.8005	1.084438	0.6370	1.335	1.056	1.541091	1.105	0.3810	0.8605	44.8	1.076833	70	70
71	71	1.170	1.061	1.155204	0.8460	0.8858	0.7722	1.076610	0.6049	1.426	1.029	1.515881	1.088	0.4000	0.8460	43.3	1.069453	71	71
72	72	1.167	1.046	1.143653	0.8121	0.9414	0.7435	1.069197	0.5726	1.528	1.002	1.491148	1.070	0.4202	0.8308	41.8	1.062484	72	72
73	73	1.165	1.029	1.132528	0.7775	1.003	0.7146	1.062190	0.5399	1.642	0.9745	1.466849	1.051	0.4420	0.8148	40.3	1.055915	73	73
74	74	1.162	1.013	1.121816	0.7421	1.071	0.6855	1.055577	0.5069	1.771	0.9464	1.442942	1.032	0.4655	0.7979	38.7	1.049736	74	74
75	75	1.159	0.9950	1.111505	0.7059	1.147	0.6560	1.049351	0.4735	1.918	0.9177	1.419386	1.011	0.4911	0.7801	37.2	1.043940	75	75
76	76	1.155	0.9767	1.101583	0.6684	1.233	0.6262	1.043503	0.4399	2.088	0.8884	1.396133	0.9886	0.5189	0.7613	35.6	1.038518	76	76
77	77	1.152	0.9576	1.092039	0.6306	1.330	0.5963	1.038027	0.4059	2.286	0.8584	1.373135	0.9650	0.5496	0.7412	34.0	1.033463	77	77
78	78	1.148	0.9374	1.082864	0.5914	1.442	0.5661	1.032917	0.3717	2.522	0.8276	1.350338	0.9398	0.5836	0.7197	32.4	1.028769	78	78
79	79	1.144	0.9162	1.074049	0.5510	1.573	0.5357	1.028167	0.3371	2.806	0.7961	1.327682	0.9127	0.6216	0.6967	30.7	1.024431	79	79
80	80	1.139	0.8937	1.065586	0.5094	1.729	0.5051	1.023774	0.3023	3.156	0.7638	1.305098	0.8833	0.6647	0.6717	29.0	1.020444	80	80
81	81	1.134	0.8697	1.057469	0.4664	1.918	0.4744	1.019736	0.2673	3.598	0.7305	1.282504	0.8513	0.7142	0.6445	27.2	1.016805	81	81
82	82	1.128	0.8438	1.049693	0.4217	2.152	0.4436	1.016050	0.2321	4.174	0.6961	1.259798	0.8160	0.7722	0.6146	25.3	1.013513	82	82
83	83	1.122	0.8156	1.042254	0.3753	2.453	0.4129	1.012717	0.1968	4.954	0.6606	1.236854	0.7767	0.8417	0.5813	23.3	1.010565	83	83
84	84	1.114	0.7845	1.035150	0.3268	2.856	0.3824	1.009739	0.1616	6.070	0.6238	1.213501	0.7321	0.9276	0.5436	21.3	1.007963	84	84
85	85	1.105	0.7494	1.028382	0.2758	3.428	0.3525	1.007120	0.1267	7.781	0.5856	1.189500	0.6805	1.039	0.5001	19.0	1.005708	85	85
θ	T	C_8	L_7	Ω_6	C_6	L_6	L_5	Ω_4	C_4	L_4	L_3	Ω_2	L_2	C_2	L_1	A_{min}	Ω_s		θ

* Reprinted from R. Saal and E. Ulbrich, "On the Design of Filters by Synthesis," IRE Transactions on Circuit Theory, December 1958.

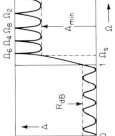

CO9 20*

θ	Ω_s	A_min	C_1	C_2	L_2	Ω_2	C_3	C_4	L_4	Ω_4	C_5	C_6	L_6	Ω_6	C_7	C_8	L_8	Ω_8	C_9
T	∞	∞	1.349	0.0000	1.405	∞	2.274	0.0000	1.551	∞	2.339	0.0000	1.551	∞	2.274	0.0000	1.405	∞	1.349
36	1.701302	116.1	1.318	0.0354	1.367	4.543865	2.067	0.2078	1.310	1.916452	1.934	0.2703	1.247	1.722454	1.949	0.1273	1.263	2.494683	1.233
37	1.661640	113.8	1.316	0.0376	1.365	4.414407	2.055	0.2207	1.297	1.869159	1.912	0.2873	1.230	1.682023	1.931	0.1352	1.254	2.428228	1.226
38	1.624269	111.5	1.315	0.0399	1.362	4.291507	2.043	0.2341	1.283	1.824497	1.889	0.3050	1.213	1.643916	1.912	0.1435	1.246	2.365290	1.219
39	1.589016	109.3	1.313	0.0422	1.360	4.174652	2.030	0.2481	1.269	1.782266	1.866	0.3234	1.196	1.607957	1.893	0.1521	1.237	2.305598	1.212
40	1.555724	107.2	1.310	0.0446	1.357	4.063382	2.017	0.2626	1.254	1.742285	1.842	0.3426	1.178	1.573989	1.874	0.1610	1.228	2.248907	1.204
41	1.524253	105.1	1.308	0.0471	1.355	3.957281	2.004	0.2777	1.240	1.704392	1.817	0.3626	1.160	1.541869	1.854	0.1703	1.219	2.194997	1.197
42	1.494477	103.0	1.306	0.0498	1.352	3.855969	1.991	0.2934	1.224	1.668439	1.792	0.3834	1.142	1.511468	1.834	0.1800	1.209	2.143669	1.189
43	1.466279	100.9	1.304	0.0525	1.349	3.759105	1.977	0.3097	1.209	1.634294	1.767	0.4052	1.123	1.482668	1.813	0.1901	1.199	2.094742	1.180
44	1.439557	98.9	1.301	0.0553	1.346	3.666376	1.963	0.3267	1.193	1.601835	1.741	0.4278	1.104	1.455364	1.792	0.2005	1.189	2.048051	1.172
45	1.414214	97.0	1.299	0.0582	1.343	3.577497	1.948	0.3444	1.177	1.570952	1.714	0.4515	1.084	1.429460	1.770	0.2114	1.178	2.003447	1.163
46	1.390164	95.0	1.296	0.0612	1.340	3.492207	1.934	0.3628	1.160	1.541544	1.687	0.4762	1.064	1.404867	1.748	0.2228	1.168	1.960793	1.154
47	1.367327	93.1	1.294	0.0643	1.336	3.410268	1.918	0.3820	1.143	1.513520	1.659	0.5020	1.044	1.381504	1.725	0.2346	1.157	1.919963	1.145
48	1.345633	91.2	1.291	0.0676	1.333	3.331459	1.903	0.4019	1.126	1.486796	1.631	0.5289	1.023	1.359299	1.702	0.2468	1.145	1.880842	1.135
49	1.325013	89.3	1.288	0.0710	1.329	3.255578	1.887	0.4227	1.108	1.461293	1.603	0.5571	1.002	1.338183	1.679	0.2596	1.134	1.843326	1.126
50	1.305407	87.5	1.285	0.0745	1.326	3.182438	1.871	0.4444	1.090	1.436942	1.574	0.5867	0.9811	1.318096	1.655	0.2730	1.121	1.807315	1.116
51	1.286760	85.7	1.282	0.0781	1.322	3.111865	1.854	0.4671	1.071	1.413677	1.544	0.6176	0.9595	1.298979	1.631	0.2869	1.109	1.772722	1.105
52	1.269018	83.9	1.279	0.0819	1.318	3.043699	1.837	0.4908	1.052	1.391438	1.514	0.6501	0.9377	1.280780	1.606	0.3014	1.096	1.739462	1.094
53	1.252136	82.1	1.275	0.0858	1.314	2.977790	1.820	0.5155	1.033	1.370170	1.484	0.6843	0.9155	1.263452	1.581	0.3166	1.083	1.707460	1.083
54	1.236068	80.4	1.272	0.0899	1.310	2.914000	1.802	0.5414	1.014	1.349821	1.453	0.7202	0.8930	1.246949	1.555	0.3324	1.070	1.676644	1.072
55	1.220775	78.6	1.268	0.0942	1.305	2.852198	1.784	0.5685	0.9939	1.330344	1.421	0.7580	0.8703	1.231230	1.529	0.3490	1.056	1.646949	1.060
56	1.206218	76.9	1.265	0.0986	1.301	2.792263	1.765	0.5969	0.9737	1.311695	1.389	0.7979	0.8472	1.216257	1.502	0.3664	1.042	1.618313	1.048
57	1.192363	75.2	1.261	0.1032	1.296	2.734079	1.746	0.6268	0.9531	1.293834	1.357	0.8401	0.8239	1.201995	1.476	0.3846	1.028	1.590678	1.036
58	1.179178	73.5	1.257	0.1080	1.291	2.677540	1.726	0.6586	0.9321	1.276723	1.324	0.8847	0.8003	1.188411	1.448	0.4037	1.013	1.563993	1.023
59	1.166633	71.8	1.253	0.1131	1.286	2.622544	1.707	0.6912	0.9108	1.260327	1.291	0.9321	0.7764	1.175475	1.420	0.4238	0.9974	1.538206	1.010
60	1.154701	70.1	1.248	0.1183	1.281	2.568993	1.686	0.7261	0.8891	1.244613	1.257	0.9825	0.7523	1.163158	1.392	0.4449	0.9816	1.513271	0.9959

TABLE 12-56 (Continued)

θ / T	C_9	Ω_8	L_8	C_8	C_7	Ω_6	L_6	C_6	C_5	Ω_4	L_4	C_4	C_3	Ω_2	L_2	C_2	C_1	A_{min}	Ω_s	θ
T	1.349	∞	1.405	0.0000	2.274	∞	1.551	0.0000	2.339	∞	1.551	0.0000	2.274	∞	1.405	0.0000	1.349	∞	∞	T
61	0.9817	1.489144	0.9654	0.4671	1.363	1.151435	0.7279	1.036	1.223	1.229551	0.8670	0.7629	1.666	2.516797	1.275	0.1238	1.244	68.5	1.143354	61
62	0.9671	1.465786	0.9487	0.4906	1.334	1.140280	0.7033	1.093	1.189	1.215114	0.8446	0.8019	1.644	2.465867	1.269	0.1296	1.239	66.8	1.132570	62
63	0.9520	1.443156	0.9315	0.5155	1.305	1.129672	0.6785	1.155	1.154	1.201275	0.8217	0.8433	1.623	2.416121	1.263	0.1356	1.234	65.2	1.122926	63
64	0.9364	1.421219	0.9138	0.5418	1.275	1.119590	0.6534	1.221	1.119	1.188009	0.7985	0.8873	1.600	2.367476	1.257	0.1420	1.229	63.5	1.112602	64
65	0.9202	1.399940	0.8956	0.5698	1.244	1.110013	0.6281	1.292	1.083	1.175295	0.7749	0.9342	1.578	2.319854	1.250	0.1487	1.223	61.9	1.103378	65
66	0.9034	1.379288	0.8768	0.5995	1.213	1.100924	0.6026	1.369	1.047	1.163112	0.7509	0.9844	1.554	2.273180	1.243	0.1557	1.217	60.2	1.094636	66
67	0.8860	1.359230	0.8574	0.6313	1.182	1.092306	0.5769	1.453	1.011	1.151440	0.7265	1.038	1.531	2.227378	1.236	0.1631	1.211	58.6	1.086360	67
68	0.8679	1.339739	0.8374	0.6653	1.150	1.084144	0.5510	1.544	0.9738	1.140260	0.7017	1.096	1.506	2.182375	1.228	0.1710	1.205	56.9	1.078535	68
69	0.8491	1.320787	0.8167	0.7019	1.118	1.076422	0.5250	1.644	0.9367	1.129558	0.6764	1.159	1.481	2.138097	1.220	0.1793	1.198	55.2	1.071145	69
70	0.8294	1.302346	0.7953	0.7413	1.085	1.069128	0.4987	1.754	0.8992	1.119316	0.6507	1.227	1.455	2.094470	1.211	0.1882	1.191	53.6	1.064178	70
71	0.8089	1.284392	0.7732	0.7840	1.052	1.062248	0.4723	1.876	0.8614	1.109521	0.6245	1.301	1.429	2.051420	1.202	0.1977	1.184	51.9	1.057621	71
72	0.7875	1.266900	0.7503	0.8304	1.018	1.055772	0.4457	2.013	0.8233	1.100160	0.5979	1.382	1.401	2.008869	1.192	0.2078	1.176	50.2	1.051462	72
73	0.7650	1.249847	0.7265	0.8812	0.9841	1.049689	0.4190	2.166	0.7849	1.091222	0.5708	1.471	1.373	1.966738	1.182	0.2187	1.167	48.5	1.045692	73
74	0.7414	1.233209	0.7017	0.9370	0.9494	1.043989	0.3922	2.339	0.7463	1.082695	0.5432	1.571	1.344	1.924942	1.171	0.2305	1.158	46.8	1.040299	74
75	0.7165	1.216956	0.6760	0.9988	0.9141	1.038663	0.3652	2.538	0.7073	1.074570	0.5150	1.682	1.314	1.883393	1.159	0.2433	1.148	45.1	1.035276	75
76	0.6902	1.201093	0.6491	1.068	0.8782	1.033703	0.3381	2.768	0.6681	1.066839	0.4862	1.807	1.283	1.841992	1.146	0.2572	1.137	43.3	1.030614	76
77	0.6622	1.185571	0.6211	1.146	0.8418	1.029101	0.3110	3.036	0.6287	1.059494	0.4569	1.950	1.251	1.800631	1.132	0.2724	1.126	41.5	1.026304	77
78	0.6323	1.170376	0.5917	1.234	0.8048	1.024852	0.2838	3.355	0.5891	1.052530	0.4268	2.115	1.218	1.759188	1.117	0.2893	1.113	39.6	1.022341	78
79	0.6004	1.155487	0.5607	1.336	0.7669	1.020948	0.2564	3.741	0.5493	1.045943	0.3961	2.308	1.183	1.717524	1.100	0.3081	1.099	37.7	1.018717	79
80	0.5658	1.140881	0.5281	1.455	0.7286	1.017385	0.2291	4.216	0.5094	1.039728	0.3645	2.538	1.146	1.675471	1.082	0.3292	1.084	35.8	1.015427	80
81	0.5281	1.126534	0.4935	1.597	0.6895	1.014158	0.2019	4.817	0.4693	1.033885	0.3321	2.817	1.108	1.632828	1.061	0.3534	1.067	33.8	1.012465	81
82	0.4868	1.112418	0.4567	1.770	0.6497	1.011264	0.1746	5.599	0.4293	1.028414	0.2986	3.166	1.067	1.589344	1.038	0.3814	1.047	31.7	1.009828	82
83	0.4407	1.098505	0.4172	1.986	0.6090	1.008700	0.1476	6.660	0.3894	1.023319	0.2641	3.616	1.024	1.544692	1.011	0.4145	1.025	29.5	1.007510	83
84	0.3886	1.084760	0.3747	2.268	0.5676	1.006464	0.1208	8.175	0.3496	1.018605	0.2282	4.223	0.9784	1.498431	0.9794	0.4548	0.9995	27.1	1.005508	84
85	0.3281	1.071141	0.3283	2.655	0.5253	1.004554	0.0944	10.50	0.3103	1.014284	0.1909	5.093	0.9284	1.449932	0.9411	0.5054	0.9688	24.6	1.003820	85
θ	L_9	C_8	Ω_8	L_8	C_8	L_7	Ω_6	L_6	C_6	L_5	Ω_4	L_4	C_4	L_3	Ω_2	L_2	C_2	L_1	A_{min}	Ω_s	θ

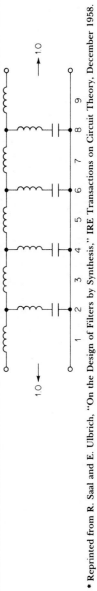

* Reprinted from R. Saal and E. Ulbrich, "On the Design of Filters by Synthesis," IRE Transactions on Circuit Theory, December 1958.

C10 20 b*

→ 0.6667

1.0

Ω₆ Ω₄ Ω₈ Ω₂

A_min R_dB A

0 1 Ω_s Ω ∞

θ	Ω_s	A_min	C₁	C₂	L₂	Ω₂	C₃	C₄	L₄	Ω₄	C₅	C₆	L₆	Ω₆	C₇	C₈	L₈	Ω₈	C₉	L₁₀
T	∞	∞	1.353	0.0000	1.410	∞	2.283	0.0000	1.559	∞	2.357	0.0000	1.571	∞	2.338	0.0000	1.522	∞	2.114	0.9019
46	1.402036	108.4	1.267	0.0953	1.304	2.837237	1.908	0.3703	1.158	1.526953	1.702	0.4577	1.092	1.414442	1.787	0.2327	1.258	1.848119	1.913	0.9056
47	1.378775	106.3	1.263	0.1002	1.298	2.772462	1.891	0.3898	1.141	1.499419	1.674	0.4823	1.072	1.390741	1.763	0.2448	1.246	1.810618	1.904	0.9058
48	1.356667	104.2	1.258	0.1053	1.293	2.710218	1.874	0.4102	1.123	1.473170	1.646	0.5079	1.052	1.368207	1.738	0.2574	1.233	1.774719	1.894	0.9059
49	1.335647	102.1	1.254	0.1106	1.287	2.650344	1.857	0.4315	1.105	1.448130	1.617	0.5347	1.031	1.346772	1.714	0.2705	1.221	1.740324	1.885	0.9061
50	1.315651	100.1	1.249	0.1161	1.282	2.592690	1.839	0.4536	1.087	1.424228	1.588	0.5628	1.010	1.326374	1.688	0.2841	1.208	1.707343	1.874	0.9063
51	1.296624	98.1	1.244	0.1218	1.276	2.537118	1.821	0.4767	1.068	1.401400	1.558	0.5921	0.9887	1.306956	1.663	0.2983	1.194	1.675693	1.864	0.9064
52	1.278514	96.1	1.239	0.1277	1.270	2.483501	1.802	0.5009	1.049	1.379588	1.528	0.6229	0.9670	1.288463	1.636	0.3130	1.180	1.645296	1.853	0.9066
53	1.261271	94.1	1.234	0.1339	1.263	2.431720	1.783	0.5262	1.029	1.358736	1.497	0.6552	0.9450	1.270848	1.610	0.3284	1.166	1.616083	1.842	0.9068
54	1.244851	92.2	1.229	0.1403	1.257	2.381665	1.763	0.5526	1.009	1.338794	1.466	0.6891	0.9227	1.254066	1.582	0.3444	1.151	1.587987	1.831	0.9069
55	1.229214	90.2	1.223	0.1470	1.250	2.333233	1.743	0.5803	0.9895	1.319714	1.434	0.7247	0.9002	1.238073	1.555	0.3611	1.136	1.560947	1.820	0.9071
56	1.214321	88.3	1.217	0.1539	1.243	2.286327	1.723	0.6093	0.9689	1.301455	1.402	0.7623	0.8773	1.222833	1.527	0.3786	1.121	1.534907	1.808	0.9073
57	1.200137	86.4	1.211	0.1612	1.236	2.240858	1.702	0.6398	0.9481	1.283976	1.369	0.8019	0.8541	1.208310	1.498	0.3969	1.105	1.509814	1.795	0.9075
58	1.186629	84.5	1.205	0.1687	1.228	2.196740	1.681	0.6719	0.9268	1.267239	1.336	0.8437	0.8307	1.194469	1.469	0.4160	1.089	1.485619	1.783	0.9077
59	1.173768	82.7	1.198	0.1767	1.220	2.153895	1.659	0.7056	0.9052	1.251211	1.303	0.8880	0.8070	1.181281	1.439	0.4360	1.073	1.462275	1.770	0.9079
60	1.161525	80.8	1.192	0.1849	1.212	2.112247	1.637	0.7412	0.8833	1.235859	1.269	0.9350	0.7830	1.168718	1.409	0.4570	1.056	1.439742	1.757	0.9081
61	1.149873	79.0	1.185	0.1936	1.204	2.071724	1.614	0.7789	0.8610	1.221153	1.234	0.9850	0.7587	1.156752	1.378	0.4790	1.038	1.417977	1.743	0.9083
62	1.138790	77.1	1.177	0.2026	1.195	2.032259	1.591	0.8187	0.8383	1.207066	1.199	1.038	0.7342	1.145359	1.347	0.5022	1.020	1.396945	1.729	0.9085
63	1.128252	75.3	1.170	0.2121	1.186	1.993787	1.567	0.8610	0.8153	1.193573	1.164	1.095	0.7095	1.134517	1.316	0.5267	1.002	1.376609	1.715	0.9087
64	1.118238	73.4	1.162	0.2221	1.176	1.956247	1.543	0.9059	0.7919	1.180648	1.129	1.156	0.6845	1.124204	1.284	0.5525	0.9831	1.356938	1.700	0.9089
65	1.108729	71.6	1.154	0.2327	1.166	1.919579	1.518	0.9538	0.7682	1.168271	1.093	1.221	0.6593	1.114401	1.251	0.5797	0.9636	1.337900	1.684	0.9091

TABLE 12-56 (Continued)

θ	Ω_s	A_{min}	C_1	C_2	L_2	Ω_2	C_3	C_4	L_4	Ω_4	C_5	C_6	L_6	Ω_6	C_7	C_8	L_8	Ω_8	C_9	L_{10}	θ
T	∞	∞	1.353	0.0000	1.410	∞	2.283	0.0000	1.559	∞	2.357	0.0000	1.571	∞	2.338	0.0000	1.522	∞	2.114	0.9019	T
66	1.099707	69.8	1.145	0.2438	1.156	1.883725	1.493	1.005	0.7440	1.156420	1.056	1.292	0.6338	1.105088	1.218	0.6087	0.9437	1.319465	1.669	0.9093	66
67	1.091154	67.9	1.136	0.2555	1.145	1.848630	1.467	1.060	0.7195	1.145077	1.019	1.368	0.6082	1.096250	1.184	0.6394	0.9231	1.301607	1.652	0.9095	67
68	1.083056	66.1	1.127	0.2679	1.134	1.814240	1.440	1.119	0.6946	1.134223	0.9820	1.451	0.5823	1.087871	1.150	0.6721	0.9020	1.284300	1.635	0.9097	68
69	1.075399	64.3	1.117	0.2810	1.122	1.780499	1.413	1.183	0.6694	1.123842	0.9444	1.542	0.5562	1.079935	1.115	0.7071	0.8802	1.267519	1.618	0.9099	69
70	1.068167	62.4	1.106	0.2951	1.110	1.747356	1.385	1.252	0.6437	1.113920	0.9063	1.641	0.5299	1.072430	1.079	0.7446	0.8578	1.251241	1.600	0.9101	70
71	1.061350	60.5	1.095	0.3100	1.097	1.714757	1.356	1.327	0.6176	1.104442	0.8680	1.750	0.5034	1.065342	1.043	0.7850	0.8346	1.235443	1.581	0.9103	71
72	1.054935	58.7	1.084	0.3260	1.083	1.682648	1.327	1.410	0.5911	1.095394	0.8292	1.871	0.4768	1.058661	1.007	0.8287	0.8106	1.220106	1.562	0.9105	72
73	1.048912	56.8	1.071	0.3432	1.069	1.650973	1.297	1.501	0.5642	1.086767	0.7901	2.007	0.4500	1.052375	0.9696	0.8760	0.7859	1.205209	1.542	0.9107	73
74	1.043271	54.9	1.058	0.3618	1.054	1.619676	1.265	1.601	0.5368	1.078548	0.7506	2.159	0.4229	1.046476	0.9317	0.9277	0.7602	1.190733	1.521	0.9109	74
75	1.038003	52.9	1.078	0.3870	1.031	1.583010	1.245	1.716	0.5086	1.070450	0.7126	2.324	0.3972	1.040840	0.8909	0.9766	0.7408	1.175676	1.494	0.9111	75
θ	Ω_s	A_{min}	L_1	C_2	L_2	Ω_2	L_3	C_4	L_4	Ω_4	L_5	C_6	L_6	Ω_6	L_7	C_8	L_8	Ω_8	L_9	C_{10}	θ

* Reprinted from R. Saal and E. Ulbrich, "On the Design of Filters by Synthesis," IRE Transactions on Circuit Theory, December 1958.

C10 20 c°

θ	Ω_s	A_min	C_1	C_2	L_2	Ω_2	C_3	C_4	L_4	Ω_4	C_5	C_6	L_6	Ω_6	C_7	C_8	L_8	Ω_8	C_9	L_10	θ
		∞	1.248	0.0000	1.513	∞	2.033	0.0000	1.809	∞	1.957	0.0000	1.957	∞	1.809	0.0000	2.033	∞	1.513	1.248	
46	1.414011	108.4	1.154	0.0853	1.413	2.880291	1.692	0.3083	1.364	1.542102	1.421	0.3593	1.367	1.426739	1.401	0.1712	1.669	1.870799	1.382	1.236	46
47	1.390318	106.3	1.149	0.0896	1.408	2.814916	1.677	0.3244	1.344	1.514114	1.398	0.3787	1.342	1.402604	1.383	0.1802	1.652	1.832802	1.375	1.235	47
48	1.367792	104.2	1.144	0.0941	1.403	2.752107	1.661	0.3412	1.325	1.487423	1.375	0.3988	1.317	1.379650	1.365	0.1895	1.635	1.796425	1.369	1.234	48
49	1.346966	102.1	1.139	0.0987	1.398	2.691699	1.645	0.3587	1.304	1.461955	1.352	0.4199	1.292	1.357807	1.346	0.1992	1.618	1.761569	1.363	1.233	49
50	1.325976	100.1	1.134	0.1035	1.392	2.633543	1.629	0.3769	1.284	1.437637	1.328	0.4419	1.266	1.337012	1.327	0.2093	1.600	1.728142	1.356	1.233	50
51	1.305565	98.1	1.129	0.1085	1.387	2.577498	1.613	0.3959	1.263	1.414404	1.303	0.4650	1.240	1.317206	1.308	0.2198	1.582	1.696060	1.349	1.232	51
52	1.288080	96.1	1.125	0.1137	1.381	2.523435	1.596	0.4157	1.241	1.392196	1.279	0.4891	1.213	1.298337	1.289	0.2307	1.563	1.665246	1.342	1.232	52
53	1.270472	94.1	1.118	0.1191	1.375	2.471234	1.579	0.4364	1.219	1.370959	1.253	0.5145	1.186	1.280355	1.269	0.2422	1.544	1.635627	1.335	1.230	53
54	1.253696	92.2	1.112	0.1247	1.369	2.420784	1.561	0.4580	1.197	1.350640	1.228	0.5411	1.158	1.263213	1.248	0.2541	1.524	1.607137	1.328	1.229	54
55	1.237711	90.2	1.106	0.1305	1.362	2.371980	1.543	0.4806	1.174	1.331192	1.202	0.5691	1.130	1.246870	1.228	0.2665	1.503	1.579712	1.320	1.228	55
56	1.222478	88.3	1.100	0.1365	1.356	2.324726	1.525	0.5043	1.151	1.312572	1.175	0.5986	1.102	1.231287	1.206	0.2795	1.483	1.553297	1.312	1.227	56
57	1.207961	86.4	1.093	0.1428	1.349	2.278929	1.506	0.5292	1.127	1.294738	1.149	0.6297	1.073	1.216427	1.185	0.2931	1.461	1.527838	1.304	1.226	57
58	1.194127	84.5	1.087	0.1493	1.342	2.234505	1.487	0.5553	1.103	1.277654	1.121	0.6626	1.044	1.202258	1.163	0.3074	1.440	1.503284	1.296	1.225	58
59	1.180946	82.7	1.080	0.1561	1.334	2.191372	1.467	0.5827	1.079	1.261283	1.094	0.6973	1.015	1.188747	1.141	0.3223	1.417	1.479589	1.287	1.224	59
60	1.168389	80.8	1.073	0.1632	1.326	2.149455	1.447	0.6116	1.054	1.245594	1.066	0.7342	0.9851	1.175867	1.118	0.3380	1.394	1.456709	1.278	1.223	60
61	1.156430	79.0	1.065	0.1706	1.318	2.108681	1.427	0.6420	1.029	1.230556	1.038	0.7733	0.9551	1.163590	1.095	0.3544	1.371	1.434604	1.269	1.221	61
62	1.145044	77.1	1.057	0.1783	1.310	2.068982	1.406	0.6743	1.003	1.216141	1.009	0.8150	0.9247	1.151891	1.072	0.3717	1.347	1.413236	1.260	1.220	62
63	1.134209	75.3	1.049	0.1864	1.301	2.030293	1.385	0.7084	0.9766	1.202323	0.9801	0.8595	0.8940	1.140748	1.048	0.3900	1.322	1.392568	1.251	1.219	63
64	1.123902	73.4	1.041	0.1949	1.292	1.992550	1.363	0.7446	0.9499	1.189078	0.9508	0.9073	0.8630	1.130139	1.024	0.4093	1.297	1.372567	1.241	1.217	64
65	1.114106	71.6	1.032	0.2038	1.283	1.955693	1.341	0.7831	0.9227	1.176383	0.9211	0.9585	0.8317	1.120044	0.9991	0.4297	1.271	1.353202	1.230	1.216	65
66	1.104800	69.8	1.023	0.2131	1.273	1.919665	1.318	0.8242	0.8951	1.164217	0.8911	1.014	0.8001	1.110444	0.9740	0.4514	1.244	1.334441	1.220	1.214	66
67	1.095969	67.9	1.014	0.2230	1.263	1.884408	1.295	0.8682	0.8670	1.152560	0.8607	1.073	0.7682	1.101323	0.9485	0.4744	1.217	1.316257	1.209	1.212	67
68	1.087597	66.1	1.004	0.2334	1.252	1.849867	1.272	0.9155	0.8384	1.141395	0.8300	1.138	0.7361	1.092664	0.9226	0.4989	1.189	1.298624	1.198	1.211	68
69	1.079669	64.3	0.9932	0.2444	1.241	1.815988	1.247	0.9664	0.8094	1.130704	0.7989	1.209	0.7036	1.084453	0.8962	0.5252	1.159	1.281515	1.186	1.209	69
70	1.072171	62.4	0.9823	0.2560	1.229	1.782715	1.223	1.021	0.7798	1.120472	0.7676	1.286	0.6709	1.076676	0.8693	0.5533	1.130	1.264906	1.174	1.207	70

TABLE 12-56 *(Continued)*

θ	A_{min}	C_1	C_2	L_2	Ω_2	C_3	C_4	L_4	Ω_4	C_5	C_6	L_6	Ω_6	C_7	C_8	L_8	Ω_8	C_9	L_{10}	θ
T	∞	1.248	0.0000	1.513	∞	2.033	0.0000	1.809	∞	1.957	0.0000	1.957	∞	1.809	0.0000	2.033	∞	1.513	1.248	T
71	60.5	0.9709	0.2684	1.216	1.749995	1.197	1.081	0.7496	1.110685	0.7358	1.371	0.6378	1.069320	0.8419	0.5836	1.099	1.248774	1.162	1.205	71
72	58.7	0.9588	0.2817	1.203	1.717773	1.171	1.147	0.7189	1.101329	0.7038	1.465	0.6046	1.062375	0.8141	0.6163	1.067	1.233098	1.149	1.202	72
73	56.8	0.9461	0.2958	1.189	1.685990	1.144	1.219	0.6877	1.092393	0.6715	1.570	0.5714	1.055828	0.7857	0.6518	1.034	1.217855	1.135	1.200	73
74	54.9	0.9326	0.3110	1.174	1.654590	1.117	1.298	0.6558	1.083865	0.6387	1.692	0.5365	1.049671	0.7567	0.6906	1.000	1.203026	1.121	1.197	74
75	52.9	0.9215	0.3329	1.151	1.615393	1.087	1.396	0.6196	1.075199	0.6033	1.848	0.4968	1.043669	0.7292	0.7369	0.9633	1.186858	1.101	1.194	75
θ	Ω_s	L_1	L_2	C_2	Ω_2	L_3	C_4	L_4	Ω_4	L_5	L_6	C_6	Ω_6	L_7	C_8	C_8	Ω_8	L_9	C_{10}	θ

θ	Ω_s
71	1.065092
72	1.058418
73	1.052141
74	1.046250
75	1.040737

1.0

1.0 → 1.0

1 2 3 4 5 6 7 8 9 10

* Reprinted from R. Saal and E. Ulbrich, "On the Design of Filters by Synthesis," **IRE** Transactions on Circuit Theory, December 1958.

TABLE 12-56   *(Continued)*

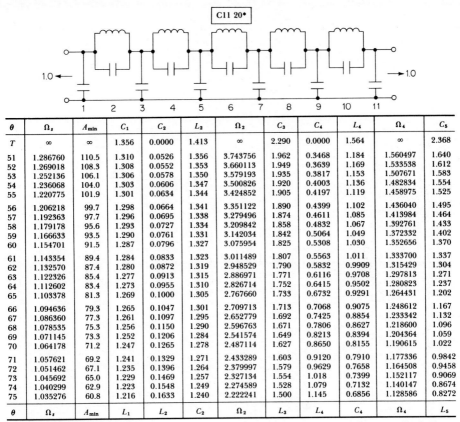

θ	Ω_s	A_{min}	C_1	C_2	L_2	Ω_2	C_3	C_4	L_4	Ω_4	C_5
T	∞	∞	1.356	0.0000	1.413	∞	2.290	0.0000	1.564	∞	2.368
51	1.286760	110.5	1.310	0.0526	1.356	3.743756	1.962	0.3468	1.184	1.560497	1.640
52	1.269018	108.3	1.308	0.0552	1.353	3.660113	1.949	0.3639	1.169	1.533538	1.612
53	1.252136	106.1	1.306	0.0578	1.350	3.579193	1.935	0.3817	1.153	1.507671	1.583
54	1.236068	104.0	1.303	0.0606	1.347	3.500826	1.920	0.4003	1.136	1.482834	1.554
55	1.220775	101.9	1.301	0.0634	1.344	3.424852	1.905	0.4197	1.119	1.458975	1.525
56	1.206218	99.7	1.298	0.0664	1.341	3.351122	1.890	0.4399	1.102	1.436040	1.495
57	1.192363	97.7	1.296	0.0695	1.338	3.279496	1.874	0.4611	1.085	1.413984	1.464
58	1.179178	95.6	1.293	0.0727	1.334	3.209842	1.858	0.4832	1.067	1.392761	1.433
59	1.166633	93.5	1.290	0.0761	1.331	3.142034	1.842	0.5064	1.049	1.372332	1.402
60	1.154701	91.5	1.287	0.0796	1.327	3.075954	1.825	0.5308	1.030	1.352656	1.370
61	1.143354	89.4	1.284	0.0833	1.323	3.011489	1.807	0.5563	1.011	1.333700	1.337
62	1.132570	87.4	1.280	0.0872	1.319	2.948529	1.790	0.5832	0.9909	1.315429	1.304
63	1.122326	85.4	1.277	0.0913	1.315	2.886971	1.771	0.6116	0.9708	1.297813	1.271
64	1.112602	83.4	1.273	0.0955	1.310	2.826714	1.752	0.6415	0.9502	1.280823	1.237
65	1.103378	81.3	1.269	0.1000	1.305	2.767660	1.733	0.6732	0.9291	1.264431	1.202
66	1.094636	79.3	1.265	0.1047	1.301	2.709713	1.713	0.7068	0.9075	1.248612	1.167
67	1.086360	77.3	1.261	0.1097	1.295	2.652779	1.692	0.7425	0.8854	1.233342	1.132
68	1.078535	75.3	1.256	0.1150	1.290	2.596763	1.671	0.7806	0.8627	1.218600	1.096
69	1.071145	73.3	1.252	0.1206	1.284	2.541574	1.649	0.8213	0.8394	1.204364	1.059
70	1.064178	71.2	1.247	0.1265	1.278	2.487114	1.627	0.8650	0.8155	1.190615	1.022
71	1.057621	69.2	1.241	0.1329	1.271	2.433289	1.603	0.9120	0.7910	1.177336	0.9842
72	1.051462	67.1	1.235	0.1396	1.264	2.379997	1.579	0.9629	0.7658	1.164508	0.9458
73	1.045692	65.0	1.229	0.1469	1.257	2.327134	1.554	1.018	0.7399	1.152117	0.9069
74	1.040299	62.9	1.223	0.1548	1.249	2.274589	1.528	1.079	0.7132	1.140147	0.8674
75	1.035276	60.8	1.216	0.1633	1.240	2.222241	1.500	1.145	0.6856	1.128586	0.8272
θ	Ω_s	A_{min}	L_1	L_2	C_2	Ω_2	L_3	L_4	C_4	Ω_4	L_5

* Reprinted from R. Saal and E. Ulbrich, "On the Design of Filters by Synthesis," IRE Transactions on Circuit Theory, December 1958.

TABLE 12-56   *(Continued)*

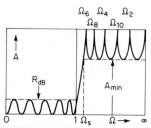

C_6	L_6	Ω_6	C_7	C_8	L_8	Ω_8	C_9	C_{10}	L_{10}	Ω_{10}	C_{11}	θ
0.0000	1.583	∞	2.368	0.0000	1.564	∞	2.290	0.0000	1.413	∞	1.356	T
0.6038	0.9877	1.294892	1.536	0.5121	1.045	1.367227	1.752	0.1984	1.198	2.051040	1.180	51
0.6351	0.9657	1.276846	1.505	0.5382	1.025	1.346576	1.730	0.2083	1.188	2.010050	1.172	52
0.6680	0.9434	1.259666	1.474	0.5656	1.004	1.326857	1.708	0.2185	1.179	1.970531	1.164	53
0.7025	0.9209	1.243308	1.441	0.5942	0.9837	1.308019	1.686	0.2292	1.168	1.932394	1.155	54
0.7388	0.8980	1.227731	1.409	0.6242	0.9628	1.290019	1.663	0.2403	1.158	1.895562	1.147	55
0.7770	0.8748	1.212897	1.376	0.6556	0.9415	1.272815	1.640	0.2520	1.147	1.859959	1.137	56
0.8173	0.8514	1.198770	1.342	0.6887	0.9198	1.256369	1.616	0.2641	1.136	1.825515	1.128	57
0.8599	0.8277	1.185319	1.308	0.7236	0.8979	1.240644	1.592	0.2769	1.125	1.792167	1.118	58
0.9050	0.8037	1.172513	1.274	0.7603	0.8756	1.225607	1.567	0.2902	1.113	1.759852	1.108	59
0.9528	0.7795	1.160324	1.239	0.7991	0.8530	1.211229	1.542	0.3042	1.100	1.728513	1.098	60
1.004	0.7551	1.148726	1.204	0.8402	0.8300	1.197480	1.516	0.3188	1.088	1.698097	1.087	61
1.058	0.7304	1.137695	1.169	0.8837	0.8067	1.184335	1.490	0.3342	1.075	1.668552	1.076	62
1.116	0.7055	1.127208	1.133	0.9300	0.7831	1.171767	1.463	0.3505	1.061	1.639830	1.065	63
1.178	0.6803	1.117245	1.096	0.9793	0.7592	1.159755	1.436	0.3676	1.047	1.611885	1.053	64
1.244	0.6549	1.107786	1.059	1.032	0.7349	1.148278	1.408	0.3857	1.032	1.584674	1.040	65
1.316	0.6294	1.098813	1.022	1.088	0.7103	1.137315	1.380	0.4049	1.017	1.558153	1.027	66
1.394	0.6036	1.090309	0.9849	1.149	0.6854	1.126849	1.351	0.4252	1.002	1.532284	1.014	67
1.478	0.5777	1.082259	0.9470	1.214	0.6602	1.116862	1.321	0.4468	0.9854	1.507028	0.9994	68
1.570	0.5515	1.074648	0.9089	1.285	0.6346	1.107340	1.291	0.4699	0.9685	1.482346	0.9846	69
1.671	0.5252	1.067462	0.8704	1.362	0.6087	1.098266	1.260	0.4946	0.9508	1.458204	0.9692	70
1.782	0.4988	1.060691	0.8316	1.446	0.5824	1.089629	1.228	0.5212	0.9323	1.434564	0.9530	71
1.905	0.4721	1.054321	0.7924	1.538	0.5558	1.081416	1.196	0.5498	0.9131	1.411392	0.9359	72
2.043	0.4454	1.048342	0.7530	1.640	0.5289	1.073617	1.162	0.5808	0.8929	1.388652	0.9179	73
2.198	0.4185	1.042744	0.7133	1.754	0.5016	1.066220	1.128	0.6145	0.8717	1.366309	0.8986	74
2.373	0.3914	1.037520	0.6733	1.881	0.4739	1.059218	1.093	0.6515	0.8493	1.344327	0.8788	75
L_6	C_6	Ω_6	L_7	L_8	C_8	Ω_8	L_9	L_{10}	C_{10}	Ω_{10}	L_{11}	θ

TABLE 12-57 Elliptic-Function Active Low-Pass Values

CO3 05

	R_1	R_2	R_3	R_4	R_5	C_1	C_2	C_3	C_4	C_5	K	Gain	α	β	ω_∞
$\theta = 3$ $\Omega_s = 19.107$ $A_{min} = 74.93$	0.3355	0.6710	34.12	153.6	1.0000	2.5030	0.5562	0.0109	0.00547	0.6381	1.0920	1.0850	0.7784 1.5670	1.6070	22.060
$\theta = 5$ $\Omega_s = 11.4737$ $A_{min} = 61.61$	0.3394	0.6788	12.46	56.11	1.0000	2.4780	0.5506	0.0300	0.0150	0.6354	1.1090	1.0890	0.7726 1.5740	1.6080	13.240
$\theta = 8$ $\Omega_s = 7.1853$ $A_{min} = 49.33$	0.3487	0.6974	5.0460	22.71	1.0000	2.4190	0.5375	0.0743	0.0371	0.6289	1.1480	1.0980	0.7584 1.5900	1.6090	8.2870
$\theta = 11$ $\Omega_s = 5.2408$ $A_{min} = 41.00$	0.3620	0.7240	2.8050	12.62	1.0000	2.3400	0.5199	0.1342	0.0671	0.6193	1.2060	1.1111	0.7378 1.6150	1.6100	6.0380
$\theta = 16$ $\Omega_s = 3.6280$ $A_{min} = 31.14$	0.3922	0.7844	1.4810	6.6620	1.0000	2.1830	0.4851	0.2570	0.1285	0.5968	1.3430	1.1410	0.6898 1.6760	1.6100	4.1690
$\theta = 24$ $\Omega_s = 2.4586$ $A_{min} = 20.40$	0.4561	0.9121	0.8258	3.7160	1.0000	1.9300	0.4290	0.4738	0.2369	0.5438	1.6580	1.2120	0.5809 1.8390	1.6020	2.8080

TABLE 12-57 *(Continued)*

CO3 15

	R_1	R_2	R_3	R_4	R_5	C_1	C_2	C_3	C_4	C_5	K	Gain	α	β	ω_∞
$\theta = 5$ $\Omega_s = 11.474$ $A_{min} = 71.24$	0.3366	0.6731	23.23	104.5	1.0000	3.4250	0.7610	0.0221	0.0110	1.0240	1.1830	1.1710	0.4832 0.9757	1.2080	13.24
$\theta = 7$ $\Omega_s = 8.2055$ $A_{min} = 62.45$	0.3396	0.6793	11.99	53.95	1.0000	3.3950	0.7545	0.0427	0.0214	1.0210	1.1960	1.1740	0.4804 0.9792	1.2080	9.4660
$\theta = 11$ $\Omega_s = 5.2408$ $A_{min} = 50.63$	0.3488	0.6975	5.0220	22.60	1.0000	3.3100	0.7356	0.1021	0.0511	1.0100	1.2370	1.1820	0.4720 0.9899	1.2100	6.0380
$\theta = 16$ $\Omega_s = 3.6280$ $A_{min} = 40.77$	0.3655	0.7311	2.5220	11.34	1.0000	3.1660	0.7036	0.2040	0.1020	0.9897	1.3130	1.1970	0.4564 1.0100	1.2130	4.1690
$\theta = 23$ $\Omega_s = 2.5593$ $A_{min} = 31.13$	0.3980	0.7961	1.3670	6.1520	1.0000	2.9240	0.6497	0.3783	0.1892	0.9484	1.4650	1.2270	0.4249 1.0540	1.2160	2.9260
$\theta = 34$ $\Omega_s = 1.7883$ $A_{min} = 20.53$	0.4651	0.9301	0.7845	3.5300	1.0000	2.5400	0.5645	0.6692	0.3346	0.8544	1.8070	1.2950	0.3539 1.1700	1.2190	2.0200

CO3 20

	R_1	R_2	R_3	R_4	R_5	C_1	C_2	C_3	C_4	C_5	K	Gain	α	β	ω_∞
$\theta = 5$ $\Omega_s = 11.474$ $A_{min} = 73.81$	0.3361	0.6722	26.98	121.4	1.0000	3.6980	0.8219	0.0205	0.0102	1.1850	1.2140	1.2040	0.4182 0.8434	1.1310	13.24
$\theta = 8$ $\Omega_s = 7.1850$ $A_{min} = 61.54$	0.3404	0.6808	11.70	48.18	1.0000	3.6530	0.8117	0.0516	0.0258	1.1790	1.2330	1.2080	0.4149 0.8478	1.1320	8.2870
$\theta = 12$ $\Omega_s = 4.8097$ $A_{min} = 50.92$	0.3491	0.6983	4.9130	22.11	1.0000	3.5640	0.7920	0.1125	0.0563	1.1670	1.2730	1.2150	0.4081 0.8567	1.1340	5.5390
$\theta = 18$ $\Omega_s = 3.2361$ $A_{min} = 40.23$	0.3683	0.7367	2.3380	10.52	1.0000	3.3830	0.7519	0.2369	0.1184	1.1390	1.3610	1.2310	0.3927 0.8776	1.1370	3.7140
$\theta = 26$ $\Omega_s = 2.2812$ $A_{min} = 30.41$	0.4042	0.8084	1.2670	5.7030	1.0000	3.0950	0.6877	0.4387	0.2194	1.0850	1.5320	1.2630	0.3627 0.9213	1.1420	2.6000
$\theta = 38$ $\Omega_s = 1.6243$ $A_{min} = 20.00$	0.4743	0.9485	0.7479	3.3650	1.0000	2.6660	0.5925	0.7514	0.3757	0.9689	1.8950	1.3320	0.2979 1.0320	1.1480	1.8240

TABLE 12-57 *(Continued)*

CO3 25

	R_1	R_2	R_3	R_4	R_5	C_1	C_2	C_3	C_4	C_5	K	Gain	α	β	ω_∞
$\theta = 6$ $\Omega_s = 9.5668$ $A_{min} = 71.10$	0.3369	0.6738	21.02	94.57	1.0000	3.9010	0.8670	0.0278	0.0139	1.3400	1.2480	1.2350	0.3693 0.7465	1.0790	11.03
$\theta = 9$ $\Omega_s = 6.3925$ $A_{min} = 60.50$	0.3413	0.6826	9.4920	42.72	1.0000	3.8510	0.8558	0.0616	0.0308	1.3320	1.2680	1.2380	0.3663 0.7505	1.0300	7.3700
$\theta = 13$ $\Omega_s = 4.4454$ $A_{min} = 50.86$	0.3499	0.6998	4.6920	21.11	1.0000	3.7580	0.8350	0.1245	0.0623	1.3180	1.3070	1.2450	0.3604 0.7585	1.0820	5.1170
$\theta = 20$ $\Omega_s = 2.9238$ $A_{min} = 39.48$	0.3719	0.7438	2.1420	9.6380	1.0000	3.5380	0.7861	0.2730	0.1365	1.2820	1.4100	1.2630	0.3449 0.7801	1.0860	3.3500
$\theta = 28$ $\Omega_s = 2.1301$ $A_{min} = 30.44$	0.4069	0.8137	1.2290	5.5330	1.0000	3.2390	0.7198	0.4764	0.2382	1.2210	1.5780	1.2930	0.3191 0.8188	1.0920	2.4230
$\theta = 40$ $\Omega_s = 1.5557$ $A_{min} = 20.58$	0.4739	0.9479	0.7490	3.3710	1.0000	2.7970	0.6215	0.7865	0.3933	1.0930	1.9270	1.3550	0.2643 0.9142	1.1000	1.7420

CO5 05

	R_1	R_2	R_3	R_4	R_5	C_1	C_2	C_3	C_4	C_5	K	Gain	α	β	ω_∞
$\theta = 18$	0.3451	0.6903	6.5020	29.26		4.2650	0.9478	0.1006	0.0503		0.8492	0.8202	0.6469	0.7873	5.4170
$\Omega_s = 3.2361$	0.3774	0.7547	1.9050	8.5720		3.2220	0.7160	0.2837	0.1418	1.1890	1.6070	1.4200	0.2286	1.2120	3.3950
$A_{min} = 72.08$					1.0000								0.8405		
$\theta = 23$	0.3531	0.7062	3.9700	17.86		4.1140	0.9143	0.1627	0.0813		0.8866	0.8370	0.6424	0.8082	4.2400
$\Omega_s = 2.5593$	0.4030	0.8059	1.2860	5.7870		3.0380	0.6751	0.4231	0.2116	1.1570	1.7310	1.4320	0.2160	1.2060	2.6810
$A_{min} = 61.20$					1.0000								0.8636		
$\theta = 30$	0.3686	0.7371	2.3250	10.46		3.8510	0.8557	0.2713	0.1357		0.9609	0.8690	0.6322	0.8469	3.2510
$\Omega_s = 2.0000$	0.4452	0.8904	0.8843	3.9790		2.7840	0.6186	0.6228	0.3114	1.1000	1.9410	1.4530	0.1940	1.1940	2.0890
$A_{min} = 49.22$					1.0000								0.9087		
$\theta = 36$	0.3866	0.7732	1.6130	7.2590		3.5830	0.7963	0.3817	0.1908		1.0500	0.9057	0.6180	0.8891	2.7090
$\Omega_s = 1.7013$	0.4848	0.9695	0.7114	3.2010		2.5900	0.5756	0.7845	0.3923	1.0390	2.1450	1.4750	0.1717	1.1820	1.7720
$A_{min} = 40.81$					1.0000								0.9623		
$\theta = 45$	0.4239	0.8479	1.0390	4.6790		3.1330	0.6962	0.5678	0.2839		1.2470	0.9806	0.5811	0.9683	2.1660
$\Omega_s = 1.4142$	0.5443	1.0880	0.5733	2.5800		2.3640	0.5253	0.9974	0.4987	0.9260	2.4710	1.5130	0.1341	1.1580	1.4650
$A_{min} = 30.17$					1.0000								1.0790		
$\theta = 55$	0.4832	0.9664	0.7164	3.2240		2.6170	0.5815	0.7845	0.3922		1.5950	1.1000	0.5039	1.0730	1.7690
$\Omega_s = 1.2208$	0.6021	1.2040	0.4979	2.2400		2.2090	0.4909	1.1870	0.5936	0.7689	2.8190	1.5610	0.0894	1.124	1.2560
$A_{min} = 20.04$					1.0000								1.3010		

TABLE 12-57 *(Continued)*

CO5 15

	R_1	R_2	R_3	R_4	R_5	C_1	C_2	C_3	C_4	C_5	K	Gain	α	β	ω_∞
$\theta = 23$	0.3459	0.6918	6.1240	27.56		5.2710	1.1710	0.1323	0.0662			0.9639	0.4336	0.6991	4.2400
$\Omega_s = 2.5593$	0.3882	0.7764	1.5730	7.0770		3.5530	0.7896	0.3898	0.1949	1.7550	1.0000	1.4820	0.1505	1.0770	2.6810
$A_{min} = 70.83$					1.0000						1.7260		0.5698		
$\theta = 29$	0.3540	0.7080	3.8060	17.13		5.0600	1.1240	0.2092	0.1045			0.9808	0.4307	0.7181	3.3630
$\Omega_s = 2.0627$	0.4176	0.8352	1.1010	4.9560		3.3150	0.7366	0.5585	0.2792	1.6980	1.0410	1.4930	0.1410	1.0740	2.1560
$A_{min} = 60.39$					1.0000						1.8710		0.5887		
$\theta = 36$	0.3669	0.7338	2.4300	10.93		4.7570	1.0570	0.3192	0.1596			1.0060	0.4252	0.7468	2.7090
$\Omega_s = 1.7013$	0.4567	0.9135	0.8225	3.7010		3.0460	0.6768	0.7517	0.3758	1.6160	1.1080	1.5090	0.1274	1.0700	1.7720
$A_{min} = 50.44$					1.0000						2.0680		0.6186		
$\theta = 44$	0.3873	0.7745	1.5960	7.1810		4.3460	0.9659	0.4688	0.2344			1.0460	0.4144	0.7889	2.2150
$\Omega_s = 1.4396$	0.5048	1.0090	0.6543	2.9450		2.7760	0.6170	0.9518	0.4759	1.5010	1.2150	1.5310	0.1090	1.0640	1.4930
$A_{min} = 40.90$					1.0000						2.3180		0.6662		
$\theta = 54$	0.4243	0.8485	1.0360	4.6660		3.7560	0.8347	0.6831	0.3415			1.1150	0.3891	0.8570	1.8020
$\Omega_s = 1.2361$	0.5600	1.1200	0.5490	2.4700		2.5310	0.5625	1.1470	0.5739	1.3230	1.4200	1.5630	0.0825	1.0540	1.2830
$A_{min} = 30.59$					1.0000						2.6260		0.7558		
$\theta = 60$	0.4546	0.9092	0.8331	3.7490		3.3800	0.7511	0.8197	0.4099			1.1700	0.3635	0.9060	1.6180
$\Omega_s = 1.1547$	0.5954	1.1900	0.5049	2.2720		2.4010	0.5335	1.2580	0.6292	1.1960	1.5960	1.5840	0.0653	1.0470	1.1830
$A_{min} = 24.92$					1.0000						2.8290		0.8355		

CO5 20

	R_1	R_2	R_3	R_4	R_5	C_1	C_2	C_3	C_4	C_5	K	Gain	α	β	ω_∞
$\theta = 24$ $\Omega_s = 2.4586$ $A_{\min} = 71.50$	0.3456 0.3897	0.6912 0.7794	6.2520 1.5360	28.13 6.9120	 1.0000	5.5640 3.6350	1.2360 0.8078	0.1367 0.4099	0.0684 0.2050	 1.9940	1.0520 1.7540	1.0150 1.5000	0.3812 0.1320 0.5016	0.6805 1.0500	4.0630 2.5740
$\theta = 30$ $\Omega_s = 2.0000$ $A_{\min} = 61.43$	0.3532 0.4186	0.7065 0.8371	3.9450 1.0910	17.75 4.9110	 1.0000	5.3470 3.3920	1.1880 0.7538	0.2128 0.5782	0.1063 0.2891	 1.9300	1.0920 1.8960	1.0300 1.5100	0.3789 0.1238 0.5180	0.6980 1.0490	3.2510 2.0890
$\theta = 38$ $\Omega_s = 1.6243$ $A_{\min} = 50.49$	0.3673 0.4626	0.7345 0.9251	2.4050 0.7954	10.82 3.5790	 1.0000	4.9880 3.0820	1.1080 0.6848	0.3385 0.7964	0.1693 0.3982	 1.8240	1.1650 2.1190	1.0580 1.5270	0.3737 0.1105 0.5481	0.7285 1.0460	2.5660 1.6900
$\theta = 47$ $\Omega_s = 1.3673$ $A_{\min} = 40.23$	0.3900 0.5156	0.7800 1.0310	1.5300 0.6285	6.8850 2.8280	 1.0000	4.5000 2.7800	1.0000 0.6177	0.5098 1.0130	0.2549 0.5068	 1.6760	1.2880 2.3960	1.1000 1.5490	0.3629 0.0923 0.5968	0.7738 1.0420	2.0730 1.4150
$\theta = 52$ $\Omega_s = 1.2690$ $A_{\min} = 35.13$	0.4067 0.5448	0.8135 1.0890	1.2310 0.5725	5.5410 2.5760	 1.0000	4.1970 2.6410	0.9327 0.5869	0.6162 1.1170	0.3081 0.5586	 1.5790	1.3800 2.5550	1.1300 1.5630	0.3533 0.0809 0.6334	0.8045 1.0390	1.8720 1.3090
$\theta = 57$ $\Omega_s = 1.1924$ $A_{\min} = 30.30$	0.4273 0.5727	0.8546 1.1450	1.0100 0.5317	4.5470 2.3930	 1.0000	3.8760 2.5230	0.8613 0.5607	0.7285 1.2070	0.3642 0.6039	 1.4710	1.4950 2.7120	1.1660 1.5780	0.3398 0.0688 0.6796	0.8395 1.0350	1.7060 1.2250

TABLE 12-57 *(Continued)*

CO5 25

	R_1	R_2	R_3	R_4	R_5	C_1	C_2	C_3	C_4	C_5	K	Gain	α	β	ω_∞
$\theta = 25$	0.3457	0.6913	6.2320	28.04		5.7850	1.2860	0.1426	0.0713		1.0990	1.0600	0.3410	0.6681	3.9010
$\Omega_s = 2.3662$	0.3920	0.7840	1.4850	6.6830	1.0000	3.6830	0.8185	0.4321	0.2160		1.7820	1.5160	0.1176	1.0320	2.4770
$A_{\min} = 71.71$										2.2240			0.4497		
$\theta = 32$	0.3544	0.7089	3.7330	16.80		5.1190	1.2260	0.2329	0.1164		1.1460	1.0770	0.3386	0.6879	3.0480
$\Omega_s = 1.8871$	0.4258	0.8515	1.0230	4.6060	1.0000	3.3970	0.7548	0.6280	0.3140		1.9500	1.5260	0.1091	1.0310	1.9690
$A_{\min} = 60.51$										2.1400			0.4673		
$\theta = 40$	0.3685	0.7370	2.3300	10.48		5.1420	1.1420	0.3615	0.1807		1.2200	1.1040	0.3339	0.7177	2.4380
$\Omega_s = 1.5557$	0.4698	0.9396	0.7651	3.4430	1.0000	3.0860	0.6858	0.8422	0.4211		2.1730	1.5420	0.0970	1.0300	1.6170
$A_{\min} = 50.10$										2.0190			0.4952		
$\theta = 50$	0.3942	0.7884	1.4390	6.4770		4.5720	1.0150	0.5564	0.2782		1.3600	1.1490	0.3226	0.7673	1.9480
$\Omega_s = 1.3054$	0.5280	1.0560	0.6027	2.7120	1.0000	2.7570	0.6127	1.0730	0.5368		2.4790	1.5650	0.0786	1.0270	1.3480
$A_{\min} = 39.17$										1.8310			0.5462		
$\theta = 54$	0.4078	0.8157	1.2160	5.4740		4.3170	0.9592	0.6433	0.3216		1.4350	1.1730	0.3154	0.7915	1.8020
$\Omega_s = 1.2361$	0.5509	1.1010	0.5627	2.5320	1.0000	2.6480	0.5884	1.1520	0.5761		2.6040	1.5750	0.0704	1.0250	1.2730
$A_{\min} = 35.21$										1.7430			0.5737		
$\theta = 59$	0.4284	0.8568	1.0010	4.5070		3.9820	0.8848	0.7570	0.3785		1.5510	1.2070	0.3031	0.8255	1.6470
$\Omega_s = 1.1666$	0.5781	1.1560	0.5249	2.3620	1.0000	2.5310	0.5624	1.2380	0.6194		2.7560	1.5890	0.0595	1.0230	1.1960
$A_{\min} = 30.46$										1.6230			0.6163		

CO7 05

	R_1	R_2	R_3	R_4	R_5	C_1	C_2	C_3	C_4	C_5	K	Gain	α	β	ω_∞
$\theta = 37$	0.3499	0.6998	4.6990	21.44		5.4850	1.2180	0.1815	0.0908		0.8162	0.7776	0.5212	0.5825	3.5080
$\Omega_s = 1.6616$	0.4114	0.8229	1.1700	5.2680	1.0000	3.6920	0.8205	0.5767	0.2883	1.5680	1.5680	1.2700	0.2938	0.9427	2.0400
$A_{min} = 70.56$	0.4722	0.9443	0.7558	3.4010		2.9030	0.6451	0.8060	0.4030		2.2060	1.5570	0.0900	1.0900	1.6960
													0.6378		
$\theta = 43$	0.3579	0.7157	3.2420	14.59		5.1550	1.1450	0.2529	0.1265		0.8599	0.8009	0.5279	0.6183	2.9970
$\Omega_s = 1.4663$	0.4387	0.8773	0.9255	4.1650	1.0000	3.4290	0.7620	0.7223	0.3612	1.4820	1.7110	1.3000	0.2740	0.9588	1.7740
$A_{min} = 60.55$	0.5094	1.0180	0.6430	2.8930		2.7130	0.6028	0.9551	0.4776		2.3980	1.5690	0.0793	1.0820	1.4940
													0.6746		
$\theta = 50$	0.3712	0.7423	2.1810	9.8130		4.7060	1.0450	0.3560	0.1780		0.9318	0.8369	0.5345	0.6722	2.5490
$\Omega_s = 1.3054$	0.4751	0.9501	0.7449	3.3520	1.0000	3.1280	0.6950	0.8865	0.4433	1.3660	1.9130	1.3420	0.2456	0.9792	1.5480
$A_{min} = 50.11$	0.5517	1.1030	0.5614	2.5260		2.5320	0.5626	1.1050	0.5529		2.6240	1.5850	0.0657	1.0710	1.3270
													0.7322		
$\theta = 53$	0.3787	0.7573	1.8560	8.3530		4.4920	0.9982	0.4072	0.2036		0.9725	0.8561	0.5361	0.7002	2.3910
$\Omega_s = 1.2521$	0.4920	0.9840	0.6890	3.1010	1.0000	3.0040	0.6675	0.9533	0.4767	1.3100	2.0110	1.3630	0.2314	0.9881	1.4710
$A_{min} = 45.93$	0.5688	1.1370	0.5368	2.4150		2.4680	0.5485	1.1620	0.5812		2.7170	1.5920	0.0595	1.0660	1.2710
													0.7634		
$\theta = 57$	0.3909	0.7819	1.5080	6.7860		4.1870	0.9305	0.4825	0.2412		1.0390	0.8862	0.5363	0.7429	2.2040
$\Omega_s = 1.1924$	0.5157	1.0310	0.6285	2.8280	1.0000	2.8460	0.6325	1.0380	0.5190	1.2290	2.1530	1.3920	0.2106	1.0000	1.3820
$A_{min} = 40.54$	0.5902	1.1800	0.5107	2.2980		2.3960	0.5325	1.2300	0.6153		2.8370	1.6020	0.0512	1.0590	1.2080
													0.8130		
$\theta = 60$	0.4023	0.8046	1.2960	5.8340		3.9460	0.8768	0.5442	0.2721		1.1010	0.9129	0.5342	0.7795	2.0780
$\Omega_s = 1.1547$	0.5340	1.0670	0.5914	2.6610	1.0000	2.7360	0.6079	1.0970	0.5489	1.1650	2.2670	1.4150	0.1934	1.0080	1.3230
$A_{min} = 36.61$	0.6048	1.2090	0.4951	2.2280		2.3510	0.5225	1.2760	0.6382		2.9210	1.6100	0.0448	1.0530	1.1680
													0.8581		

TABLE 12-57 (Continued)

CO7 15

	R_1	R_2	R_3	R_4	R_5	C_1	C_2	C_3	C_4	C_5	K	Gain	α	β	ω_∞
$\theta = 46$	0.3526	0.7052	4.0690	18.31		6.3450	1.4100	0.2444	0.1221		0.9973	0.9429	0.3640	0.5631	2.7890
$\Omega_s = 1.3902$	0.4343	0.8686	0.9557	4.3010		3.7610	0.8358	0.7596	0.3798		1.8120	1.3910	0.1899	0.8984	1.6680
$A_{\min} = 65.57$	0.5108	1.0210	0.6397	2.8790	1.0000	2.8450	0.6322	1.0090	0.5048	2.1640	2.4450	1.5960	0.0548	1.0300	1.4150
													0.4620		
$\theta = 49$	0.3564	0.7128	3.4350	15.46		6.1410	1.3650	0.2832	0.1416		1.0190	0.9536	0.3665	0.5791	2.6060
$\Omega_s = 1.3250$	0.4481	0.8962	0.8677	3.9050		3.6190	0.8043	0.8307	0.4153		1.8860	1.4030	0.1827	0.9066	1.5760
$A_{\min} = 61.17$	0.5284	1.0560	0.6020	2.7090	1.0000	2.7550	0.6122	1.0740	0.5374	2.0990	2.5370	1.6010	0.0511	1.0290	1.3470
													0.4764		
$\theta = 53$	0.3625	0.7251	2.7590	12.41		5.8460	1.2990	0.3414	0.1707		1.0550	0.9701	0.3697	0.6035	2.3910
$\Omega_s = 1.2521$	0.4678	0.9356	0.7730	3.4790		3.4320	0.7626	0.9230	0.4615		1.9940	1.4210	0.1721	0.9183	1.4710
$A_{\min} = 55.56$	0.5513	1.1020	0.5621	2.5290	1.0000	2.6470	0.5882	1.1530	0.5770	2.0060	2.6580	1.6070	0.0459	1.0260	1.2710
													0.4985		
$\theta = 57$	0.3702	0.7405	2.2290	10.03		5.5240	1.2270	0.4078	0.2039		1.0990	0.9897	0.3724	0.6318	2.2040
$\Omega_s = 1.1924$	0.4890	0.9780	0.6980	3.1410		3.2480	0.7219	1.0110	0.5057		2.1140	1.4410	0.1600	0.9306	1.3820
$A_{\min} = 50.1$	0.5732	1.1460	0.5311	2.3900	1.0000	2.5530	0.5673	1.2240	0.6123	1.9040	2.7760	1.6140	0.0404	1.0240	1.2080
													0.5251		
$\theta = 60$	0.3774	0.7547	1.9050	8.5710		5.2640	1.1690	0.4635	0.2318		1.1390	1.0060	0.3737	0.6561	2.0780
$\Omega_s = 1.1547$	0.5058	1.0110	0.6517	2.9330		3.1150	0.6922	1.0740	0.5372		2.2110	1.4570	0.1499	0.9402	1.3230
$A_{\min} = 46.24$	0.5886	1.1770	0.5124	2.3060	1.0000	2.4910	0.5536	1.2720	0.6359	1.8230	2.8600	1.6200	0.0361	1.0220	1.1680
													0.5486		

CO7 20

	R_1	R_2	R_3	R_4	R_5	C_1	C_2	C_3	C_4	C_5	K	Gain	α	β	ω_∞
$\theta = 44$	0.3487	0.6973	5.0550	22.75		6.8590	1.5240	0.2103	0.1051		1.0320	0.9870	0.3197	0.5396	2.9250
$\Omega_s = 1.4396$	0.4220	0.8441	1.0570	4.7580		3.9660	0.8814	0.7036	0.3518		1.7840	1.4090	0.1731	0.8792	1.7370
$A_{min} = 71.20$	0.4955	0.9910	0.6790	3.0550	1.0000	2.9600	0.6578	0.9602	0.4801	2.5120	2.3780	1.6000	0.0512	1.0210	1.4660
													0.3980		
$\theta = 48$	0.3528	0.7057	4.0180	18.08		6.5930	1.4650	0.2573	0.1287		1.0570	0.9989	0.3229	0.5581	2.6650
$\Omega_s = 1.3456$	0.4392	0.8784	0.9218	4.1480		3.7740	0.8386	0.7992	0.3996		1.8750	1.4230	0.1654	0.8897	1.6060
$A_{min} = 65.20$	0.5189	1.0370	0.6214	2.7960	1.0000	2.8310	0.6291	1.0500	0.5253	2.4200	2.4990	1.6050	0.0470	1.0200	1.3680
													0.4132		
$\theta = 51$	0.3566	0.7133	3.4010	15.30		6.3760	1.4170	0.2972	0.1486		1.0790	1.0090	0.3252	0.5739	2.4950
$\Omega_s = 1.2868$	0.4531	0.9063	0.8405	3.7820		3.6290	0.8065	0.8696	0.4348		1.9500	1.4340	0.1589	0.8981	1.5210
$A_{min} = 60.91$	0.5362	1.0720	0.5873	2.6430	1.0000	2.7430	0.6095	1.1130	0.5565	2.3450	2.5890	1.6100	0.0437	1.0180	1.3070
													0.4264		
$\theta = 55$	0.3628	0.7255	2.7380	12.32		6.0630	1.3470	0.3570	0.1785		1.1150	1.0250	0.3281	0.5978	2.2950
$\Omega_s = 1.2208$	0.4730	0.9461	0.7525	3.3860		3.4390	0.7641	0.9607	0.4803		2.0590	1.4510	0.1493	0.9100	1.4240
$A_{min} = 55.42$	0.5587	1.1170	0.5509	2.4790	1.0000	2.6370	0.5860	1.1880	0.5942	2.2380	2.7080	1.6160	0.0391	1.0170	1.2380
													0.4468		
$\theta = 60$	0.3728	0.7455	2.1010	0.9456		5.6320	1.2520	0.4440	0.2220		1.1730	1.0490	0.3309	0.6332	2.0780
$\Omega_s = 1.1547$	0.4999	0.9999	0.6669	3.0010		3.2070	0.7126	1.0680	0.5342		2.2100	1.4740	0.1354	0.9258	1.3230
$A_{min} = 48.81$	0.5851	1.1700	0.5165	2.3240	1.0000	2.5240	0.5608	1.2710	0.6353	2.0920	2.8500	1.6230	0.0329	1.0150	1.1680
													0.4781		

12-115

TABLE 12-57 *(Continued)*

CO7 25

	R_1	R_2	R_3	R_4	R_5	C_1	C_2	C_3	C_4	C_5	K	Gain	α	β	ω_∞
$\theta = 46$	0.3493	0.6986	4.8560	21.85		7.0290	1.5620	0.2247	0.1123		1.0870	1.0370	0.2882	0.5386	2.7890
$\Omega_s = 1.3902$	0.4277	0.8555	1.0060	4.5310		3.9490	0.8777	0.7458	0.3729		1.8440	1.4370	0.1530	0.8746	1.6680
$A_{min} = 70.19$	0.5047	1.0090	0.6544	2.9450	1.0000	2.9290	0.6509	1.0040	0.5021	2.7620	2.4350	1.6080	0.0446	1.0130	1.4150
													0.3621		
$\theta = 49$	0.3525	0.7049	4.0970	18.44		6.8190	1.5150	0.2607	0.1304		1.1060	1.0460	0.2904	0.5524	2.6060
$\Omega_s = 1.3250$	0.4407	0.8815	0.9119	4.1030		3.8030	0.8452	0.8170	0.4085		1.9130	1.4460	0.1478	0.8825	1.5760
$A_{min} = 65.79$	0.5221	1.0440	0.6146	2.7650	1.0000	2.8340	0.6297	1.0700	0.5350	2.6840	2.5250	1.6120	0.0418	1.0120	1.3470
													0.3726		
$\theta = 53$	0.3575	0.7150	3.2890	14.80		6.5180	1.4480	0.3149	0.1574		1.1360	1.0590	0.2932	0.5731	2.3910
$\Omega_s = 1.2521$	0.4595	0.9189	0.8095	3.6430		3.6080	0.8018	0.9102	0.4551		2.0130	1.4610	0.1399	0.8939	1.4710
$A_{min} = 60.18$	0.5449	1.0890	0.5723	2.5750	1.0000	2.7180	0.6040	1.1500	0.5751	2.5710	2.6430	1.6170	0.0377	1.0120	1.2710
													0.3889		
$\theta = 56$	0.3621	0.7242	2.7980	12.59		6.2720	1.3940	0.3608	0.1804		1.1630	1.0700	0.2952	0.5908	2.2480
$\Omega_s = 1.2062$	0.4745	0.9490	0.7470	3.3620		3.4630	0.7696	0.9777	0.4888		2.0950	1.4720	0.1333	0.9030	1.4030
$A_{min} = 56.12$	0.5615	1.1230	0.5468	2.4610	1.0000	2.6400	0.5866	1.2040	0.6024	2.4810	2.7310	1.6210	0.0346	1.0110	1.2230
													0.4030		
$\theta = 60$	0.3696	0.7393	2.2630	10.18		5.9190	1.3150	0.4297	0.2148		1.2060	1.0880	0.2974	0.6177	2.0780
$\Omega_s = 1.1547$	0.4958	0.9916	0.6782	3.0520		3.2740	0.7276	1.0630	0.5319		2.2140	1.4890	0.1234	0.9157	1.3230
$A_{min} = 50.86$	0.5825	1.1650	0.5195	2.3380	1.0000	2.5480	0.5661	1.2700	0.6349	2.3520	2.8430	1.6270	0.0302	1.0100	1.1680
													0.4252		

Index